GEOMETRY OF PROJECTIVE ALGEBRAIC CURVES

PURE AND APPLIED MATHEMATICS

A Program of Monographs, Textbooks, and Lecture Notes

EXECUTIVE EDITORS

Earl J. Taft
Rutgers University
New Brunswick, New Jersey

Zuhair Nashed
University of Delaware
Newark, Delaware

CHAIRMEN OF THE EDITORIAL BOARD

S. Kobayashi
University of California, Berkeley
Berkeley, California

Edwin Hewitt
University of Washington
Seattle, Washington

EDITORIAL BOARD

M. S. Baouendi
Purdue University

Jack K. Hale
Brown University

Marvin Marcus
University of California, Santa Barbara

W. S. Massey
Yale University

Leopoldo Nachbin
Centro Brasileiro de Pesquisas Físicas
and University of Rochester

Anil Nerode
Cornell University

Donald Passman
University of Wisconsin

Fred S. Roberts
Rutgers University

Gian-Carlo Rota
Massachusetts Institute of Technology

David Russell
University of Wisconsin-Madison

Jane Cronin Scanlon
Rutgers University

Walter Schempp
Universität Siegen

Mark Teply
University of Florida

MONOGRAPHS AND TEXTBOOKS IN PURE AND APPLIED MATHEMATICS

1. *K. Yano,* Integral Formulas in Riemannian Geometry (1970) *(out of print)*
2. *S. Kobayashi,* Hyperbolic Manifolds and Holomorphic Mappings (1970) *(out of print)*
3. *V. S. Vladimirov,* Equations of Mathematical Physics (A. Jeffrey, editor; A. Littlewood, translator) (1970) *(out of print)*
4. *B. N. Pshenichnyi,* Necessary Conditions for an Extremum (L. Neustadt, translation editor; K. Makowski, translator) (1971)
5. *L. Narici, E. Beckenstein, and G. Bachman,* Functional Analysis and Valuation Theory (1971)
6. *D. S. Passman,* Infinite Group Rings (1971)
7. *L. Dornhoff,* Group Representation Theory (in two parts). Part A: Ordinary Representation Theory. Part B: Modular Representation Theory (1971, 1972)
8. *W. Boothby and G. L. Weiss (eds.),* Symmetric Spaces: Short Courses Presented at Washington University (1972)
9. *Y. Matsushima,* Differentiable Manifolds (E. T. Kobayashi, translator) (1972)
10. *L. E. Ward, Jr.,* Topology: An Outline for a First Course (1972) *(out of print)*
11. *A. Babakhanian,* Cohomological Methods in Group Theory (1972)
12. *R. Gilmer,* Multiplicative Ideal Theory (1972)
13. *J. Yeh,* Stochastic Processes and the Wiener Integral (1973) *(out of print)*
14. *J. Barros-Neto,* Introduction to the Theory of Distributions (1973) *(out of print)*
15. *R. Larsen,* Functional Analysis: An Introduction (1973) *(out of print)*
16. *K. Yano and S. Ishihara,* Tangent and Cotangent Bundles: Differential Geometry (1973) *(out of print)*
17. *C. Procesi,* Rings with Polynomial Identities (1973)
18. *R. Hermann,* Geometry, Physics, and Systems (1973)
19. *N. R. Wallach,* Harmonic Analysis on Homogeneous Spaces (1973) *(out of print)*
20. *J. Dieudonné,* Introduction to the Theory of Formal Groups (1973)
21. *I. Vaisman,* Cohomology and Differential Forms (1973)
22. *B.-Y. Chen,* Geometry of Submanifolds (1973)
23. *M. Marcus,* Finite Dimensional Multilinear Algebra (in two parts) (1973, 1975)
24. *R. Larsen,* Banach Algebras: An Introduction (1973)
25. *R. O. Kujala and A. L. Vitter (eds.),* Value Distribution Theory: Part A; Part B: Deficit and Bezout Estimates by Wilhelm Stoll (1973)
26. *K. B. Stolarsky,* Algebraic Numbers and Diophantine Approximation (1974)
27. *A. R. Magid,* The Separable Galois Theory of Commutative Rings (1974)
28. *B. R. McDonald,* Finite Rings with Identity (1974)
29. *J. Satake,* Linear Algebra (S. Koh, T. A. Akiba, and S. Ihara, translators) (1975)

30. *J. S. Golan,* Localization of Noncommutative Rings (1975)
31. *G. Klambauer,* Mathematical Analysis (1975)
32. *M. K. Agoston,* Algebraic Topology: A First Course (1976)
33. *K. R. Goodearl,* Ring Theory: Nonsingular Rings and Modules (1976)
34. *L. E. Mansfield,* Linear Algebra with Geometric Applications: Selected Topics (1976)
35. *N. J. Pullman,* Matrix Theory and Its Applications (1976)
36. *B. R. McDonald,* Geometric Algebra Over Local Rings (1976)
37. *C. W. Groetsch,* Generalized Inverses of Linear Operators: Representation and Approximation (1977)
38. *J. E. Kuczkowski and J. L. Gersting,* Abstract Algebra: A First Look (1977)
39. *C. O. Christenson and W. L. Voxman,* Aspects of Topology (1977)
40. *M. Nagata,* Field Theory (1977)
41. *R. L. Long,* Algebraic Number Theory (1977)
42. *W. F. Pfeffer,* Integrals and Measures (1977)
43. *R. L. Wheeden and A. Zygmund,* Measure and Integral: An Introduction to Real Analysis (1977)
44. *J. H. Curtiss,* Introduction to Functions of a Complex Variable (1978)
45. *K. Hrbacek and T. Jech,* Introduction to Set Theory (1978)
46. *W. S. Massey,* Homology and Cohomology Theory (1978)
47. *M. Marcus,* Introduction to Modern Algebra (1978)
48. *E. C. Young,* Vector and Tensor Analysis (1978)
49. *S. B. Nadler, Jr.,* Hyperspaces of Sets (1978)
50. *S. K. Segal,* Topics in Group Rings (1978)
51. *A. C. M. van Rooij,* Non-Archimedean Functional Analysis (1978)
54. *L. Corwin and R. Szczarba,* Calculus in Vector Spaces (1979)
53. *C. Sadosky,* Interpolation of Operators and Singular Integrals: An Introduction to Harmonic Analysis (1979)
54. *J. Cronin,* Differential Equations: Introduction and Quantitative Theory (1980)
55. *C. W. Groetsch,* Elements of Applicable Functional Analysis (1980)
56. *I. Vaisman,* Foundations of Three-Dimensional Euclidean Geometry (1980)
57. *H. I. Freedman,* Deterministic Mathematical Models in Population Ecology (1980)
58. *S. B. Chae,* Lebesgue Integration (1980)
59. *C. S. Rees, S. M. Shah, and C. V. Stanojević,* Theory and Applications of Fourier Analysis (1981)
60. *L. Nachbin,* Introduction to Functional Analysis: Banach Spaces and Differential Calculus (R. M. Aron, translator) (1981)
61. *G. Orzech and M. Orzech,* Plane Algebraic Curves: An Introduction Via Valuations (1981)
62. *R. Johnsonbaugh and W. E. Pfaffenberger,* Foundations of Mathematical Analysis (1981)
63. *W. L. Voxman and R. H. Goetschel,* Advanced Calculus: An Introduction to Modern Analysis (1981)
64. *L. J. Corwin and R. H. Szcarba,* Multivariable Calculus (1982)
65. *V. I. Istrătescu,* Introduction to Linear Operator Theory (1981)
66. *R. D. Järvinen,* Finite and Infinite Dimensional Linear Spaces: A Comparative Study in Algebraic and Analytic Settings (1981)

67. *J. K. Beem and P. E. Ehrlich,* Global Lorentzian Geometry (1981)
68. *D. L. Armacost,* The Structure of Locally Compact Abelian Groups (1981)
69. *J. W. Brewer and M. K. Smith, eds.,* Emmy Noether: A Tribute to Her Life and Work
70. *K. H. Kim,* Boolean Matrix Theory and Applications (1982)
71. *T. W. Wieting,* The Mathematical Theory of Chromatic Plane Ornaments (1982)
72. *D. B. Gauld,* Differential Topology: An Introduction (1982)
73. *R. L. Faber,* Foundations of Euclidean and Non-Euclidean Geometry (1983)
74. *M. Carmeli,* Statistical Theory and Random Matrices (1983)
75. *J. H. Carruth, J. A. Hildebrant, and R. J. Koch,* The Theory of Topological Semigroups (1983)
76. *R. L. Faber,* Differential Geometry and Relativity Theory: An Introduction (1983)
77. *S. Barnett,* Polynomials and Linear Control Systems (1983)
78. *G. Karpilovsky,* Commutative Group Algebras (1983)
79. *F. Van Oystaeyen and A. Verschoren,* Relative Invariants of Rings: The Commutative Theory (1983)
80. *I. Vaisman,* A First Course in Differential Geometry (1984)
81. *G. W. Swan,* Applications of Optimal Control Theory in Biomedicine (1984)
82. *T. Petrie and J. D. Randall,* Transformation Groups on Manifolds (1984)
83. *K. Goebel and S. Reich,* Uniform Convexity, Hyperbolic Geometry, and Nonexpansive Mappings (1984)
84. *T. Albu and C. Năstăsescu,* Relative Finiteness in Module Theory (1984)
85. *K. Hrbacek and T. Jech,* Introduction to Set Theory, Second Edition, Revised and Expanded (1984)
86. *F. Van Oystaeyen and A. Verschoren,* Relative Invariants of Rings: The Noncommutative Theory (1984)
87. *B. R. McDonald,* Linear Algebra Over Commutative Rings (1984)
88. *M. Namba,* Geometry of Projective Algebraic Curves (1984)

Other Volumes in Preparation

GEOMETRY OF PROJECTIVE ALGEBRAIC CURVES

MAKOTO NAMBA
Mathematical Institute
Tôhoku University
Sendai, Japan

MARCEL DEKKER, INC. *New York and Basel*

Library of Congress Cataloging in Publication Data

Namba, Makoto, [date]
Geometry of projective algebraic curves.

(Monographs and textbooks in pure and applied mathematics ; v. 88)
Includes index.
1. Curves, Algebraic. I. Title. II. Series.
QA565.N36 1984 516.3'52 84-17636
ISBN 0-8247-7222-9

MARCEL DEKKER, INC.
270 Madison Avenue, New York, New York 10016

Current printing (last digit):
10 9 8 7 6 5 4 3 2 1

PRINTED IN THE UNITED STATES OF AMERICA

PREFACE

In this book I present the theory of projective algebraic curves from the geometric point of view. I begin from the classical geometry of conics and develop the theory up to the frontier of present-day research.

The main readers of this book will be graduate students and nonspecialists: This book can be used as a textbook for first- and second-year graduate students and as an introductory book on algebraic geometry for nonspecialists.

This book has originated on the basis of lectures I presented for first- and second-year graduate students at Tôhoku University and the University of Tokyo in 1980 and 1981. In the academic year of 1981-1982, I had an opportunity to work at the Institute for Advanced Study in Princeton. The first draft of this book was written during that period. I express my gratitude to the Institute for Advanced Study and the Sloan Foundation, supporting me during my stay in Princeton.

I also express my thanks to Professor Borel, Professor Griffiths, Professor Fulton, and the members of the Institute for Advanced Study for many suggestions and stimulating discussions.

In writing this book, the books by Mumford [70], Fulton [29], Griffiths-Harris [33], and Kawai [51] (in Japanese) were particularly helpful.

Last but not least, I extend my thanks to Messrs. Shimizu and Konno, who read the manuscript and gave me a lot of suggestions, and to Mrs. Kazuko Kawauchi for her beautiful job of typing the manuscript.

Makoto Namba

INTRODUCTION

The purpose of this book is to introduce the curve theory from the geometric point of view to graduate students and nonspecialists. Because of this, we always work over the complex number field $\mathbb{C}$.

Roughly speaking, geometry of curves can be divided into two kinds: extrinsic geometry and intrinsic geometry. Extrinsic geometry is concerned with curves and their ambient spaces. Thus their positions and mutual relations in an ambient space are important in extrinsic geometry. On the other hand, intrinsic geometry is concerned with the properties of a curve itself, disregarding how it is imbedded in an ambient space. Thus its topological and analytic (algebraic) structures are important in intrinsic geometry.

Modern curve theory is a part of algebraic geometry and is highly sophisticated. Intrinsic study on curves is so successful that people sometimes neglect extrinsic study. It should not, however, be forgotten that intrinsic and extrinsic studies have intertwined and developed the curve theory. Also, there are many beautiful jewels in extrinsic geometry which move our heart.

For this reason, we take up extrinsic geometry as one of the subjects. We divide the book into two parts. In Part I, we discuss extrinsic geometry. In Part II, we discuss intrinsic geometry.

The construction of this book is dialectic. That is, Part I is the thesis, Part II the antithesis, and Chapter 5, Sec. 4, the synthesis. Thus this book is not written in a linear fashion: In Part I, we use freely the language and theorems of Part II. A few results in Part I are used in Part II.

In Chapter 1, we discuss projective geometry of curves. Bezout's theorem and linear systems of plane curves are the central subjects in this chapter.

In Chapter 2, we consider the following problem. "What singular irreducible plane or space curves of given degree and genus exist?" This is a naive but difficult problem. Following the idea of Veronese [99], we solve the problem for curves of lower degree as an application of the theory of linear systems on a curve developed in Chapter 5. Our method is intuitive and projective geometric.

In Chapter 3, we discuss the concepts and properties of complex manifolds and projective varieties which are used in other chapters.

In Chapter 4, we discuss compact Riemann surfaces. We explain the so-called trinity; the three categories are equivalent:

(a) compact Riemann surfaces (analysis),
(b) nonsingular irreducible projective curves (geometry),
(c) algebraic function fields of one variable (algebra).

In Chapter 5, we discuss the Riemann-Roch theorem, Jacobian varieties, Abel's theorem, linear systems, etc., which are the subjects of the usual curve theory.

In Chapter 5, Sec. 4, we discuss recent topics of linear systems on curves in order to give the reader a higher view.

It is impossible to discuss everything. Many important topics are not developed at all: Cremona transformations, correspondences, automorphism groups, theta functions, Torelli's theorem, Schottky problem, moduli spaces of curves, etc. We give some references for these topics.

In this book, we present a lot of figures. They are intended to provide a better understanding, by appealing to the reader's intuition. Real understanding, however, will be obtained by watching with the eyes in his mind.

SUGGESTION TO THE READER

1. This book is not written in a linear fashion. The relations among various parts of the book are given by the following diagram:

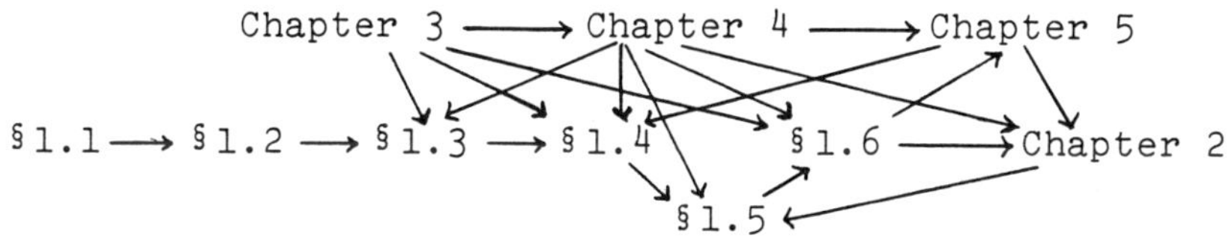

1-a. The reader who is familiar with the concepts of complex manifolds and algebraic varieties may read this book linearly.

1-b. The reader who knows some of the concepts may also read Part I first, by referring to the material in Part II.

1-c. The reader who knows little is advised to read Part II first, and then Part I.

2. This book is not self-contained. In Chapter 3, some important theorems which are used in other parts of the book are stated without proof: Hilbert Nullstellensatz (Theorem 3.2.12), Remmert-Stein continuation theorem (Theorem 3.2.13), the proper mapping theorem (Theorem 3.2.14), Hironaka's theorem (Theorem 3.3.11), and the theorem of finite dimensionality of linear systems (Theorem 3.4.4). For the latter three theorems, proofs for the case of one dimension are given in Chapter 4. In Chapter 4, Riemann's existence theorem (Theorem 4.1.7) is stated without proof. We give suitable reference books for the assertions without proofs. All other parts of the book can be read within the basic knowledge of undergraduate (or first year graduate) algebra, geometry, general topology, and

the function theory of one variable, provided the reader accepts the assertions without proofs in Chapters 3 and 4.

3. Almost all exercises at the end of each section are easy applications of the text material. But some of them are propositions which could be given in the text, and a few of them are a little difficult to solve.

CONTENTS

PART II INTRINSIC GEOMETRY OF CURVES

GEOMETRY OF PROJECTIVE ALGEBRAIC CURVES

PART I

EXTRINSIC GEOMETRY OF CURVES

1

PROJECTIVE GEOMETRY OF CURVES

1.1. PROJECTIVE SPACES AND PROJECTIVE TRANSFORMATIONS

Let n be a positive integer and $\mathbb{C}^{n+1}$ be the (n + 1)-fold Cartesian product of the complex plane $\mathbb{C}$. Let $0 = (0, \ldots, 0)$ be the origin of $\mathbb{C}^{n+1}$. We introduce an equivalence relation $\sim$ in $\mathbb{C}^{n+1} - \{0\}$ as follows

$$(z_0, z_1, \ldots, z_n) \sim (w_0, w_1, \ldots, w_n)$$

if there is a nonzero $c \in \mathbb{C}$ such that

$$w_0 = cz_0, \quad w_1 = cz_1, \ldots, \quad w_n = cz_n$$

We denote the equivalence class in which $(z_0, z_1, \ldots, z_n)$ belongs by the ratio $(z_0 : z_1 : \cdots : z_n)$.

The <u>n-dimensional (complex) projective space</u> $\mathbb{P}^n = \mathbb{P}^n(\mathbb{C})$ is, by definition, the quotient space $(\mathbb{C}^{n+1} - \{0\})/\sim$. We may write

$$\mathbb{P}^n = \frac{\mathbb{C}^{n+1} - \{0\}}{\mathbb{C}^*}$$

where $\mathbb{C}^* = \mathbb{C} - \{0\}$ is the multiplicative group of nonzero complex numbers acting on $\mathbb{C}^{n+1} - \{0\}$ as follows

$$c(z_0, z_1, \ldots, z_n) = (cz_0, cz_1, \ldots, cz_n) \quad \text{for} \quad c \in \mathbb{C}^*$$

Then we may write the above relation $\sim$ as

$$(z_0, z_1, \ldots, z_n) \sim (w_0, w_1, \ldots, w_n) \pmod{\mathbb{C}^*}$$

The topology of $\mathbb{P}^n$ is the quotient topology. That is, $W \subset \mathbb{P}^n$ is open if and only if $\pi^{-1}(W)$ is open in $\mathbb{C}^{n+1} - \{0\}$, where $\pi: \mathbb{C}^{n+1} - \{0\} \to \mathbb{P}^n$ is the natural projection. Since $\mathbb{P}^n$ is the quotient space by a group action, π is an open map, that is, $\pi(U)$ is open, if U is open.

$\mathbb{P}^n$ is compact. In fact, $\pi(S^{2n+1}) = \mathbb{P}^n$ where

$$S^{2n+1} = \{(z_0, z_1, \ldots, z_n) \in \mathbb{C}^{n+1} \mid |z_0|^2 + |z_1|^2 + \cdots + |z_n|^2 = 1\}$$

the $(2n + 1)$-unit sphere.

We call $\mathbb{P}^1$ and $\mathbb{P}^2$ the projective line and the projective plane, respectively.

$\mathbb{P}^1$ is the set of all ratios $(z_0 : z_1)$. If we set $z_0 = 0$, then we get the point $(0: 1)$, called the point of infinity and denoted by ∞. Other points satisfy $z_0 \neq 0$. The map

$$(z_0 : z_1) \in \mathbb{P}^1 - \{\infty\} \to z = z_1/z_0 \in \mathbb{C}$$

is well-defined and homeomorphic. Thus $\mathbb{P}^1$ can be identified with the Riemann sphere $\hat{\mathbb{C}} = \mathbb{C} \cup \{\infty\}$. We project a point p on the sphere to $z \in \mathbb{C}$ with the center being the south pole s as in Fig. 1.1.

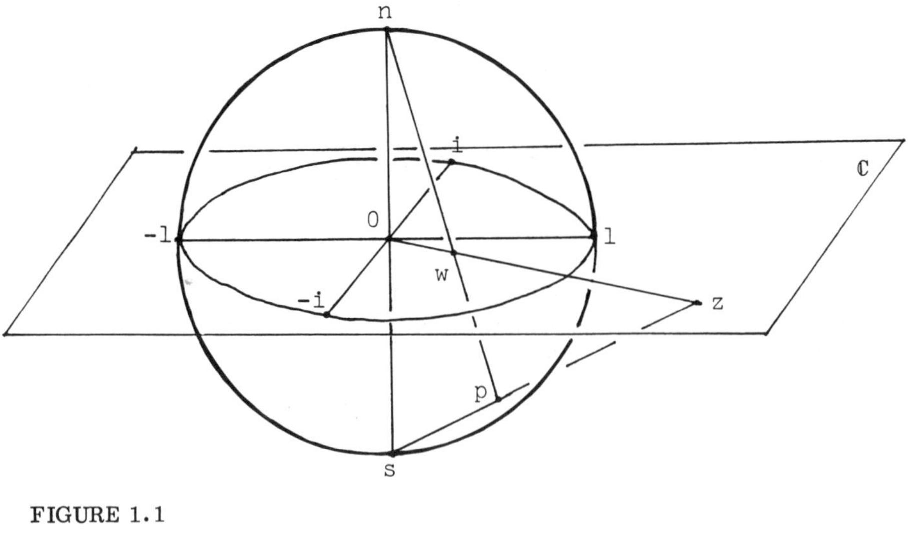

FIGURE 1.1

If we project p with the center n, which is the north pole, to $w \in \mathbb{C}$, then the relation of z and w is

$$\bar{w}z = 1$$

$\bar{w}$ is the complex conjugate of w. Consider the homeomorphisms

$$\phi_0 \colon p \in \hat{\mathbb{C}} - \{s\} \to z \in \mathbb{C}$$

$$\phi_1 \colon p \in \hat{\mathbb{C}} - \{n\} \to \bar{w} \in \mathbb{C}$$

Then $\hat{\mathbb{C}} - \{s, n\}$ is mapped by both ϕ_0 and ϕ_1 onto $\mathbb{C}^* = \mathbb{C} - \{0\}$, and the homeomorphism $\phi_1\phi_0^{-1}$ of $\mathbb{C}^*$ onto itself is given by

$$z \to \frac{1}{z}$$

Hence, $\mathbb{P}^1$ is one of the simplest examples of complex manifolds. (See Sec. 3.1 for the definition of complex manifolds.)

Next, we consider $\mathbb{P}^2$, which is the set of all ratios $(z_0 : z_1 : z_2)$. Consider the open sets

$$U_0 = \{z_0 \neq 0\}$$

$$U_1 = \{z_1 \neq 0\}$$

$$U_2 = \{z_2 \neq 0\}$$

of $\mathbb{P}^2$. They form an open covering of $\mathbb{P}^2$. Now, consider the homeomorphisms

$$\phi_0 : (z_0 : z_1 : z_2) \in U_0 \to \left(\frac{z_1}{z_0}, \frac{z_2}{z_0}\right) \in \mathbb{C}^2$$

$$\phi_1 : (z_0 : z_1 : z_2) \in U_1 \to \left(\frac{z_0}{z_1}, \frac{z_2}{z_1}\right) \in \mathbb{C}^2$$

$$\phi_2 : (z_0 : z_1 : z_2) \in U_2 \to \left(\frac{z_0}{z_2}, \frac{z_1}{z_2}\right) \in \mathbb{C}^2$$

Then $\phi_1\phi_0^{-1}$ of $\phi_0(U_0 \cap U_1) = \{(x, y) \in \mathbb{C}^2 \mid x \neq 0\}$ onto itself is given by

$$(x, y) \to \left(\frac{1}{x}, \frac{y}{x}\right)$$

and so on. Hence, $\mathbb{P}^2$ is a two-dimensional complex manifold.

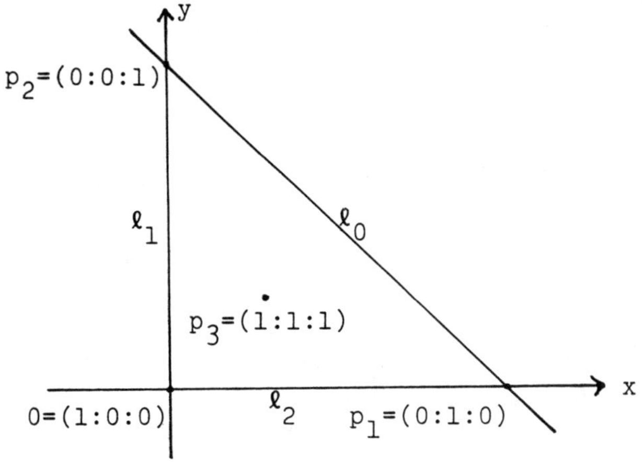

FIGURE 1.2

The coordinate system $(x,y) = (z_1/z_0, z_2/z_0)$ on U_0, and so forth, is called the affine coordinate system on U_0 or the inhomogeneous coordinate system. The ratio $(z_0 : z_1 : z_2)$ is called the homogeneous coordinate of the point. We usually write the homogeneous coordinate system $(X_0 : X_1 : X_2)$.

$\ell_0 = \mathbb{P}^2 - U_0$ is called the line of infinity. $\mathbb{P}^2$ is thus $\mathbb{C}^2$ plus the ideal point set ℓ_0. We visualize $\mathbb{P}^2$ as in Fig. 1.2. Here, the x-axis and y-axis are $\ell_2 = \mathbb{P}^2 - U_2$ and $\ell_1 = \mathbb{P}^2 - U_1$, respectively. The lines ℓ_0, ℓ_1 and ℓ_2 are the coordinate axis. The triangle $\Delta 0 p_1 p_2$ together with the point $p_3 = (1: 1: 1)$ is the coordinate triangle. Note that we are watching only the real part of whole $\mathbb{P}^2$. Topologically, $\mathbb{P}^2$ is four-dimensional.

Similar considerations work for $\mathbb{P}^n$ $(n \geq 3)$.

Projective spaces have another, more flexible, definition. Let V be a (n+1)-dimensional complex vector space. The set $\mathbb{P}(V)$ of all one-dimensional vector subspaces of V is called the projective space associated with V. Take a basis $\{v_0, v_1, \ldots, v_n\}$ of V. Any nonzero vector $v \in V$ determines a one-dimensional vector subspace $\mathbb{C}v$ containing v. Thus, we find a bijection

$$\phi: \mathbb{C}v \in \mathbb{P}(V) \to (a_0 : a_1 : \cdots : a_n) \in \mathbb{P}^n$$

where $v = a_0 v_0 + a_1 v_1 + \cdots + a_n v_n$. We regard $\mathbb{P}(V)$ as a projective space through this bijection, which is called a homogeneous coordinate system on $\mathbb{P}(V)$. (The original projective space $\mathbb{P}^n$ is nothing but $\mathbb{P}(\mathbb{C}^{n+1})$.) By abuse of notation, we sometimes identify $\mathbb{P}(V)$ with $\mathbb{P}^n$ through this bijection and call $(a_0 : a_1 : \cdots : a_n)$ a homogeneous coordinate system. (We call $(a_1/a_0, \ldots, a_n/a_0)$ an affine coordinate system.)

If we take another basis $\{w_0, w_1, \ldots, w_n\}$ of V, then we get another homogeneous coordinate system

$$\psi: \mathbb{C}v \in \mathbb{P}(V) \to (b_0 : b_1 : \cdots : b_n) \in \mathbb{P}^n$$

where $v = b_0w_0 + b_1w_1 + \cdots + b_nw_n$. ϕ and ψ give the same complex structure (see Sec. 3.1) on $\mathbb{P}(V)$. In fact,

$$\phi\psi^{-1}: (b_0 : b_1 : \cdots : b_n) \in \mathbb{P}^n \to (a_0 : a_1 : \cdots : a_n) \in \mathbb{P}^n$$

is given by the simultaneous linear equations

$$a_j = \sum_{k=0}^{n} c_{jk} b_j \qquad 0 \leq j \leq n$$

where (c_{jk}) is the nonsingular $(n+1) \times (n+1)$-matrix defined by

$$w_k = \sum_{j=0}^{n} c_{jk} v_j \qquad 0 \leq k \leq n$$

A bijection $\mathbb{P}^n \to \mathbb{P}^n$ given by simultaneous linear equations is called a projective transformation. Hence, homogeneous coordinate systems on $\mathbb{P}(V)$ and projective transformations on $\mathbb{P}^n$ have a one-to-one correspondence.

In general, a bijection $\mathbb{P}(V) \to \mathbb{P}(W)$ is called a projective transformation if it is induced by a linear isomorphism $V \to W$. One example would be if it is given by simultaneous linear equations, as above with respect to homogeneous coordinate systems on $\mathbb{P}(V)$ and $\mathbb{P}(W)$.

A s-dimensional linear subspace (or simply an s-plane) of $\mathbb{P}(V)$ is, by definition, $\pi(W - \{0\})$ for a $(s+1)$-dimensional vector subspace W of V, where

$$\pi: v \in V - \{0\} \to \mathbb{C}v \in \mathbb{P}(V)$$

is the canonical projection. If $s = 0$ (respectively -1), it is nothing but a point (respectively the empty set). If $s = 1$, it is called a line. An $(n-1)$-plane is called a hyperplane. An s-plane is itself an s-dimensional projective space. For an s-plane S of $\mathbb{P}(V)$, we denote its dimension s by dim S.

If $\phi: \mathbb{P}(V) \to \mathbb{P}(V')$ is a projective transformation and S is an s-plane of $\mathbb{P}(V)$, then $\phi(S)$ is an s-plane of $\mathbb{P}(V')$ and the restriction

$$\phi: S \to \phi(S)$$

to S is a projective transformation.

For linear subspaces S_1 and S_2 of $\mathbb{P}(V)$, $S_1 \cap S_2$ is also a linear subspace. We denote the minimal linear subspace of $\mathbb{P}(V)$ containing S_1 and S_2 by $S_1 \vee S_2$, and call it the linear subspace spanned by S_1 and S_2. The equality

$$\dim S_1 + \dim S_2 = \dim S_1 \cap S_2 + \dim S_1 \vee S_2$$

easily follows. In particular, if $\dim S_1 + \dim S_2 \geq \dim \mathbb{P}(V)$, then $S_1 \cap S_2$ is nonempty. For example, two distinct lines in $\mathbb{P}^2$ meet at one point. For a hyperplane P and a line ℓ in $\mathbb{P}(V)$, $P \cap \ell$ is one point unless $\ell \subset P$. (See Fig. 1.3.)

Now, let $(X_0 : X_1 : \cdots : X_n)$ be a homogeneous coordinate system on $\mathbb{P}^n = \mathbb{P}(V)$. For $(a_0 : a_1 : \cdots : a_n) \in \mathbb{C}^{n+1} - \{0\}$, the set

$$\{(X_0 : X_1 : \cdots : X_n) \in \mathbb{P}^n \mid a_0X_0 + a_1X_1 + \cdots + a_nX_n = 0\}$$

is well-defined and is a hyperplane of $\mathbb{P}^n$. Conversely, every hyperplane can be given in this way. $(a_0, a_1, \ldots, a_n)$ and $(b_0, b_1, \ldots, b_n)$ in $\mathbb{C}^{n+1} - \{0\}$ give the same hyperplane if and only if

$$(a_0, a_1, \ldots, a_n) \sim (b_0, b_1, \ldots, b_n) \pmod{\mathbb{C}^*}$$

Hence, the set $\mathbb{P}(V)^*$ of all hyperplanes in $\mathbb{P}^n = \mathbb{P}(V)$ can be regarded as an n-dimensional projective space. We call $\mathbb{P}(V)^*$ the dual projective space to $\mathbb{P}(V)$. It can be naturally identified with $\mathbb{P}(V^*)$, where V^* is the dual vector space to V. Note that $\mathbb{P}(V)^{**}$ is naturally identified with $\mathbb{P}(V)$.

Points $p_0, \ldots, p_r$ $(0 \leq r \leq n)$ in $\mathbb{P}^n$ are said to be in general position if $\dim p_0 \vee \cdots \vee p_r = r$. This occurs if and only if the matrix

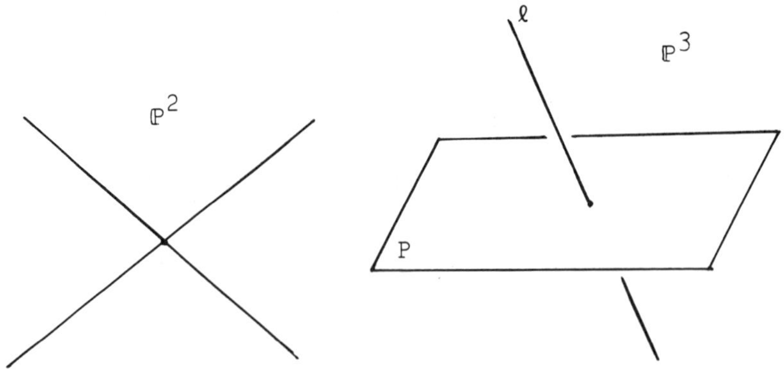

FIGURE 1.3

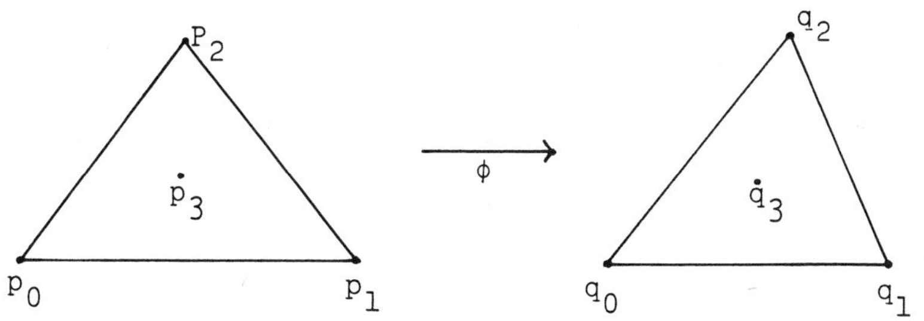

FIGURE 1.4

$$\begin{bmatrix} a_{00} & a_{01} & \cdots & a_{0n} \\ \vdots & \vdots & & \vdots \\ a_{r0} & a_{r1} & \cdots & a_{rn} \end{bmatrix}$$

has the rank r+1, where $p_j = (a_{j0}: a_{j1}: \cdots : a_{jn})$ for $0 \leq j \leq r$.

Points $p_0, p_1, \ldots, p_r$ $(r \geq n+1)$ are said to be <u>in general position</u> if any n+1 points of them span the whole space $\mathbb{P}^n$.

<u>Proposition 1.1.1</u>. Let $\{p_0, p_1, \ldots, p_{n+1}\}$ and $\{q_0, q_1, \ldots, q_{n+1}\}$ be two sets of (n+2)-points of $\mathbb{P}^n$ in general position. Then there is a unique projective transformation $\phi: \mathbb{P}^n \to \mathbb{P}^n$ such that $\phi(p_j) = q_j$ for $0 \leq j \leq n+1$. (See Fig. 1.4.)

The proof is straightforward and is left to the reader as an exercise (Exercise 1 at the end of this section). The proposition implies in particular that we may take any four points in $\mathbb{P}^2$ in general position as a coordinate triangle.

The group of all projective transformations of $\mathbb{P}^n$ onto itself is written as $PGL(n+1, \mathbb{C})$. It is isomorphic to $GL(n+1, \mathbb{C})/\Delta$, the general linear group $GL(n+1, \mathbb{C})$ of all nonsingular $(n+1) \times (n+1)$-matrices, divided by its center

$$\Delta = \left\{ \begin{bmatrix} \lambda & & 0 \\ & \ddots & \\ 0 & & \lambda \end{bmatrix} \middle| \lambda \in \mathbb{C}^* \right\}$$

The group $PGL(n+1, \mathbb{C})$ coincides with $Aut(\mathbb{P}^n)$, which is the group of all biholomorphic maps of $\mathbb{P}^n$ onto itself (see Exercise 3 of Sec. 3.4). $PGL(2, \mathbb{C}) = Aut(\mathbb{P}^1)$ is the group of all linear fractional transformations

$$\phi: z \in \hat{\mathbb{C}} \to \frac{az + b}{cz + d} \in \hat{\mathbb{C}}, \qquad ad - bc \neq 0$$

Proposition 1.1.1 implies that for any pair of distinct three points $\{p_1, p_2, p_3\}$ and $\{q_1, q_2, q_3\}$ on $\hat{\mathbb{C}}$, there is a unique linear fractional transformation ϕ such that $\phi(p_j) = q_j$ for $1 \le j \le 3$. The reader can directly check this.

It is well known that a linear fractional transformation $\phi: \hat{\mathbb{C}} \to \hat{\mathbb{C}}$ satisfies the following properties: (1) ϕ map circles on $\hat{\mathbb{C}}$ to circles (a circle on $\hat{\mathbb{C}}$ is the intersection of the sphere $\hat{\mathbb{C}}$ and a plane in the real 3-space $\mathbb{R}^3$), (2) ϕ preserves angles, that is, it is conformal (see Fig. 1.5), and (3) the fixed point set of ϕ is nonempty and consists of at most two points, unless ϕ is the identity transformation.

Let z_1, z_2, z_3 and z_4 be points in $\hat{\mathbb{C}}$. The cross ratio $(z_1 z_2, z_3 z_4)$ is defined by

$$(z_1 z_2, z_3 z_4) = \frac{(z_3 - z_1)/(z_3 - z_2)}{(z_4 - z_1)/(z_4 - z_2)}$$

(We put $\infty/\infty = 1$, $k/0 = \infty$ for $k \neq 0$, and so on, for convenience.)

If we put $(z_1 z_2, z_3 z_4) = \lambda$, then

$$\begin{aligned}(z_1 z_2, z_3 z_4) &= (z_3 z_4, z_1 z_2) = (z_2 z_1, z_4 z_3)\\ &= (z_4 z_3, z_2 z_1) = \lambda\end{aligned}$$

$$(z_2 z_1, z_3 z_4) = \frac{1}{\lambda}$$

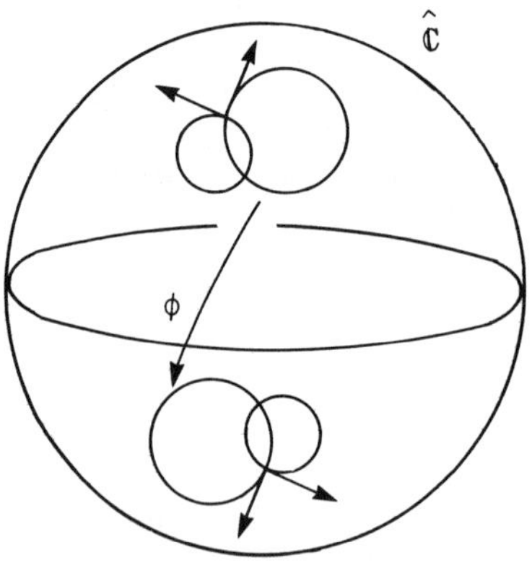

FIGURE 1.5

$$(z_1 z_3, z_2 z_4) = 1 - \lambda$$

$$(z_3 z_1, z_2 z_4) = \frac{1}{1 - \lambda}$$

$$(z_1 z_4, z_2 z_3) = \frac{\lambda - 1}{\lambda}$$

$$(z_1 z_4, z_3 z_2) = \frac{\lambda}{\lambda - 1}$$

Lemma 1.1.2. (1) If z_1, z_2, z_3, and z_4 are mutually distinct and $(z_1 z_2, z_3 z_4) = (z_1 z_2, z_3 z_4')$, then $z_4 = z_4'$. (2) $(z_1 z_2, z_3 z_4)$ is a real number if and only if z_1, z_2, z_3, and z_4 are on a circle on $\hat{\mathbb{C}}$. (3) If $\phi \in \mathrm{Aut}(\mathbb{P}^1)$, then $(\phi(z_1)\phi(z_2), \phi(z_3)\phi(z_4)) = (z_1 z_2, z_3 z_4)$ for any points z_1, z_2, z_3, z_4 on $\hat{\mathbb{C}}$. Conversely, a bijective map $\phi: \mathbb{P}^1 \to \mathbb{P}^1$ satisfies this equality, then ϕ is a linear fractional transformation of $\hat{\mathbb{C}}$.

The proof is easy and is left to the reader (Exercise 1). Let $\mathbb{P}(V)$ (dim $V = 2$) be a projective line. By the lemma, we may define the cross ratio $(p_1 p_2, p_3 p_4)$ for points p_1, p_2, p_3, and p_4 of $\mathbb{P}(V)$ by

$$(p_1 p_2, p_3 p_4) = (z_1 z_2, z_3 z_4)$$

for any affine coordinate system z, where $z = z_j$ at p_j for $1 \leq j \leq 4$.

The sequence of mutually distinct points p_1, p_2, p_3, and p_4 of $\mathbb{P}^1$ is said to be harmonic if $(p_1 p_2, p_3 p_4) = -1$.

Lemma 1.1.3. Let ℓ be a line in $\mathbb{P}^2$. Then the sequence of mutually distinct points p_1, p_2, p_3, and p_4 on ℓ is harmonic if and only if there are points a, b, c, d in $\mathbb{P}^2 - \ell$ such that the configuration in Fig. 1.6 can be drawn.

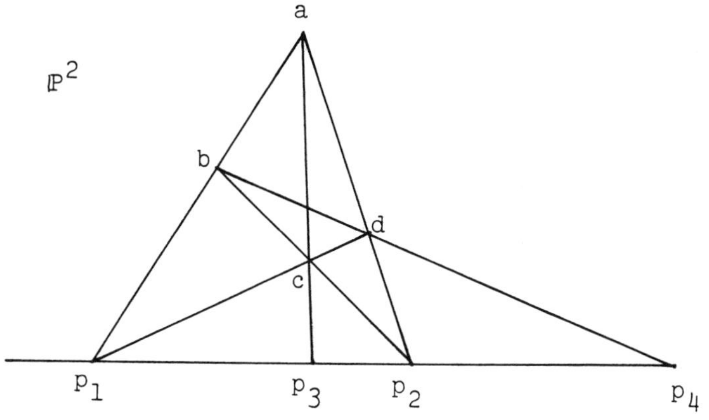

FIGURE 1.6

Proof. We first show the "if" part. We may choose a homogeneous coordinate system on $\mathbb{P}^2$ such that

$$a = (0:0:1),\ p_1 = (1:0:0),\ p_2 = (1:1:0),\ p_4 = (0:1:0)$$

$$\text{and} \quad d = (1:1:1)$$

Then,

$$b = (1:0:1),\ c = (1:\tfrac{1}{2}:\tfrac{1}{2}) \quad \text{and} \quad p_3 = (1:\tfrac{1}{2}:0)$$

Hence,

$$(p_1p_2,\ p_3p_4) = \frac{(\frac{1}{2}-0)/(\frac{1}{2}-1)}{(\infty-0)/(\infty-\frac{1}{2})} = -1$$

Conversely, from the points p_1, p_2, and p_4, we can draw a configuration as above. Put $p_3' = \ell \cap \overline{ac}$. (We denote by $\overline{ac}$ the line passing through the points a and c.) Then,

$$(p_1p_2,\ p_3'p_4) = -1 = (p_1p_2,\ p_3p_4), \quad \text{so} \quad p_3' = p_3$$

Q.E.D.

Lemma 1.1.4. Let ℓ and ℓ' be lines in $\mathbb{P}^n$ $(n \geq 2)$. A homeomorphism $\phi: \ell \to \ell'$ is a projective transformation if and only if (1) harmonic sequences of points on ℓ are mapped by ϕ to harmonic sequences of points on ℓ', and (2) there are mutually distinct points p, q, r, s on ℓ such that (pq, rs) is not a real number and $(pq,\ rs) = (\phi(p)\phi(q),\ \phi(r)\phi(s))$.

Proof. The "only if" part follows from Lemma 1.1.2. To prove the "if" part let z (respectively w) be the affine coordinate system on ℓ (respectively ℓ') such that

$$z = \infty,\ 0,\ 1 \quad \text{at} \quad p,\ q,\ r, \quad \text{respectively}$$

$$(\text{respectively } w = \infty,\ 0,\ 1 \quad \text{at} \quad \phi(p),\ \phi(q),\ \phi(r))$$

Let $z = \lambda$ at s. Then, by condition (2),

$$\lambda \in \mathbb{C} - \mathbb{R} \quad \text{and} \quad w = \lambda \quad \text{at} \quad \phi(s)$$

Let p_z (respectively q_w) be the point on ℓ (respectively ℓ') with the coordinate z (respectively w). It suffices to show that

$$\phi(p_z) = q_z \quad \text{for all} \quad z \in \mathbb{C}$$

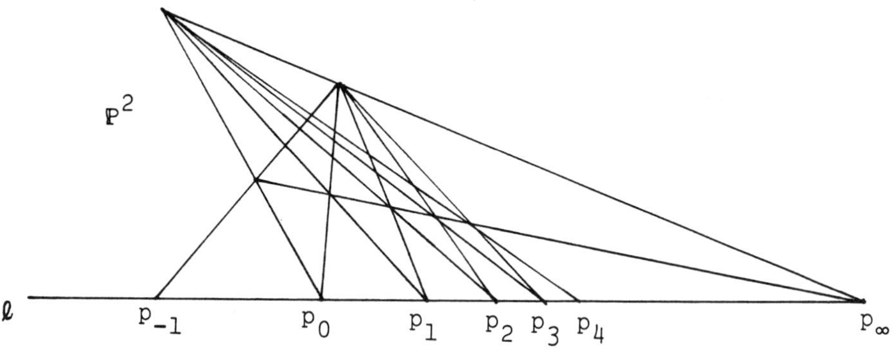

FIGURE 1.7

Now, p_2 (respectively q_2) satisfies $(p_1 p_\infty, p_0 p_2) = -1$ (respectively $(q_1 q_\infty, q_0 q_2) = -1$). Hence, $\phi(p_2) = q_2$ by Condition (1). p_3 (repectively q_3) satisfies $(p_2 p_\infty, p_1 p_3) = -1$ (respectively $(q_2 q_\infty, q_1 q_3) = -1$). Hence, $\phi(p_3) = q_3$. In a similar way, we get

$$\phi(p_n) = q_n \quad \text{for} \quad n = 4, 5, \ldots$$

Next, p_{-1} (respectively q_{-1}) satisfies $(p_0 p_\infty, p_1 p_{-1}) = -1$ (respectively $(q_0 q_\infty, q_1 q_{-1}) = -1$). Hence, $\phi(p_{-1}) = q_{-1}$. Similarly, we get

$$\phi(p_n) = q_n \quad \text{for all} \quad n \in Z$$

If we take a two-plane $\mathbb{P}^2$ in $\mathbb{P}^n$ with $\ell \subset \mathbb{P}^2$, then the configuration is as in Fig. 1.7.

Next, the point $p_{1/n}$ (respectively $q_{1/n}$) satisfies $(p_0 p_1, p_n p_{1/n}) = -1$ (respectively $(q_0 q_1, q_n q_{1/n}) = -1$). Hence, $\phi(p_{1/n}) = q_{1/n}$. Thus

$$\phi(p_r) = q_r \quad \text{for any} \quad r \in Q$$

By the continuity of ϕ,

$$\phi(p_x) = q_x \quad \text{for any} \quad x \in R$$

Finally, take any $z \in \mathbb{C} - \mathbb{R}$ with $z \neq \lambda$. Let Γ be a circle on $\mathbb{C}$ passing through λ and z and cutting the real line at two distinct points x and y. (See Fig. 1.8.) By exchanging p_x, p_y, p_λ (respectively q_x, q_y, q_λ) for p_∞, p_0, p_1 (respectively q_∞, q_0, q_1), and by repeating the above process, we get

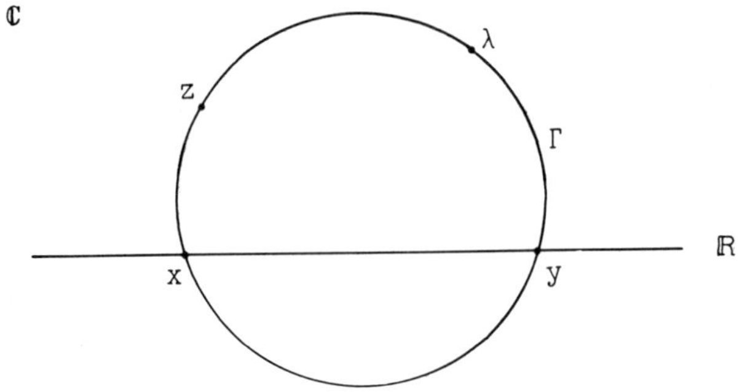

FIGURE 1.8

$$\phi(p_u) = q_u \quad \text{for all} \quad u \in \Gamma$$

In particular, $\phi(p_z) = q_z$. Q.E.D.

Remark 1.1.5. If we drop Condition (2) in Lemma 1.1.4, then ϕ is either (i) a projective transformation or (ii) the composition of a projective transformation and the map $z \to \bar{z}$ (the complex conjugate).

Lemma 1.1.4 is used to prove the following geometric characterization of projective transformation on $\mathbb{P}^n$ $(n \geq 2)$.

Theorem 1.1.6. A homeomorphism $\phi: \mathbb{P}^n \to \mathbb{P}^n$ $(n \geq 2)$ is a projective transformation if and only if (1) ϕ maps lines to lines and (2) on every line ℓ in $\mathbb{P}^n$, there are mutually distinct points p, q, r, s such that (pq, rs) is not a real number and $(pq, rs) = (\phi(p)\phi(q), \phi(r)\phi(s))$.

Proof. The "only if" part is trivial. Let us prove the "if" part. We first show the theorem in the case $n = 2$.

Take a line ℓ in $\mathbb{P}^2$. Then $\phi(\ell)$ is a line by Condition (1). We claim that

$$\phi: \ell \to \phi(\ell)$$

is a projective transformation. By Lemma 1.1.4, it suffices to show that ϕ maps harmonic sequences of points on ℓ to harmonic sequences of points on $\ell' = \phi(\ell)$. Now, take points p, q, r, s, a, b, c, d on $\mathbb{P}^2$ as in Fig. 1.9. Then, this configuration is mapped by ϕ to Fig. 1.10 ($p' = \phi(p)$, and so on). Hence, ϕ is a projective transformation.

Now, take four points p_j, $0 \leq j \leq 3$, in $\mathbb{P}^2$ in general position. Then $p_j' = \phi(p_j)$, $0 \leq j \leq 3$, are in general position. (See Fig. 1.11.) Define the points p_j and p_j' $(4 \leq j \leq 6)$ as in the figure. Then $p_j' = \phi(p_j)$ for $4 \leq j \leq 6$. By

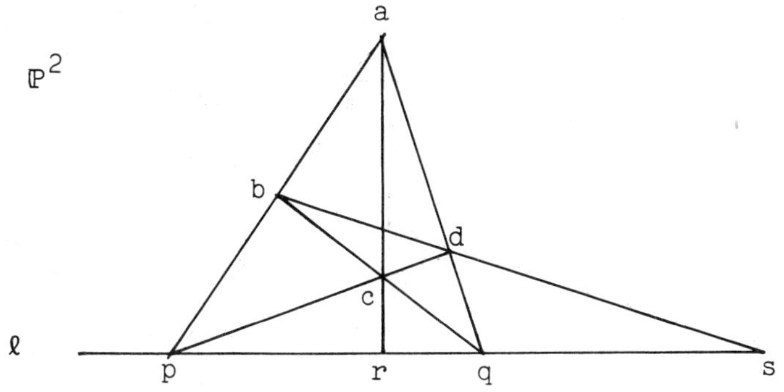

FIGURE 1.9

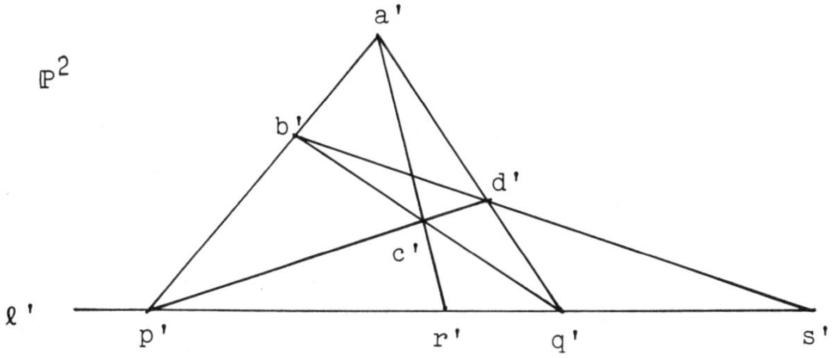

FIGURE 1.10

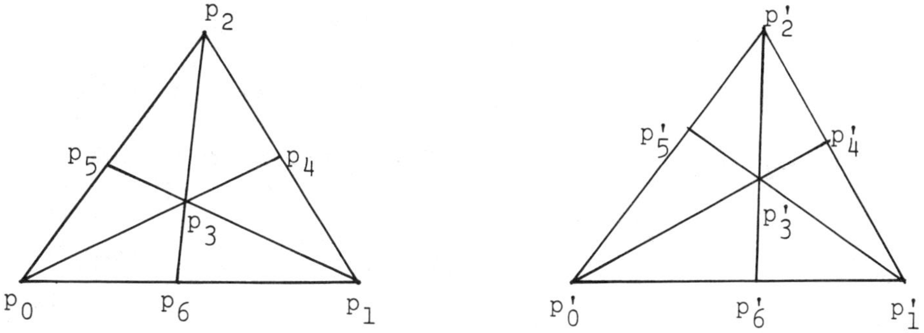

FIGURE 1.11

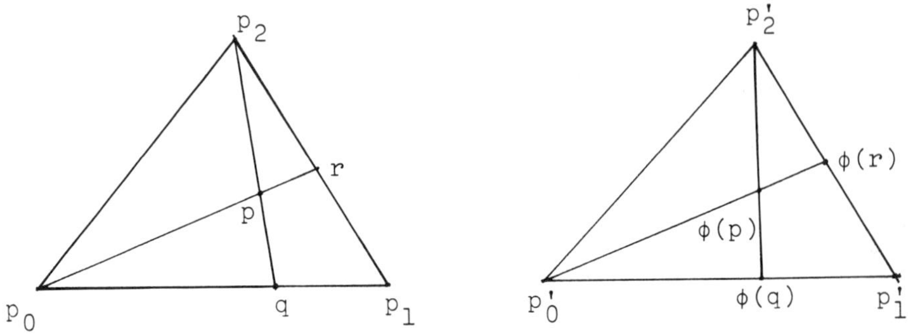

FIGURE 1.12

Proposition 1.1.1, there is a unique projective transformation ψ of $\mathbb{P}^2$ such that $\psi(p_j) = \phi(p_j)$ for $0 \leq j \leq 3$. We show that $\phi = \psi$. It is clear that $\phi(p_j) = \psi(p_j)$ for $4 \leq j \leq 6$. Now, ϕ: $\overline{p_0 p_1} \to \overline{p_0' p_1'}$ is a projective transformation. Hence, $\phi = \psi$ on $\overline{p_0 p_1}$. In a similar way, $\phi = \psi$ on $\overline{p_1 p_2}$, and so on. Take a point $p \in \mathbb{P}^2$ which is not on these lines. Put $q = \overline{p_0 p_1} \cap \overline{p_2 p}$ and $r = \overline{p_1 p_2} \cap \overline{p_0 p}$. (See Fig. 1.12.) Then

$$\phi(p) = \overline{p_2' \phi(q)} \cap \overline{p_0' \phi(r)} = \overline{\psi(p_2)\psi(q)} \cap \overline{\psi(p_0)\psi(r)} = \psi(p)$$

This proves the theorem for $n = 2$.

Next we prove the theorem for $n = 3$. (The general case is similar by induction.) It is easy to see that ϕ maps planes to planes. For a plane H in $\mathbb{P}^3$, the restriction ϕ: $H \to \phi(H)$ is a projective transformation by the case $n = 2$.

Take five points p_j $(0 \leq j \leq 4)$ in $\mathbb{P}^3$ in general position. (See Fig. 1.13.) $(p_j' = \phi(p_j)$ for $0 \leq j \leq 4$.) The rest is parallel to the case $n = 2$. Q.E.D.

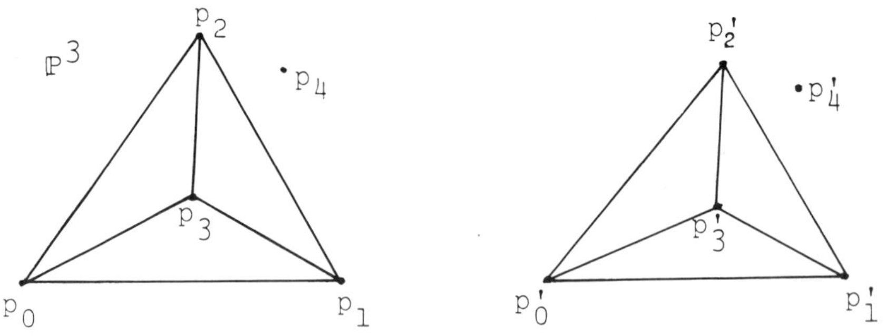

FIGURE 1.13

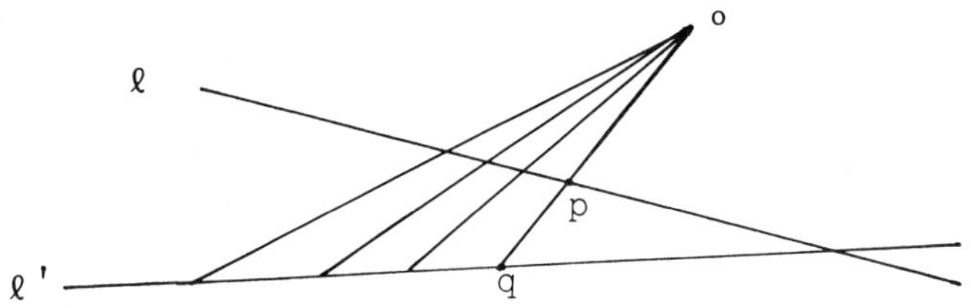

FIGURE 1.14

Now, let ℓ and ℓ' be lines in $\mathbb{P}^2$. For points $p_1, \ldots, p_m$ on ℓ and $q_1, \ldots, q_m$ on ℓ', we denote

$$p_1 \cdots p_m \barwedge q_1 \cdots q_m$$

if there is a projective transformation $\phi: \ell \to \ell'$ such that $\phi(p_j) = q_j$ for $1 \leq j \leq m$. This is a classical notation.

If $p_1p_2p_3p_4 \barwedge q_1q_2q_3q_4$, then $(p_1p_2, p_3p_4) = (q_1q_2, q_3q_4)$. The converse is also true. If $p_1p_2p_3p \barwedge p_1p_2p_3q$, then $p = q$.

Take a point $o \in \mathbb{P}^2$ such that $o \notin \ell$ and $o \notin \ell'$. The <u>projection</u> π_o <u>with the center</u> o gives a projective transformation $\ell \to \ell'$. (See Fig. 1.14.) Here $\pi_o(p) = q$, where $p \in \ell$ and $q \in \ell'$ and o, p, q are <u>collinear</u>, that is, on a line. For points $p_1, \ldots, p_m$ on ℓ and $q_1, \ldots, q_m$ on ℓ', we denote

$$p_1 \cdots p_m \overset{\underline{\underline{o}}}{\barwedge} q_1 \cdots q_m$$

if $\pi_o(p_j) = q_j$ for $1 \leq j \leq m$. (See Fig. 1.15.) This is also a classical notation. Using these notations, we prove the classical

<u>Theorem 1.17 (Desargues)</u>. Let Δpqr and $\Delta p'q'r'$ be triangles in $\mathbb{P}^2$. Then

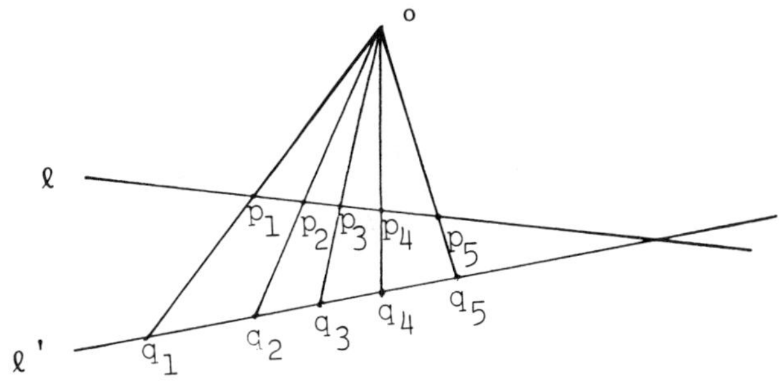

FIGURE 1.15

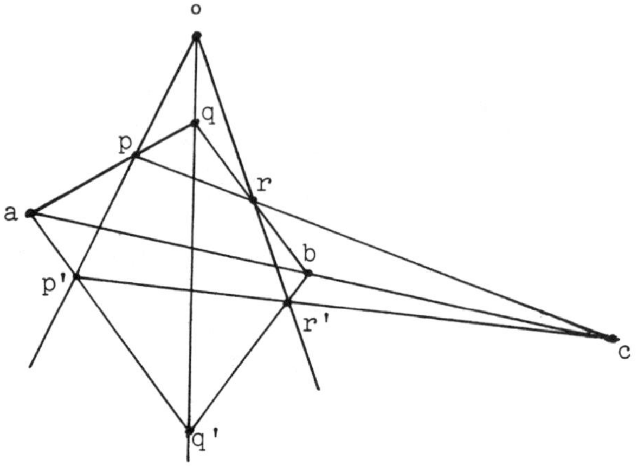

FIGURE 1.16

the lines $\overline{pp'}$, $\overline{qq'}$, and $\overline{rr'}$ meet at one point if and only if the points

$$a = \overline{pq} \cap \overline{p'q'}, \quad b = \overline{qr} \cap \overline{q'r'} \quad \text{and} \quad c = \overline{rp} \cap \overline{r'p'}$$

are collinear. (See Fig. 1.16.)

<u>Proof</u>: Suppose that $\overline{pp'}$, $\overline{qq'}$, and $\overline{rr'}$ meet at a point o. Put $s = \overline{ac} \cap \overline{op}$, $t = \overline{ac} \cap \overline{oq}$, $u = \overline{ac} \cap \overline{or}$, and $u' = \overline{bt} \cap \overline{or}$. (See Fig. 1.17.) It suffices to show that u = u'. Now,

$$orr'u' \overset{b}{\bar{\bar{\wedge}}} oqq't \overset{a}{\bar{\bar{\wedge}}} opp's \overset{c}{\bar{\bar{\wedge}}} orr'u$$

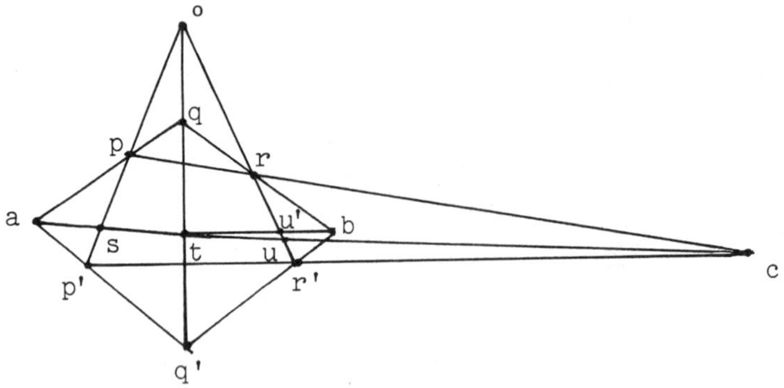

FIGURE 1.17

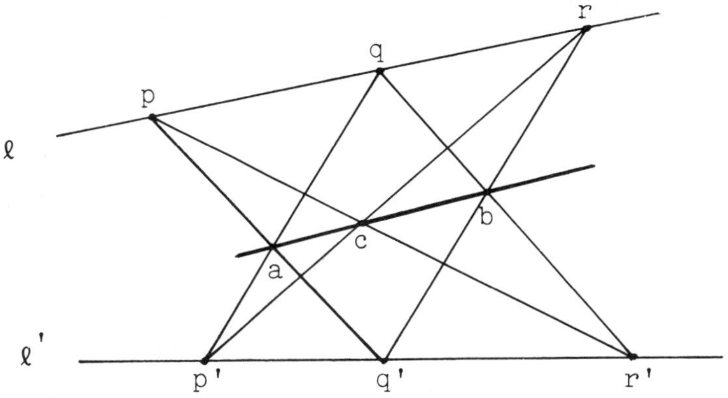

FIGURE 1.18

Hence, orr'u' $\overline{\wedge}$ orr'u, so u' = u.

The converse follows automatically if we exchange

$$o \to c,\ \Delta pqr \to \Delta pap',\ \Delta p'q'r' \to \Delta rbr'$$

$$a \to q,\ b \to q' \quad \text{and} \quad c \to o$$

(as in Figure 1.17). Q.E.D.

Theorem 1.1.8 (Pappus). Let ℓ and ℓ' be distinct lines in $\mathbb{P}^2$. Take distinct points p, q, r, on ℓ and p', q', r' on ℓ'. Suppose that none is the point ℓ ∩ ℓ'. Then the points

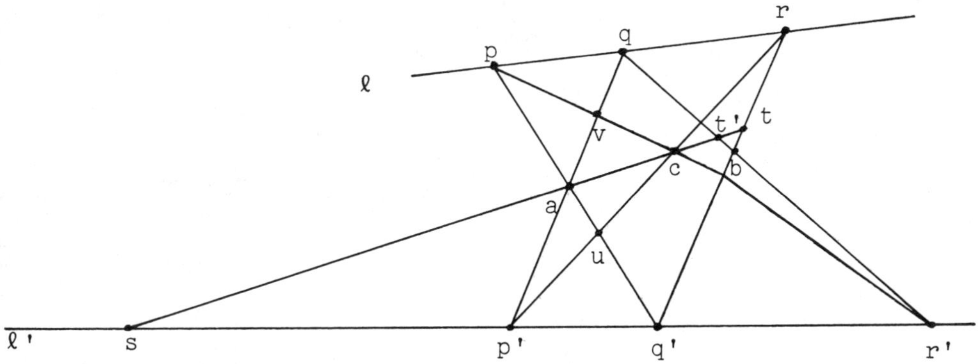

FIGURE 1.19

$$a = \overline{pq'} \cap \overline{qp'}, \quad b = \overline{qr'} \cap \overline{rq'}, \quad \text{and} \quad c = \overline{rp'} \cap \overline{pr'}$$

are collinear. (See Fig. 1.18.)

Proof: Put $s = \overline{ac} \cap \ell'$, $t = \overline{ac} \cap \overline{rq'}$, $t' = \overline{ac} \cap \overline{qr'}$, $u = \overline{pq'} \cap \overline{rp'}$ and $v = \overline{pr'} \cap \overline{qp'}$. (See Fig. 1.19.) It suffices to show that $t = t'$. We have

$$\text{sact}' \overset{\underline{\underline{r'}}}{\wedge} \text{p'avq} \overset{\underline{\underline{p}}}{\overline{\wedge}} \text{p'ucr} \overset{\underline{\underline{q'}}}{\wedge} \text{sact}$$

Hence, sact' $\overline{\wedge}$ sact, so $t' = t$. Q.E.D.

The above theorems relate incidence relations among points and lines in $\mathbb{P}^2$.

If there is a proposition about incidence relations among points and lines in $\mathbb{P}^2$, then we get a new proposition, called the dual proposition, by exchanging points for lines and lines for points. In fact, it suffices to read the proposition in the dual space $\mathbb{P}^{2*}$ to $\mathbb{P}^2$. For example, the statement "3 lines meet at a point" in $\mathbb{P}^2$ can be read as the statement "3 points are collinear" in $\mathbb{P}^{2*}$. (See Fig. 1.20.)

This principle, established by Plücker, is called the principle of duality. Note that the dual propositions to Desargues' Theorem and Pappus' Theorem are nothing but themselves, so they are self-dual propositions.

Finally, we generalize the notion of projections, which plays an important role in this book.

Let S and T be an r-plane and an $(n - r - 1)$-plane in $\mathbb{P}^n$, respectively, such that $S \cap T = \phi$. For every point $p \in \mathbb{P}^n - S$, the $(r + 1)$-plane $p \vee S$, spanned by p and S, meets T at a unique point q. Then the map

$$\pi_S: p \in \mathbb{P}^n - S \to q \in T$$

is called the projection with the center S. (See Fig. 1.21.)

The set of all $(r + 1)$-planes containing S can be naturally regarded as a $(n - r - 1)$-dimensional projective space $\mathbb{P}^{n-r-1}$. The definition of the projection π_S is sometimes

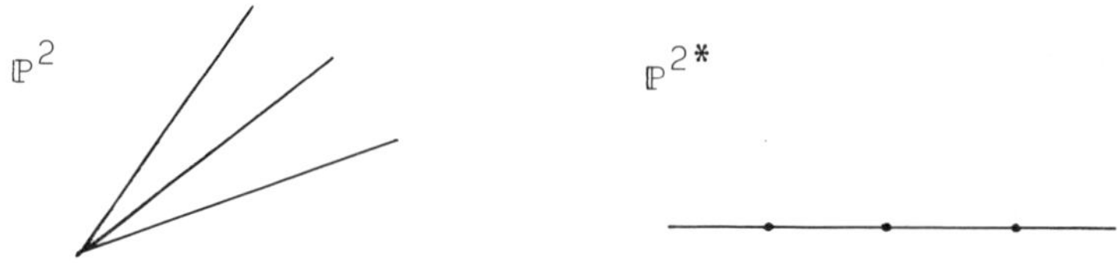

FIGURE 1.20

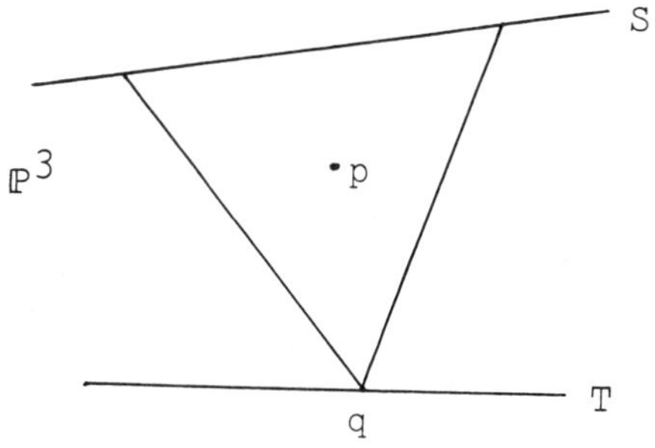

FIGURE 1.21

$$\pi_S: p \in \mathbb{P}^n - S \to p \vee S \in \mathbb{P}^{n-r-1}$$

Exercises

1. Prove Proposition 1.1.1, Lemma 1.1.2, and the assertion in Remark 1.1.5.
2. Prove that, given a s-plane S in $\mathbb{P}^n$, there are hyperplanes $H_1, \ldots, H_{n-s}$ such that $S = H_1 \cap \cdots \cap H_{n-s}$.
3. Prove the following assertion: A projective transformation ϕ of $\mathbb{P}^2$ can be written as one of the following forms under a suitable affine coordinate system: (i) $(x,y) \to (ax, by)$, (ii) $(x,y) \to (ax + by, ay)$, or (iii) $(x,y) \to (x + ay, y + a)$, where a and b are nonzero constants. Generalize this assertion to projective transformations of $\mathbb{P}^n$.
4. Let ℓ and ℓ' be distinct lines in $\mathbb{P}^2$. Take a point $0 \in \mathbb{P}^2$ such that $0 \notin \ell$ and $0 \notin \ell'$. Prove: (i) The restriction to ℓ of the projection $\pi_0: \mathbb{P}^2 - \{0\} \to \ell'$ with the center 0 can be extended to a projective transformation ϕ of the form

$$\phi: (X_0 : X_1 : X_2) \to (X_0 : X_1 : aX_2 + bX_0)$$

(a ($\neq 0$) and b are constants), under a suitable homogeneous coordinate system; (ii) the fixed point set of ϕ is $\ell_\phi \cup \{0\}$, where ℓ_ϕ is a line. ($0 \in \ell_\phi$ or $0 \notin \ell_\phi$); (iii) for any point p and q ($p \neq q$) in $\mathbb{P}^2 - \{0\}$ with $\overline{pq} \not\ni 0$, the lines $\overline{pq}$ and $\overline{\phi(p)\phi(q)}$ meet at a point on ℓ_ϕ. (iv) Using (iii), prove Desargues' Theorem.
5. Generalize Desargues' Theorem to higher dimensional cases.
6. Prove the following: Let $\alpha: V \to W$ be a surjective linear map of vector spaces V and W with $\dim V = n + 1$ and $\dim W = n - r$. Put $S = \mathbb{P}(\ker \alpha)$ and let $\hat{\alpha}: \mathbb{P}(V) - S \to \mathbb{P}(W)$ be the induced map. Then, for any $(n-r-1)$-plane T in $\mathbb{P}(V)$ with $S \cap T = \phi$, the restriction $\phi = \hat{\alpha} | T: T \to \mathbb{P}(W)$ is a projective transformation such that $\phi\pi_S = \hat{\alpha}$, where $\pi_S: \mathbb{P}(V) - S \to T$ is the projection with the center S.

1.2. PROJECTIVE GEOMETRY OF CONICS

A line ℓ in $\mathbb{P}^2$ is defined by a linear equation

$$\ell: L = a_0X_0 + a_1X_1 + a_2X_2 = 0$$

L is a linear form, that is, a homogeneous polynomial of degree 1 in the variables X_0, X_1, X_2 and is determined by ℓ up to nonzero constant multiple. ℓ is a point of the dual projective plane $\mathbb{P}^{2*}$.

Three lines ℓ_1, ℓ_2, and ℓ_3 are collinear as points in $\mathbb{P}^{2*}$ if and only if they meet at a point. Putting $\ell_j = \{L_j = 0\}$, $1 \leq j \leq 3$, this occurs if and only if there are constants c_j, $1 \leq j \leq 3$, not all zero, such that

$$c_1L_1 + c_2L_2 + c_3L_3 = 0$$

as a polynomial. The set of all lines passing through a fixed point is called a <u>linear pencil of lines in</u> $\mathbb{P}^2$. Let $\ell = \{L = 0\}$ and $\ell' = \{L' = 0\}$ be distinct lines in $\mathbb{P}^2$. Then the linear pencil of lines in $\mathbb{P}^2$ passing through $p = \ell \cap \ell'$ is given by

$$\Lambda = \{\ell_\lambda = \{\lambda_0 L + \lambda_1 L' = 0\} \mid \lambda = (\lambda_0 : \lambda_1) \in \mathbb{P}^1\}$$

(See Fig. 1.22.)

Now, let $F = F(X_0, X_1, X_2)$ be a two-form, that is, a homogeneous polynomial of degree 2, in the variables X_0, X_1, X_2. The subset

$$C = V(F) = \{(X_0 : X_1 : X_2) \in \mathbb{P}^2 \mid F(X_0, X_1, X_2) = 0\}$$

is well defined and is called a <u>conic</u>. F = 0 is called a <u>defining equation of the conic</u> C. F can be written as

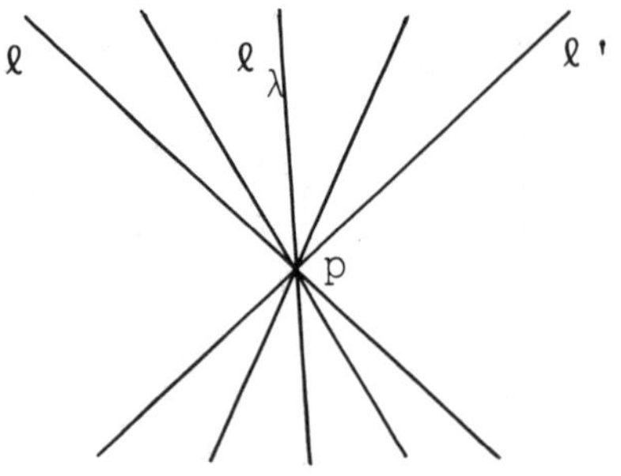

FIGURE 1.22

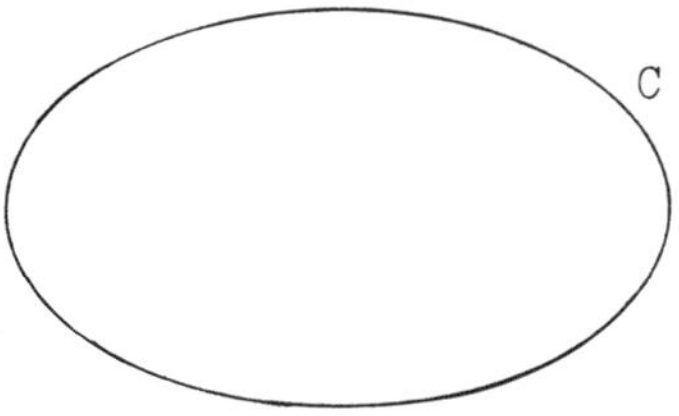

FIGURE 1.23

$$F = \sum_{j,k=0}^{2} a_{jk} X_j X_k \quad (a_{jk} = a_{kj})$$

The polynomial F is either (1) irreducible or (2) $F = L_1 L_2$, where L_1 and L_2 are linear forms. In Case (1), the conic C is said to be irreducible. (See Fig. 1.23.) In case (2), $C = \{L_1 = 0\} \cup \{L_2 = 0\}$ is the union of two lines and C is said to be reducible. The lines $\{L_1 = 0\}$ and $\{L_2 = 0\}$ are called the irreducible components of C. (See Figure 1.24.)

It can be shown that C is irreducible if and only if

$$\begin{vmatrix} a_{00} & a_{01} & a_{02} \\ a_{10} & a_{11} & a_{12} \\ a_{20} & a_{21} & a_{22} \end{vmatrix} \neq 0$$

(Exercise 1).

In Case (2), it may happen that $\{L_1 = 0\} = \{L_2 = 0\}$. In this case, C is set theoretically a line. But we consider it a double line. (See Fig. 1.25.)

For conics $V(F_1)$ and $V(F_2)$, $V(F_1) = V(F_2)$ if and only if a nonzero constant c exists such that $F_2 = cF_1$. Hence, it is more precise to define a conic to be an equivalence class of two-forms under the $\mathbb{C}^*$-action

$$(c, F) \to cF$$

Then the set of all conics is nothing but the projective space $\mathbb{P}(W)$, where

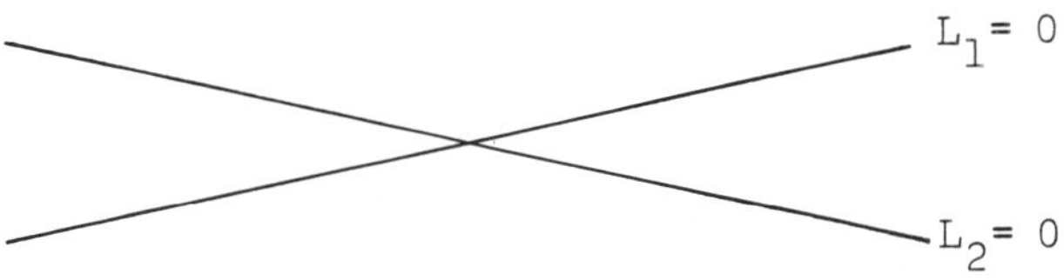

FIGURE 1.24

FIGURE 1.25

W is the vector space of all two-forms. $\mathbb{P}(W)$ is five-dimensional. A homogeneous coordinate system in $\mathbb{P}(W)$ is given by

$$F(\text{mod } \mathbb{C}^*) \in \mathbb{P}(W) \to (a_0 : \cdots : a_5) \in \mathbb{P}^5$$

where

$$F(X_0, X_1, X_2) = a_0 X_0^2 + a_1 X_0 X_1 + a_2 X_0 X_2 + a_3 X_1^2 + a_4 X_1 X_2 + a_5 X_2^2 \quad (*)$$

We sometimes identify $\mathbb{P}(W)$ and $\mathbb{P}^5$ through this bijection. The set of all irreducible conics is open in $\mathbb{P}^5$.

For an r-plane S in $\mathbb{P}^5$, the set of all conics in S is called a linear system of conics of dimension r. If $r = 1$, it is called a linear pencil of conics. If $S = \mathbb{P}^5$, it is called the complete linear system of conics.

Lemma 1.2.1. Let $C = V(F)$ be an irreducible conic and ℓ be a line in $\mathbb{P}^2$. Then $C \cap \ell$ is nonempty and consists of at most two points.

Proof. Choose a homogeneous coordinate system in $\mathbb{P}^2$ such that $\ell = \{X_0 = 0\}$. If F is given by (*), then

$$F(0, X_1, X_2) = a_3 X_1^2 + a_4 X_1 X_2 + a_5 X_2^2$$

This is not zero as a polynomial of X_1 and X_2, because C is irreducible. Hence, $F(0, x_1, X_2) = 0$ has at least one, and at most two, roots $(X_1 : X_2)$.

Q.E.D.

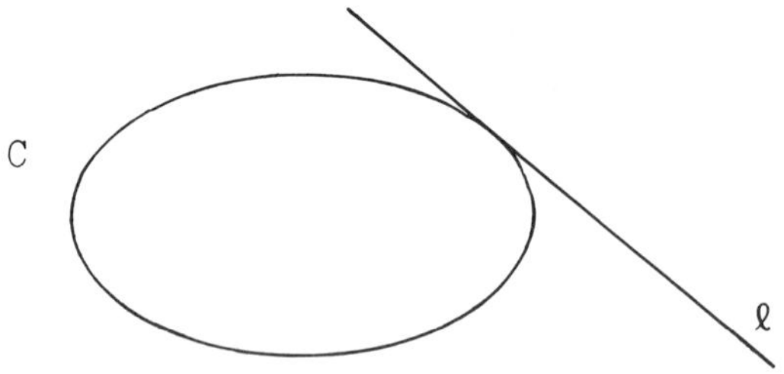

FIGURE 1.26

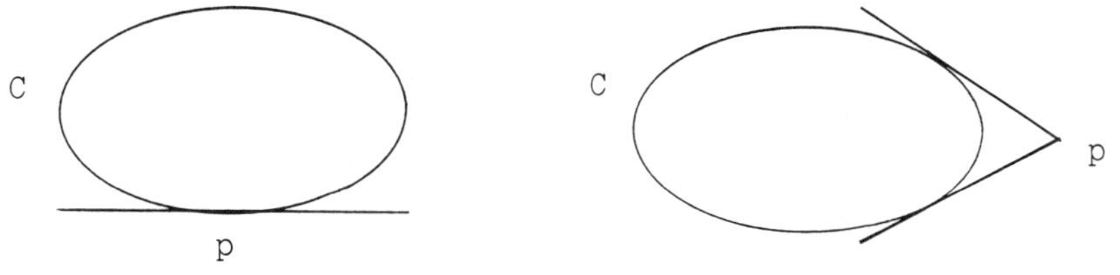

FIGURE 1.27

Using the notations in the proof, $C \cap \ell$ consists of one point if and only if $a_4^2 - 4a_3a_5 = 0$. If this is the case, ℓ is said to be tangent to C at $(0: X_1: X_2)$. (See Fig. 1.26.)

The proof of the following lemma is left to the reader (Exercise 2).

Lemma 1.2.2. Let C be an irreducible conic. (1) If $p \in C$, then C has a unique tangent line to C at p (which is denoted by T_pC). (2) If $p \in \mathbb{P}^2 - C$, then exactly two lines pass through p which are tangent to C. (See Fig. 1.27.)

Now we prove

Lemma 1.2.3. Any two irreducible conics can be mapped each other by projective transformations.

Proof. Let C be an irreducible conic. It suffices to show that the equation of C is given by $X_0X_2 - X_1^2 = 0$ in a suitable homogeneous coordinate system $(X_0 : X_1 : X_2)$. Take mutually distinct points p_0, p_2, and p_3 on C. Let p_1 be the intersection point of $T_{p_0}C$ and $T_{p_2}C$. (See Fig. 1.28.) Clearly, p_0, p_1, p_2, and p_3 are in general position. Take the homogeneous coordinate system $(X_0 : X_1 : X_2)$ in $\mathbb{P}^2$ such that

$$p_0 = (1: 0: 0), \quad p_1 = (0: 1: 0), \quad p_2 = (0: 0: 1) \quad \text{and} \quad p_3 = (1: 1: 1)$$

Then $C = \{X_0X_2 - X_1^2 = 0\}$. Q.E.D.

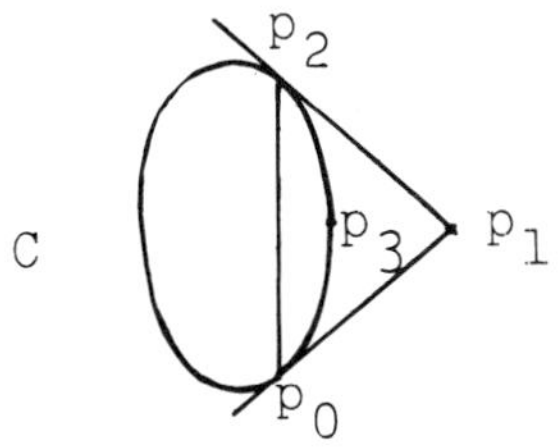

FIGURE 1.28

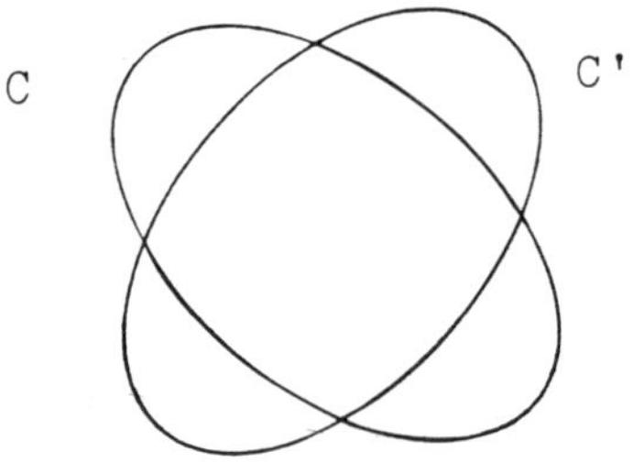

FIGURE 1.29

Remark 1.2.4. Thus, there is no distinction between an ellipse, a parabola and a hyperbola on the complex projective plane.

The following lemma is a special case of Bezout's Theorem (see Sec. 1.3). It is easy to prove it directly (Exercise 2).

Lemma 1.2.5. Let C and C' be distinct irreducible conics. Then $C \cap C'$ is non-empty and consists of at most four points. (See Fig. 1.29.)

Proposition 1.2.6. Let p_1, p_2, p_3, p_4, and p_5 be five distinct points on $\mathbb{P}^2$. Then a conic passes through all these points. Such a conic is uniquely determined, unless four or five points are collinear.

Proof. Put $p_j = (b_{j0}: b_{j1}: b_{j2})$, $1 \leq j \leq 5$. Then the simultaneous linear equations

$$a_0b_{j0}^2 + a_1b_{j0}b_{j1} + a_2b_{j0}b_{j2} + a_3b_{j1}^2 + a_4b_{j1}b_{j2} + a_5b_{j2}^2 = 0, \quad 1 \leq j \leq 5$$

have the nontrivial solution $a_0, \ldots, a_5$. Then the conic

$$C: a_0X_0^2 + a_1X_0X_1 + a_2X_0X_2 + a_3X_1^2 + a_4X_1X_2 + a_5X_2^2 = 0$$

passes through p_j, $1 \leq j \leq 5$.

If no three points of p_j are collinear, then C must be irreducible. It is unique by Lemma 1.2.5. (See Fig. 1.30.)

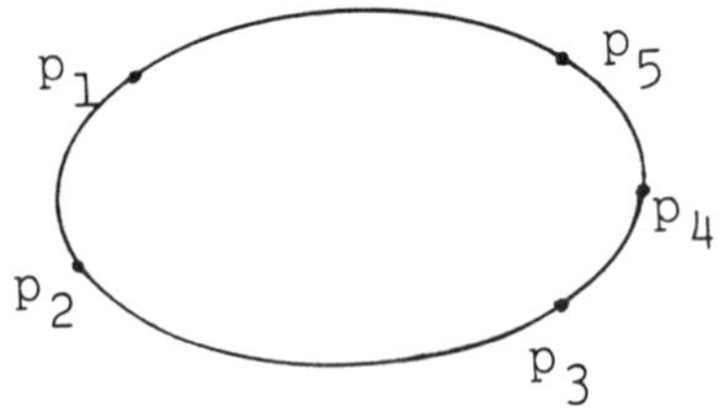

FIGURE 1.30

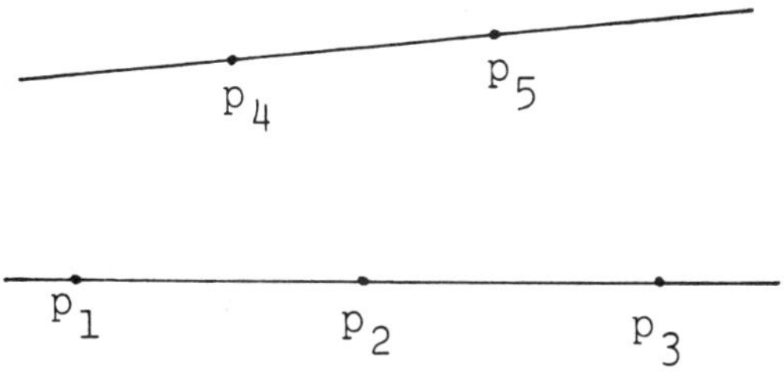

FIGURE 1.31

If p_1, p_2, and p_3 are collinear and $p_4 \notin \overline{p_1p_2}$ and $p_5 \notin \overline{p_1p_2}$, then C must be the union of two lines $C = \overline{p_1p_2} \cup \overline{p_4p_5}$. Hence, C is again uniquely determined. (See Fig. 1.31.)

If four or five points of p_i are collinear, then C is not uniquely determined. (See Fig. 1.32.) Q.E.D.

Corollary 1.2.7. Let p_1, p_2, p_3, and p_4 be not all collinear, mutually distinct points on $\mathbb{P}^2$. Then

$$\Lambda = \{C \mid C \text{ is a conic passing through } p_1,\ p_2,\ p_3, \text{ and } p_4\}$$

is a linear pencil of conics. (See Fig. 1.33.)

Proof. Take a conic $C = \{F = 0\}$ passing through p_1, p_2, p_3, and p_4. Take a point p_5' in $\mathbb{P}^2 - C$. Let $C' = \{F' = 0\}$ be a conic passing through p_1, p_2, p_3, p_4, and p_5'. Then

$$\Lambda' = \{C_\lambda = \{\lambda_0 F + \lambda_1 F' = 0\} \mid \lambda = (\lambda_0 : \lambda_1) \in \mathbb{P}^1\}$$

is a linear pencil of conics. Clearly, $\Lambda' \subset \Lambda$. To show that $\Lambda' = \Lambda$, take $C'' \in \Lambda$ and a point $p_5'' = (b_0 : b_1 : b_2)$ on C'' such that no four points of p_1, p_2, p_3, p_4, and p_5'' are collinear. Let $\lambda = (\lambda\ : \lambda_1) \in \mathbb{P}^1$ satisfy

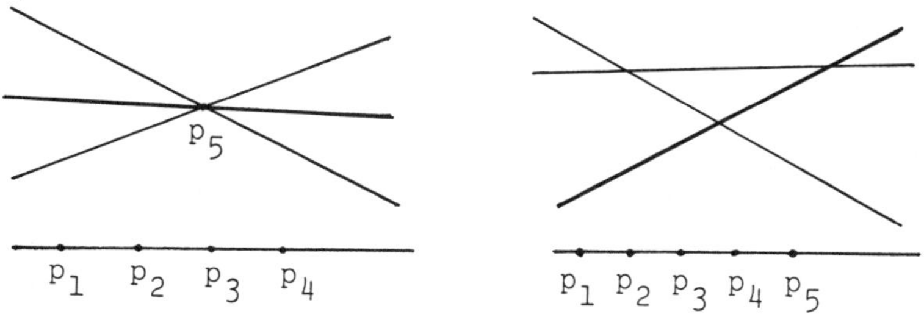

FIGURE 1.32

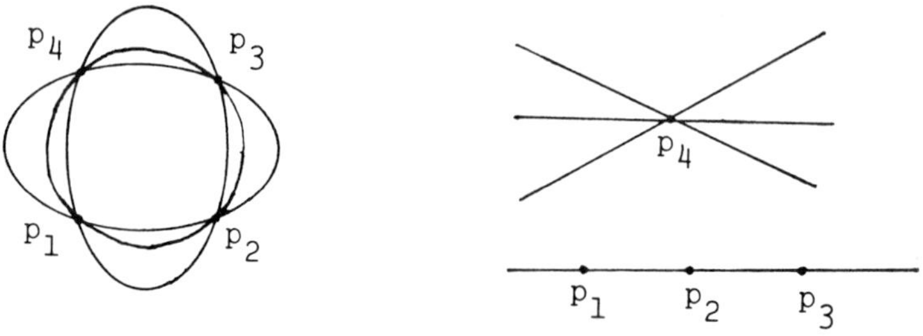

FIGURE 1.33

$$\lambda_0 F(b_0, b_1, b_2) + \lambda_1 F'(b_0, b_1, b_2) = 0$$

Then $C'' = C_\lambda$ by Proposition 1.2.6. Q.E.D.

Remark 1.2.8. Not every linear pencil of conics can be obtained in this way. For example,

$$\Lambda = \{C_\lambda = \{\lambda_0 X_0 X_2 - \lambda_1 X_1^2 = 0\} \mid \lambda = (\lambda_0 : \lambda_1) \in \mathbb{P}^1\}$$

is a linear pencil of conics as in Fig. 1.34.

Now let C be an irreducible conic and $p \in \mathbb{P}^2 - C$. Let T_qC and T_rC $(q \neq r)$ be tangent lines to C passing through p. The line $\overline{qr}$ is called the polar line of p with respect to C, and the point is called the polar point of the line $\overline{qr}$ with respect to C (p is uniquely determined by $\overline{qr}$). See Fig. 1.35.) If $p \in C$, Then T_pC itself is the polar line of p with respect to C, and p is the polar point of T_pC with respect to C.

The proof of the following lemma is left to the reader (Exercise 2).

Lemma 1.2.9. Let C be an irreducible conic defined by the equation $F = \Sigma_{j,k=0}^{2} a_{jk} X_j X_k = 0$ $(a_{jk} = a_{kj})$. Then, for $p = (b_0 : b_1 : b_2) \in \mathbb{P}^2$,

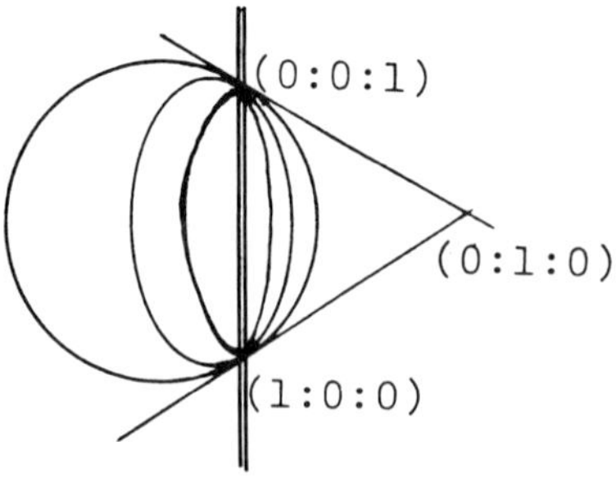

FIGURE 1.34

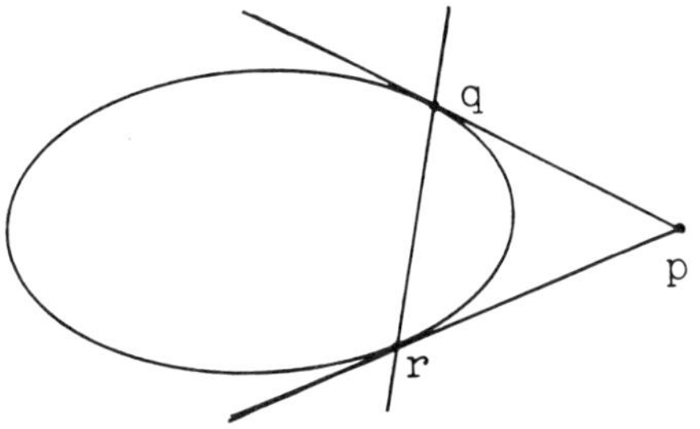

FIGURE 1.35

(1) the equation of the polar line of p with respect to C is given by

$$\sum_{j,k=0}^{2} a_{jk} b_j X_k = 0$$

(in particular, if $p \in C$, then this is the equation of T_pC), and (2) if a point q is on the polar line of p, then p is on the polar line of q.

Proposition 1.2.10. Let C be an irreducible conic. Then the map

$$\psi : p \in \mathbb{P}^2 \to \ell \in \mathbb{P}^{2*}$$

where ℓ is the polar line with respect to C, is a projective transformation.

Proof. Under the notations in Lemma 1.2.9, ψ is given by

$$\psi : p = (b_0 : b_1 : b_2) \in \mathbb{P}^2 \to (\Sigma_j a_{j0} b_j : \Sigma_j a_{j1} b_j : \Sigma_j a_{j2} b_j) \in \mathbb{P}^{2*}$$ Q.E.D.

By the proposition, $\psi(C)$ is an irreducible conic in $\mathbb{P}^{2*}$. $\psi(C)$ is the set of all tangent lines to C and is called the dual conic to C. Let $\psi^* : \mathbb{P}^{2*} \to \mathbb{P}^2$ be the projective transformation in the proposition with respect to the conic $\psi(C)$. Then, it is clear that

$$\psi^*\psi = \text{the identity transformation of } \mathbb{P}^2$$

If ℓ is a line in $\mathbb{P}^2$, then $\psi(\ell)$ is a line in $\mathbb{P}^{2*}$. $\psi(\ell)$ is the set of all lines passing through the polar point of ℓ with respect to C. (See Fig. 1.36.)

Thus, if we are given a proposition on a configuration of points and lines around an irreducible conic C, then we get a new proposition, called the dual proposition with respect to C, by exchanging

points $\to$ the polar lines of the points

lines $\to$ the polar points of the lines

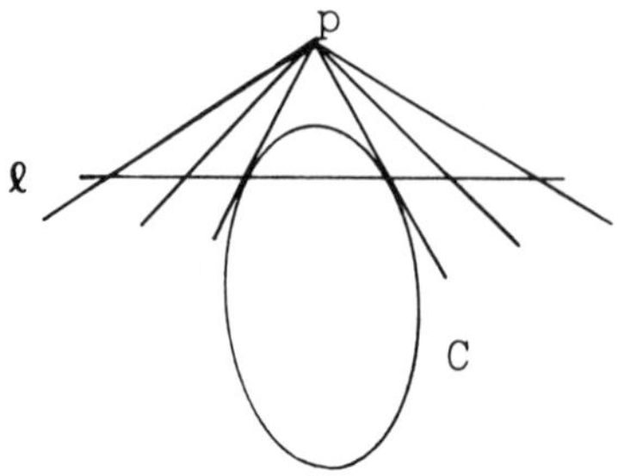

FIGURE 1.36

This principle was established by Poncelet and was called the principle of polar duality, or polar reciprocity.

Proposition 1.2.11. Let C be an irreducible conic. Let p_1, p_2, and p_3 be mutually distinct points on $\mathbb{P}^2$. Let ℓ_j, $1 \leq j \leq 3$, be the polar lines of p_j, $1 \leq j \leq 3$, with respect to C. Put

$$q_1 = \ell_2 \cap \ell_3, \quad q_2 = \ell_3 \cap \ell_1 \quad \text{and} \quad q_3 = \ell_1 \cap \ell_2$$

Then the lines $\overline{p_1 q_1}$, $\overline{p_2 q_2}$, and $\overline{p_3 q_3}$ meet at one point. (See Fig. 1.37.)

Proof. Put $p_\nu = (b_0^\nu : b_1^\nu : b_2^\nu)$, $1 \leq \nu \leq 3$. Let C be defined by

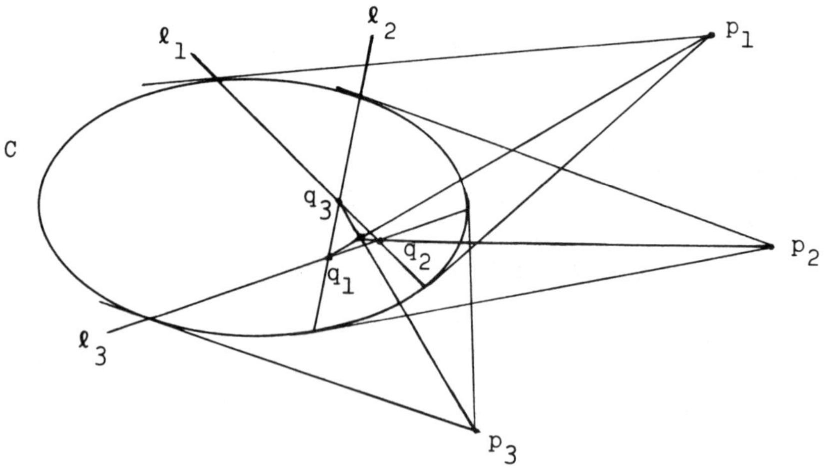

FIGURE 1.37

$$C: \sum_{j,k} a_{jk} X_j X_k = 0 \quad (a_{jk} = a_{kj})$$

Then, ℓ_ν, $1 \leq \nu \leq 3$, are given by

$$\ell_\nu: L_\nu = \sum_{j,k} a_{jk} b_j^\nu X_k = 0, \quad 1 \leq \nu \leq 3$$

Put $c_{\nu\mu} = L_\nu(p^\mu) = \Sigma_{j,k}\, a_{jk} b_j^\nu b_k^\mu$. Then $c_{\nu\mu} = c_{\mu\nu}$.

Now, the line $\overline{p_1 q_1}$ is given by the equation

$$c_{13} L_2 - c_{12} L_3 = 0 \tag{1}$$

In fact, this line passes through $q_1 = \ell_2 \cap \ell_3$. This line also passes through p_1, because

$$c_{13} L_2(p_1) - c_{12} L_3(p_1) = c_{13} c_{21} - c_{12} c_{31} = 0$$

Similarly, the lines $\overline{p_2 q_2}$ and $\overline{p_3 q_3}$ are given by the equations

$$c_{12} L_3 - c_{23} L_1 = 0 \tag{2}$$

$$c_{23} L_1 - c_{13} L_2 = 0 \tag{3}$$

respectively. Now we have (1) + (2) + (3) = 0, as a polynomial. This means that these lines meet at one point. Q.E.D.

The dual proposition to Proposition 1.2.11 is Proposition 1.2.11*.

Proposition 1.2.11*. Under the same notations as in Proposition 1.2.11, the points $r_1 = \ell_1 \cap \overline{p_2 p_3}$, $r_2 = \ell_2 \cap \overline{p_3 p_1}$, and $r_3 = \ell_3 \cap \overline{p_1 p_2}$ are collinear.

Note that Proposition 1.2.11* follows directly from Proposition 1.2.11 and Desargues' Theorem.

Corollary 1.2.12. Let $\Delta p_1 p_2 p_3$ be a triangle circumscribed to an irreducible conic C. Put

$$q_1 = C \cap \overline{p_2 p_3}, \quad q_2 = C \cap \overline{p_3 p_1}, \quad \text{and} \quad q_3 = C \cap \overline{p_1 p_2}$$

Then the lines $\overline{p_1 q_1}$, $\overline{p_2 q_2}$, and $\overline{p_3 q_3}$ meet at one point. (See Fig. 1.38.)

Proof. Proposition 1.2.11 can be applied if

$$\ell_1 = \overline{q_2 q_3}, \quad \ell_2 = \overline{q_3 q_1}, \quad \text{and} \quad \ell_3 = \overline{q_1 q_2}$$ Q.E.D.

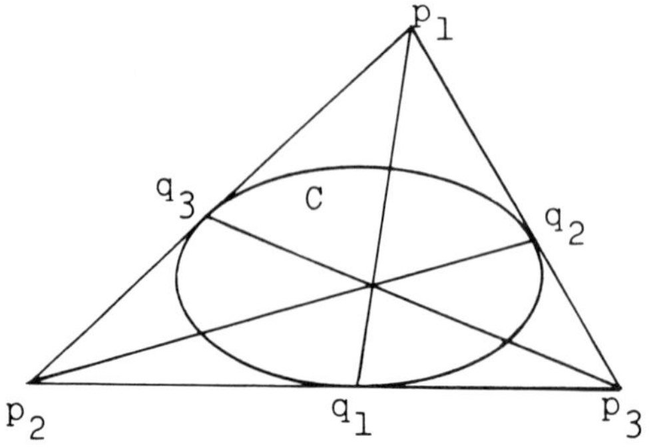

FIGURE 1.38

Now we prove the very famous and beautiful theorem by Pascal, which, it is said, he found at age sixteen.

Theorem 1.2.13 (Pascal). If a hexagon is inscribed in an irreducible conic C, then the opposite sides meet in collinear points. Conversely, if the opposite sides of a hexagon $p_1, \ldots, p_6$ meet in collinear points and if $p_1, \ldots, p_5$ are on an irreducible conic C, then $p_6 \in C$. (See Fig. 1.39.)

Proof. Put

$$q_1 = \overline{p_1p_2} \cap \overline{p_4p_5}, \quad q_2 = \overline{p_2p_3} \cap \overline{p_5p_6}, \quad \text{and} \quad q_3 = \overline{p_3p_4} \cap \overline{p_6p_1}$$

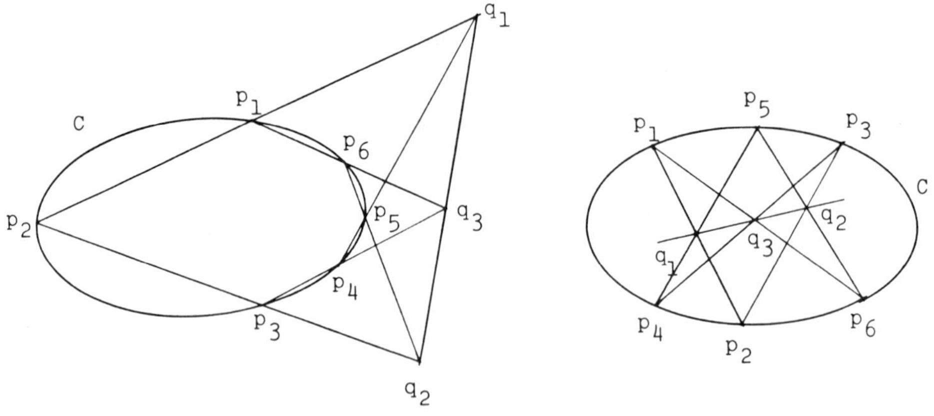

FIGURE 1.39

Let $L_{jk} = 0$ $(j \neq k)$ be the equation of the line $\overline{p_j p_k}$. Then there is a nonzero constant k such that the conic C is defined by the equation

$$L_{12} L_{34} - kL_{23} L_{14} = 0 \tag{1}$$

In fact, the conic C' defined by (1) passes through p_1, p_2, p_3, and p_4. If k is so chosen that

$$L_{12}(p_5) L_{34}(p_5) - kL_{23}(p_5) L_{14}(p_5) = 0$$

then $p_5 \in C'$. Hence, $C' = C$ by Proposition 1.2.6. In a similar way, there is a nonzero constant k' such that C is defined by the equation

$$L_{45} L_{16} - k' L_{56} L_{14} = 0 \tag{2}$$

Hence, equations (1) and (2) differ only by a nonzero constant multiple. By multiplying a suitable nonzero constant to L_{45} and L_{56}, we may assume that

$$L_{12} L_{34} - kL_{23} L_{14} = L_{45} L_{16} - k' L_{56} L_{14}$$

as polynomials. Hence, as polynomials,

$$L_{12} L_{34} - L_{45} L_{16} = L_{14}(kL_{23} - k' L_{56})$$

Consider the conic

$$C'': L_{12} L_{34} - L_{45} L_{16} = 0$$

It passes through p_1 and p_4. It also passes through q_1 and q_3, since

$$L_{12}(q_1) = L_{45}(q_1) = 0 \quad \text{and} \quad L_{34}(q_3) = L_{16}(q_3) = 0$$

On the other hand, $C'' = \overline{p_1 p_4} \cup \ell$, where

$$\ell: kL_{23} - k' L_{56} = 0$$

is a line. Since $\overline{p_1 p_4}$ does not pass through q_1 nor q_3, ℓ must pass through q_1 and q_2. Moreover, ℓ passes through q_2, because

$$L_{23}(q_2) = L_{56}(q_2) = 0$$

Hence, q_1, q_2, and q_3 are collinear. This proves the first half of the theorem.

Conversely, let p_6' be the point in $\overline{p_1 p_6} \cap C$ other than p_1. Put

$$q_2' = \overline{p_2 p_3} \cap \overline{p_5 p_6'}$$

Then, by the first half of the theorem, q_1, q_2', and q_3 are collinear. Hence, $q_2' = q_2$, so $p_6' = p_6$. Q.E.D.

The dual proposition to Pascal's Theorem is known as Brianchon's Theorem.

Theorem 1.2.13* (Brianchon). If a hexagon $p_1, \ldots, p_6$ is circumscribed to an irreducible conic, then three diagonal lines $\overline{p_1p_4}$, $\overline{p_2p_5}$, and $\overline{p_3p_6}$ meet at one point. (See Fig. 1.40.)

We can deduce this theorem from Pascal's Theorem as follows. (See Figure 1.41.)

Let q_j, $1 \le j \le 6$, and r_j, $1 \le j \le 3$, be the points as in the figure. Then r_1 is on the polar lines of p_2 and p_5. Hence, $\overline{p_2p_5}$ is the polar line of r_1 by Lemma 1.2.9. In a similar way, $\overline{p_3p_6}$ (respectively $\overline{p_1p_4}$) is the polar line of r_2 (respectively r_3). By Pascal's Theorem, r_1, r_2, and r_3 are collinear. Hence, their polar lines meet at one point. This proves Brianchon's Theorem.

Now, let C be an irreducible conic. Take a point $p \in C$ and a line ℓ such that $p \notin \ell$. Consider the restriction to C of the projection

$$\pi_p : q \in C \to \overline{pq} \cap \ell \in \ell \quad (q \ne p)$$

$$p \to T_pC \cap \ell \in \ell$$

with the center p. Then π_p is a homeomorphism of C onto ℓ. (See Fig. 1.42.) We leave the proof of the following theorem to the reader (Exercise 2).

Theorem 1.2.14 (Steiner). Let p and p' be distinct points on an irreducible conic C. Take lines ℓ and ℓ' such that $p \notin \ell$ and $p' \notin \ell'$. Then

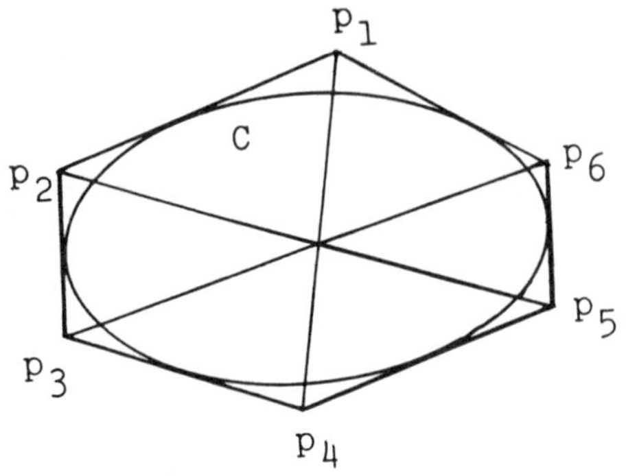

FIGURE 1.40

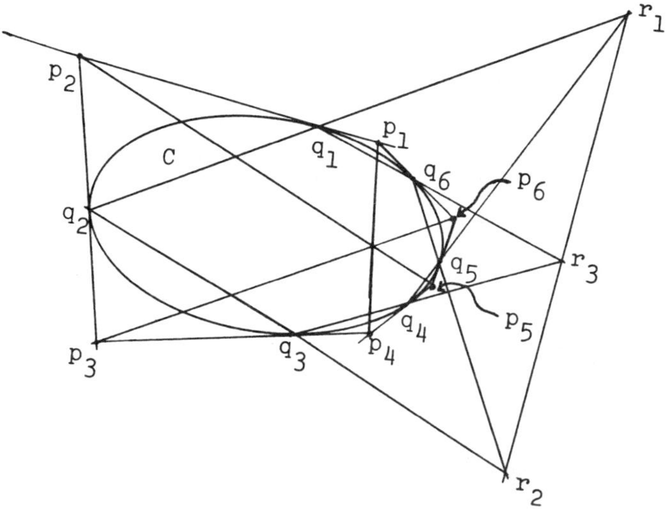

FIGURE 1.41

$$\phi = \pi_{p'}\pi_p^{-1} : \ell \to \ell'$$

is a projective transformation. Conversely, if points p, p' ($p \neq p'$), lines ℓ, ℓ' ($p \notin \ell$, $p' \notin \ell'$) and a projective transformation $\phi: \ell \to \ell'$ are given, then the closure of the set $\{q = \overline{pr} \cap \overline{p'\phi(r)} \mid r \in \ell\}$ is either an irreducible conic passing through p and p' or a line. (See Fig. 1.43.)

Steiner constructed this theorem on the basis of his study of the projective geometry of conics. For example, Pascal's Theorem can be deduced from Steiner's Theorem in the following way. Take mutually distinct points $p_1, \ldots, p_5$ on an irreducible conic C, say a line ℓ passing through p_5, but not through any other points p_j, $1 \leq j \leq 4$. Finally, put $q_1 = \ell \cap \overline{p_1 p_4}$,

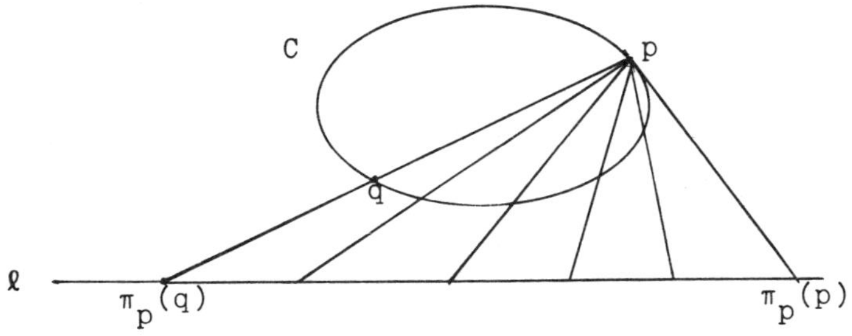

FIGURE 1.42

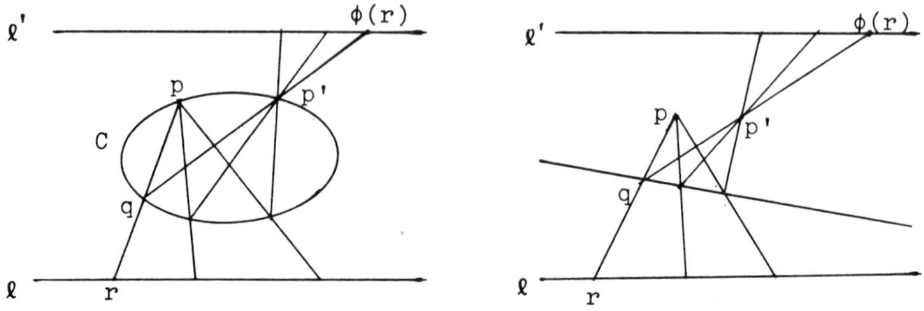

FIGURE 1.43

$q_2 = \ell \cap \overline{p_3p_4}$, $q_3 = \ell \cap \overline{p_1p_2}$, $r_1 = \ell \cap \overline{p_2p_3}$, $r_3 = \overline{p_1p_2} \cap \overline{p_4p_5}$, $r_2 = \overline{p_3p_4} \cap \overline{r_1r_3}$, $r_4 = \overline{p_1r_2} \cap \overline{p_4p_5}$, and $p_6 = \ell \cap \overline{p_1r_2}$. (See Fig. 1.44.) To prove Pascal's Theorem, it suffices to show that $p_6 \in C$. Consider the projections

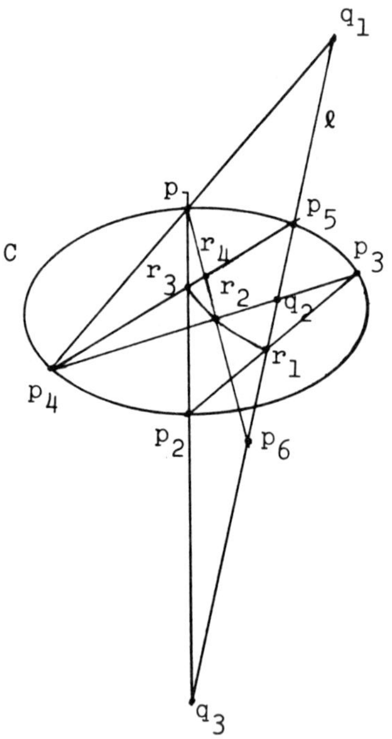

FIGURE 1.44

$$\pi = \pi_{p_4} : C \to \ell$$

$$\pi' = \pi_{p_2} : C \to \ell$$

Then

$$\phi = \pi'\pi^{-1} : \ell \to \ell$$

is a projective transformation by Steiner's Theorem. Note that $C \cap \ell$ is the fixed point set of ϕ and note also that

$$\phi: p_5 \to p_5, \quad q_1 \to q_3, \quad \text{and} \quad q_2 \to r_1$$

On the other hand,

$$p_5 q_1 q_2 p_6 \overset{p_4}{\overline{\overline{\wedge}}} r_4 p_1 r_2 p_6 \overset{r_3}{\overline{\overline{\wedge}}} p_5 q_3 r_1 p_6$$

This means that $\phi(p_6) = p_6$, that is, p_6 is the fixed point of ϕ other than p_5. Hence, $p_6 \in C$. This proves Pascal's Theorem.

The dual proposition to Steiner's Theorem is

Theorem 1.2.14*. Let ℓ_1 and ℓ_2 be distinct tangent lines to an irreducible conic C. For any tangent line ℓ to C, put $p = \ell \cap \ell_1$ and $q = \ell \cap \ell_2$. (If $\ell = \ell_1$, put $p = C \cap \ell_1$, and so on.) Then

$$\phi: p \in \ell_1 \to q \in \ell_2$$

is a projective transformation. (See Fig. 1.45.)

Conversely, if ℓ_1 and ℓ_2 $(\ell_1 \neq \ell_2)$ are lines in $\mathbb{P}^2$ and $\phi: \ell_1 \to \ell_2$ be a projective transformation, then the closure in $\mathbb{P}^{2*}$ of the set of the lines $\{\overline{p\phi(p)} \mid p \in \ell_1\}$ is either (1) the set of all tangent lines to an irreducible conic or (2) the set of all lines passing through a fixed point. (See Fig. 1.46.)

Note 1.2.15. The contents of this and the previous section are classical. See, for example, Baker [9]. The presentation given here is mainly from

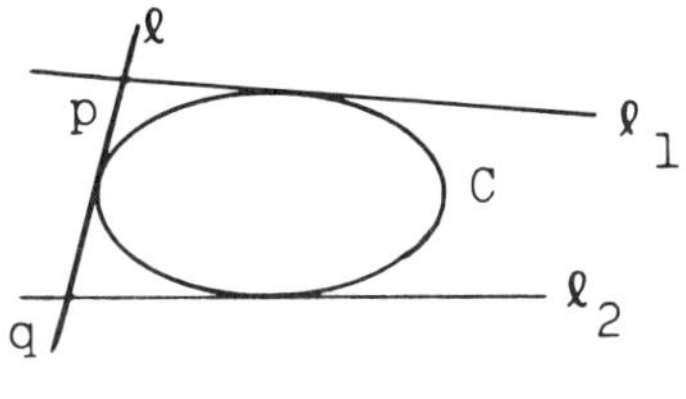

FIGURE 1.45

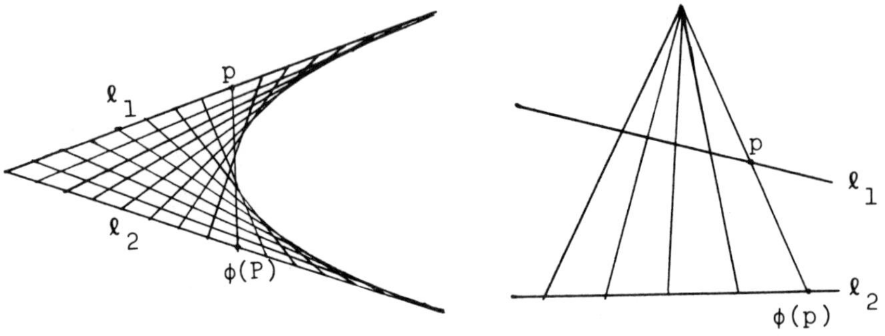

FIGURE 1.46

lectures given by Professor S. Sasaki when the author was an undergraduate student.

Exercises

1. Prove that if $F = \Sigma_{j,k=0}^{2} a_{jk}X_jX_k$ $(a_{jk} = a_{kj})$, then the conic $F = 0$ is irreducible if and only if

$$\begin{vmatrix} a_{00} & a_{01} & a_{02} \\ a_{10} & a_{11} & a_{12} \\ a_{20} & a_{21} & a_{22} \end{vmatrix} \neq 0$$

2. Prove Lemmas 1.2.2, 1.2.5, and 1.2.9. Also, prove Theorem 1.2.14.

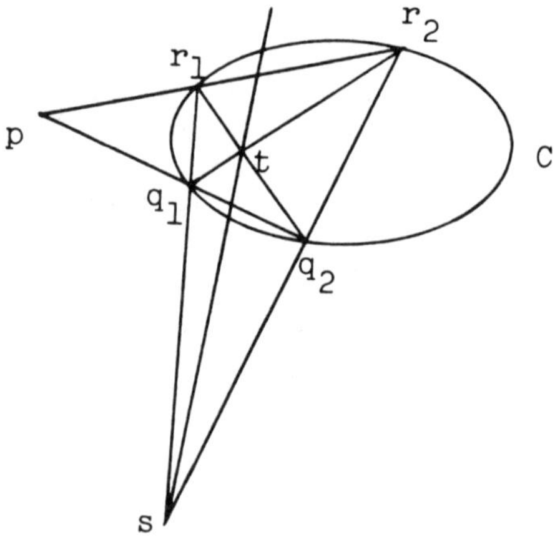

FIGURE 1.47

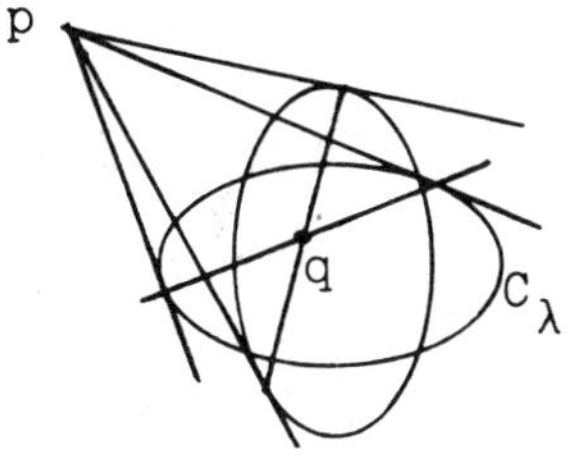

FIGURE 1.48

3. Let C be an irreducible conic. Take a point $p \in \mathbb{P}^2 - C$. Let ℓ_1 and ℓ_2 $(\ell_1 \neq \ell_2)$ be lines passing through p and are not tangent to C. Put $\ell_1 \cap C = \{q_1, q_2\}$, $\ell_2 \cap C = \{r_1, r_2\}$, $s = \overline{q_1 r_1} \cap \overline{q_2 r_2}$ and $t = \overline{q_1 r_2} \cap \overline{q_2 r_1}$. Now, prove that the line $\overline{st}$ is the polar line of p with respect to C. (See Fig. 1.47.)
4. Prove the following proposition Let $\Lambda = \{C_\lambda = \{\lambda_0 F + \lambda_1 F' = 0\} \mid \lambda = (\lambda_0 : \lambda_1) \in \mathbb{P}^1\}$ be a linear pencil of conics. For a fixed point $p \in \mathbb{P}^2$, the polar lines ℓ_λ of p with respect to irreducible C_λ's pass through a fixed point q (independent of λ).
5. Let C be an irreducible conic and p_1, p_2, p_3, and p_4 be mutually distinct points on C. Put $q_1 = \overline{p_1 p_2} \cap \overline{p_3 p_4}$, $q_2 = \overline{p_1 p_4} \cap \overline{p_2 p_3}$, $q_3 = T_{p_1} C \cap T_{p_3} C$ and $q_4 = T_{p_2} C \cap T_{p_4} C$. Prove that q_1, q_2, q_3, and q_4 are collinear and state its dual proposition. (See Fig. 1.49.)
6. Prove the following theorem of Brianchon: If two triangles $\Delta p_1 p_2 p_3$ and $\Delta q_1 q_2 q_3$ are circumscribed to an irreducible conic C, then the six points p_1, p_2, p_3, q_1, q_2, q_3 are on a conic. State the dual proposition to this theorem. (See Fig. 1.50.)

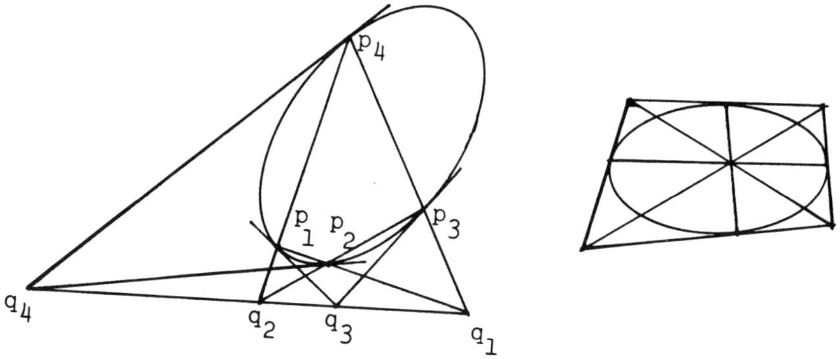

FIGURE 1.49

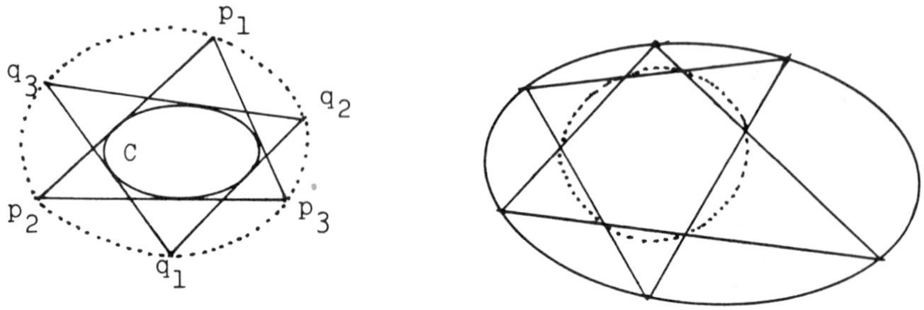

FIGURE 1.50

1.3 BEZOUT'S THEOREM

Let $F = F(X_0, X_1, X_2)$ be a homogeneous polynomial of degree n ($n \geq 1$) in the variables X_0, X_1, and X_2. The set

$$V(F) = \{(X_0 : X_1 : X_2) \in \mathbb{P}^2 \mid F(X_0, X_1, X_2) = 0\}$$

is well-defined and is called a plane algebraic curve or simply a plane curve. This is a primary definition. However, since it is more convenient to allow a plane curve to have multiple components, we define a plane curve as follows:

A plane (algebraic) curve of degree n ($n \geq 1$) is an equivalence class of homogeneous polynomials of degree n in the variables X_0, X_1, X_2 under the $\mathbb{C}^*$-action: $(c, F) \to cF$.

Using a homogeneous polynomial F, we denote a plane curve by F (mod $\mathbb{C}^*$) or simply by F.

Plane curves F and G are said to be projectively equivalent if there is a projective transformation $\phi: (X_0 : X_1 : X_2) \in \mathbb{P}^2 \to (\Sigma a_{0j}X_j : \Sigma a_{1j}X_j : \Sigma a_{2j}X_j) \in \mathbb{P}^2$, such that

$$G(\Sigma a_{0j}X_j, \Sigma a_{1j}X_j, \Sigma a_{2j}X_j) = F(X_0, X_1, X_2)$$

as polynomials. We denote this relation by $\phi(F) = G$ or $\phi^*G = F$ or $G \circ \phi = F$. A projective property (respectively invariant) of a plane curve is a property (respectively invariant) which is invariant under projective equivalence, that is, under any change of homogeneous coordinate systems. For example, the degree, deg F, of a plane curve F is a projective invariant.

Plane curves of degree 1, 2, 3, 4, 5, 6, ... are called lines, conics, cubics, quartics, quintics, sextics, and so forth, respectively. (See Fig. 1.51.)

As is well known, $\mathbb{C}[X_0, X_1, X_2]$ is a unique factorization domain (UFD), so a homogeneous polynomial F can be decomposed into irreducible divisors

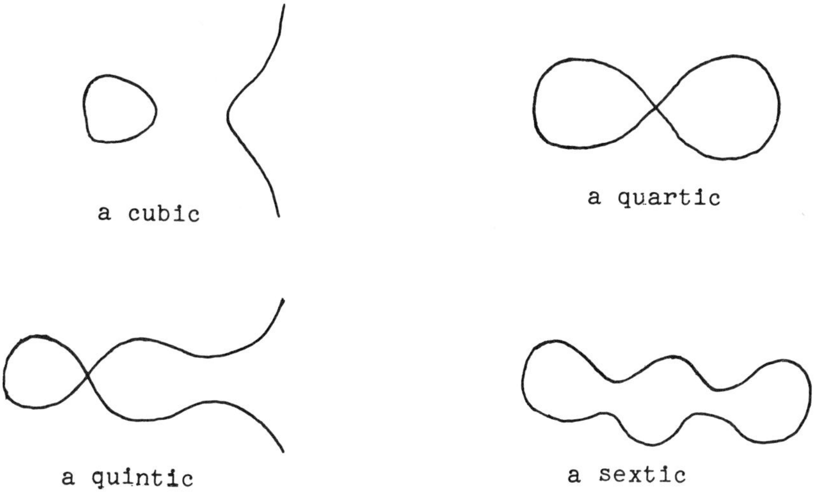

FIGURE 1.51

$$F = F_1^{\nu_1} \cdots F_s^{\nu_s} \qquad (\nu_j \geq 1)$$

Note that each F_j is homogeneous. The curve F_j is called an irreducible component of the plane curve F with the multiplicity ν_j. If $\nu_j \geq 2$, then the curve $F_j^{\nu_j}$ is called a multiple component. It is a ν_j-ple curve. For example, $F = F_1^2 F_2^2$ (deg $F_1 = 1$, deg $F_2 = 2$) is a sextic curve as in Fig. 1.52. F is said to be irreducible if F is irreducible as a polynomial. Otherwise, F is said to be reducible. Put

$$V(F) = \{(X_0 : X_1 : X_2) \in \mathbb{P}^2 \mid F(X_0, X_1, X_2) = 0\}$$

Then, as a set,

$$V(F) = V(F_1) \cup \cdots \cup V(F_s) = V(F_1 \cdots F_s)$$

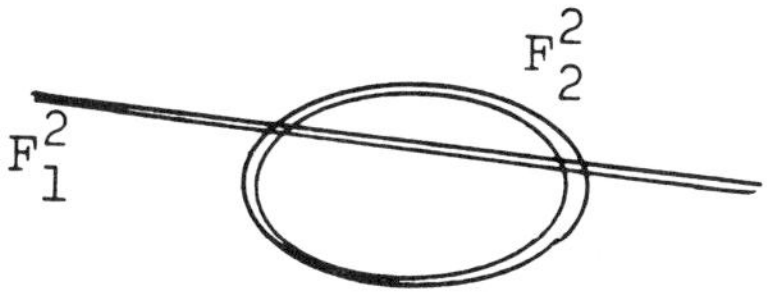

FIGURE 1.52

The following proposition is a special case of the Projective Nullstellensatz (see Sec. 3.3).

Proposition 1.3.1. For any polynomial F of degree n ($n \geq 1$), (1) $V(F) \neq \mathbb{P}^2$ and (2) V(F) is nonempty.

Proof.

1. Write

$$F = a_0 X_2^n + a_1(X_0, X_1)X_2^{n-1} + \cdots + a_n(X_0, X_1)$$

where $a_j(X_0, X_1)$ is a homogeneous polynomial in X_0, X_1 of degree j. Suppose that $V(F) = \mathbb{P}^2$. Take any $(\lambda_0, \lambda_1) \in \mathbb{C}^2$. Then

$$a_0 X_2^n + a_1(\lambda_0, \lambda_1)X_2^{n-1} + \cdots + a_n(\lambda_0, \lambda_1) = 0$$

as a polynomial in X_2. Hence,

$$a_0 = a_1 = \cdots = a_n = 0, \quad \text{i.e.,} \quad F = 0$$

2. Suppose that $(0: 0: 1) \notin V(F)$. Writing F as above, $a_0 \neq 0$. Hence, for any $(\lambda_0, \lambda_1) \in \mathbb{C}^2 - \{0\}$, there is a solution of the equation

$$a_0 X^n + a_1(\lambda_0, \lambda_1)X^{n-1} + \cdots + a_n(\lambda_0, \lambda_1) = 0$$ Q.E.D.

Proposition 1.3.2. Let F and G be plane curves. Suppose that (1) F has no multiple component and (2) $V(F) \subset V(G)$. Then the polynomial G can be divided by F.

Proof. We may choose a homogeneous coordinate system $(X_0: X_1: X_2)$ such that $(0: 0: 1) \notin V(F)$. Write

$$F = a_0 X_2^n + a_1(X_0, X_1)X_2^{n-1} + \cdots + a_n(X_0, X_1), \quad a_0 \neq 0$$

$$G = b_0 X_2^m + b_1(X_0, X_1)X_2^{m-1} + \cdots + b_m(X_0, X_1)$$

Consider F and G polynomials in X_2 over the ring $\mathbb{C}[X_0, X_1]$ and divide G by F

$$G = AF + B$$

A and B are homogeneous polynomials, because $a_0 \neq 0$. Write

$$B = c_0X_2^q + c_1(X_0,X_1)X_2^{q-1} + \cdots + c_q(X_0,X_1), \qquad q \leq n - 1$$

The discriminant $D(X_0, X_1)$ of F is not zero as a polynomial in X_0, X_1, because F has no multiple component. Take any $(\lambda_0 : \lambda_1) \in \mathbb{P}^1$ such that $D(\lambda_0, \lambda_1) \neq 0$. Then there are mutually distinct complex numbers $\lambda_{21}, \ldots, \lambda_{2n}$ such that $F(\lambda_0, \lambda_1, \lambda_{2j}) = 0$, $1 \leq j \leq n$. By assumption (2), $G(\lambda_0, \lambda_1, \lambda_{2j}) = 0$, $1 \leq j \leq n$. Hence,

$$B(\lambda_0, \lambda_1, \lambda_{2j}) = 0, \qquad 1 \leq j \leq n$$

Thus,

$$c_0 = c_1(\lambda_0, \lambda_1) = \cdots = c_q(\lambda_0, \lambda_1) = 0$$

This is true for any $(\lambda_0 : \lambda_1) \in \mathbb{P}^1$ except finite points. Hence,

$$c_0 = c_1 = \cdots = c_q = 0, \qquad \text{that is,} \qquad B = 0$$

Q.E.D.

Corollary 1.3.3. For plane curves F and G without multiple component, $V(F) = V(G)$ if and only if $F = G \pmod{\mathbb{C}^*}$.

Thus we may identify a plane curve F without multiple component with the subset V(F) of $\mathbb{P}^2$. $F = 0$ is then called the defining equation of $C = V(F)$.

Let F be a plane curve. Suppose that $p = (0: 0: 1) \notin V(F)$. As in the proof of Proposition 1.3.1, write

$$F = a_0X_2^n + a_1(X_0,X_1)X_2^{n-1} + \cdots + a_n(X_0,X_1), \qquad (a_0 \neq 0)$$

Then, for any point $q = (\lambda_0 : \lambda_1 : 0)$ on the x-axis, the equation

$$a_0X^n + a_1(\lambda_0,\lambda_1)X^{n-1} + \cdots + a_n(\lambda_0,\lambda_1) = 0 \qquad (*)$$

has n solutions counting multiplicity. This means that V(F) and $\overline{pq}$ meet at least one and at most n times. (See Fig. 1.53.) This property is a projective property. As in the proof of Proposition 1.3.2, F and $\overline{pq}$ meet at just n points if (1) F has no multiple component and (2) $q \in \ell_2 - \Delta$ (ℓ_2 = the x-axis), where

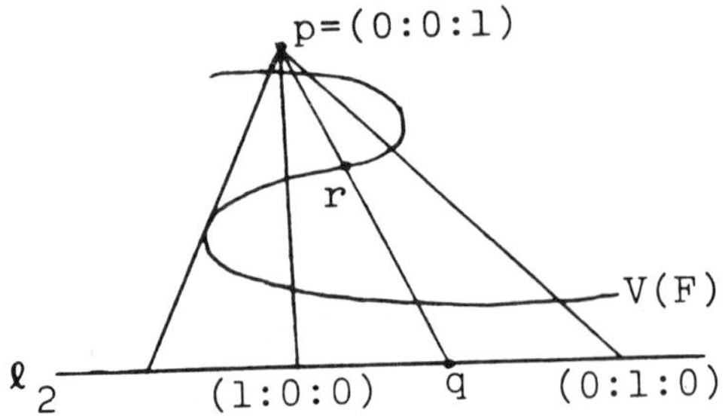

FIGURE 1.53

$$\Delta = \{(\lambda_0 : \lambda_1 : 0) \in \ell_2 \mid D(\lambda_0, \lambda_1) = 0\}$$

is a finite set.

Proposition 1.3.4. (1) A plane curve V(F) is closed and nowhere dense in $\mathbb{P}^2$. (2) $\mathbb{P}^2$ - V(F) is arcwise connected.

Proof. The closedness of V(F) is clear. Take $p \in \mathbb{P}^2$ - V(F) and $r \in$ V(F). The line $\overline{pr}$ is homeomorphic to a sphere S^2 and V(F) $\cap$ $\overline{pr}$ is a finite set. Hence, the proposition is proved. Q.E.D.

If $X = \lambda_2$ is a solution of the equation (*) with the multiplicity ν, then the intersection number of F and $\overline{pq}$ at $r = (\lambda_0 : \lambda_1 : \lambda_2)$ is ν and

$$I_r(F, \overline{pq}) = \nu$$

Also, if $r \notin$ V(F) $\cap$ $\overline{pq}$, then the intersection number $I_r(F, \overline{pq})$ is zero. Then

$$\Sigma_r I_r(F, \overline{pq}) = \deg F$$

where Σ is extended over all points $r \in$ V(F) $\cap$ $\overline{pq}$ or all points $r \in \mathbb{P}^2$. This is a special case of the theorem below.

Theorem 1.3.5 (Bezout). Let F and G be plane curves. Suppose that they have no common irreducible component. Then

$$\Sigma_p I_p(F,G) = \deg F \cdot \deg G$$

where Σ is extended over all points $p \in$ V(F) $\cap$ V(G). In particular, V(F) $\cap$ V(G) is nonempty.

Before proving the theorem, of course, we have to define the inter-section number $I_p(F,G)$ of F and G at $p \in V(F) \cap V(G)$. We do it analytically, using languages and propositions in Part II. (For a purely algebraic definition and discussion, see Fulton [29].)

Let M be a two-dimensional complex manifold. For a point $p \in M$, let (x, y) be a local coordinate system (see Sec. 3.1) in a neighborhood U of p in M such that p = (0, 0). We replace U by a smaller neighborhood if necessary.

A (germ of) analytic curve at p is, by definition, an equivalence class of holomorphic functions (see Sec. 3.1) f with f(p) = 0 defined on a neighborhood of p, under the equivalence relation $f \sim g$ (p) if there is a holomorphic function h on a neighborhood of p with $h(p) \neq 0$ such that g = fh. We denote such an analytic curve by $\{f = 0\}$.

We may consider f to be an element of the ring $\mathbb{C}\{x,y\}$ of convergent power series (see Sec. 3.2). Since $C\{x,y\}$ is a UFD (see Theorem 3.2.6), so f can be decomposed into the irreducible divisors

$$f = f_1^{\nu_1} \cdots f_s^{\nu_s}$$

Correspondingly, the analytic curve $\{f = 0\}$ at p has the irreducible branches $\{f_1 = 0\}, \ldots, \{f_s = 0\}$ with the multiplicity $\nu_1, \ldots, \nu_s$, respectively. If every multiplicity is 1, then $\{f = 0\}$ is an analytic curve without multiple component. Such an analytic curve can be identified with a (closed, nowhere dense) subset of a neighborhood of p, which we also denote by $\{f = 0\}$. If f is irreducible, then $\{f = 0\}$ is said to be irreducible. These notions are independent of the choice of local coordinate systems.

For an irreducible analytic curve $\{f = 0\}$ at p, there is a so-called local uniformizing parameter t of $\{f = 0\}$. That is, there are a disc $\Delta(0,\epsilon) = \{t \in \mathbb{C} \mid |t| < \epsilon\}$ and a holomorphic map ϕ: $\Delta(0,\epsilon) \to M$ such that (1) $\phi(0) = p$ and $\phi^{-1}(p) = \{0\}$ and (2) ϕ maps $\Delta(0,\epsilon)$ onto $\{f = 0\}$ bijectively (see Definition 4.1.9). The map ϕ is called a local uniformizing parameter of $\{f = 0\}$.

For example, if $f = x^a - y^b$ with (a, b) = 1 (coprime), then we may take (see Fig. 1.54)

$$\phi: t \to (x,y) = (t^b, t^a)$$

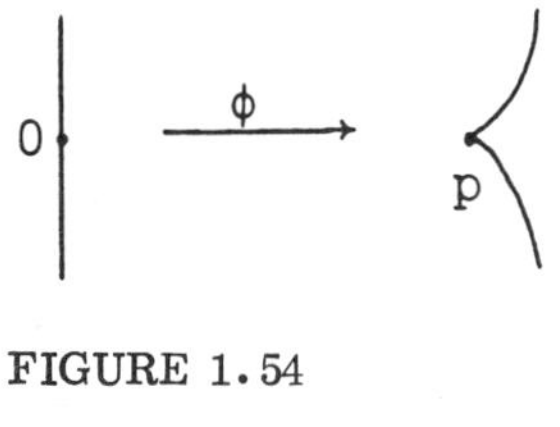

FIGURE 1.54

Lemma 1.3.6. For an irreducible analytic curve $\{f = 0\}$ at p, (1) there is a local uniformizing parameter ϕ such that $\phi(t) = (t^m, v(t))$, where $v(t)$ is a holomorphic function such that $v(0) = 0$ and (2) there is an irreducible polynomial

$$R(x,y) = y^m + a_1(x)y^{m-1} + \cdots + a_m(x), \qquad (a_j(x) \in \mathbb{C}\{x\})$$

over the convergent power series ring $\mathbb{C}\{x\}$ such that $\{f = 0\} = \{R = 0\}$.

Proof.

1. Put $\phi: t \to (u(t), v(t))$, where $u(0) = v(0) = 0$. Put $m = \mathrm{ord}_0 u(t)$, the order of zero of $u(t)$ at $t = 0$. Let

$$u(t) = a_m t^m + a_{m+1} t^{m+1} + \cdots \qquad (a_m \neq 0)$$

be the power series expansion. Put

$$s = t\sqrt[m]{a_m + a_{m+1}t + \cdots}.$$

(Here, we take a branch of $\sqrt[m]{}$.) Then t can be solved by s. That is, t is a holomorphic function $t = t(s)$ of s. Hence, s is a local uniformizing parameter of $\{f = 0\}$ such that

$$(\phi \circ t)(s) = (s^m, (v \circ t)(s))$$

2. Let $\phi: t \to (t^m, v(t))$ be a local uniformizing parameter. Put

$$R(x,y) = \prod_{j=0}^{m-1} (y - v(\epsilon^j x^{1/m})), \qquad (\epsilon = \exp 2\pi\sqrt{-1}/m)$$

$$= y^m + a_1(x)y^{m-1} + \cdots + a_m(x)$$

$x^{1/m}$ is a m-th root of x. Note that $a_1(x), \ldots, a_m(x)$ are holomorphic functions of x. In fact, say

$$a_1(x) = -v(x^{1/m}) - v(\epsilon x^{1/m}) - \cdots - v(\epsilon^{m-1}x^{1/m})$$

is a holomorphic function of $x^{1/m}$ invariant under the action

$x^{1/m} \to \epsilon^j x^{1/m}$. Hence, it is a holomorphic function of x. It is clear that $\{f = 0\} = \{R = 0\}$. If $R = R_1 R_2$ over $\mathbb{C}\{x\}$, then the analytic curve $\{f = 0\} = \{R = 0\}$ is not irreducible, which is a contradiction. Hence, R is irreducible over $\mathbb{C}\{x\}$. Q.E.D.

Now, we are ready to define the intersection number.

First, let $C = \{f = 0\}$ and $D = \{g = 0\}$ be irreducible analytic curves at p. Let $t \to (x(t), y(t))$ be a local uniformizing parameter of C. The order, $\mathrm{ord}_0\, g(t)$, of zero at $t = 0$ of the holomorphic function $g(t) = g(x(t), y(t))$ is called the intersection number, $I_p(C, D)$, of C and D (at p). (If $C = D$, then we put $I_p(C, D) = +\infty$.)

Clearly, it is independent of the choice of local uniformizing parameters and local coordinate systems. But it is not clear that the definition is symmetric, that is, $I_p(C, D) = I_p(D, C)$. To show this, we may assume that

$$C = \{R(x,y) = y^m + a_1(x)y^{m-1} + \cdots + a_m(x) = 0\}$$

$$D = \{S(x,y) = y^r + b_1(x)y^{r-1} + \cdots + b_r(x) = 0\}$$

by Lemma 1.3.6.

Lemma 1.3.7. Let $T = T(x) \in \mathbb{C}\{x\}$ be the resultant of the polynomials R and S over $\mathbb{C}\{x\}$. Then $I_p(C, D) = \mathrm{ord}_0\, T(x)$. In particular, $I_p(C, D) = I_p(D, C)$.

Proof. Using the notations in Lemma 1.3.6,

$$S(t^m, v(t)) = c_\nu t^\nu + c_{\nu+1} t^{\nu+1} + \cdots, \qquad (c_\nu \neq 0)$$

where $\nu = I_p(C, D)$. Note that

$$T(x) = S(x, v(x^{1/m}))\, S(x, v(\epsilon x^{1/m})) \cdots S(x, v(\epsilon^{m-1} x^{1/m}))$$

Here the equality holds as holomorphic function of x. But the right hand side is

$$= \{c_\nu (x^{1/m})^\nu + \cdots\} \cdots \{c_\nu (\epsilon^{m-1} x^{1/m})^\nu + \cdots\}$$

$$= c_\nu^m x^\nu + \cdots$$

Q.E.D.

We give some examples of the intersection numbers $I_p(C, D)$ of two analytic curves $C = \{f = 0\}$ and $D = \{g = 0\}$ at $p = (0, 0)$. (See Figs. 1.55 to 1.60.)

1°

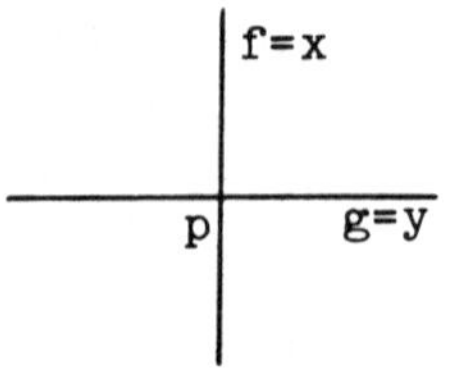

$I_p(C, D) = 1.$

FIGURE 1.55

2°

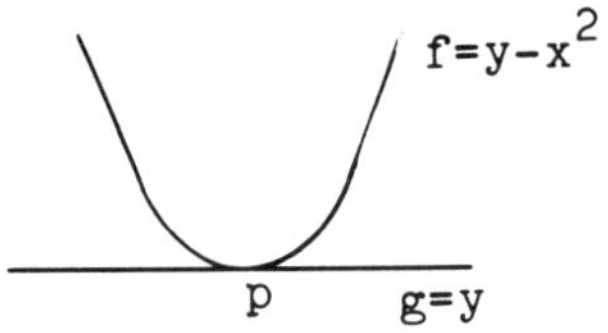

$I_p(C, D) = 2.$

FIGURE 1.56

3°

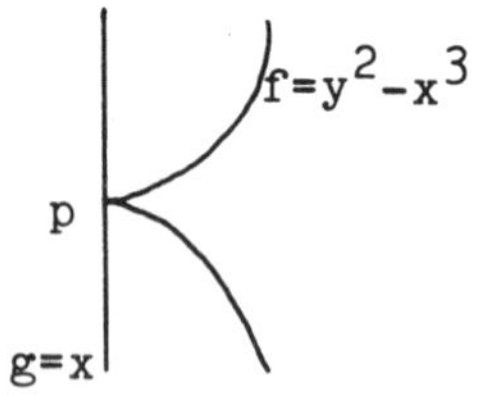

$I_p(C, D) = 2.$

FIGURE 1.57

4°

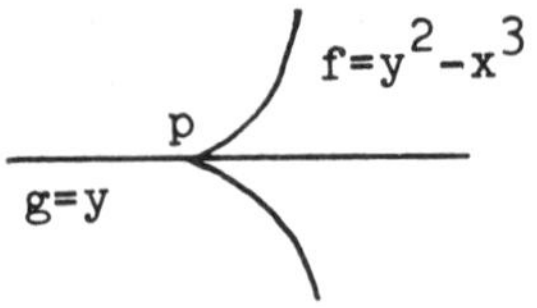

$I_p(C, D) = 3.$

FIGURE 1.58

5°

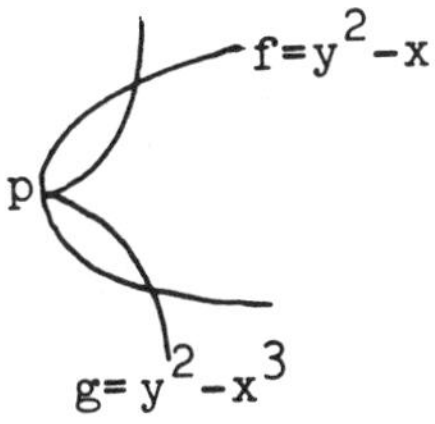

$I_p(C, D) = 2.$

FIGURE 1.59

6°

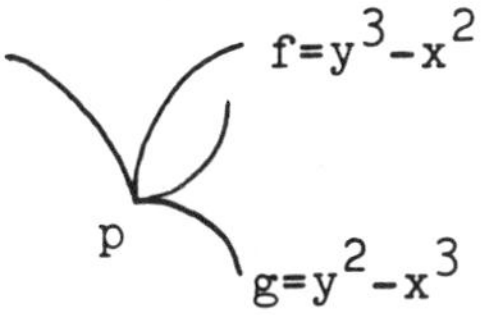

$I_p(C, D) = 4.$

FIGURE 1.60

An analytic curve $C = \{f = 0\}$ at p is said to be nonsingular (respectively singular) if

$$f(x,y) = f_1(x,y) + f_2(x,y) + \cdots$$

is the power series expansion such that $f_1(x,y) \neq 0$ (respectively $f_1(x,y) = 0$), where $f_k(x,y)$ is the k-th homogeneous part. We say also that p is a nonsingular (respectively singular) point of C. These notions are independent of the choice of local coordinate systems. If C is nonsingular, then C is necessarily irreducible. The analytic curves $C = \{f = 0\}$ at p in the examples 1°, 2°, and 5° above are nonsingular, while C in 3°, 4°, and 6° are singular. For a nonsingular $C = \{f = 0\}$, write

$$f_1(x,y) = ax + by$$

we may assume that, say, $b \neq 0$. Then we may take x a local uniformizing parameter. In this case, $I_p(C,D)$ is easy to compute.

Now, for analytic curves $C = \{f = 0\}$ and $D = \{g = 0\}$ at p such that $f = f_1^{\nu_1} \cdots f_s^{\nu_s}$ and $g = g_1^{\mu_1} \cdots g_t^{\mu_t}$ (the irreducible decompositions), we define the intersection number, $I_p(C,D)$, of C and D (at p) by

$$I_p(C,D) = \sum_{j,k} \nu_j \mu_k I_p(C_j, D_k)$$

where $C_j = \{f_j\}$ and $D_k = \{g_k = 0\}$.

It is clear that this definition is independent of the choice of local coordinate systems. Also, we have

$$I_p(C, D) = I_p(D, C)$$

For the later use, we prove

Lemma 1.3.8. Let C, D, and E be analytic curves at p. Suppose that C is nonsingular. Then

$$\min\{I_p(C, D), I_p(C, E)\} \leq I_p(D, E)$$

Proof. (Namba [71, p. 74]). We may take a local coordinate system (x, y) such that $p = (0, 0)$ and $C = \{x = 0\}$. Then, y is a local uniformizing parameter of C. Let $D = \{f = 0\}$ and $E = \{g = 0\}$. Expand f(0, y) and g(0, y) as follows

$$f(0, y) = y^s(c_0 + c_1 y + \cdots), \qquad c_0 \neq 0$$

$$g(0, y) = y^t(d_0 + d_1 y + \cdots), \qquad d_0 \neq 0$$

Then $I_p(C, D) = s$ and $I_p(C, E) = t$. (If $C \subset D$, then $s = +\infty$.) Assume that $t \geq s \geq 1$. By the Weierstrass Division Theorem (see Theorem 3.2.4), g can be written as

$$g = af + (b_1 y^{s-1} + \cdots + b_s) \tag{*}$$

where $a = a(x, y)$ is a holomorphic function around p and $b_1, \ldots, b_s$ are holomorphic functions of x. In (*), put $x = 0$. Then

$$y^t(d_0 + d_1 y + \cdots) = a(0, y)y^s(c_0 + c_1 y + \cdots) + b_1(0)y^{s-1} + \cdots + b_s(0)$$

Hence, $b_1(0) = \cdots = b_s(0) = 0$, so we can write

$$b_j(x) = xe_j(x), \qquad 1 \leq j \leq s$$

where $e_j = e_j(x)$ are holomorphic functions of X. Consider the analytic curves

$$B = \{b_1 y^{s-1} + \cdots + b_s = 0\}$$

$$B' = \{e_1 y^{s-1} + \cdots + e_s = 0\}$$

at p. Then

$$I_p(D, E) = I_p(D, B) = I_p(D, C) + I_p(D, B') \geq I_p(D, C) \qquad \text{Q.E.D.}$$

Now, let F and G be plane curves and $p \in V(F) \cap V(G)$. We define $I_p(F, G)$ by considering F and G as analytic curves at p.

More precisely, take a homogeneous coordinate system $(X_0 : X_1 : X_2)$ such that (1) $p = (1: 0: 0) \in V(F) \cap V(G)$ and (2) $(0: 0: 1) \notin V(F) \cap V(G)$. Put $x = X_1/X_0$ and $y = X_2/X_0$ and

$$f(x,y) = F(1,x,y), \qquad g(x,y) = G(1,x,y)$$

The <u>intersection number</u> $I_p(F, G)$ <u>of</u> F <u>and</u> G <u>at</u> p is, by definition,

$$I_p(F, G) = I_p(C, D)$$

where $C = \{f = 0\}$ and $D = \{g = 0\}$. (If $p \notin V(F) \cap V(G)$, then we put $I_p(F, G) = 0$.) This definition does not depend on the choice of homogeneous coordinate systems.

The following lemma is easy to prove by the definition of the intersection number.

<u>Lemma 1.3.9.</u>

1. $I_p(F, G) = I_p(G, F)$.
2. $I_p(F, G) = +\infty$ if and only if F and G have a common irreducible component containing p.
3. If $F = F_1 F_2$, then $I_p(F, G) = I_p(F_1, G) + I_p(F_2, G)$.
4. If A is a homogeneous polynomial with deg A = deg G - deg F, then $I_p(F, G) = I_p(F, G + AF)$.
5. If deg G = deg G' and a and a' are constants, then

$$I_p(F, aG + a'G') \geq \min\{I_p(F, G), I_p(F, G')\}$$

<u>Lemma 1.3.10.</u> Let $C = \{F = 0\}$ be an irreducible plane curve and $\Phi: M \to C$ be a <u>nonsingular model of</u> C (Theorem 4.1.11). Let G and H be homogeneous polynomials with deg G = deg H such that $C \cap V(G) \cap V(H) = \phi$.

Then the degree of the meromorphic function $(G/H)\cdot\Phi$ on M (see Proposition 4.1.5) is equal to $\Sigma I_p(F, G)$, where Σ is taken over all $p \in C \cap V(G)$.

Proof. Take $p \in C \cap V(G)$ and $p_1 \in \Phi^{-1}(p)$. Let F_1 be the irreducible branch of F at p corresponding to p_1. Write

$$\Phi: t \to (x(t), y(t))$$

where t (respectively (x,y)) is a local coordinate system around p_1 (respectively p) such that $t = 0$ at p_1 (respectively $(x,y) = (0,0)$ at p). Write $G = \{g = 0\}$ and $H = \{h = 0\}$ at p. Then $\phi = (G/H)\cdot\Phi$ is given by

$$t \to \frac{g(x(t), y(t))}{h(x(t), y(t))}$$

Since $h(0,0) \neq 0$ by the assumption, $I_p(F_1, G)$ is the order, $\operatorname{ord}_{p_1} \phi$, of zero of ϕ at $p_1 \in M$. Hence, $\Sigma I_p(F, G)$ is the order of zeros of ϕ, which is equal to the degree of ϕ (see Sec. 4.1). Q.E.D.

Now we are ready to prove Bezout's Theorem.

Proof of Theorem 1.3.5. We have already shown the theorem in the case $\deg G = 1$. Suppose that $n = \deg G \geq 2$. By (3) of Lemma 1.3.9, we may assume that the curve F is irreducible. Let $\Phi: M \to V(F)$ be a nonsingular model. Let

$$H = L_1 \cdots L_n$$

be a product of 1-forms L_j. If we choose general lines L_j, then $\Sigma I_p(F, G)$ is equal to the degree of $\phi = (G/H)\cdot\Phi$ by Lemma 1.3.10. But the degree of ϕ is equal to the order of poles of ϕ (see Sec. 4.1). Hence,

$$\Sigma I_p(F, G) = \Sigma I_p(F, H) = \Sigma_p \Sigma_j I_p(F, L_j) = n \deg F = \deg F \deg G$$

Q.E.D.

Let F be a plane curve of degree n. Suppose that F does not contain the line $\ell_0 = \{X_0 = 0\}$ of infinity as an irreducible component. Then

$$f(x,y) = F(1,x,y)$$

is a polynomial of degree n. An equivalence class of polynomials $f(x,y)$ of degree n under $\mathbb{C}^*$-action is called a (plane) affine curve of degree n. (Irreducible components, and so on, can be defined similarly for affine curves.) Thus there is a correspondence

$$F \to f$$

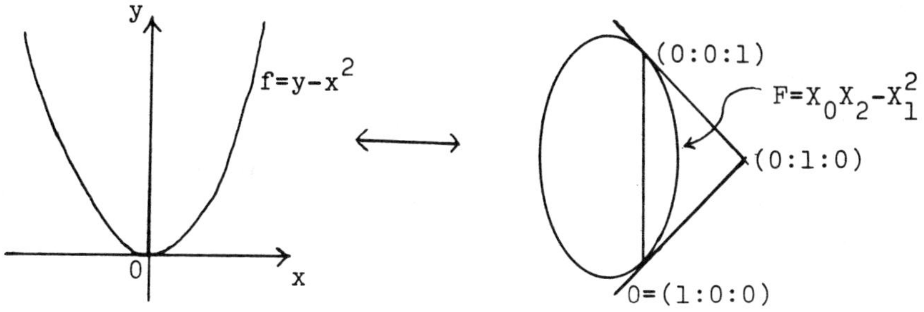

FIGURE 1.61

of plane curves F of degree n with $\ell_0 \not\subset V(F)$ and affine curves of degree n. This is in fact a bijection. Given a polynomial $f = f(x,y)$ of degree n, put

$$F(X_0, X_1, X_2) = X_0^n f(X_1/X_0, X_2/X_0)$$

F is called the closure in $\mathbb{P}^2$ of the affine curve f. If f has no multiple component, then $\{F = 0\}$ is in fact the closure of the set $\{f = 0\}$ in the topology of $\mathbb{P}^2$. For example, see Fig. 1.61. We sometimes use affine curves to indicate their closure in $\mathbb{P}^2$.

Let F be a plane curve and $p \in V(F)$. Choose a homogeneous coordinate system $(X_0 : X_1 : X_2)$ such that $\ell_0 = \{X_0 = 0\} \not\subset V(F)$ and $p = (1: 0: 0)$. Let $f = f(x,y) = F(1,x,y)$ be the affine curve corresponding to F. Write

$$f(x,y) = f_m(x,y) + f_{m+1}(x,y) + \cdots + f_n(x,y), \qquad (m \geq 1)$$

where $f_k(x,y)$ is the k-th homogeneous part of f and $f_m(x,y) \neq 0$ as a polynomial. The integer m is called the multiplicity of the curve F at p and is denoted by $m_p(F)$. This is independent of the choice of homogeneous coordinate systems. As in the case of analytic curves, F is said to be nonsingular (respectively singular) at p if $m_p(F) = 1$ (respectively $m_p(F) \geq 2$). p is then called a nonsingular (respectively singular) point of F. If $m_p(F) = 2, 3, \ldots, m$, then p is called a double point, a triple point, ..., an m-ple point, respectively. A singular point is sometimes called a multiple point. Every point on a multiple component of F is a singular point. We denote by Sing F the set of all singular points of F. Sing F is a finite set if and only if F has no multiple component. A plane curve without singular point is called a nonsingular plane curve. Otherwise, it is called a singular plane curve.

We decompose the above $f_m(x,y)$ into 1-forms

$$f_m(x,y) = (a_1x + b_1y)^{\nu_1} \cdots (a_sx + b_sy)^{\nu_s}, \qquad (\Sigma \nu_j = m)$$

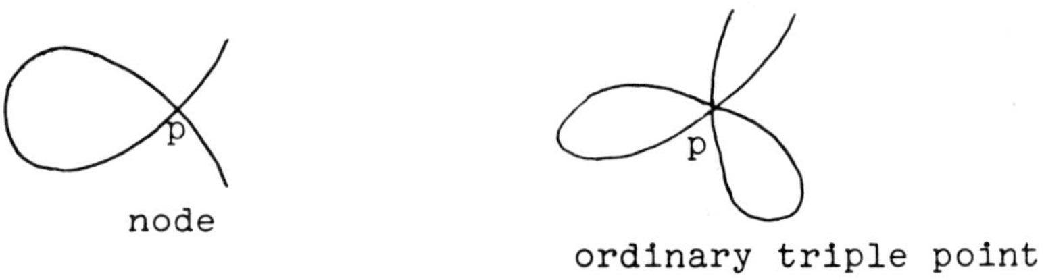

FIGURE 1.62

where $a_1x + b_1y \neq a_2x + b_2y$, and so forth, as affine lines. The closure of the affine line $\{a_jx + b_jy = 0\}$ in $\mathbb{P}^2$ is called a *tangent line to* F *at* p. ν_j is called the *multiplicity of the tangent line*. If p is a nonsingular point of F, the tangent line to F at p is uniquely determined and is denoted by T_pF. If $m_p(F) \geq 2$ and every $\nu_j = 1$, then p is called an *ordinary m-ple point*. In this case, there are m distinct tangent lines at p. An ordinary double point is called a *node*. (See Fig. 1.62.)

In the other extremal case, if f is irreducible at p and $m = m_p(F) \geq 2$, then p is called a *cusp*. In this case, $f_m(x, y) = (ax + by)^m$, so there is a unique tangent line, T_pF, at p with the multiplicity m. (See Fig. 1.63.)

It should be noted that, even if $f_m(x, y) = (ax + by)^m$, p may not be a cusp, that is, f may not be irreducible at p. For example, consider $f = y(y - x^2)$ at $p = 0$. (See Fig. 1.64.)

These notions can be easily extended to analytic curves and affine curves.

The following lemmas and corrolaries, Lemma 1.3.11 to Corollary 1.3.19, are easy to prove and are left to the reader as exercises (Exercise 1).

Lemma 1.3.11 (Euler's Equality). Let F be a homogeneous polynomial of degree n in the variables $X_0, \ldots, X_r$. Then

1. $X_0 \dfrac{\partial F}{\partial X_0} + \cdots + X_r \dfrac{\partial F}{\partial X_r} = nF$

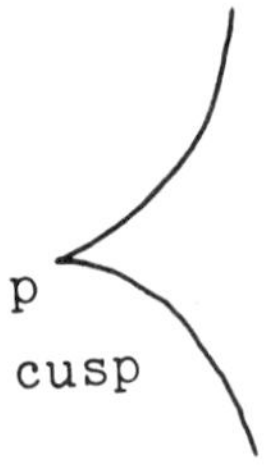

FIGURE 1.63

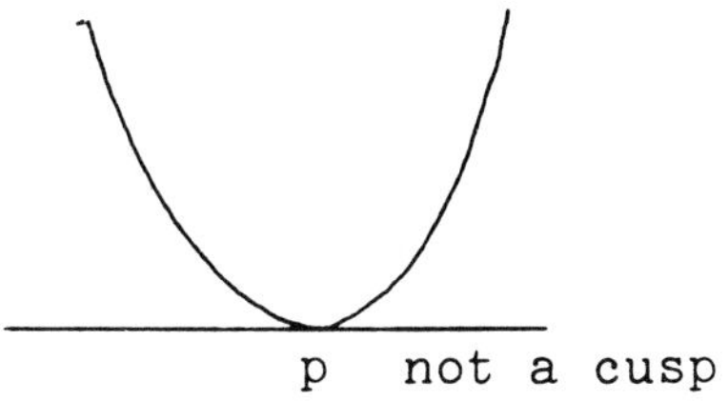

FIGURE 1.64

2. $\sum_{j,k} X_j X_k \frac{\partial^2 F}{\partial X_j \partial X_k} = n(n-1)F$ and

3. in general, $\left(X_0 \frac{\partial}{\partial X_0} + \cdots + X_r \frac{\partial}{\partial X_r}\right)^k F = \binom{n}{k} F$

Proposition 1.3.12. Let F be a plane curve and $p \in \mathbb{P}^2$. Then p is a singular point of F if and only if

$$\frac{\partial F}{\partial X_0}(p) = \frac{\partial F}{\partial X_1}(p) = \frac{\partial F}{\partial X_2}(p) = 0$$

Proposition 1.3.13. For $\lambda = (\lambda_0, \lambda_1, \lambda_2) \in \mathbb{C}^3 - 0$, suppose that $p = (\lambda : \lambda_1 : \lambda_2)$ is a nonsingular point of F. Then

$$T_p F = \frac{\partial F}{\partial X_0}(\lambda) X_0 + \frac{\partial F}{\partial X_1}(\lambda) X_1 + \frac{\partial F}{\partial X_2}(\lambda) X_2$$

Proposition 1.3.14. A plane curve F is irreducible if and only if (1) F has no multiple component and (2) V(F)-Sing F is connected. (See Fig. 1.65.)

Proposition 1.3.15. If $F = F_1 F_2$, then every point of $V(F_1) \cap V(F_2)$ is a singular point of F. In particular (by Bezout's Theorem), a nonsingular plane curve is necessarily irreducible. (See Fig. 1.66.)

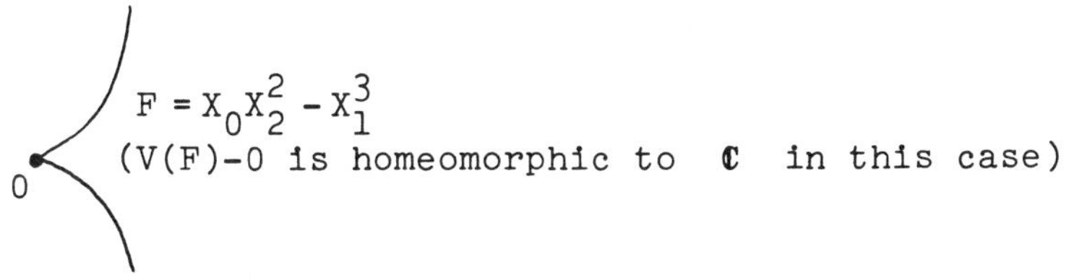

FIGURE 1.65

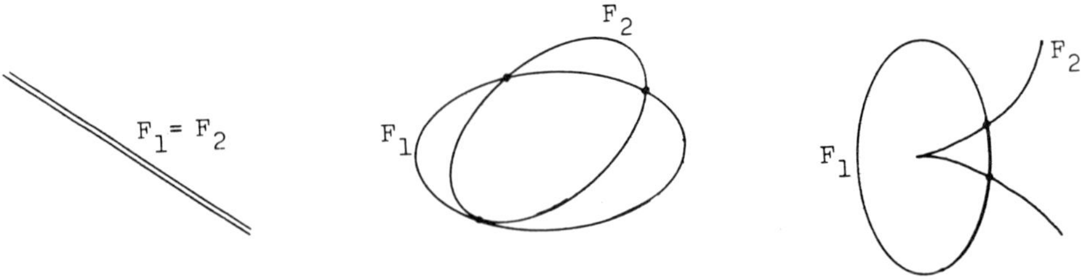

FIGURE 1.66

Proposition 1.3.16. For $p \in V(F)$, (1) $m_p(F) = \min \{I_p(F, L) \mid L$ is a line passing through $p\}$ and (2) $I_p(F, L) > m_p(F)$ if and only if L is a tangent line to F at p. (See Fig. 1.67.)

Proposition 1.3.17. For plane curves F and G and $p \in V(F) \cap V(G)$

$$I_p(F, G) \geq m_p(F)m_p(G)$$

The equality holds if and only if F and G have no tangent line at p in common. (If this is the case, F and G are said to intersect at p transversally.) (See Fig. 1.68.)

Corollary 1.3.18. For plane curves F and G and $p \in V(F) \cap V(G)$, $I_p(F, G) = 1$ if and only if (1) F and G are nonsingular at p and (2) $T_pF \neq T_pG$.

Corollary 1.3.19. For plane curves F and G, $V(F) \cap V(G)$ consists of distinct deg F deg G points if and only if every point $p \in V(G) \cap V(G)$ is a nonsingular point of both F and G such that $T_pF \neq T_pG$. (If this is the case, then neither F nor G has a multiple component.) (See Fig. 1.69.)

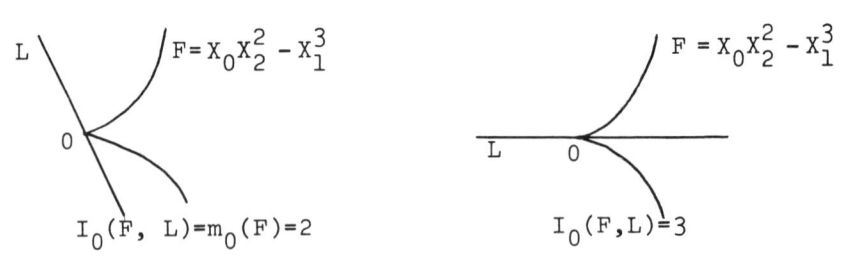

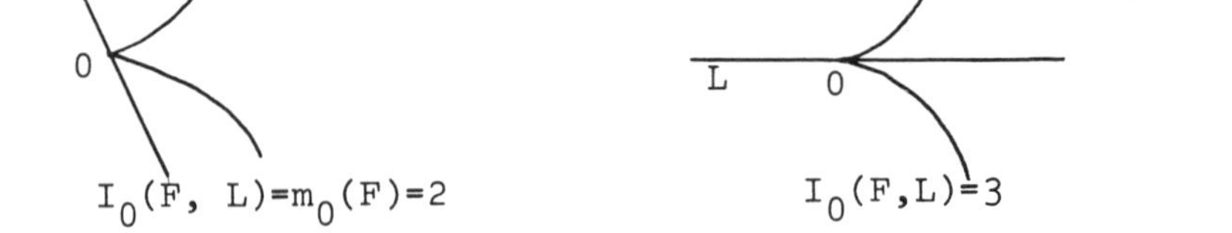

FIGURE 1.67

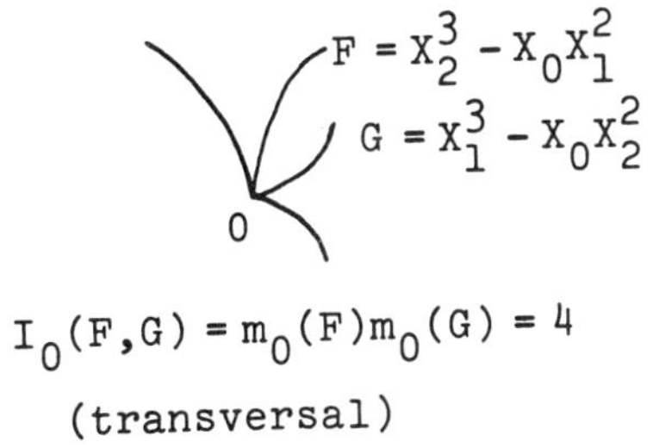

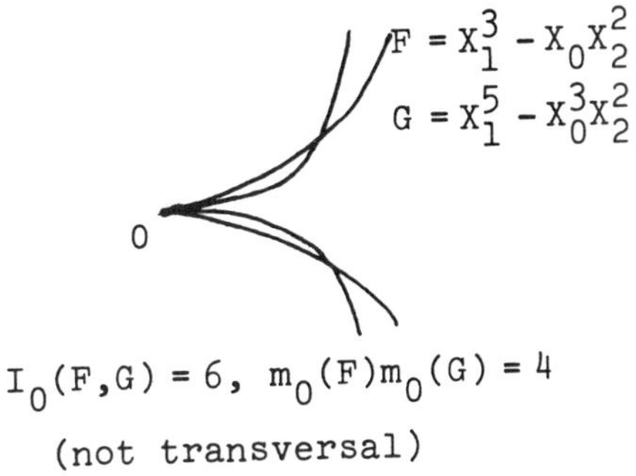

FIGURE 1.68

Now, a nonsingular point p on an irreducible plane curve F of degree n (≥ 3) is called a flex if $I_p(F, T_pF) \geq 3$. It is called an ordinary (respectively higher) flex if $I_p(F, T_pF) = 3$ (respectively ≥ 4). $r = I_p(F, T_pF) - 2$ is called the order of the flex p. (See Fig. 1.70.)

Put

$$\text{Hess}(F) = \begin{vmatrix} \dfrac{\partial^2 F}{\partial X_0\, \partial X_0} & \dfrac{\partial^2 F}{\partial X_1\, \partial X_0} & \dfrac{\partial^2 F}{\partial X_2\, \partial X_0} \\[2ex] \dfrac{\partial^2 F}{\partial X_0\, \partial X_1} & \dfrac{\partial^2 F}{\partial X_1\, \partial X_1} & \dfrac{\partial^2 F}{\partial X_2\, \partial X_1} \\[2ex] \dfrac{\partial^2 F}{\partial X_0\, \partial X_2} & \dfrac{\partial^2 F}{\partial X_1\, \partial X_2} & \dfrac{\partial^2 F}{\partial X_2\, \partial X_2} \end{vmatrix}$$

This is a homogeneous polynomial of degree $3(n - 2)$, and is called the Hessian of F.

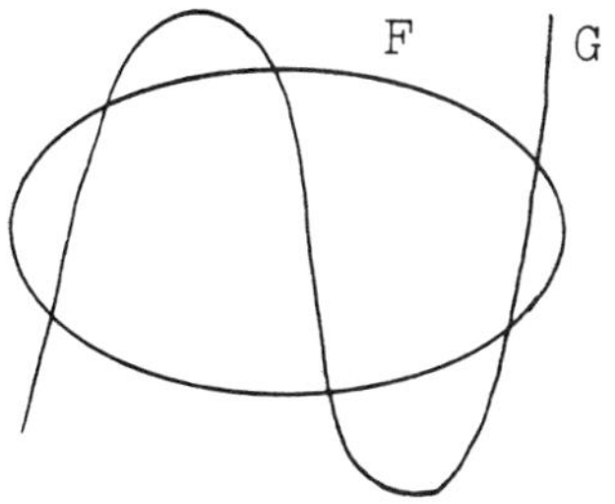

FIGURE 1.69

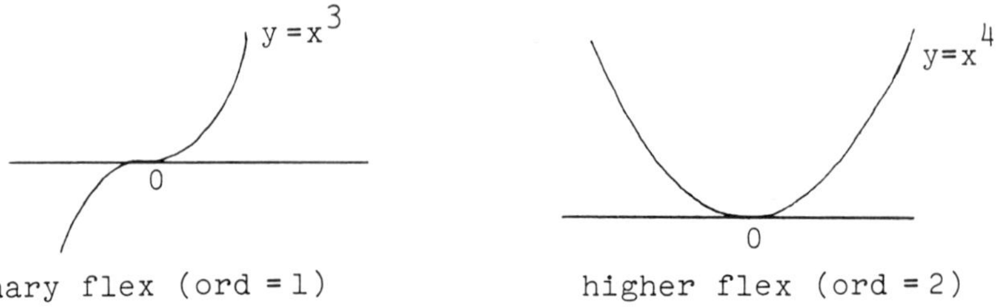

FIGURE 1.70

Proposition 1.3.20. Let F be an irreducible plane curve of degree n (≥ 3). A nonsingular point $p \in V(F)$ is a flex if and only if $\text{Hess}(F)(p) = 0$.

Proof. Note first that, if $\phi: (X_0: X_1: X_2) \to (\Sigma a_{0j} X_j: \Sigma a_{1j} X_j: \Sigma a_{2j} X_j)$ is a projective transformation, then

$$\text{Hess}(\phi(F)) = \det(a_{jk})^{-2}\phi(\text{Hess}(F))$$

Hence, we may choose a homogeneous coordinate system $(X_0 : X_1 : X_2)$ such that $p = (1: 0: 0)$ and $T_pF = \{X_2 = 0\}$. Put $x = X_1/X_0$, $y = X_2/X_0$ and $f(x,y) = F(1,x,y)$. Then $f(x,y)$ can be written as

$$f(x,y) = y(a + bx + cy + (\text{terms of degree} \geq 2)) + dx^2 + h(x)$$

where $a\ (\neq 0)$, b, c, d are constants and $h(x)$ is a polynomial in x with no term of degree ≤ 2. Since $F(X_0, X_1, X_2) = X_0^n f(X_1/X_0,\ X_2/X_0)$, we get

$$\text{Hess}(F)(1: 0: 0) = \begin{vmatrix} 0 & 0 & (n-1)a \\ 0 & 2d & b \\ (n-1)a & b & 2c \end{vmatrix} = -2(n-1)^2a^2d$$

Hence, $\text{Hess}(F)(1: 0: 0) = 0$ if and only if $d = 0$. But, it is clear that $d = 0$ if and only if $(1: 0: 0)$ is a flex. Q.E.D.

By Bezout's Theorem,

Corollary 1.3.21. On a nonsingular plane curve of degree n (≥ 3), there are at least one and at most $3n(n-2)$ flexes.

The proof of the following proposition is left to the reader as an exercise (Exercise 1).

Proposition 1.3.22. Let p be a flex of a nonsingular plane cubic F. Then F can be written as

$$y^2 = 4x^3 - g_2x - g_3$$

in a suitable affine coordinate system $(x,y) = (X_1/X_0, X_2/X_0)$ such that $p = (0: 0: 1)$, where g_2 and g_3 are constants which satisfy $g_2^3 - 27g_3^2 \neq 0$.

The equation in the proposition is called the Weierstrass canonical form of a nonsingular plane cubic.

A property (respectively an invariant) concerning a point on a (plane) curve is called an analytic property (respectively invariant) if it is invariant under any change of local coordinate systems. For example, the singularity (respectively nonsingularity) of a point on a curve is an analytic property, while "A point is a flex" is not (though it is of course a projective property). The multiplicity, the number of irreducible branches and the intersection number are analytic invariants. "A point is an ordinary m-ple point (respectively a cusp)," "to intersect transversally," and so on, are analytic properties.

Note 1.3.23. Good beginning books on curve theory are Orzech [78], Fulton [29], Walker [100], and Kawai [51] (in Japanese). Our definition of the intersection number is different from that of Fulton [29]. See Fulton [29, pp. 182-183].

Exercises

1. Prove Lemma 1.3.11 to Corollary 1.3.19 and Proposition 1.3.22.
2. Compute the singular points of the plane quartics (1) $F = (X_0X_2 - X_1^2)^2 - X_1X_2^3$ and (2) $F = (X_0X_2 - X_1^2)^2 - X_1^3X_2$.
3. Let F be a plane curve of degree n (≥ 3) containing no line as an irreducible component. Suppose that there is a point $p \in V(F)$ of multiplicity $n - 1$. Then p is a unique singular point of F.
4. Let F be a plane curve of degree n without multiple component. For any singular point $p \in V(F)$ of multiplicity r, there is a line L passing through p such that L intersects F at just $n - r$ distinct points other than p.
5. Let F be a plane curve of degree n without multiple component. Let $p_1, \ldots, p_s$ be singular points on F of multiplicity $r_1, \ldots, r_s$, respectively. Then $n(n - 1) \geq \Sigma_j r_j(r_j - 1)$. Moreover, if F is irreducible, then $(n - 1)(n - 2) \geq \Sigma_j r_j(r_j - 1)$.
6. Let F be an irreducible plane curve. Then (1) $p \in V(F) \cap V(\mathrm{Hess}(F))$ if and only if p is either a flex or a singular point of F, and (2) $I_p(F, \mathrm{Hess}(F)) = 1$ if and only if p is an ordinary flex of F.
7. There are just nine flexes on a nonsingular plane cubic.

8. In a suitable affine (respectively homogeneous) coordinate system (x, y) (respectively $(X_0 : X_1 : X_2)$), a nonsingular plane cubic can be written as

$$y^2 = x(x - 1)(x - \lambda) \tag{1}$$

where λ is a constant with $\lambda(\lambda - 1) \neq 0$ (Riemann's canonical form of a nonsingular plane cubic) and

$$X_0^3 + X_1^3 + X_2^3 + 3mX_0X_1X_2 = 0 \tag{2}$$

where m is a constant with $m^3 \neq -1$ (Hessian canonical form).

9. Let F be a plane curve of degree n without multiple component. Take distinct points $p = (X_0 : X_1 : X_2)$ and $q = (Y_0 : Y_1 : Y_2)$ on $\mathbb{P}^2$. Then the coordinate of a point on the line $\overline{pq}$ can be written as $\lambda X + \mu Y$, where $(\lambda : \mu) \in \mathbb{P}^1$, $X = (X_0, X_1, X_2)$ and $Y = (Y_0, Y_1, Y_2)$. By Taylor's Formula, $F(\lambda X + \mu Y)$ can be written as

$$F(\lambda X + \mu Y) = \lambda^n F(X) + \lambda^{n-1}\mu\Delta_Y^{(1)}F(X) + \dots + \frac{\lambda\mu^{n-1}}{(n-1)!}\Delta_Y^{(n-1)}F(X) + \frac{\mu^n}{n!}\Delta_Y^{(n)}F(X)$$

where

$$\Delta_Y^{(k)}F(X) = \left(Y_0\frac{\partial}{\partial X_0} + Y_1\frac{\partial}{\partial X_1} + Y_2\frac{\partial}{\partial X_2}\right)^k F(X)$$

Fixing q and moving p, the curve

$$P^{(k)}(q, F)(X) = \Delta_Y^{(k)}F(X)$$

is called the k-th polar of q with respect to the curve F. Prove that (1) $\deg P^{(k)}(q, F) = n - k$ and (2) if $p \in V(P^{(k)}(q, F))$, then $q \in V(P^{(n-k)}(p, F))$.

1.4. LINEAR SYSTEMS OF PLANE CURVES

Let W_n be the vector space of all homogeneous polynomials of degree n (≥ 1) in the variables X_0, X_1, and X_2. W_n has the basis consisting of all monomials

$$\{X_0^a X_1^b X_2^c \quad 0 \leq a,\ b,\ c, \quad a + b + c = n\}$$

so W_n is $\frac{1}{2}n(n + 3) + 1$-dimensional. The set of all plane curves of degree n can be identified with the projective space $\mathbb{P}(W_n)$ of dimension

$$N = \frac{1}{2}n(n + 3)$$

A projective transformation $\phi: \mathbb{P}^2 \to \mathbb{P}^2$ naturally induces a projective transformation

$$\tilde{\phi}: F \in \mathbb{P}(W_n) \to \phi(F) = F \cdot \phi^{-1} \in \mathbb{P}(W_n)$$

For an r-plane S in $\mathbb{P}(W_n)$, the set of all plane curves in S is called a linear system of plane curves of degree n and dimension r. If $r = 1$, it is called a linear pencil. If $S = \mathbb{P}(W_m)$, then it is said to be complete.

Proposition 1.4.1. Let $p_1, \ldots, p_m$ be distinct points on $\mathbb{P}^2$.

1. $\Lambda(n; p_1, \ldots, p_m) = \{F \in \mathbb{P}(W_n) \mid F$ passes through every $p_j\}$ is a linear system of dimension $\geq N - m$. (It may be empty.)
2. If $m \leq N$, then $\Lambda(n; p_1, \ldots, p_m)$ is nonempty.
3. If $m \leq n + 1$, then $\dim \Lambda(n; p_1, \ldots, p_m) = N - m$.

Proof. We first show (1) and (2). It is clear that, for $p \in \mathbb{P}^2$

$$\Lambda(n; p) = \{F \in \mathbb{P}(W_n) \mid F \text{ passes through } p\}$$

is a hyperplane in $\mathbb{P}(W_n)$. Then

$$\Lambda(n; p_1, \ldots, p_m) = \Lambda(n; p_1) \cap \cdots \cap \Lambda(n; p_m)$$

is a linear subspace of $\mathbb{P}(W_n)$ of dimension $\geq N - m$, and is nonempty if $m \leq N$. Next, we show (3) by induction on m. (3) is clear for $m = 1$. Suppose that (3) holds for m with $m \leq n$. It suffices to show that

$$\Lambda(n; p_1, \ldots, p_m) \neq \Lambda(n; p_1, \ldots, p_{m+1})$$

Let L_j be a line passing through p_j but not p_k $(k \neq j)$ and L_0 be a line passing through none of p_j. Put

$$F = L_0^{n-m} L_1 \cdots L_m$$

Then $F \in V(n; p_1, \ldots, p_m)$, but $F \notin V(n; p_1, \ldots, p_{m+1})$. Q.E.D.

(2) of the proposition says that, given any distinct N points on $\mathbb{P}^2$, there is a plane curve F of degree n passing through all these points. We can tell when F is uniquely determined by the following.

Let $\{X_0^a X_1^b X_2^c \mid 0 \leq a,b,c,\ a+b+c=n\}$ be the set of all monomials of degree n. Arranging them in a fixed order, we get the following map, called the n-th Veronese map

$$\Phi = \Phi_n : (X_0 : X_1 : X_2) \in \mathbb{P}^2 \to (\cdots : X_0^a X_1^b X_2^c : \cdots) \in \mathbb{P}^N$$

This is a holomorphic imbedding (see Sec. 3.1) and nondegenerate, that is, $\Phi(\mathbb{P}^2)$ is not contained in any hyperplane of $\mathbb{P}^N$. (In fact, $\Phi = \Phi_{|nL|}$, the map associated with the complete linear system $|nL| = \mathbb{P}(W_n)$ on $\mathbb{P}^2$, where L is a line (see Sec. 3.4).) Here $\mathbb{P}^N$ can be considered as the dual space to $\mathbb{P}(W_n)$. Hyperplanes H in $\mathbb{P}^N$ and plane curves F of degree n are naturally in one-to-one correspondence

$$H \to F = \Phi^{-1}(H) = \Phi^{-1}(H \cap \Phi(\mathbb{P}^2))$$

Linear pencils of plane curves of degree n and (N - 2)-planes in $\mathbb{P}^N$ are in one-to-one correspondence.

Proposition 1.4.2. (1) For distinct points $p_1, \ldots, p_N$ on $\mathbb{P}^2$, a plane curve F passing through every p_j is uniquely determined if an only if $\Phi(p_1), \ldots, \Phi(p_N)$ are in general position in $\mathbb{P}^N$. (2) There is an algebraic set (see Sec. 3.3) T in $(\mathbb{P}^2)^N = \mathbb{P}^2 \times \cdots \times \mathbb{P}^2$ (N-times) with $T \neq (\mathbb{P}^2)^N$ such that, for every $(p_1, \ldots, p_N) \in (\mathbb{P}^2)^N - T$, there is a uniquely determined plane curve of degree n passing through every p_j.

Proof. (1) is clear from the above consideration. We show (2). Put $p_j = (\xi_{j0} : \xi_{j1} : \xi_{j2})$, $1 \leq j \leq N$. Put

$$T = \{(p_1, \ldots, p_N) \in (\mathbb{P}^2)^N \mid \operatorname{rank}(\xi_{j0}^a \xi_{j1}^b \xi_{j2}^c)_{j,(abc)} \leq N - 1\}$$

Then T satisfies the condition. ($T \neq (\mathbb{P}^2)^N$, for Φ is nondegenerate.) Q.E.D.

Since $T \neq (\mathbb{P}^2)^N$, T is closed and nowhere dense in $(\mathbb{P}^2)^N$ (see Proposition 3.2.10). Hence, (2) of the proposition says that for "general" points $p_1, \ldots, p_N$ on $\mathbb{P}^2$, there is a unique plane curve of degree n passing through them.

A "general" plane curve of degree n is nonsingular in the following sense.

Proposition 1.4.3. $S = \{F \in \mathbb{P}(W_n) \mid F \text{ is a singular curve}\}$ is an algebraic set in $\mathbb{P}(W_n)$ such that $S \neq \mathbb{P}(W_n)$ such that $S \neq \mathbb{P}(W_n)$ (so S is closed and nowhere dense in $\mathbb{P}(W_n)$).

Proof. Put

$$\tilde{S} = \left\{ (p, F) \in \mathbb{P}^2 \times \mathbb{P}(W_n) \;\middle|\; \frac{\partial F}{\partial X_j}(p) = 0, \; j = 0, 1, 2 \right\}$$

Then $\tilde{S}$ is an algebraic set in $\mathbb{P}^2 \times \mathbb{P}(W)$. Let $\pi: \mathbb{P}^2 \times \mathbb{P}(W_n) \to \mathbb{P}(W_n)$ be the second projection. Then $S = \pi(\tilde{S})$. By the main theorem of elimination theory (see Theorem 3.3.13), S is an algebraic set of $\mathbb{P}(W_n)$. $S \neq \mathbb{P}(W_n)$, for $F = X_0^n + X_1^n + X_2^n$, say, is not in S. Q.E.D.

Proposition 1.4.4. Let F be a plane curve of degree n without multiple component. Let m be a positive integer. Then there is an algebraic set Y in $\mathbb{P}(W_m)$ with $Y \neq \mathbb{P}(W_m)$ such that every curve $G \in \mathbb{P}(W_m) - Y$ meets F at mn distinct points.

Proof. Put

$$\tilde{Y} = \left\{ (p, G) \in \mathbb{P}^2 \times \mathbb{P}(W_m) \mid F(p) = G(p) = 0, \; \left(\frac{\partial F}{\partial X_j}\frac{\partial G}{\partial X_k} - \frac{\partial F}{\partial X_k}\frac{\partial G}{\partial X_j}\right)(p) = 0, \; 0 \leq j, k \leq 2 \right\}$$

(The last condition means that $T_pF = T_pG$ if p is a nonsingular point of both F and G.) Then $\tilde{Y}$ is an algebraic set in $\mathbb{P}^2 \times \mathbb{P}(W_m)$. Put $Y = \pi(\tilde{Y})$, where $\pi: \mathbb{P}^2 \times \mathbb{P}(W_m) \to \mathbb{P}(W_m)$ is the second projection. Then Y is an algebraic set which satisfies the condition. Note that $Y \neq \mathbb{P}(W_m)$. In fact, if $G = L_1 \cdots L_m$, say, where L_j are general lines passing through a fixed point $q \in \mathbb{P}^2 - V(F)$, then $G \notin Y$. (See Fig. 1.71.) Q.E.D.

Now, let Λ be a linear system of plane curves of degree n. A plane curve G of degree m ($\leq$ n) is called a fixed component of Λ if the polynomial G divides every member of Λ as a polynomial. If there is a fixed component, then there is the maximal fixed component G_0 in the sense that every fixed

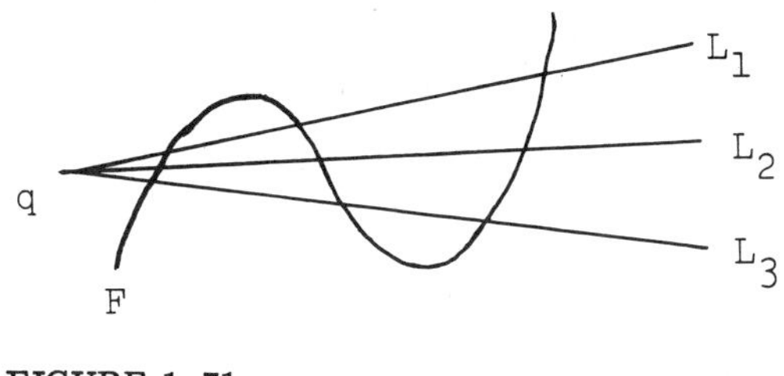

FIGURE 1.71

component G divides G_0. G_0 is uniquely determined and is called the fixed part of Λ.

Proposition 1.4.5. Let G be a plane curve of degree m. Let $n \geq m$ and put

$$\Lambda = \{F = GH \mid H \in \mathbb{P}(W_{n-m})\}, \quad (W_0 = \mathbb{C})$$

Then Λ is a linear system of plane curves of degree n and of dimension $N' = \frac{1}{2}(n-m)(n-m+3)$ such that (1) the fixed part of Λ is G and (2) any linear system Λ' of degree n having G as a fixed component is contained in Λ.

Proof. Λ is $\mathbb{P}(S)$, where S is the image of the injective linear map

$$H \in W_{n-m} \to GH \in W_n$$

Hence, (1) and (2) are clear. Q.E.D.

Let Λ be a linear system of plane curves of degree n without fixed component. A point $p \in \mathbb{P}^2$ is called a base point of Λ if every member F of Λ passes through p. The set $\mathrm{Bs}(\Lambda)$ of all base points of Λ is called the base locus of Λ. It is easy to see that $\mathrm{Bs}(\Lambda)$ is a finite set.

Let F and G be plane curves of degree n. Then

$$\Lambda = \{F_\lambda = \lambda_0 F + \lambda_1 G \mid \lambda = (\lambda_0 : \lambda_1) \in \mathbb{P}^1\}$$

is a linear pencil, called the linear pencil generated by F and G. Conversely, any linear pencil can be obtained in this way. Λ has no fixed component if and only if F and G have no common irreducible component. In this case, $\mathrm{Bs}(\Lambda) = V(F) \cap V(G)$.

The following proposition is a special case of Bertini's Theorem (see Theorem 3.4.14).

Proposition 1.4.6. Let $\Lambda = \{F_\lambda \mid \lambda \in \mathbb{P}^1\}$ be a linear pencil of plane curves of degree n without fixed component. Then there is a finite set Γ in $\mathbb{P}^1$ such that every F_λ, $\lambda \in \mathbb{P}^1 - \Gamma$, is nonsingular at every point of $V(F_\lambda) - \mathrm{Bs}(\Lambda)$. In other words, a general member of Λ is nonsingular outside the base locus.

Next, we introduce the notion of zero-cycles on $\mathbb{P}^2$. A zero-cycle on $\mathbb{P}^2$ is a formal finite sum

$$Z = \nu_1 p_1 + \cdots + \nu_r p_r$$

where $p_1, \ldots, p_r$ are (distinct) points of $\mathbb{P}^2$ and $\nu_1, \ldots, \nu_r$ are integers. The set of all zero-cycles forms a free abelian group. The degree, deg Z, of Z is, by definition, $\nu_1 + \cdots + \nu_r$. deg: $Z \to \deg Z$ is a homomorphism. We write $Z \geq 0$ if $\nu_j \geq 0$ for all j. Z is said to be positive, $Z > 0$, if $Z \geq 0$ and $\nu_j > 0$ for some j. We write

$$\nu_1 p_1 + \cdots + \nu_r p_r \geq \mu_1 p_1 + \cdots + \mu_r p_r$$

if $\nu_j \geq \mu_j$ for all j. The support of $Z = \nu_1 p_1 + \cdots + \nu_r p_r$ ($\nu_j \neq 0 \quad 1 \leq j \leq r$) is, by definition, the set $\{p_1, \ldots, p_r\}$.

For plane curves F and G without common irreducible component, the intersection zero-cycle F.G of F and G is, by definition, the zero-cycle

$$F.G = \Sigma_p I_p(F,G)p$$

where Σ is extended over all points $p \in V(F) \cap V(G)$. By Bezout's Theorem, F.G is a positive zero-cycle of degree deg F deg G. Note that

1. $F.G = G.F$,
2. $F.(GH) = F.G + F.H$, and
3. $F.(AF+G) = F.G$, where A satisfies $\deg A = \deg G - \deg F$.

Proposition 1.4.7. Let F and G be plane curves of degree n without common irreducible component. Let H be an irreducible plane curve of degree k ($<n$) such that (1) $V(H) \not\subset V(F)$, (2) every point of $V(F) \cap V(H)$ is a nonsingular point of F, and (3) $F.H \leq F.G$. Then there is a plane curve H' of degree n - k such that $F.G = F.H + F.H'$.

Proof. Take a point $p \in V(H) - V(F)$. Put

$$\lambda = -G(p)/F(p) \quad \text{and} \quad F_\lambda = \lambda F + G$$

Then F_λ passes through p. We claim that $V(H) \subset V(F_\lambda)$. Suppose that $V(H) \not\subset V(F_\lambda)$. Since H is irreducible, the intersection zero-cycle $F_\lambda.H$ is defined and has degree nk. For every point $q \in V(F) \cap V(H)$, $I_q(F,H) = \min\{I_q(F,H), I_q(F,G)\} = \min\{I_q(F,H), I_q(F,F_\lambda)\} \leq I_q(F_\lambda, H)$, by Lemma 1.3.8. Hence, $F.H \leq F_\lambda.H$. On the other hand, $p \in V(F_\lambda) \cap V(H)$, so $F.H + p \leq F_\lambda.H$. Hence, $\deg(F_\lambda.H) \geq nk + 1$, a contradiction. Thus, we have $V(H) \subset V(F_\lambda)$. By Proposition 1.3.2, there is a plane curve H' of degree n - k such that $HH' = F_\lambda$. Then

$$F.G = F.F_\lambda = F.H + F.H' \qquad \text{Q.E.D.}$$

Remark 1.4.8. It is possible to drop the assumption that H is irreducible. See Corollary 1.4.14 below.

Theorem 1.4.9. Let F be a nonsingular plane curve of degree n. Let m be a positive integer and D be a positive divisor of degree mn on F (see Sec. 4.1). D is regarded as a positive zero-cycle on $\mathbb{P}^2$. Then there is a plane curve G of degree m such that F.G = D if and only if D is linearly equivalent to mE (see Sec. 4.1), where E is a divisor on F defined by E = F.L for a line L, called a line section of F.

Proof. If G is a plane curve of degree m, then G is linearly equivalent to mL as divisors on $\mathbb{P}^2$, where L is a line. Hence, F.G is linearly equivalent to F.mL = mF.L = mE.

Next, we prove the converse.

Case 1. $m \geq n$. Put $\Lambda = \{F.G \mid G \in \mathbb{P}(W_m)\}$. We may regard Λ as a linear system on F (see Sec. 3.4). We compute its dimension. Let L(mE) be the vector space of (the zero function and) all meromorphic functions f on F such that $D_\infty(f) \leq mE$, where $D_\infty(f)$ is the polar divisor of f (see Sec. 4.1). Then

$$\alpha\colon G \in W_m \to f = \left(\frac{G}{mL}\right)\Big|_F \in L(mE)$$

is a linear map, whose kernel is

$$\ker(\alpha) = \{FH \mid H \in W_{m-n}\}$$

Hence,

$$\dim \Lambda = \dim \alpha(W_m) - 1 = \left\{\frac{m(m+3)}{2} + 1\right\} - \left\{\frac{(m-n)(m-n+3)}{2} + 1\right\} - 1$$

$$= mn - \tfrac{1}{2}(n-1)(n-2)$$

Note that the genus of F is $g = \frac{1}{2}(n-1)(n-2)$ (see Theorem 2.1.9). Hence, $mn \geq n^2 \geq 2g + 1$. By the Riemann-Roch Theorem (Theorem 5.1.5),

$$\dim |mE| = mn - g$$

This means that $\Lambda = |mE|$, that is, Λ is a complete linear system on F. Hence, if $D \in |mE|$, then there is $G \in \mathbb{P}(W_m)$ such that D = F.G.

Case 2. $m < n$. Let G' be a (general) curve of degree n - m such that (1) G' is nonsingular, (2) F and G' meet at n(n - m) distinct points, and (3) $V(F) \cap V(G) \cap V(G') = \phi$ (see Propositions 1.4.3 and 1.4.4). Put D' = F.G'. Then $D' \sim (n-m)E$ (linearly equivalent). Hence, if $D \sim mE$, then $D + D' \sim nE$. By Case 1, there is a plane curve H of degree n such that

$D + D' = F.H$. Then, by Proposition 1.4.7, there is a plane curve G of degree m such that $D = F.G$. Q.E.D.

Remark 1.4.10. If one knows cohomology, Theorem 1.4.9 follows from the exact sequence

$$0 \to H^0(\mathbb{P}^2, \mathcal{O}((m-n)L)) \to H^0(\mathbb{P}^2, \mathcal{O}(mL))$$

$$\to H^0(F, \mathcal{O}(mE)) \to H^1(\mathbb{P}^2, \mathcal{O}((m-n)L)) \to \cdots$$

and from the fact that $H^1(\mathbb{P}^2, \mathcal{O}((m-n)L)) = 0$ (see Hartshorne [39, p. 225]). If $m < n$, then $H^0(\mathbb{P}^2, \mathcal{O}((m-n)L)) = 0$, so G is unique. This also follows from Max Noether's Theorem (Theorem 1.4.12) below.

Proposition 1.4.11. Let F be a plane curve of degree n and $Z = \nu_1 p_1 + \cdots + \nu_s p_s$ ($p_j \neq p_k$ for $j \neq k$) be a positive zero-cycle on $\mathbb{P}^2$ such that $p_1, \ldots, p_s$ are nonsingular points on F. Let m be a positive integer. Then,

1. $\Lambda = \Lambda(m; F; Z) = \{G \in \mathbb{P}(W_m) \mid I_{p_j}(F, G) \geq \nu_j, \ 1 \leq j \leq s\}$ is a linear system of plane curves of degree m such that

$$\dim \Lambda \geq \tfrac{1}{2}m(m+3) - \deg Z.$$

2. If $\deg Z \leq \frac{1}{2}m(m+3)$, then Λ is nonempty.
3. If $\deg Z \leq m + 1$, then $\dim \Lambda = \frac{1}{2}m(m+3) - \deg Z$.

Proof. (A modification of the proof of Proposition 1.4.1.) First we prove (1) and (2). By (5) of Lemma 1.3.9, Λ is a linear system. If $Z = p_1$, then $\Lambda = \Lambda(m; p_1)$ is a hyperplane of $\mathbb{P}(W_m)$ (see Proposition 1.4.1). Consider the case $Z = 2p_1$. We may assume that $p_1 = (1: 0: 0)$. Put

$$G = \Sigma a_{\alpha\beta\gamma} X_0^{\alpha} X_1^{\beta} X_2^{\gamma}, \qquad (\alpha + \beta + \gamma = m)$$

$$g(x, y) = G(1, x, y) = \Sigma a_{\alpha\beta\gamma} x^{\beta} y^{\gamma}$$

Let $t \to (x(t), y(t))$ be a local uniformizing parameter of F at p_1 (see Definition 4.1.9). Then $I_{p_1}(F, G) \geq 2$ if and only if

$$g(0, 0) = 0$$

$$\frac{\partial g}{\partial x}(0, 0)\frac{dx}{dt}(0) + \frac{\partial g}{\partial y}(0, 0)\frac{dy}{dt}(0) = 0$$

These are linear conditions on the coefficients $a_{\alpha\beta\gamma}$'s. Hence,

$$\dim \Lambda \geq \tfrac{1}{2}m(m+3) - 2$$

In a similar way, if $Z = \nu_1 p_1$, then $\dim \Lambda \geq \frac{1}{2}m(m+3) - \nu_1$. In general,

$$\Lambda(m;\, F;\, \nu_1 p_1 + \cdots + \nu_s p_s) = \Lambda(m;\, F;\, \nu_1 p_1) \cap \cdots \cap \Lambda(m;\, F;\, \nu_s p_s)$$

This proves (1) and (2).

Now we prove (3) (Induction on deg Z). Let p_{s+1} be a nonsingular point of F such that $p_{s+1} \neq p_j$ for $1 \leq j \leq s$. Let L_j $(1 \leq j \leq s)$ be lines such that (1) $p_j \in L_j$, (2) $L_j \neq T_{p_j}F$, (3) $p_k \notin L_j$ for $k \neq j$, and (4) $p_{s+1} \notin L_j$. Let L_0 be a line such that $p_j \notin L_0$ for $1 \leq j \leq s+1$. Suppose that $Z = \nu_1 p_1 + \cdots + \nu_s p_s$ satisfy $\deg Z = \nu_1 + \cdots + \nu_s \leq m$. Put

$$G = L_1^{\nu_1} \cdots L_s^{\nu_s} L_0^{m-d} \qquad (d = \deg Z)$$

Then $G \in \Lambda(m;\, F;\, Z)$. But $G \notin \Lambda(m;\, Z + p_s)$, for p_s is a nonsingular point of F. Also, $G \notin \Lambda(m;\, F;\, Z + p_{s+1})$. Q.E.D.

Now, we prove the famous theorem of Max Noether.

Theorem 1.4.12 (Max Noether). Let F, G, and H be plane curves of degree n, m, and ℓ $(\ell \geq m)$, respectively. Suppose that (1) F and G have no common irreducible component, (2) every point of $V(F) \cap V(G)$ is a nonsingular point of F, and (3) $I_p(F, H) \geq I_p(F, G)$ for all $p \in V(F) \cap V(G)$. Then there are homogenous polynomials A of degree $\ell - n$ $(A = 0$ if $\ell < n)$ and B of degree $\ell - m$ such that

$$H = AF + BG$$

Proof.

Case 1. $\ell \geq mn$. Consider the linear map

$$\alpha:\ (A, B) \in W_{\ell-n} \times W_{\ell-m} \to AF + BG \in W_\ell$$

Since F and G have no common irreducible component,

$$\ker(\alpha) = \{(GE, -FE) \mid E \in W_{\ell-m-n}\}$$

which is isomorphic to $W_{\ell-m-n}$. (For convenience, we put $W_0 = \mathbb{C}$ and $W_{-1} = (0)$.) Hence,

$\dim \alpha\,(W_{\ell-n} \times W_{\ell-m})$

$= \frac{1}{2}(\ell-n)(\ell-n+3)+1+\frac{1}{2}(\ell-m)(\ell-m+3)+1-\frac{1}{2}(\ell-m-n)(\ell-m-n+3)-1$

$= \frac{1}{2}\ell(\ell+3)+1-mn$

Hence, $\Lambda = \mathbb{P}(\alpha(W_{\ell-n} \times W_{\ell-m}))$ is a linear system of plane curves of degree ℓ and of dimension $\frac{1}{2}\ell(\ell+3) - mn$. On the other hand,

$$\Lambda' = \{H \in \mathbb{P}(W_\ell) \mid I_p(F,H) \geq I_p(F,G) \quad \text{for all} \quad p \in V(F) \cap V(G)\}$$

is a linear system of dimension $\frac{1}{2}\ell(\ell+3) - mn$ by (3) of Proposition 1.4.11. Clearly, Λ' contains Λ, so $\Lambda' = \Lambda$.

Case 2. $mn \geq \ell \geq \max\{m,n\}$. Suppose that the theorem is true for ℓ such that $\ell - 1 \geq \max\{m,n\}$. We prove it for $\ell - 1$. Let H be a plane curve of degree $\ell - 1$ satisfying the condition of the theorem. Take a general line L such that (a) $V(F) \cap V(L)$ consists of n distinct points and (b) $V(F) \cap V(L) \cap V(G) = \phi$. Then $LH \in \mathbb{P}(W_\ell)$ satisfies the condition (3). Hence, there are $A \in \mathbb{P}(W_{\ell-n})$ and $B \in \mathbb{P}(W_{\ell-m})$ such that

$$LH = AF + BG \tag{*}$$

Case 2-i. L divides B. In this case, B can be written as $B = LB_1$, where B_1 is of degree $\ell - m - 1$. Then

$$L(H - B_1G) = AF$$

Hence, A can be written as $A = LA_1$, where A_1 is of degree $\ell - n - 1$. Thus,

$$H = A_1F + B_1G$$

Case 2-ii. L does not divide B. In this case, by (*) and by Assumption (b) on L, B passes through every point of $V(F) \cap V(L)$. Hence, $\deg B = \ell - m \geq n$. By Case 1 applied to F, L, and B, there are A_1 of degree $\ell - m - n$ and B_1 of degree $\ell - m - 1$ such that $B = A_1F + B_1L$. Substituting this for B in (*), we get

$$L(H - B_1G) = (A + A_1G)F$$

Hence, $A + A_1G$ can be written as $A + A_1G = LA_2$, where A_2 is of degree $\ell - n - 1$. Then $H = A_2F + B_1G$.

Case 3. $n > \ell \geq m$. Let G_1 be an irreducible component of G such that $V(G_1) \not\subset V(H)$. Then, for every point $p \in V(F) \cap V(G_1)$,

$$I_p(F,G_1) = \min\{I_p(F,H),\ I_p(F,G_1)\} \le I_p(H,G_1)$$

by Lemma 1.3.8. Hence, $F.G_1 \le H.G_1$, so

$$n \deg G_1 = \deg F.G_1 \le \deg H.G_1 = \ell \deg G_1$$

Hence, $n \le \ell$, a contradiction. Hence every irreducible component of G divides H. Let

$$G = G_1^{\nu_1} \cdots G_s^{\nu_s} \quad (\nu_j \ge 1)$$

be the irreducible decomposition of G. Write

$$H = G_1^{\mu_1} \cdots G_s^{\mu_s} H_0 \quad (\mu_j \ge 1)$$

where H_0 contains no G_j. Then, for every point $p \in V(F) \cap V(G)$,

$$I_p(F,H) = \Sigma \mu_j I_p(F,G_j) + I_p(F,H_0) \ge I_p(F,G) = \Sigma \nu_j I_p(F,G_j) \tag{**}$$

Put

$$I = \{1,2,\dots,s\}$$

$$I' = \{j \in I \mid \mu_j - \nu_j \ge 0\} \quad \text{and} \quad I'' = I - I'$$

Claim that I'' is empty. Take $i \in I''$. Put

$$H' = \prod_{j \in I'} G_j^{\mu_j - \nu_j} H_0$$

Then $\deg H' \le \ell$. For every point $p \in V(F) \cap V(G_i)$,

$$I_p(F,H') = \sum_{j \in I'} (\mu_j - \nu_j) I_p(F,G_j) + I_p(F,H_0)$$

$$\ge \sum_{k \in I''} (\nu_k - \mu_k) I_p(F,G_k) \ge I_p(F,G_i)$$

by (**). Hence, by the same reason as above, $F.G_i \le H'.G_i$, so

$$n \deg G_i \leq \deg H' \deg G_i \leq \ell \deg G_i$$

Hence, $n \leq \ell$, a contradiction. Thus, I'' is empty. That is, $\nu_j \leq \mu_j$ for all j. Hence, G divides H. Q.E.D.

Remark 1.4.13. Our statement of Max Noether's Theorem is a restricted one. For a general form, see Fulton [29] and Walker [100].

Corollary 1.4.14. Let F, G, and H be plane curves of degree n, m, and ℓ ($\ell \geq m$), respectively. Suppose that (1) F and H (respectively F and G) have no common irreducible component, (2) every point of $V(F) \cap V(G)$ is a nonsingular point of F, and (3) $F.G \leq F.H$. Then there is a plane curve B of degree $\ell - m$ such that $F.H = F.G + F.B$.

Proof. By the theorem, H can be written as $H = AF + BG$. By (1), $B \neq 0$. Hence,

$$F.H = F.(AF + BG) = F.G + F.B \qquad \text{Q.E.D.}$$

Corollary 1.4.15. Let F and H be plane cubics and G be a conic such that (1) $F.H = p_1 + \cdots + p_9$, (2) $F.G = p_1 + \cdots + p_6$, and (3) $p_1, \ldots, p_6$ are nonsingular points of F. Then p_7, p_8, and p_9 are collinear.

This corollary implies both Pascal's Theorem and Pappus' Theorem by putting $F = L_1 L_2 L_3$ (three lines) and $H = L_4 L_5 L_6$ (three lines). Moreover, it should be noted that the points $p_1, \ldots, p_9$ in the corollary are not necessarily distinct. Hence, Pascal's Theorem holds even if, say, $p_1 = p_2$ and $\overline{P_1 P_2} = T_{p_1}C$. (See Fig. 1.72.)

Theorem 1.4.16. Let F, G, and H be plane curves of degree n (≥ 3) such that (1) F and G (F and H) have no common irreducible component and (2) every point of $V(F) \cap V(G)$ is a nonsingular point of F. Suppose that there is a positive zero-cycle Z on $\mathbb{P}^2$ such that (3) $\deg Z = n^2 - n + 2$ and (4) $F.G \geq Z$ and $F.H \geq Z$. Then H is a member of the linear pencil

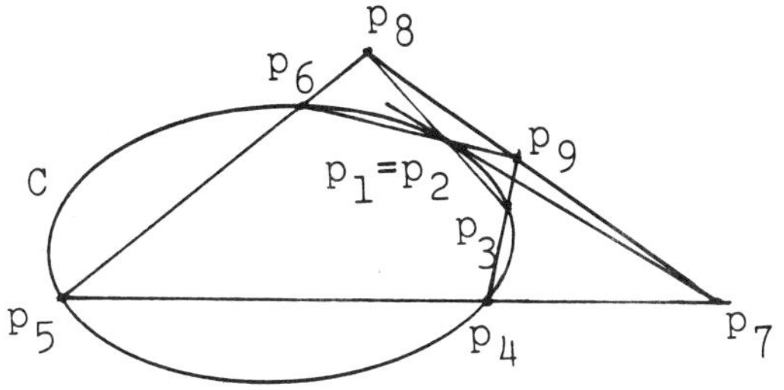

FIGURE 1.72

$$\Lambda = \{F_\lambda = \lambda_0 F + \lambda_1 G \mid \lambda = (\lambda_0 : \lambda_1) \in \mathbb{P}^1\}$$

In particular, F.G = F.H.

Proof. A general member F_λ of Λ is nonsingular outside $V(F) \cap V(G)$, by Proposition 1.4.6. Hence, by assumption (2), there is $\lambda = (\lambda_0 : \lambda_1) \in \mathbb{P}^1$ ($\lambda \neq (1:0)$, $\lambda \neq (0:1)$) near (1: 0) such that F_λ is nonsingular. Since $F_\lambda \neq G$, we get

$$F_\lambda . G = (\lambda_0 F + \lambda_1 G).G = F.G \geq Z$$

In a similar way, we get $F_\lambda . F = F.G$.

If $F_\lambda = H$, then there is nothing to prove. Suppose that $F_\lambda \neq H$. For every point p in the support of Z,

$$\min\{I_p(F,G), I_p(F,H)\} = \min\{I_p(F,F_\lambda), I_p(F,H)\} \leq I_p(F_\lambda, H)$$

by Lemma 1.3.8. Hence, $Z \leq F_\lambda . H$.

Now, the meromorphic function $f = (G/H)|_{F_\lambda}$ on F_λ has degree at most n - 2. Hence, by Theorem 5.3.17, f must be a constant. Thus,

$$F_\lambda . G = F_\lambda . H$$

We regard $D = F_\lambda . G = F_\lambda . H$ as a positive divisor on F_λ. By the proof of Theorem 1.4.9 (m = n), the linear system

$$\Sigma = \{F_\lambda . G' \mid G' \in \mathbb{P}(W_n)\}$$

is complete and of dimension N - 1 ($N = \frac{1}{2}n(n+3)$). Moreover the surjective map

$$\pi: G' \in \mathbb{P}(W_n) - \{F_\lambda\} \to F_\lambda . G' \in \Sigma = |D|$$

is induced by the surjective linear map

$$\alpha: G' \in W_n \to (G'/nL)|_{F_\lambda} \in L(nE)$$

where L is a line and $E = F_\lambda . L$ (see the proof of Theorem 1.4.9). Note that $\ker(\alpha) = \{cF_\lambda \mid c \in \mathbb{C}\}$. Hence, π is the projection with the center the point $F_\lambda \in \mathbb{P}(W_n)$ (see Exercise 6 of Sec. 1.1). Since $D = F_\lambda . G = F_\lambda . H$, H is on the line in $\mathbb{P}(W_n)$ passing through F_λ and G, so $H \in \Lambda$. Since $H \neq F$, we may write $H = a_0 F + a_1 G$ with $a_1 \neq 0$. Then,

$$F.H = F.(a_0 F + a_1 G) = F.G \qquad \text{Q.E.D.}$$

Putting n = 3, we get the classically well-known

Corollary 1.4.17. Let F, G, and H be plane cubics such that (1) $F.G = p_1 + \cdots + p_9$, (2) $p_1, \ldots, p_9$ are nonsingular points of F, and (3) $F.H = p_1 + \cdots + p_8 + q$. Then $q = p_9$.

Remark 1.4.18. The degree $n^2 - n + 2$ of Z in the theorem cannot be replaced by a smaller number. In fact, put

$$F = L_1 \cdots L_n \text{ (n-lines)} \quad \text{and} \quad G = L'_1 \cdots L'_n \text{ (n-lines)}$$

and suppose that

$$F.G = \Sigma p_{jk} \quad (p_{jk} = L_j . L'_k)$$

is a positive zero-cycle consisting of distinct n^2 points. Let L''_n be a line passing through p_{nn}, but no other p_{jk}. Put

$$Z = \Sigma' p_{jk} + p_{nn}$$

where Σ' is the sum taken over all j, k with $1 \leq j \leq n$ and $1 \leq k \leq n - 1$. Put

$$H = L'_1 \cdots L'_{n-1} L''_n$$

Then F, G, H, and Z satisfy all the conditions in the theorem except

(n=3)

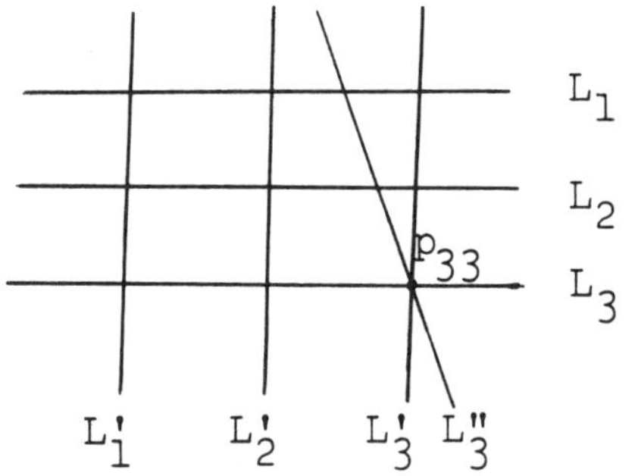

FIGURE 1.73

$$\deg Z = n^2 - n + 1$$

We get, in this case, $F.H \neq F.G$. (See Fig. 1.73.)

A far-reaching theorem in this direction is

Theorem 1.4.19 (Cayley-Bacharach). Let F and G be plane curves of degree n and m, respectively, such that they meet at nm distinct points. Then any plane curve H of degree n + m - 3 passing through all but one point of $V(F) \cap V(G)$ necessarily passes through the remaining point also.

For the proof of the theorem, see Griffiths-Harris [33, p. 671].

Note 1.4.20. For various classical results on linear systems of plane curves, see Coolidge [22] and Severi [89].

Exercises

1. Let $Z = \nu_1 p_1 + \cdots + \nu_s p_s$ ($p_j \neq p_k$ for $j \neq k$) be a zero-cycle on $\mathbb{P}^2$. Then, for positive integer n,
   (i) $\Lambda = \Lambda(n; Z) = \{F \in \mathbb{P}(W_n) \mid m_{p_j}(F) \geq \nu_j,\ 1 \leq j \leq s\}$ is a linear system of dimension $\geq N - \Sigma \frac{1}{2}\nu_j(\nu_j + 1)$ $(N = \frac{1}{2}n(n+3))$.
   (ii) If $\Sigma \frac{1}{2}\nu_j(\nu_j + 1) \leq N$, then Λ is nonempty.
   (iii) If $\deg Z \leq n + 1$, then $\dim \Lambda = N - \Sigma \frac{1}{2}\nu_j(\nu_j + 1)$.
2. Let C be a plane cubic and L and L' be lines in $\mathbb{P}^2$. Put
$$C.L = p_1 + p_2 + p_3 \quad \text{and} \quad C.L' = q_1 + q_2 + q_3$$
   Suppose that p_1, p_2, p_3, q_1, q_2, and q_3 are nonsingular points of C. Put

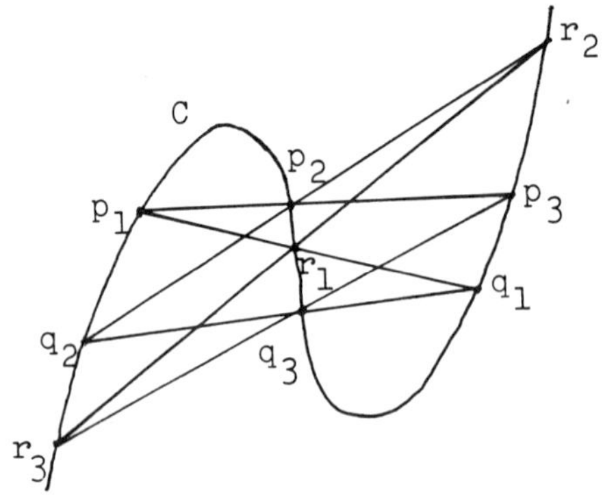

FIGURE 1.74

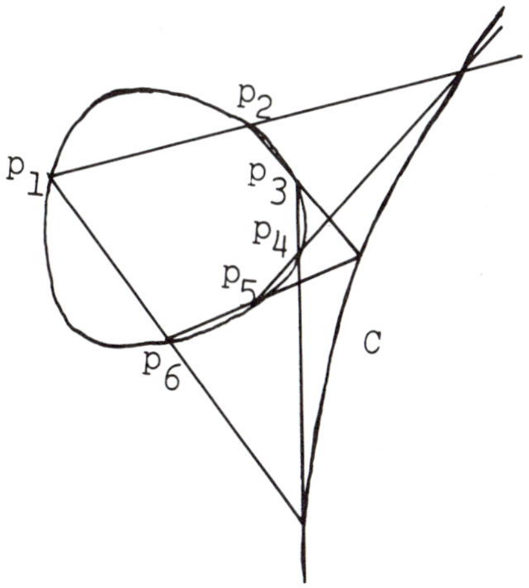

FIGURE 1.75

$C.\overline{p_1q_1} = p_1 + q_1 + r_1$, $C.\overline{p_2q_2} = p_2 + q_2 + r_2$, and $C.\overline{p_3q_3} = p_3 + q_3 + r_3$

Then r_1, r_2, and r_3 are collinear. In particular, a line passing through two flexes of C passes through another flex. (See Fig. 1.74.)

3. Let a, b, c, and d be noncollinear fixed nonsingular points on a plane cubic C without multiple component. Let D be a conic passing through a, b, c, and d without common component with C. Put

$$C.D = a + b + c + d + p + q$$

Then the line $\overline{pq}$ passes through a fixed point of C.

4. If a hexagon is inscribed in a plane cubic C and if two pairs of the opposite sides meet on C, then another pair meets also on C. (See Fig. 1.75.)

5. Let C, D, and E be irreducible conics meeting at distinct three points p, q, and r with mutually distinct tangent lines. Put

$$C.D = p + q + r + s, \quad D.E = p + q + r + t, \quad \text{and} \quad E.C = p + r + s + u$$

Let L be a line passing through s. Put $C.L = s + v$ and $D.L = s + w$. Then the lines $\overline{tw}$ and $\overline{uv}$ meet at a point on E. (See Fig. 1.76.)

6. Let G and H be plane curves of degree m. Suppose that there are a plane curve F of degree n with $n > m$ and a positive zero-cycle Z on $\mathbb{P}^2$ of degree $mn - n + 2$ such that (1) F has a common irreducible compo-

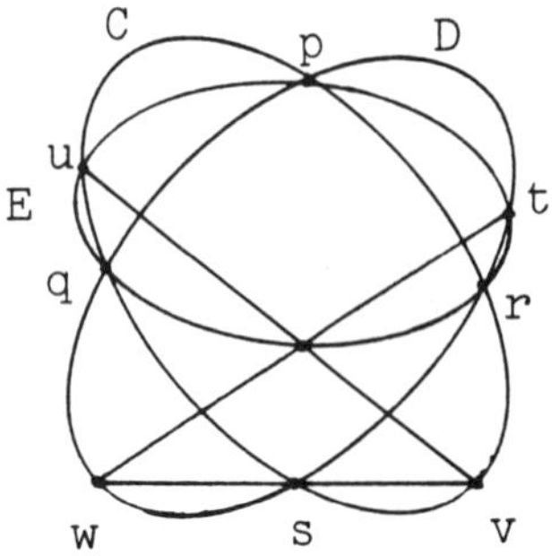

FIGURE 1.76

nent with neither G nor H, (2) every point of $V(F) \cap V(G)$ is a nonsingular point of F, and (3) $Z \leq F \cdot G$ and $Z \leq F \cdot H$. Then $G = H$.

1.5. DUAL CURVES AND PLÜCKER'S FORMULA

In this section, we treat irreducible plane curves. Let $C = \{F = 0\}$ be an irreducible plane curve of degree n ($n \geq 2$). Let

$$\phi\colon M \to C$$

be a nonsingular model of C (see Theorem 4.1.11). Take a point $p \in \mathbb{P}^2 - C$. The projection $\pi_p\colon M \to \mathbb{P}^1$ with the center p is a meromorphic function on M defined by

$$\pi_p\colon m \to \overline{p\phi(m)} \in \mathbb{P}^1$$

where $\mathbb{P}^1$ is the one-dimensional projective space of all lines in $\mathbb{P}^2$ passing through p. The degree of π_p is clearly n. (See Fig. 1.77.)

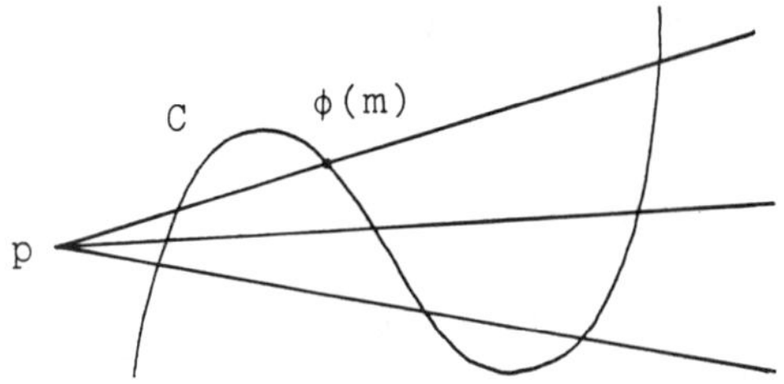

FIGURE 1.77

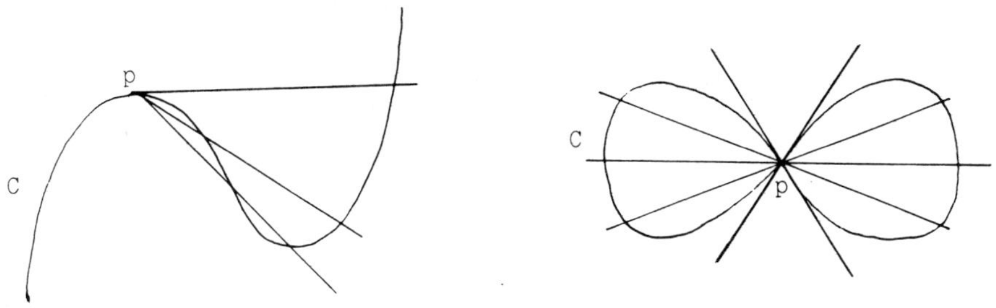

FIGURE 1.78

Even if p is on C, we can define a meromorphic function $\pi_p: M \to \mathbb{P}^1$ called the projection with the center p again. The degree of π_p is $n - m_p$, where m_p is the multiplicity of C at p. (See Fig. 1.78.)

Now, for a point $m \in M$, let C_1 be the irreducible branch at $\phi(m)$, corresponding to m. Consider the map

$$\psi: m \in M \to T_{\phi(m)}C_1 \in \mathbb{P}^{2*}$$

The image of ψ is the set of all tangent lines to C.

Lemma 1.5.1. ψ is a holomorphic map.

Proof. The problem is local. Take $m \in M$ and an open neighborhood U of m in M with a local coordinate t on U such that $t = 0$ at m and $\phi: U \to C$ is injective. (t is called a local uniformizing parameter of C at m (see Definition 4.1.9).) We may put $\phi(m) = (1: 0: 0)$. ϕ is locally written as

$$\phi: t \to (x(t), y(t))$$

where $x = X_1/X_0$ and $y = X_2/X_0$. The affine tangent line at $\phi(t)$ is given by

$$y - y(t) = (y'(t)/x'(t))(x - x(t))$$

where $x'(t) = \frac{dx}{dt}(t)$ and $y'(t) = \frac{dy}{dt}(t)$. Its closure in $\mathbb{P}^2$ is the line

$$(x(t)y'(t) - y(t)x'(t))X_0 - y'(t)X_1 + x'(t)X_2 = 0$$

Hence, ψ is locally given by

$$\psi: t \to (Y_0: Y_1: Y_2) = (x(t)y'(t) - y(t)x'(t): -y'(t): x'(t))$$

Put $u = Y_1/Y_0$ and $v = Y_2/Y_0$. Then ψ is given by

$$\psi : t \to (u(t), v(t))$$

where

$$u(t) = \frac{-y'(t)}{x(t)y'(t) - y(t)x'(t)}$$

$$v(t) = \frac{x'(t)}{x(t)y'(t) - y(t)x'(t)}$$

These are meromorphic functions of t, so

$$\psi : t \in U \to (1: u(t): v(t)) \in \mathbb{P}^{2*}$$

can be extended to a holomorphic map (see Proposition 4.1.6), which is clearly equal to the map ψ defined above. Q.E.D.

The image curve $\psi(M)$ in $\mathbb{P}^{2*}$ is called the dual curve of C and is denoted by C*. For example, the dual curve of an irreducible conic is an irreducible conic in $\mathbb{P}^{2*}$ (see Sec. 1.2).

Lemma 1.5.2. C* is an irreducible curve and is not a line.

Proof. C* is irreducible, because it is the image of M under ψ. If C* is a line in $\mathbb{P}^{2*}$, then there is a point $q \in \mathbb{P}^2$ such that every tangent line to C passes through q. This is impossible, because a general line passing through q cuts C at just n distinct points (see the discussion before Proposition 1.3.4). Q.E.D.

Theorem 1.5.3 (Reciprocity of dual curves). The dual curve to the dual curve C* to C is C itself, that is, C** = C. In particular, $\psi : M \to C^*$ is a nonsingular model of C*.

Proof. Let p be a nonsingular point of C. Then p can be identified with the point $\phi^{-1}(p)$ of M. In a neighborhood U of p in C, the map ψ is given by

$$\psi : q \in U \to T_q C \in \mathbb{P}^{2*}$$

Take an affine coordinate system (x, y) such that p = (0, 0) and $T_p C$ = x-axis. Then the coordinate x can be taken as a local uniformizing parameter at p. Hence, ϕ can be locally given by

$$x \to (x, y) = (x, f(x))$$

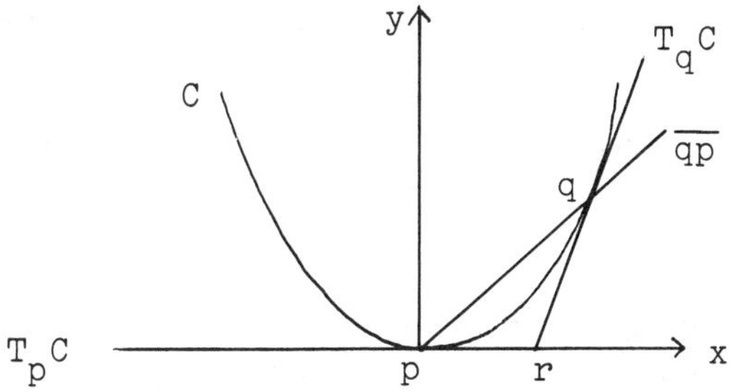

FIGURE 1.79

where f(x) is a holomorphic function of x having the power series expansion

$$f(x) = a_m x^m + a_{m+1} x^{m+1} + \cdots \qquad (a_m \neq 0,\ m \geq 2)$$

Using this expansion, an elementary calculus shows that

1. $\overline{qp} \to T_pC$ in the topology of $\mathbb{P}^{2*}$ as $q \to p$ on C and
2. putting $r = T_qC \cap T_pC$, $r \to p$ as $q \to p$ on C. (See Fig. 1.79.)

Now, assume that $\psi(p) = T_pC$ is also a nonsingular point of C*. The line $\overline{\psi(q)\psi(p)}$ in $\mathbb{P}^{2*}$ corresponds to the point $r = T_qC \cap T_pC$ in $\mathbb{P}^2$. By (1) and (2), the tangent line $T_{\psi(p)}C^*$ corresponds to p. Hence,

$$\psi(\psi(p)) = p$$

This holds for any nonsingular point $p \in C$ such that $\psi(p)$ is also a nonsingular point of C*. Thus C** = C. In particular,

$$\psi: M \to C^* \subset \mathbb{P}^{2*}$$

is a bimeromorphic map. Q.E.D.

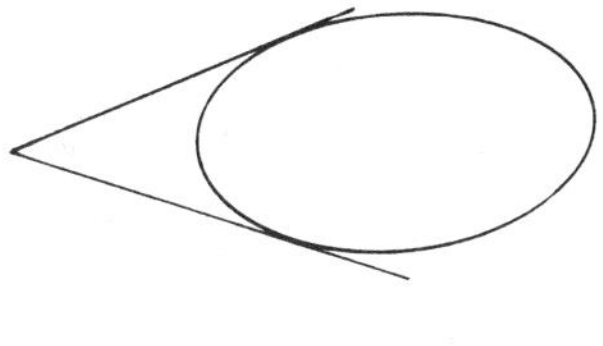

FIGURE 1.80

The degree of C^* is called the class of C. It is the number of tangent lines passing through a general point of $\mathbb{P}^2 - C$. (See Fig. 1.80.)

Theorem 1.5.4 (The Class Formula). The class of C is given by

$$c = 2(g - 1 + n) - \Sigma(m_q - s_q)$$

where n = the degree of C, g = the genus of a nonsingular model of C, m_q = the multiplicity of C at q, s_q = the number of irreducible branches at q and Σ is extended over all singular points $q \in C$.

Proof. This is a simple application of the Riemann-Hurwitz Formula (see Theorem 4.1.4). Let $\phi: M \to C$ be as before a nonsingular model of C. Take a general point $p \in \mathbb{P}^2$ such that (1) p is not on any tangent lines at singular points of C and (2) the class c is the number of tangent lines passing through p. Consider the projection

$$\pi_p : M \to \mathbb{P}^1$$

with the center p. Applying the Riemann Hurwitz Formula to π_p,

$$2g - 2 = -2n + \Sigma(e_m - 1)$$

where Σ is extended over all points $m \in M$ of ramification of π_p and e_m is the ramification index (see Lemma 4.1.2). Singular points $q \in C$ gives, in general, a point $m(\phi(m) = q)$ of ramification. By the assumption,

$$\Sigma_{m \in \phi^{-1}(q)} (e_m - 1) = m_q - s_q$$

On the other hand, a nonsingular point $q \in C$ with $p \in T_qC$ also gives a point m $(\phi(m) = q)$ of ramification with $e_m = I_q(C, T_qC) = 2$. Hence, the sum $\Sigma(e_m - 1)$ for such q's is the class of C. Thus,

$$2g - 2 = -2n + \Sigma(m_q - s_q) + c \qquad \text{Q.E.D.}$$

Let C and D be irreducible plane curves of degree ≥ 2. Suppose that a line L is tangent to both C and D. (See Fig. 1.81.) This means that the dual curves C^* and D^* meet at $p^* = L \in \mathbb{P}^{2*}$. Hence, we may define the tangent number $c_L(C, D)$ of C and D at L by

$$c_L(C, D) = I_{p^*}(C^*, D^*) \qquad (p^* = L \in \mathbb{P}^{2*})$$

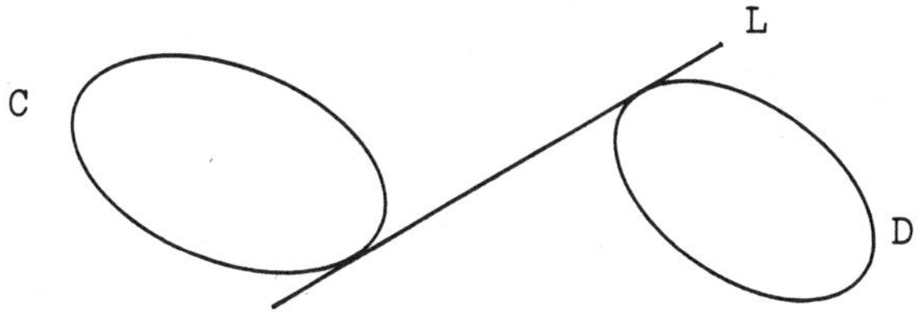

FIGURE 1.81

Then the dual to Bezout's Theorem is

Theorem 1.3.5*. Let C and D ($C \neq D$) be irreducible plane curves of degree ≥ 2. Then

$$\Sigma_L c_L(C, D) = c(C)c(D)$$

where c(C) (respectively c(D)) is the class of C (respectively D). In particular, there is a line tangent to both C and D. (See Fig. 1.82.)

Let C be as before an irreducible plane curve of degree n (≥ 2). For a point $p \in C$, put as before

m_p = the multiplicity of C at p

s_p = the number of irreducible branches of C at p

Let $C_1, \ldots, C_s$ ($s = s_p$) be the irreducible branches at p. Put

$$\lambda_{p1} = I_p(C_1, T_pC_1), \ldots, \lambda_{ps} = I_p(C_s, T_pC_s)$$

$$\lambda_p = (\lambda_{p1}, \ldots, \lambda_{ps})$$

$$|\lambda_p| = \lambda_{p1} + \cdots + \lambda_{ps}$$

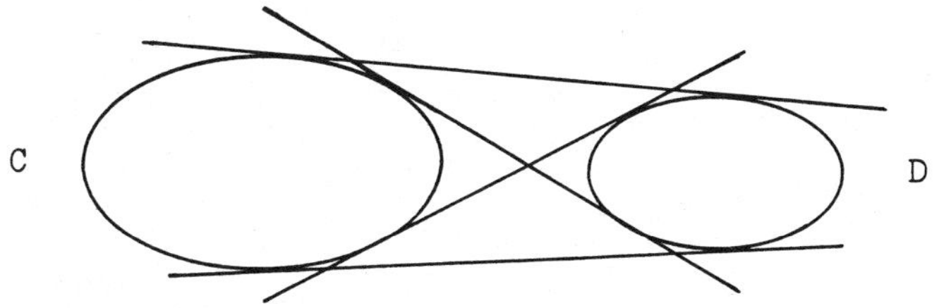

FIGURE 1.82

FIGURE 1.83

If p is a nonsingular point or a cusp, that is, $s_p = 1$, then we put simply

$$\lambda_p = |\lambda_p| = I_p(C, T_pC)$$

Note that $s_p = 1$ and $\lambda_p = 2$ if and only if p is a nonsingular point which is not a flex. Note that λ_p is not an analytic invariant, but a projective invariant.

An ordinary m-ple point p is called a regular m-ple point if none of irreducible branches of C at p has p as a flex, that is, $\lambda_p = (2, \ldots, 2)$. A node $p \in C$ is called a flecnode (repectively biflecnode) if one (respectively both) of two branches at p has p as a flex. (See Figure 1.83.)

A line L is said to be m-multitangent to C ($m \geq 2$) if L is tangent to C at m distinct nonsingular points of C which are not flexes. A two-multitangent (respectively three-multitangent) line is called a bitangent (respectively tritangent) line. (See Fig. 1.84.)

The following lemma is easy to show.

Lemma 1.5.5. If L is a m-multitangent line to C, then $p^* = L \in \mathbb{P}^{2*}$ is a regular m-ple point of C^*. In particular, there are only finitely many multitangent lines to C. Conversely, if $p^* = L \in \mathbb{P}^{2*}$ is a regular m-ple point of C^* such that every tangent line at p^* is not tangent to C at another point, then L is a m-multitangent line to C. (See Fig. 1.85.)

FIGURE 1.84

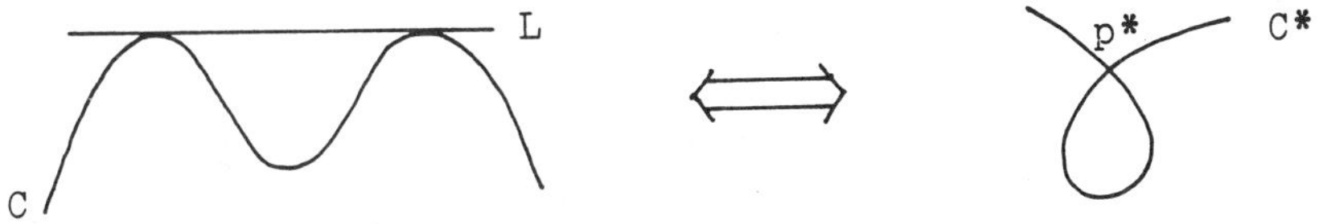

FIGURE 1.85

We leave the proof of the following lemma to the reader (Exercise 1).

Lemma 1.5.6. Let C_1 be an irreducible branch of C at p. Then there are an affine coordinate system (x, y) with p = (0, 0) and a local uniformizing parameter t at p such that C_1 is given by the image of

$$t \to (x, y) = (t^m, t^\lambda + a_{\lambda+1}t^{\lambda+1} + \cdots)$$

(Here, $\cdots$ means the higher order terms.) In the equation above, $m = m_p(C_1)$, $\lambda = I_p(C_1, T_pC_1)$, and $a_{\lambda+1}, \ldots$ are constants.

Note that irreducible branches at $p \in C$ (respectively $p^* \in C^*$) are in one-to-one correspondence to the points in $\phi^{-1}(p)$ (respectively $\psi^{-1}(p^*)$), where $\phi: M \to C$ (respectively $\psi: M \to C^*$) is a nonsingular model of C (respectively C^*) (see Theorem 4.1.11). We say that an irreducible branch C_1 at a point $p \in C$ and an irreducible branch C_1^* at a point $p^* \in C^*$ correspond if they correspond to the same point of M.

Lemma 1.5.7. Let p, C_1, (x, y), t and

$$t \to (x, y) = (t^m, t^\lambda + a_{\lambda+1}t^{\lambda+1} + \cdots)$$

be as in Lemma 1.5.6. Then the irreducible branch C_1^* of C^* at p^* corresponding to C_1 is given by the image of

$$s \to (u, v) = (s^{\lambda-m}, s^\lambda + b_{\lambda+1}s^{\lambda+1} + \cdots)$$

for a suitable affine coordinate system (u, v) in $\mathbb{P}^{2*}$ and a local uniformizing parameter s.

Proof. Using the notations in the proof of Lemma 5.1, C_1^* is given by the image of

$$\begin{aligned} t \to (Y_0 : Y_1 : Y_2) &= (x(t)y'(t) - y(t)x'(t) : -y'(t) : x'(t)) \\ &= ((\lambda - m)t^{m+\lambda-1} + \cdots : -\lambda t^{\lambda-1} + \cdots : mt^{m-1} + \cdots) \\ &= \left(\frac{\lambda - m}{m} t^{\lambda} + \cdots : \frac{-\lambda}{m} t^{\lambda-m} + \cdots : 1\right) \end{aligned}$$

From this expression, the lemma follows. Q.E.D.

In this lemma, if we put $m_p = 1$, $s_p = 1$, and $\lambda_p \geq 3$, that is, p is a flex of C, then C* is locally given by the image of

$$s \to (s^{\lambda-1}, s^{\lambda} + b_{\lambda+1}s^{\lambda+1} + \cdots)$$

A cusp p of C is said to be a <u>simple cusp of multiplicity</u> m if (1) $m_p = m$ and (2) $\lambda_p = I_p(C, T_pC) = m + 1$. The proof of the following lemma is easy and is left to the reader (Exercise 1).

<u>Lemma 1.5.8.</u> A point $p \in C$ is a simple cusp of multiplicity m if and only if there are a local (or affine) coordinate system (x, y) around p with p = (0, 0) and a uniformizing parameter t at p such that C is locally given by the image of

$$t \to (x, y) = (t^m, t^{m+1} + a_{m+2}t^{m+2} + \cdots)$$

In particular, the notion of simple cusps is analytic.

From the above consideration,

<u>Lemma 1.5.9.</u> Let p be a flex of C of order k, that is, $\lambda_p = I_p(C, T_pC) = k + 2$. Suppose that T_pC is tangent to C at no other point. Then $p^* = T_pC \in \mathbb{P}^{2*}$

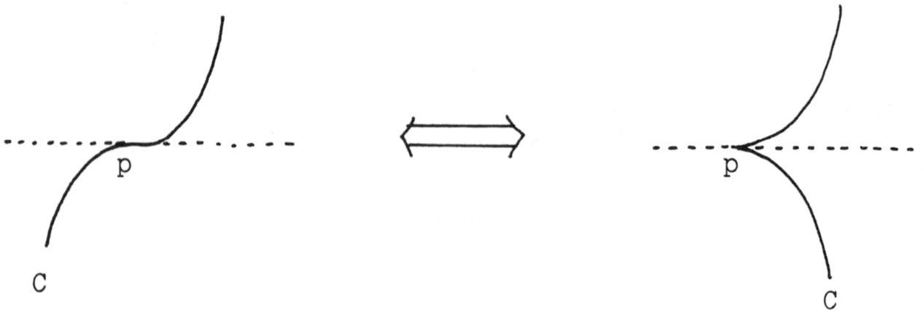

FIGURE 1.86

is a simple cusp of multiplicity $k + 1$ of C^*. Conversely, if $p^* = T_pC \in \mathbb{P}^{2*}$ is a simple cusp of multiplicity $k + 1$ of C^* such that $T_{p^*}C^*$ is tangent to C^* at no other point, then p is a flex of C of order k. (See Fig. 1.86.)

In general, a complicated tangent line to C is a complicated singular point of C^* (see Figs. 1.87 through 1.92).

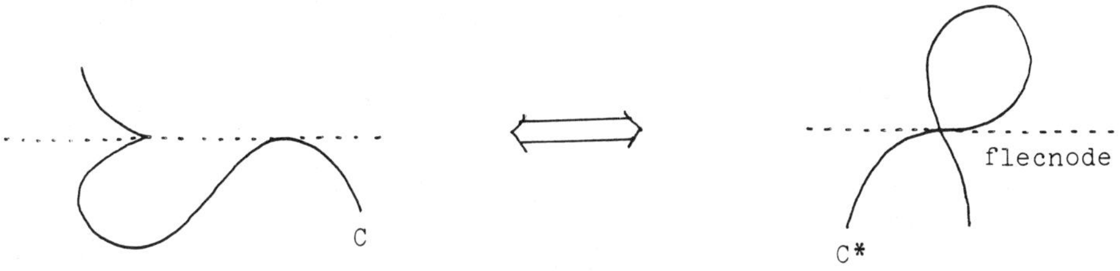

FIGURE 1.87

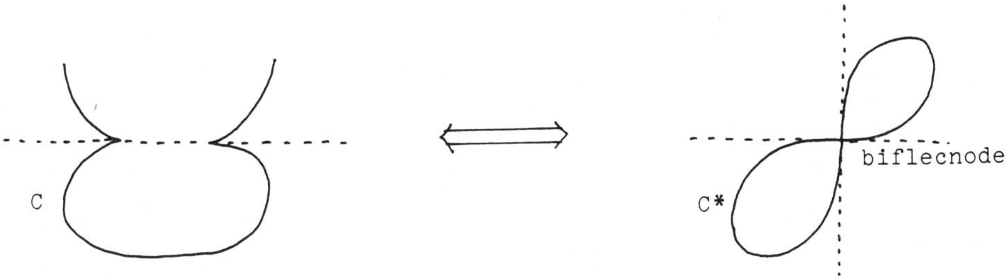

FIGURE 1.88

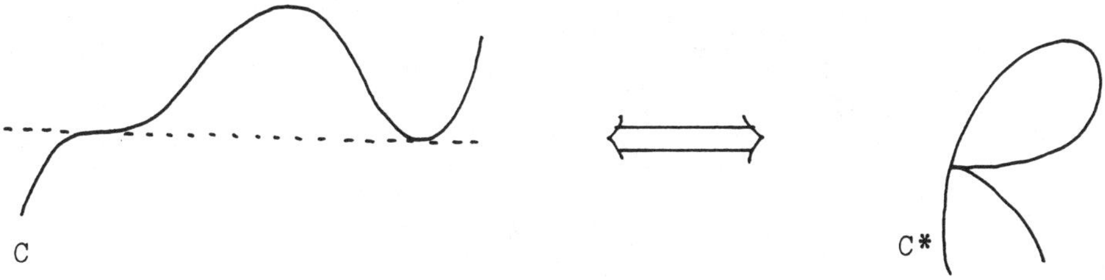

FIGURE 1.89

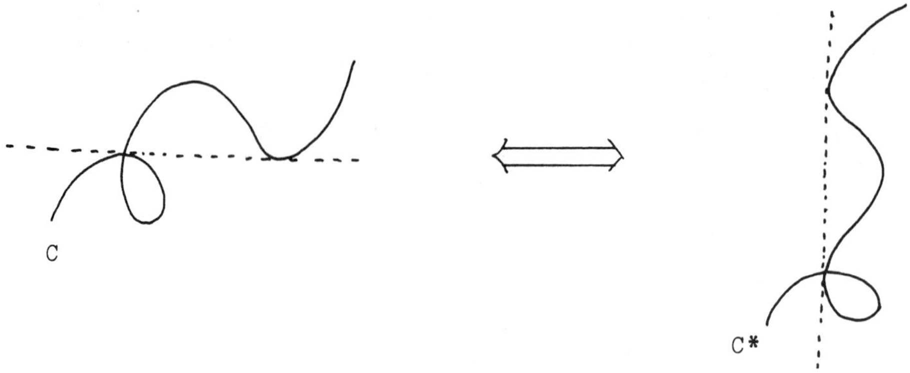

FIGURE 1.90

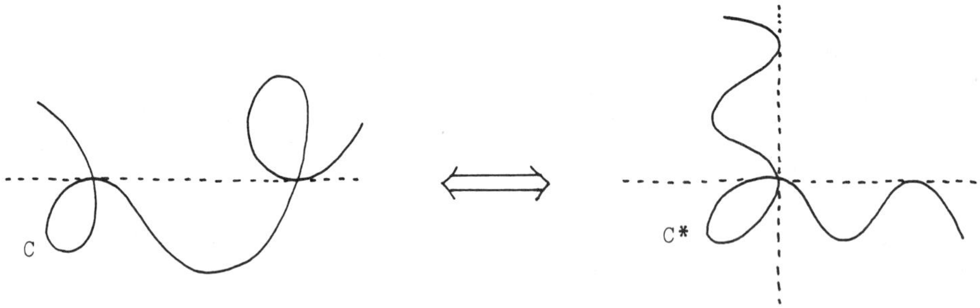

FIGURE 1.91

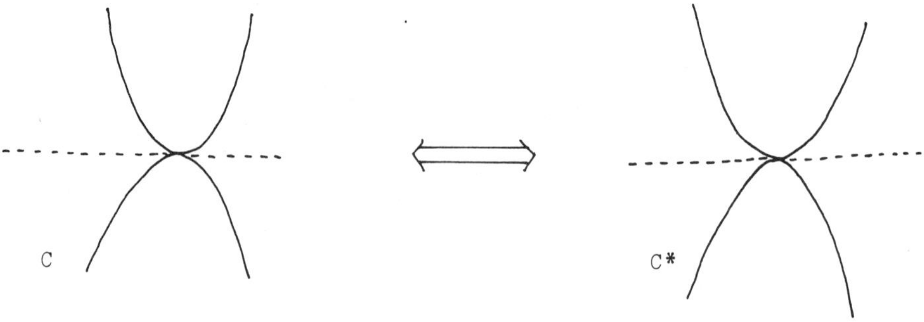

FIGURE 1.92

Applying Theorem 1.5.4 to C*, we get

<u>Theorem 1.5.4*.</u>

$$n = 2(g - 1 + c) - \sum_{q \in C} (|\lambda_q| - m_q - s_q)$$

The proof is easy (use Lemma 1.5.7). It should be noted that, if p is a nonsingular point of C which is not a flex, then $|\lambda_p| - m_p - s_p = 2 - 1 - 1 = 0$. Hence, Σ is in fact a finite sum.

For a flex q, $|\lambda_q| - m_q - s_q = \lambda_q - 2$ is the order of q. Put

$$f = f_C = \sum_q (\lambda_q - 2)$$

where Σ is extended over all flexes q of C. f is called the <u>total order of flexes on</u> C. By Theorems 1.5.4 and 1.5.4*,

<u>Theorem 1.5.10 (The flex formula).</u>

$$f = 2(g - 1 + c) - n - \sum_q (|\lambda_q| - m_q - s_q)$$

$$= 6g - 6 + 3n - \sum_q (|\lambda_q| + m_q - 3s_q)$$

where Σ is extended over all singular points q on C.

<u>Example 1.5.11.</u> For a nonsingular plane curve of degree n (≥ 2),

$$g = \tfrac{1}{2}(n - 1)(n - 2) \quad \text{(see Theorem 2.1.9)}$$

$$c = n(n - 1)$$

$$f = 3n(n - 2)$$

An irreducible plane curve C is called a <u>Plücker curve</u> if all the following conditions are satisfied:

1. C has only simple cusps of multiplicity 2 and regular nodes as singular points,
2. every flex of C is an ordinary flex,
3. every multitangent line is a bitangent line,
4. the tangent line at a singular point or a flex is a tangent line at no other point.

The following lemma is clear from the definition.

Lemma 1.5.12. If C is a Plücker curve, then C* is also a Plücker curve.

For a Plücker curve C, put

g = the genus of C

n = the degree of C

c = the class of C

δ = the number of nodes on C

κ = the number of simple cusps of multiplicity 2 on C

b = the number of bitangent lines to C

f = the number of flexes on C

Then, clearly,

g = the genus of C*

c = the degree of C*

n = the class of C*

b = the number of nodes on C*

f = the number of simple cusps of multiplicity 2 on C*

δ = the number of bitangent lines on C* and

κ = the number of flexes on C*

By the genus formula (see Theorem 2.1.9),

$$g = \tfrac{1}{2}(n - 1)(n - 2) - \delta - \kappa$$

Hence we get the classical

Theorem 1.5.13 (Plücker's Formula). For a Plücker curve,

1. $g = \frac{1}{2}(n - 1)(n - 2) - \delta - \kappa$,
2. $g = \frac{1}{2}(c - 1)(c - 2) - b - f$,
3. $c = n(n - 1) - 2\delta - 3\kappa$, and
4. $n = c(c - 1) - 2b - 3f$.

Proof. (2) is the genus formula for C*. (3) follows from (1) and Theorem 1.5.4. (4) is dual to (3). Q.E.D.

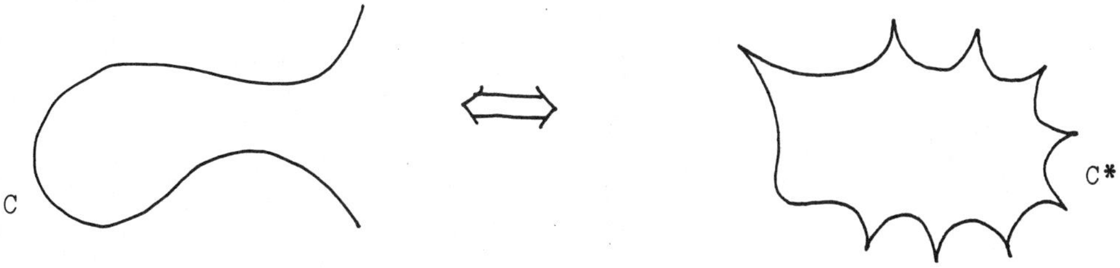

FIGURE 1.93

Example 1.5.14. Any irreducible plane cubic C is a Plücker curve. This is easily checked when C is nonsingular. If C has a singular point, it is a unique singular point and is either a node or a simple cusp of multiplicity 2 (see Corollary 2.1.6). From this, C is a Plücker curve in this case also. Thus there are three types of irreducible plane cubics:

	n	δ	κ	g	c	b	f
I	3	0	0	1	6	0	9
II	3	1	0	0	4	0	3
III	3	0	1	0	3	0	1

First, the dual curve C* to a nonsingular plane cubic C is an irreducible plane sextic with nine simple cusps of multiplicity 2. (See Fig. 1.93.)

Second, the dual curve C* to an irreducible plane cubic C with a node is an irreducible plane quartic with three simple cusps of multiplicity 2 and one bitangent line. (See Fig. 1.94.)

Third, the dual curve C* to an irreducible plane cubic C with a simple cusp of multiplicity 2 is projectively equivalent to C itself. (See Fig. 1.95.)

An irreducible plane curve C is said to be self-dual if C* is projectively equivalent to C. For example, the closure in $\mathbb{P}^2$ of the affine curve

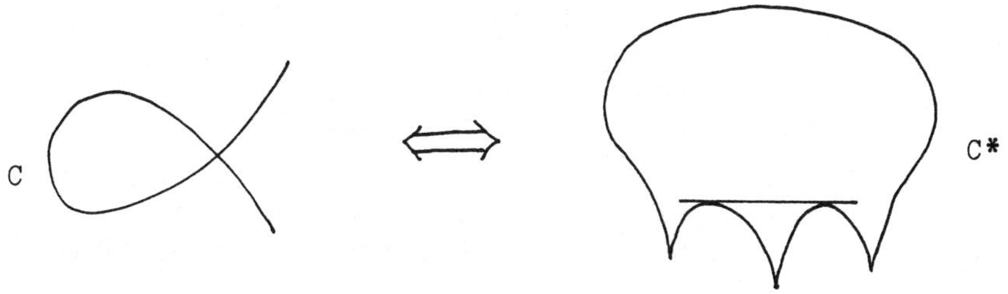

FIGURE 1.94

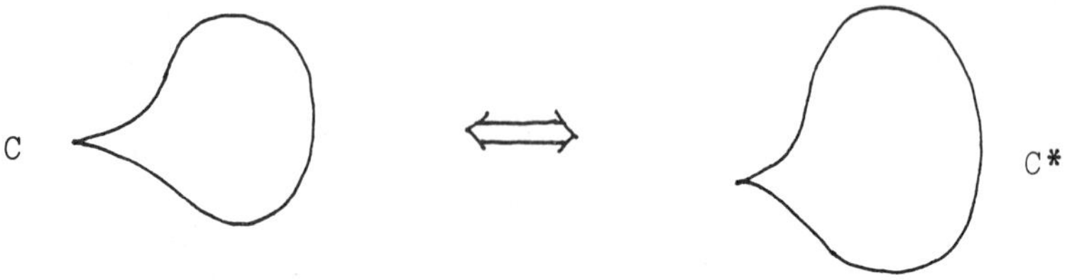

FIGURE 1.95

$$y^a - x^b = 0, \quad (a, b) = 1$$

is self-dual. It is an unsolved problem to determine all self-dual plane curves up to projective equivalence.

Example 1.5.15. An irreducible plane quartic C is not necessarily a Plücker curve. It may have another kind of singularity (see Sec. 2.2) or have a higher flex of order 2. The curve

$$y - x^4 = 0$$

is such an example. If a nonsingular plane quartic C has no higher flex, then it is a Plücker curve. For such C,

$$g = 3, \ c = 12, \ f = 24, \ \text{and} \ b = 28$$

It can be shown that a general nonsingular plane curve is a Plücker curve. For a nonsingular Plücker curve of degree n, we have

$$b = \tfrac{1}{2} n(n+1)(n-1)(n-2) - 4n(n-2)$$

Note 1.5.16. See Iitaka [49] for further discussion on dual curves and the generalized Plücker's Formula.

Exercises

1. Prove Lemmas 1.5.6 and 1.5.8.
2. Let $L_1, \ldots, L_5$ be distinct 5 lines in $\mathbb{P}^2$ no three of which meet at one point. Then there is a unique irreducible conic tangent to all L_j.
3. For a nonsingular plane quartic C, let b be the number of bitangent lines and f_1 be the number of ordinary flexes on C. Then (1) f_1 is even and $0 \leq f_1 \leq 24$, (2) $b = 16 + \frac{1}{2} f_1$, and (3) $16 \leq b \leq 28$. For the Fermat quartic $C = \{X_0^4 + X_1^4 + X_2^4 = 0\}$, $b = 16$.
4. The plane quartic C: $(X_0 X_2 - X_1^2)^2 - X_1^3 X_2 = 0$ is self-dual.

5. There exists no curve with one of the following conditions.

   i. An irreducible plane quintic with six simple cusps of multiplicity 2.
   ii. An irreducible plane quintic with five simple cusps of multiplicity 2 and one node.

6. Let C be an irreducible plane curve of degree n (≥ 2). Give a formula for the number of tangent lines (except T_pC) to C from a general point $p \in C$.

1.6. SPACE CURVES

By a projective algebraic curve or simply a curve, we mean a projective algebraic set of dimension 1 in a projective space $\mathbb{P}^r$ ($r \geq 2$) (see Sec. 3.3). In this section, we treat irreducible curves. A curve is said to be nondegenerate if it is not contained in any hyperplane. If $r \geq 3$, then a nondegenerate irreducible curve in $\mathbb{P}^r$ is simply called a space curve (in $\mathbb{P}^r$).

One of the simplest but most typical examples of space curves is the rational normal curve. It is, by definition, the image of the holomorphic imbedding

$$\begin{array}{cccc} \Phi: (t_0 : t_1) \in \mathbb{P}^1 & \to & (t_0^r : t_0^{r-1} t_1 : \cdots : t_1^r) \in \mathbb{P}^r \\ \| & & \| \\ t & \to & (1 : t : \cdots : t^r) \end{array}$$

($t = t_1/t_0$). Φ is nothing but the map $\Phi_{|r(\infty)|}$ associated with the complete linear system $|r(\infty)|$, where (∞) is the point divisor of the point ∞ of infinity on $\mathbb{P}^1$ (see Sec. 4.1). (It is an irreducible conic if $r = 2$.) The rational normal curve in $\mathbb{P}^3$ is called the twisted cubic.

Another interesting example of space curves is the image of the holomorphic imbedding

$$\Phi_{|n(0)|}: M \to \mathbb{P}^{n-1} \qquad (n \geq 4)$$

where M is a complex 1-torus (see Sec. 3.1) and (0) is the point divisor of the zero 0 of the additive group M. If $n = 4$, then $C = \Phi_{|4(0)|}(M)$ is the complete intersection of two quadric surfaces in $\mathbb{P}^3$ (see Example 5.3.25). We call it an elliptic quatic curve in $\mathbb{P}^3$.

One of the most important examples of space curves is the canonical curve $C_K = \Phi_K(M) \subset \mathbb{P}^{g-1}$ of a non-hyperelliptic compact Riemann surface M of genus g (≥ 4) (see Sec. 5.1).

Henceforth, let C be a space curve in $\mathbb{P}^r$ ($r \geq 3$) and $\phi: M \to C$ be a nonsingular model of C.

Take a point $m \in M$ and put $p = \phi(m) \in C$. Take a local coordinate t in a neighborhood U of m in M such that $t = 0$ at p, and an affine coordinate system $(x_1, \ldots, x_r)$ in $\mathbb{P}^r$ such that $(x_1, \ldots, x_r) = (0, \ldots, 0)$ at p. Then ϕ is locally given by

$$\phi: t \in U \to (x_1, \ldots, x_r) = (x_1(t), \ldots, x_r(t))$$

where $x_j(t)$ are holomorphic functions of t such that $x_j(0) = 0$. The proof of the following lemma is left to the reader as an exercise (Exercise 1).

Lemma 1.6.1. There are t and an affine coordinate system $(x_1, \ldots, x_r)$ such that

$$\phi: t \to (x_1, \ldots, x_r) = (t^{\alpha_1}, t^{\alpha_2} + a_{22}t^{\alpha_2+1} + \ldots, \ldots, t^{\alpha_r} + a_{r2}t^{\alpha_r+1} + \ldots)$$

where $1 \leq \alpha_1 < \alpha_2 < \cdots < \alpha_r$. $(t^{\alpha_2} + a_{22}t^{\alpha_2+} + \cdots$ means the power series expansion of the function $x_2(t)$, and so on. Moreover, $(\alpha\ , \ldots, \alpha_r)$ are independent of the choices of t and $(x\ , \ldots, x_r)$.

$C_1 = \phi(U)$ is called the irreducible branch at p of C corresponding to m. α_1 is called the multiplicity of C_1 at p and is denoted by $m_p(C\ )$. It is an analytic invariant, that is, it is independent of the choice of local coordinate system, while $\alpha_2, \ldots, \alpha_r$ are not necessarily analytic invariants. Put

$$m_p = m_p(C) = \sum_{m \in \phi^{-1}(p)} m_p(C_1)$$

and call it the multiplicity of C at p. (Here Σ runs over $m \in \phi^{-1}(p)$ and C_1 is the irreducible branch corresponding to m.)

C is said to be nonsingular at p if ($\phi^{-1}(p)$ is one point and) $m_p(C) = 1$. Otherwise p is called a singular point. Only finitely many singular points exist.

Using t and $(x_1, \ldots, x_r)$ in Lemma 1.6.1, the closure in $\mathbb{P}^r$ of the affine line: $x_2 = \cdots = x_r = 0$ is called the tangent line to the irreducible branch C_1 at p and is denoted by T_pC_1. It is called a tangent line to C at p. If C has a unique irreducible branch at p (in particular, if p is a nonsingular point), then T_pC_1 is written T_pC and is called the tangent line to C at p.

An easy calculus shows (Exercise 1)

Lemma 1.6.2. Under the above notations,

$$\overline{p\phi(t)} \to T_pC_1 \quad \text{as} \quad t \to 0$$

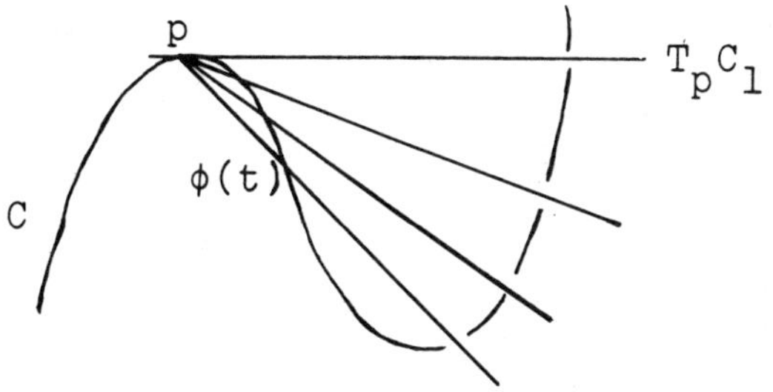

FIGURE 1.96

in the topology of G(1, r), the Grassmann variety (see Sec. 3.3) of all lines in $\mathbb{P}^r$. (See Fig. 1.96.)

Lemma 1.6.3. For a hyperplane H, $C \cap H$ is a nonempty finite set.

Proof. If $C \cap H$ is empty, then C is a compact connected analytic set in $\mathbb{P}^r - H$, which is biholomorphic to $\mathbb{C}^r$. Hence, C is a point (see Exercise 2 of Sec. 3.2), a contradiction. Hence, $C \cap H$ is a nonempty algebraic subset of the irreducible C with $C \cap H \neq C$, so is a finite point set. Q.E.D.

Take $m \in M$ and put $p = \phi(m) \in C$ as before. Let $\phi: t \to (x_1(t), \ldots, x_r(t))$ be as before. Let $\phi: t \to (x_1(t), \ldots, x_r(t))$ be as before. Let H be a hyperplane in $\mathbb{P}^r$ defined by the equation

$$H: h(x_1, \ldots, x_r) = 0$$

We then define the intersection number $I_m(C, H)$ (respectively $I_p(C, H)$) of C and H at m (respectively p) by

$$I_m(C, H) = \text{the order of zero of } h(x_1(t), \ldots, x_r(t)) \text{ at } t = 0$$

$$I_p(C, H) = \sum_{m \in \phi^{-1}(p)} I_m(C, H)$$

It is an analytic invariant. The intersection number $I_p(C, S)$ of C and a hypersurface S at p can be defined in a similar way. Put

$$D_H = \sum_{m \in M} I_m(C, H)m$$

This is a positive divisor on M, called a hyperplane divisor, or a hyperplane cut or a hyperplane section.

We show that $\deg D_H = \Sigma_m I_m(C, H)$ is constant for H. For this purpose, we need some preparations.

Lemma 1.6.4. Let $G(k, r)$ be the Grassman variety of all k-planes in $\mathbb{P}^r$ (see Sec. 3.3). Then

$$I = \{(p, P) \in C \times G(k, r) \mid p \in P\}$$

is an irreducible algebraic set in $\mathbb{P}^r \times G(k, r)$ of dimension $k(r - k) + 1$. (I is called the incidence correspondence.)

Proof. For simplicity, we put $r = 3$ and $k = 1$. The general case can be treated in a similar way.

Take an open set W of $G(1, 3)$ such that there are holomorphic maps $\sigma_1, \sigma_2 : W \to \mathbb{P}^3$ with the following properties: (1) $\sigma_1(\ell) \neq \sigma_2(\ell)$ for all $\ell \in W$ and (2) $\sigma_1(\ell)\sigma_2(\ell) = \ell$ for all $\ell \in W$. (This is equivalent to choose a holomorphic local section of the principal bundle $V \to G(1, 3)$ in Lemma 3.3.15.)

We may write

$$\sigma_1(\ell) = (X_{10}(\ell) : X_{11}(\ell) : X_{12}(\ell) : X_{13}(\ell))$$

$$\sigma_2(\ell) = (X_{20}(\ell) : X_{21}(\ell) : X_{22}(\ell) : X_{23}(\ell))$$

where $X_{jk}(\ell)$ are holomorphic functions of $\ell \in W$. Let C be defined by

$$F_1 = \cdots = F_m = 0$$

where $F_j = F_j(X_0, X_1, X_2, X_3)$ are homogeneous polynomials. Then I is locally defined by

$$I = \{(p, \ell) \in \mathbb{P}^3 \times W \mid F_j(p) = 0,\ 1 \leq j \leq m \quad \text{and} \quad \operatorname{rank}(X Y_1(\ell) Y_2(\ell)) \leq 2\}$$

where $p = (X_0 : X_1 : X_2 : X_3)$, $X = {}^t(X_0, X_1, X_2, X_3)$, $Y_j(\ell) = {}^t(X_{j0}(\ell), X_{j1}(\ell), X_{j2}(\ell), X_{j3}(\ell))$, $j = 1, 2$, the column vectors. (${}^tA$ is the transpose of A.) Hence, I is an algebraic set.

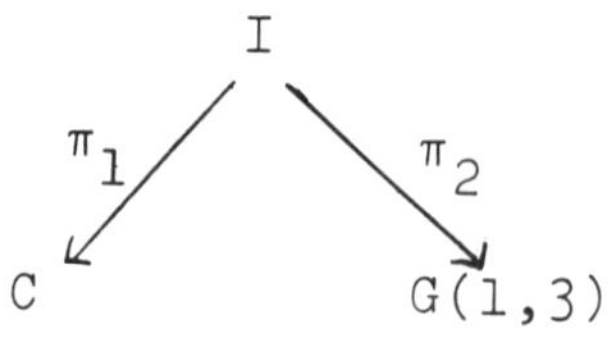

FIGURE 1.97

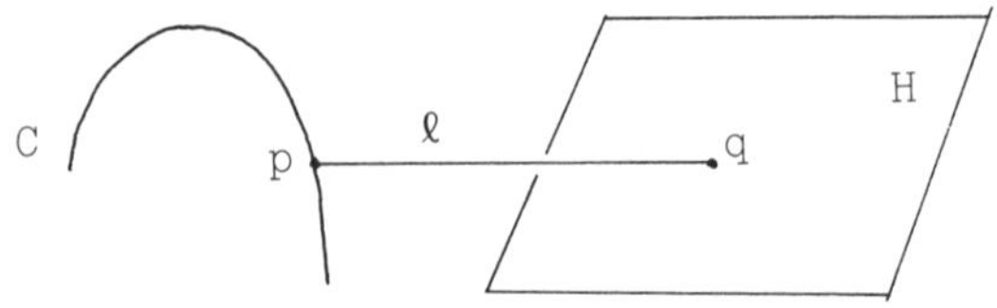

FIGURE 1.98

Next, consider the natural projections in Fig. 1.97. Each fiber of π_1 is the set of all lines passing through a fixed point $p \in C$ and is a two-dimensional projective space. In fact, take a plane H with $p \notin H$. Then there is an identification $\ell \to \{q\} = \ell \cap H$. The set Sing C of all singular points of C is a finite set. Put U = C - Sing C. On U, $\pi_1^{-1}(U) \to U$ is a $\mathbb{P}^2$-bundle. In fact, if U' is a small open set in U, then $\pi_1^{-1}(U')$ is biholomorphic to $U' \times H$. (See Fig. 1.98.)

This shows that I is irreducible and dim I = 2 + 1 = 3. Q.E.D.

Lemma 1.6.5. Let $k \leq r - 1$ and I be the incidence correspondence in Lemma 1.6.4. Then the image $\pi_2(I)$ of the projection $\pi_2: I \to G(k, r)$ is an irreducible algebraic set of G(k, r) of dimension $k(r - k) + 1$.

Proof. Since I is irreducible, $\pi_2(I)$ is also irreducible. Every fiber $\pi_2^{-1}(P)$ of $P \in G(k, r)$ is either empty or a finite set by Lemma 1.6.3. Hence, by Theorem 3.2.14, $\dim \pi_2(I) = \dim I = k(r - k) + 1$. Q.E.D.

Corollary 1.6.6. Let $k \leq r - 2$. Then a general k-plane does not intersect with C.

Proof. $k(r - k) + 1 < (k + 1)(r - k) = \dim G(k, r)$ if $k \leq r - 2$. Hence, $\pi_2(I)$ is closed and nowhere dense in G(k, r). Q.E.D.

Now, let $k \leq r - 2$ and P be a k-plane in $\mathbb{P}^r$. The set

$$\mathbb{P} = \{P' \mid P' \text{ is a } (k + 1)\text{-plane such that } P' \supset P\}$$

is a $(r - k - 1)$-dimensional projective space (compare the proof of Lemma 1.6.4). Consider the map

$$\pi_P: m \in M \to \phi(m) \vee P \in \mathbb{P}$$

This is clearly well-defined and holomorphic for m with $\phi(m) \notin P$. Hence, it is uniquely extended to a holomorphic map $\pi_P: M \to \mathbb{P}$. We call it the projection with the center P. (See Fig. 1.99.)

Now, let H and H' $(H \neq H')$ be hyperplanes in $\mathbb{P}^r$. Then $P = H \cap H'$ is an $(r - 2)$-plane. Hence,

$$\pi_P: M \to \mathbb{P}^1$$

is a meromorphic function on M. Suppose that $P \cap C = \phi$. It is then easy to see that

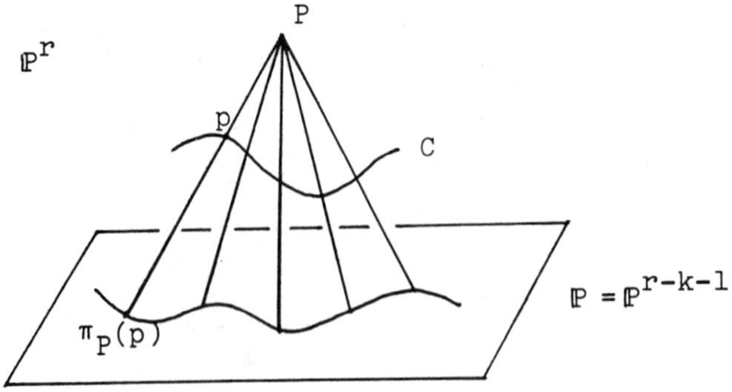

FIGURE 1.99

$$\pi_P^{-1}(t) = D_H \quad \text{and} \quad \pi_P^{-1}(t') = D_{H'}$$

where t (respectively t') corresponds to H (respectively H') and D_H (respectively $D_{H'}$) is the hyperplane cut by H (respectively H'). Hence,

$$D_H \sim D_{H'} \quad \text{(linearly equivalent)}$$

This holds even if $P \cap C \neq \phi$, for we can take H'' such that $H \cap H'' \cap C = \phi$ and $H' \cap H'' \cap C = \phi$. In particular,

$$\deg D_H = \deg D_{H'} \quad \text{for all} \quad H, H' \in \mathbb{P}^{r*}$$

Considering the function π_P above, it can be easily checked that, if $H \neq H'$, then $D_H \neq D_{H'}$.

Proposition 1.6.7.

1. The hyperplane divisors form a linear system Λ on M of dimension r without fixed point (see Sec. 4.1). In particular, $n = \deg D_H$ is constant for hyperplanes H.
2. If P is an (r - 2)-plane in $\mathbb{P}^r$ such that $P \cap C = \phi$, then a general hyperplane H with $H \supset P$ meets C at n distinct points.
3. A general hyperplane meets C at n distinct points.

Proof. (1) was shown above. (2) is clear because

$$\pi_P: M \to \mathbb{P}^1$$

is a ramified covering of degree n (see Sec. 4.1). (3) is a special case of Bertini's Theorem (see Theorem 3.4.14). Q.E.D.

The constant $n = \deg D_H$ in the proposition is called the degree of C and is denoted by deg C. It is also called the degree of the linear system Λ of all hyperplane cuts.

The degree of the rational normal curve in $\mathbb{P}^r$ is clearly r.

Proposition 1.6.8. Let C be a space curve in $\mathbb{P}^r$. Then (1) $\deg C \geq r$ and (2) $\deg C = r$ if and only if C is projectively equivalent to the rational normal curve, that is, there is $\sigma \in \mathrm{Aut}(\mathbb{P}^r)$ such that $\sigma(C)$ = the rational normal curve. (Such a C is called a rational normal curve.)

Proof. (1) C is nondegenerate, so there are distinct r points $p_1, \cdots, p_r$ on C in general position. Then they span a hyperplane H. We have then $\deg C = \deg D_H \geq r$. (2) Suppose that $\deg C = r$. Let $p_1, \cdots, p_r$ be as above. Let P be the (r - 2)-plane spanned by $p_1, \cdots, p_{r-1}$. Consider the meromorphic function

$$\pi_P : M \to \mathbb{P}^1$$

where M is a nonsingular model of C. By the assumption, $\deg \pi_P = 1$. (See Fig. 1.100.) This means that $\pi_P : M \to \mathbb{P}^1$ is a biholomorphic map. We identify M with $\mathbb{P}^1$ through π_P. The linear system Λ on $M = \mathbb{P}^1$ of all hyperplane cuts has dimension r and degree r, so $\Lambda = |r(\infty)|$ (see Example 5.2.18). Hence, $C = \Phi_\Lambda(M)$ is projectively equivalent to the rational normal curve. Q.E.D.

Now let $C \subset \mathbb{P}^r$ and $\phi: M \to C$ be as before a space curve and its nonsingular model. We fix an integer k with $0 \leq k \leq r - 1$. Let Λ be the linear system on M of all hyperplane cuts. For a k-plane P, consider the linear subsystem Λ_P of Λ defined by

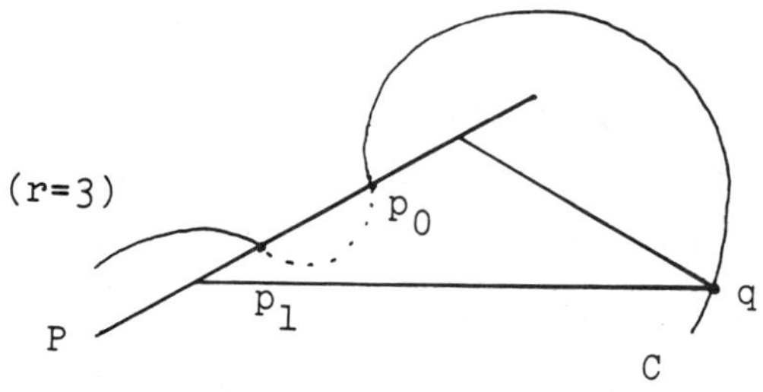

FIGURE 1.100

$$\Lambda_P = \{D_H \mid H \text{ is a hyperplane such that } H \supset P\}$$

It has the dimension r - k - 1. Λ_P has a fixed point (see Sec. 4.1) if and only if $C \cap P$ is nonempty. The fixed part of Λ_P (see Sec. 3.4) is denoted by D_P and is called the k-plane cut by P.

By Corollary 1.6.6, if $k \leq r - 2$, then Λ_P has no fixed point for a general P.

If $k \leq r - 2$ and $C \cap P$ is nonempty, then P is called a secant k-plane to C. It is said to be m-secant if $\deg D_P = m$. If $m \geq 2$ (respectively $m = 2$ or 3), then P is said to be multisecant (respectively bisecant or trisecant).

We now fix a positive integer j. Let S^jM be the j-th symmetric product of M (see Sec. 5.2). It is a j-dimensional compact complex manifold. Put

$$\tilde{S}_j^{\#} = \tilde{S}_j^{\#}(k, r) = \{(D, P) \in S^jM \times G(k, r) \mid D \leq D_P\}$$

Lemma 1.6.9. $\tilde{S}_j^{\#}$ is an analytic set in $S^jM \times G(k, r)$ (see Sec. 3.2).

Proof. The problem is local. Take $(D_0, P_0) \in S^jM \times G(k, r)$. Write

$$D_0 = i_1 m_1^0 + \cdots + i_s m_s^0 \quad (m_\nu^0 \neq m_\mu^0 \text{ for } \nu \neq \mu)$$

$(i_1 + \cdots + i_s = j)$. Take a local coordinate t in a neighborhood U of m_1^0 in M such that $t = 0$ at m_1^0. Then we can take a part of local coordinate system on a neighborhood of D_0 in S^jM as follows

$$u_1 = t_1 + \cdots + t_i$$

$$u_2 = t_1^2 + \cdots + t_i^2$$

$$\cdots$$

$$u_i = t_1^i + \cdots + t_i^i$$

where $i = i_1$, $(t_1, \ldots, t_i) \in U \times \cdots \times U$ and $t_\nu = t$ on each U (see Proposition 5.2.10).

On the other hand, there is a neighborhood W of P_0 in G(k, r) and r - k hyperplanes $H_1(P), \ldots, H_{r-k}(P)$ depending holomorphically on $P \in W$ such that

$$H_1(P) \cap \cdots \cap H_{r-k}(P) = P \quad \text{for} \quad P \in W$$

We may assume that every $H_\nu(P)$ is defined by the equation

$$h_\nu(x_1,\ldots,x_r, P) = a_{\nu 0}(P) + a_{\nu 1}(P)x_1 + \cdots + a_{\nu r}(P)x_r = 0$$

for an affine coordinate system $(x_1,\ldots,x_r)$, where $a_{\nu\mu}(P)$ are holomorphic functions of $P \in W$. Put

$$\phi\colon t \in U \to (x_1(t), \ldots, x_r(t)) \in \mathbb{P}^r$$

and

$$g_\nu(t,P) = h_\nu(x_1(t),\ldots,x_r(t), P)$$

Then

$$\tilde{g}_{\nu 1}(u,P) = g_\nu(t_1,P) + \cdots + g_\nu(t_i,P)$$

$$\tilde{g}_{\nu 2}(u,P) = g_\nu(t_1,P)^2 + \cdots + g_\nu(t_i,P)^2$$

$$\tilde{g}_{\nu i}(u,P) = g_\nu(t_1,P)^i + \cdots + g_\nu(t_i,P)^i$$

are holomorphic functions of $(u,P) = (u_1,\ldots,u_i, P)$.

For simplicity, assume that $i = i_1 = j$ (the extremal case). (The general case is similar.) Then $\tilde{S}_j^\#$ is clearly defined by

$$\tilde{g}_{\nu\mu} = 0, \text{ for } 1 \leq \nu \leq r-k,\ 1 \leq \mu \leq j$$ Q.E.D.

Let the case as in Fig. 1.101 be the natural projections. Put

$$S_j^\# = S_j^\#(k,r) = \pi_2(\tilde{S}_j^\#)$$

By the <u>Proper Mapping Theorem</u> (Theorem 3.2.14), $S_j^\#$ is an analytic set in $G(k,r)$. By <u>Chow's Theorem</u> (Theorem 3.3.1), $S_j^\#$ is an algebraic set. Note that

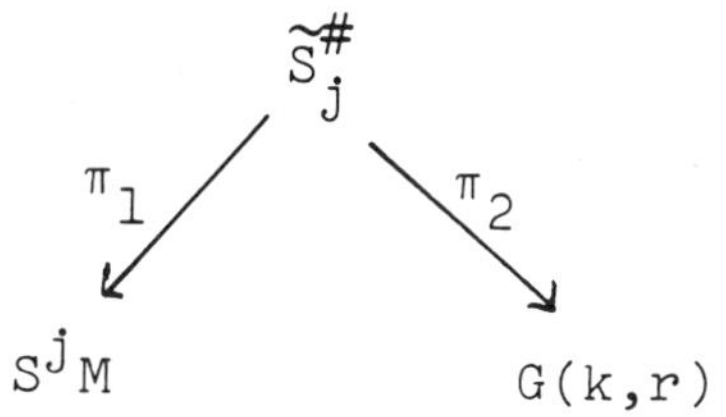

FIGURE 1.101

$$S_j^{\#} = \{P \in G(k,r) \mid \deg D_P \geq j\}$$

so

$$\cdots \supset S_{j-1}^{\#} \supset S_j^{\#} \supset S_{j+1}^{\#} \supset \cdots$$

Lemma 1.6.10. $\pi_2 : \tilde{S}_j^{\#} \to S_j^{\#}$ is a finite map, that is, every fiber of π_2 is a finite set.

Proof. This is clear, for there are only finitely many $D \in S^j M$ such that $D \leq D_P$. Q.E.D.

The proof of the following lemma is left to the reader as an exercise (Exercise 1).

Lemma 1.6.11. For $j \leq r+1$, the set

$$B_j = \{D = m_1 + \cdots + m_j \in S^j M \mid \phi(m_1), \ldots, \phi(m_j) \text{ are not in general position}\}$$

is an analytic set of $S^j M$ such that $B_j \neq S^j M$.

What we actually need is not $S_j^{\#}(k,r)$ but its algebraic subset $S_j(k,r)$ which is, by definition, the Zariski closure (see Sec. 3.3) in $G(k,r)$ of the set

$$\{P \in G(k,r) \mid P \text{ contains distinct } j \text{ points on } C\}$$

It can be shown that $S_j(k,r)$ is equal to the closure of this set in the usual topology. Note that

$$\cdots S_{j-1}(k,r) \supset S_j(k,r) \supset S_{j+1}(k,r) \supset \cdots$$

is a decreasing sequence of algebraic sets in $G(k,r)$. Put

$$\tilde{S}_j(k,r) = \pi_2^{-1}(S_j(k,r)) \quad (\subset \tilde{S}_j^{\#}(k,r))$$

and consider again the projections in Fig. 1.102. Then π_2 is a finite surjective map.

Proposition 1.6.12.

1. $\pi_1 : \tilde{S}_j(k,r) \to S^j M$ is surjective for $j \leq k+1$.

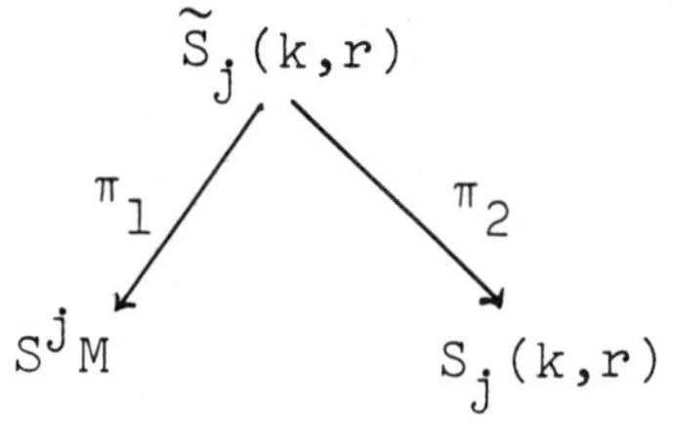

FIGURE 1.102

2. $\pi_1: \tilde{S}_{k+1}(k,r) \to S^{k+1}M$ is a bimeromorphic map (see Sec. 3.2).
3. $S_{k+1}(k,r)$ is irreducible and $\dim S_{k+1}(k,r) = k+1$.

Proof. (1) Let $j \leq k+1$. For any j points $m_1, \ldots, m_j$ on M with $\phi(m_\nu) \neq \phi(m_\mu)$ for $\nu \neq \mu$, there is a k-plane P such that $\phi(m_\nu) \in P$ for $1 \leq \nu \leq j$. Then $(m_1 + \cdots + m_j, P) \in \tilde{S}_j(k,r)$. Such $D = m_1 + \cdots + m_j$ form an open dense set in S^jM, so $\pi_1: \tilde{S}_j(k,r) \to S^jM$ is surjective.

(2) By Lemma 1.6.11, B_{k+1} is an analytic set in $S^{k+1}M$ such that $B_{k+1} \neq S^{k+1}M$. Hence, B_{k+1} is closed and nowhere dense in $S^{k+1}M$. Every fiber $\pi_1^{-1}(D)$ for $D = m_1 + \cdots + m_{k+1} \in S^{k+1}M - B_{k+1}$ consists of the unique point (D, P) where P is the k-plane spanned by $\phi(m_1), \ldots, \phi(m_{k+1})$.

(3) follows from (2) and Lemma 1.6.10. Q.E.D.

In particular, put k = 1. Then (3) of Proposition 1.6.12 says that the set of all multisecant lines to C is a two-dimensional irreducible algebraic set in G(1, r). This is intuitively clear. (See Fig. 1.103.)

The following lemma is also intuitively clear.

Lemma 1.6.13. $S_3(1,r)$ is an algebraic set in G(1, r) such that $\dim S_3(1,r) \leq 1$. In particular, a general multisecant line is bisecant.

Proof. Assume the contrary. Then, for any general points p and q on C, there is another $r \in C$ such that $r \in \overline{pq}$. Since our discussion can be done

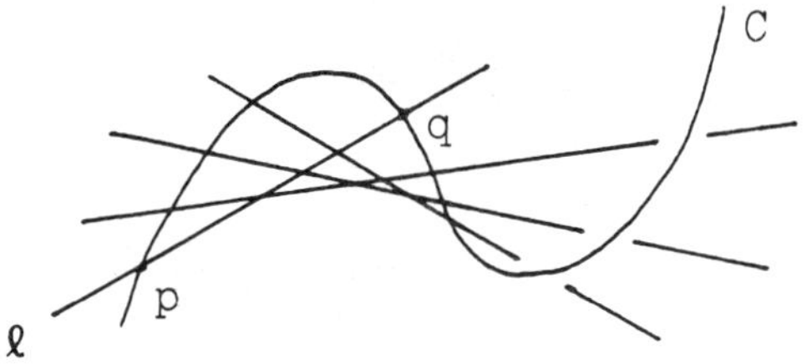

FIGURE 1.103

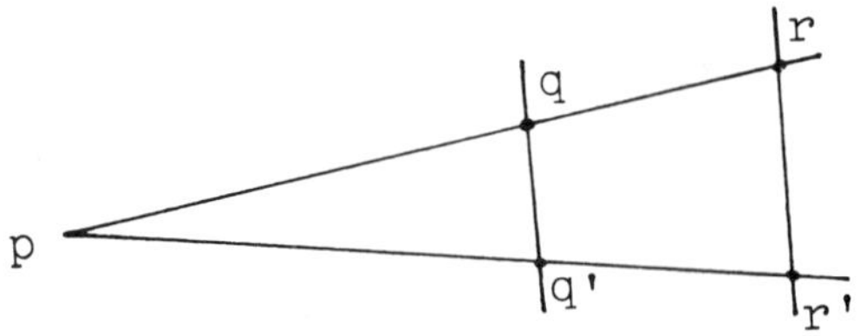

FIGURE 1.104

locally, we may assume that r depends holomorphically on (p, q). If p is fixed, then r depends holomorphically on q. Take q' near from q. Then r' corresponding to q' (that is, $r' \in \overline{pq'}$) is near from r. (See Fig. 1.104.) The lines $\overline{qq'}$ and $\overline{rr'}$ meet, so p is contained in the 2-plane $\overline{qq'} \vee \overline{rr'}$. By taking the limit $q' \to q$, we conclude that the tangent lines T_qC and T_rC meet and p is contained in the 2-plane $T_qC \vee T_rC$. In the same way, T_pC and T_rC meet and $q \in T_pC \vee T_rC$. This means that, for general p and q in C, T_pC and T_qC meet.

Take another general point $s \in C$ such that $s \notin T_pC \vee T_qC$. T_sC meet both T_pC and T_qC. Hence, T_pC, T_qC, and T_sC meet at a point $p_0 \in \mathbb{P}^r$. This means that, for every general point $s \in C$, T_sC passes through a fixed point p_0.

Then the projection $\pi_{p_0}: M \to \mathbb{P}^{r-1}$ with the center p_0 has the identically zero differential; $d\pi_{p_0} = 0$ on M. Hence, C is a line passing through p_0, a contradiction. Q.E.D.

Remark 1.6.14. If C is a rational normal curve in $\mathbb{P}^r$, then $S_3(1, r)$ is empty. On the other hand, there are examples of nonsingular C such that $\dim S_3(1, r) = 1$ (compare Exercise 2).

Lemma 1.6.15. For a general hyperplane H, $C \cap H$ consists of $n = \deg C$ distinct points no three of which are collinear.

Proof. Using the above notations, consider the cases $(k, j) = (r - 1, 2)$ and $(k, j) = (1, 2)$, and the projections in Fig. 1.105. Note that μ_1 is surjective.

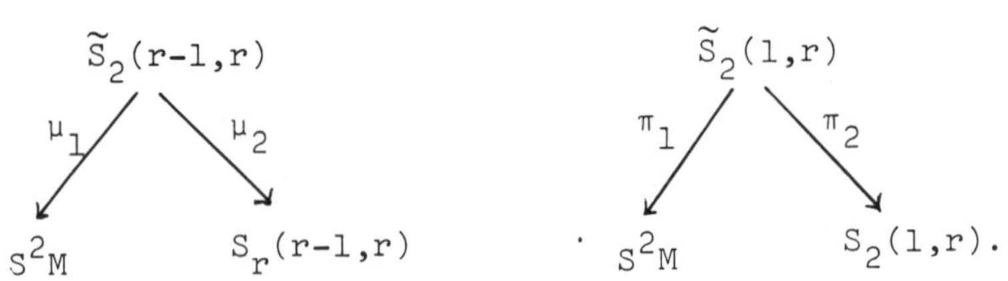

FIGURE 1.105

Moreover, if $\phi(m_1) \neq \phi(m_2)$, then $\mu_1^{-1}(m_1 + m_2)$ is the set of all hyperplanes containing the line $\overline{\phi(m_1)\phi(m_2)}$, so is an (r - 2)-dimensional projective space. On the curve (of S^2M)

$$B_2 = \{m_1 + m_2 \in S^2M \mid \phi(m_1) = \phi(m_2)\} \quad \text{(see Lemma 1.6.11)}$$

the fibers of μ_1 are (r - 1)-dimensional projective spaces.

On the other hand, since π_2 is a finite map, $\pi_2^{-1}(S_3(1, r))$ is at most one-dimensional by Lemma 1.6.13. Note that $\pi_1\pi_2^{-1}(S_3(1, r))$ meets B_2 at at most finitely many points. Hence, $S_3' = \mu_2\mu_1^{-1}\pi_1\pi_2^{-1}(S_3(1, r))$ is an algebraic set in $\mathbb{P}^{r*}$ of at most (r - 1)-dimension. This, together with (3) of Proposition 1.6.7 proves the lemma. Q.E.D.

Now, we are ready to prove

Theorem 1.6.16 (General Position Theorem). Let C be a space curve in $\mathbb{P}^r$. Then, for a general hyperplane H, $C \cap H$ consists of $n = \deg C$ distinct points in general position, that is, any r points of $C \cap H$ span H.

Proof (following Arbarello-Cornalba-Griffiths-Harris [6]). Using the above notations, consider the cases (k, j) = (r - 1, r) and (k, j) = (r - 1, 1), and the projections in Fig. 1.106. The assertion holds if and only if

$$\mu_2\mu_1^{-1}(B_r) \neq \mathbb{P}^{r*}$$

where B_r is the set in Lemma 1.6.11 (for j = r).

Suppose that $\mu_2\mu_1^{-1}(B_r) = \mathbb{P}^{r*}$.

Since every fiber of π_1 is a (r - 1)-dimensional projective space, $\tilde{S}_1(r - 1, r)$ is an irreducible analytic set of dimension r in $M \times \mathbb{P}^{r*}$ (compare Lemma 1.6.4).

Consider the set

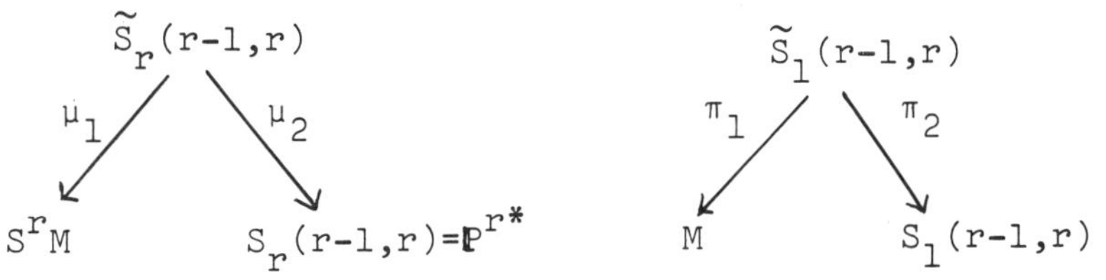

FIGURE 1.106

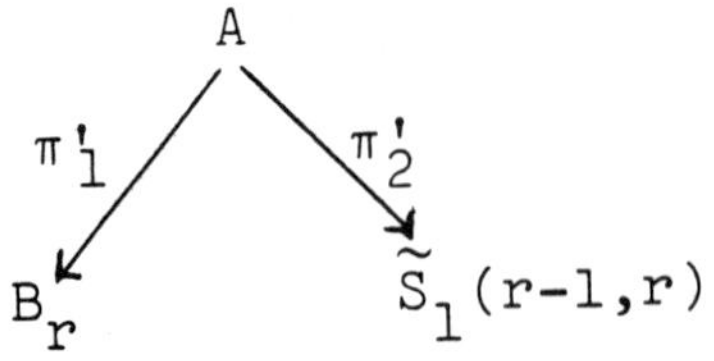

FIGURE 1.107

$$A = \{(D, m, H) \in S^r M \times M \times \mathbb{P}^{r*} \mid D \in B_r \text{ and } m \leq D \leq D_H\}$$

$$= \{(D, m, H) \in B_r \times \tilde{S}_1(r - 1, r) \mid m \leq D \leq D_H\}$$

This is an analytic set in $S^r M \times M \times \mathbb{P}^{r*}$ as is easily seen.

Consider the projections in Fig. 1.107. We show that π'_2 is a surjective finite map. In fact, for a given $(m, H) \in \tilde{S}_1(r - 1, r)$, there are at most finitely many $D \in S^r M$ such that $m \leq D \leq D_H$. Since we have assumed that $\mu_2\mu_1^{-1}(B_r) = \mathbb{P}^{r*}$, for <u>any given</u> general H, there is $D \in B_r$ such that $D \leq D_H$. Taking $m \in M$ with $m \leq D \leq D_H$, we conclude that $\dim A \geq r$. Since $\tilde{S}_1(r - 1, r)$ is irreducible and of dimension r, π'_2 is a surjective finite map.

Now, take a general point $m \in M$. Put $p = \phi(m)$. Consider the projection

$$\pi_p : M \to \mathbb{P}^{r-1}$$

with the center p. It can be shown that, π_p is a <u>birational map</u> (see Sec. 3.3) of M onto its image $C' = \pi_p(M)$ (see the proof of Theorem 1.6.22 below).

Since $\pi'_2(A) = \tilde{S}_1(r - 1, r)$, the theorem fails for $C' \subset \mathbb{P}^{r-1}$. (Draw a picture!) We can continue this process. But the theorem holds for $r = 3$ by Lemma 1.6.15, a contradiction. Q.E.D.

We will use the general position theorem to prove Clifford's Theorem (Theorem 5.3.6).

<u>Corollary 1.6.17.</u> Let C be a space curve in $\mathbb{P}^r$ and k and j be $0 \leq k \leq r - 2$ and $k + 2 \leq j$. Then $\dim S_j(k, r) \leq k$. ($S_j(k, r)$ may be empty.)

<u>Proof.</u> By Proposition 1.6.12, it suffices to show that there is a k-plane P such that $\deg D_P = k + 1$. Take a general hyperplane H and put $C \cap H = \{p_1, \ldots, p_n\}$ ($n = \deg C$). By the theorem, $p_1, \ldots, p_{k+1}$ span a k-plane P such that

$$D_P = m_1 + \cdots + m_{k+1}$$

where $\phi(m_\nu) = p_\nu$ for $1 \le \nu \le k+1$. Q.E.D.

Next, we discuss "generic" projections of space curves. For this purpose, we need some preparation.

Lemma 1.6.18. $T_C = \{\ell \mid \ell \text{ is a tangent line to } C\}$ is an irreducible curve in $G(1, r)$. Moreover, there is a natural bimeromorphic map ψ: $M \to T_C$. (See Fig. 1.108.)

Proof. For simplicity, assume that r = 3. The general case is similar. Using the notations in Lemma 1.6.1, the equations of the tangent line $T_{\phi(t)}C_1$ to the irreducible branch C_1 of C at $\phi(t)$ are given by

$$X_j = \left(\frac{\alpha_j}{\alpha_1}t^{\alpha_j-\alpha_1} + \cdots\right)X_1 + \left(\frac{\alpha_1-\alpha_j}{\alpha_1}t^{\alpha_j} + \cdots\right)X_0, \quad j = 2, 3$$

Its Plücker coordinates in $G(1, 3)$ are given by all the (2×2)-minors of the (2×4)-matrix

$$\begin{pmatrix} \frac{\alpha_1 - \alpha_2}{\alpha_1}t^{\alpha_2} + \cdots, & \frac{\alpha_2}{\alpha_1}t^{\alpha_2-\alpha_1} + \cdots, & -1, & 0 \\ \frac{\alpha_1 - \alpha_3}{\alpha_1}t^{\alpha_3} + \cdots, & \frac{\alpha_3}{\alpha_1}t^{\alpha_3-\alpha_1} + \cdots, & 0, & -1 \end{pmatrix}$$

(see Sec. 3.3), so they are holomorphic functions of t. Hence,

$$\psi: t \in M \to T_{\phi(t)}C_1 \in G(1, 3)$$

is holomorphic.

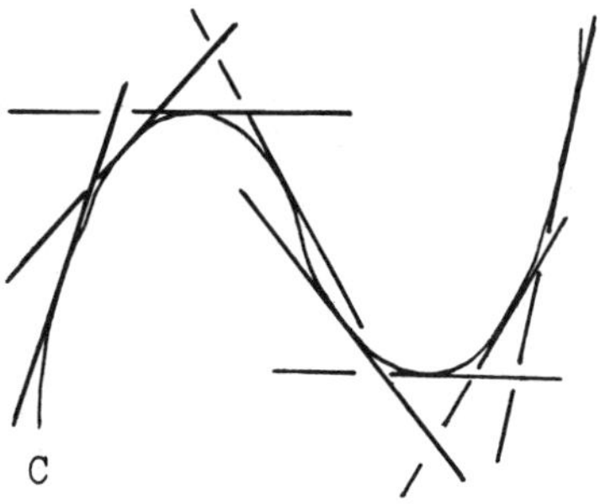

FIGURE 1.108

Suppose that ψ is not bimeromorphic. Then, for any general $p \in C$, there is $p' \in C$ ($p' \neq p$) such that $T_{p'}C = T_pC$. Take a point $r \in \mathbb{P}^3 - C$ and consider the projection π_r with the center r. As in the proof of Lemma 1.6.13, at most finitely many tangent lines pass through the point r. Other tangent lines are mapped to tangent lines to the curve $\pi_r(M)$. Hence, the irreducible plane curve $\pi_r(M)$ (of degree ≥ 2) has infinitely many multitangent lines. This contradicts Lemma 1.5.5. Thus ψ is bimeromorphic. Q.E.D.

Lemma 1.6.19. $V_C = \{\ell \in S_2(1,r) \mid \text{there is a } (m_1 + m_2, \ell) \in \tilde{S}_2(1,r) \text{ such that } \psi(m_1) \text{ and } \psi(m_2) \text{ meet}\}$ (ψ is the map in Lemma 1.6.18) is an algebraic set in $G(1,r)$ of dimension 1 and contains T_C as an irreducible component. (See Fig. 1.109.)

Proof. Consider the projections in Fig. 1.110. Consider also the set

$$\tilde{J} = \{(m_1 + m_2, \ell) \in \tilde{S}_2(1,r) \mid \psi(m_1) \text{ and } \psi(m_2) \text{ meet}\}$$

Then it is easy to see that $\tilde{J}$ is an analytic set in $S^2M \times G(1,r)$ and $\pi_2(\tilde{J}) = V_C$. Hence, it is enough to show that $\dim \tilde{J} \leq 1$.

Suppose that $\dim \tilde{J} = 2$. π_1 is bimeromorphic by Proposition 1.6.12, so $\dim \pi_1(\tilde{J}) = 2$, that is, $\pi_1(\tilde{J}) = S^2M$. This means that, for any general p and q ($p \neq q$) of C, T_pC and T_qC meet.

Fix p and put $\ell = T_pC$. Consider the projection

$$\pi_\ell : M \to \mathbb{P}^{r-2}$$

Then it has the identically zero differential; $d\pi_\ell = 0$ on M. Hence, C is on a plane which contains ℓ, a contradiction.

Note that $T_C \subset V_C$ and T_C is one-dimensional and irreducible. Q.E.D.

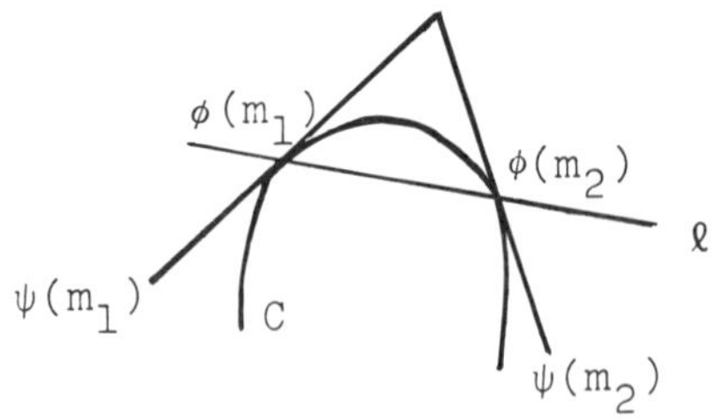

FIGURE 1.109

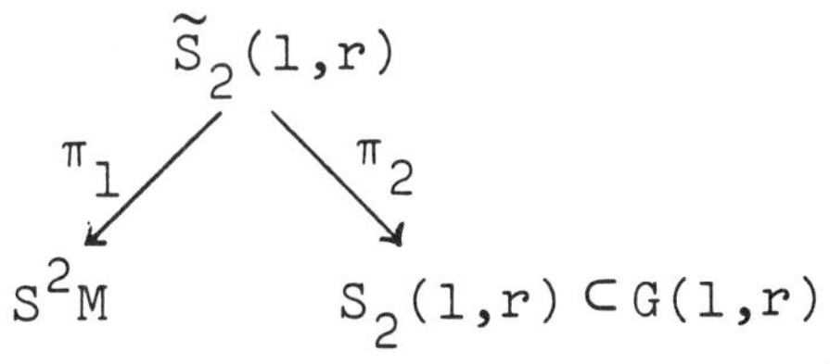

FIGURE 1.110

Lemma 1.6.20. Let S be a (respectively irreducible) algebraic set in G(1, r). Then

$$\mathrm{Ch}(S) = \{p \in \mathbb{P}^r \mid \text{there is } \ell \in S \text{ such that } p \in \ell\}$$

is a (respectively irreducible) algebraic set such that dim Ch(S) $\leq$ dim S + 1.

Proof. Consider the incidence correspondence

$$\tilde{I} = \{(p, \ell) \in \mathbb{P}^r \times G(1, r) \mid p \in \ell\}$$

and the projections in Fig. 1.111 (compare Lemma 1.6.4). Note that π_2 is a $\mathbb{P}^1$-bundle (see Sec. 3.1) and

$$\mathrm{Ch}(S) = \pi_1 \pi_2^{-1}(S) \qquad \text{Q.E.D.}$$

The irreducible algebraic set $\mathrm{Ch}(S_2(1, r))$ in $\mathbb{P}^r$ is called the chodal variety of C and is denoted by Ch(C). (Some authors write it Sec(C) and call it the secant variety of C.) The irreducible algebraic set $\mathrm{Ch}(T_C)$ in $\mathbb{P}^r$ is called the tangent variety of C and is denoted by Tan(C). Note that Tan(C) $\subset$ Ch(C). They are important to the study of the projective geometric property of C.

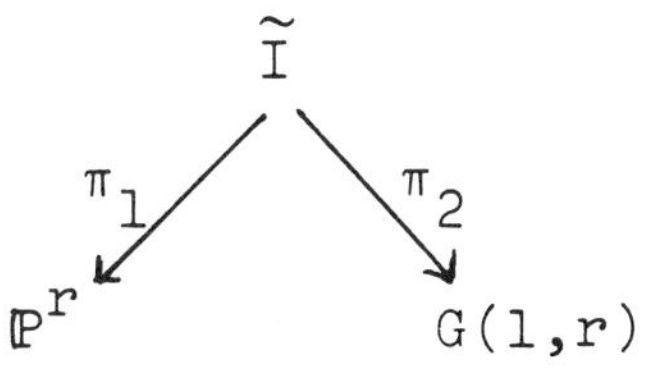

FIGURE 1.111

Lemma 1.6.21.

1. Ch(C) is an irreducible algebraic set in $\mathbb{P}^r$ of dimension 3. In particular, if r = 3, then Ch(C) = $\mathbb{P}^3$.
2. Tan(C) is an irreducible algebraic set in $\mathbb{P}^r$ of dimension 2. (Hence, we call Tan(C) the tangent surface of C.)

Proof. To prove the first part, suppose that dim Ch(C) = 2. Take a general nonsingular point $p \in C$ and put

$$S_p = \{\overline{pq} \in G(1,r) \mid q \in C\}$$

(If q = p, then put $\overline{pq} = T_pC$.) This is an algebraic set in G(1, r) which is birational to C by Lemma 1.6.13. Hence, $Ch(S_p)$ is irreducible and dim $Ch(S_p) \leq 2$. If dim $Ch(S_p) \leq 1$, then $Ch(S_p)$ must be a line passing through p because of the irreducibility of $Ch(S_p)$. But this is impossible. Hence, dim $Ch(S_p) = 2$. Note that $S_p \subset S_2(1, r)$, so $Ch(S_p) \subset Ch(C)$. Hence, by the assumption,

$$Ch(C) = Ch(S_p)$$

This means that every multisecant line of C must pass through p, a contradiction.

To prove the second part, take a nonsingular point $p \in C$. If dim Tan(C) = 1, then Tan(C) = T_pC by the irreducibility of Tan(C). But this is impossible. Q.E.D.

Now, we are ready to prove

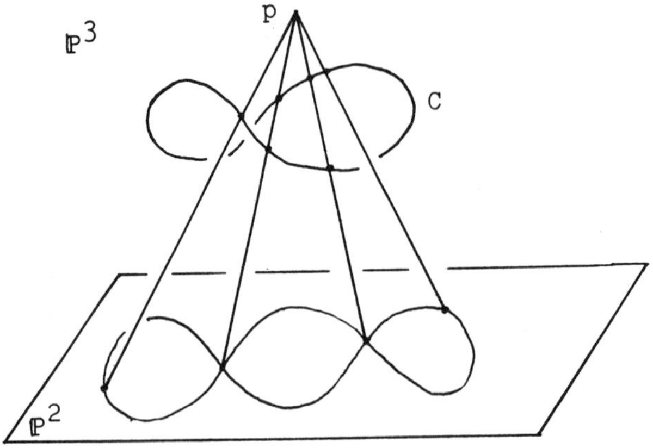

FIGURE 1.112

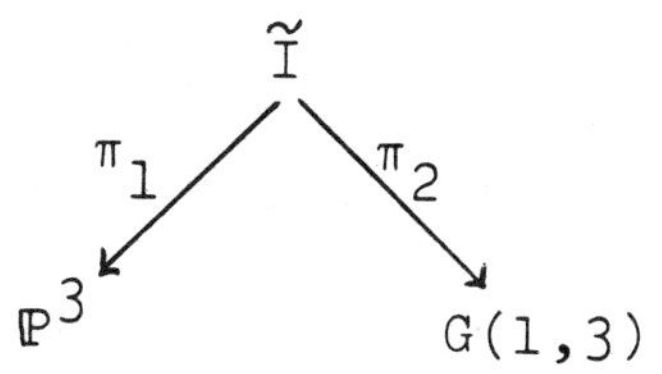

FIGURE 1.113

Theorem 1.6.22. Let C be a nonsingular space curve in $\mathbb{P}^r$.

1. If $r \geq 4$, then the projection

$$\pi_p : C \to \mathbb{P}^{r-1}$$

with the center a general point $p \in \mathbb{P}^r - C$ is a holomorphic imbedding (see Sec. 3.1).

2. If $r = 3$, then the projection

$$\pi_p : C \to \mathbb{P}^2$$

with the center a general point $p \in \mathbb{P}^3 - C$ is birational onto its image $\pi_p(C)$ which is a plane nodal curve, that is, has only nodes as its singular points. (See Fig. 1.112.)

Proof. To prove the first part, note that Ch(C) is three-dimensional and so is closed and nowhere dense in $\mathbb{P}^r$. Take $p \in \mathbb{P}^r - \mathrm{Ch}(C)$. Then π_p satisfies the condition.

To prove the second part, consider the projections in Fig. 1.113 (see the proof of Lemma 1.6.20). Let

$$\pi' : \pi_2^{-1}(S_2(1,3)) \to \mathrm{Ch}(C)$$

be the restriction of π_1 to $\pi_2^{-1}(S_2(1,3))$. Then π' is surjective and $\dim \pi_2^{-1}(S_2(1,3)) = \dim \mathrm{Ch}(C) = 3$, so π' is generically finite, that is, there is a nowhere dense algebraic set Δ in $\mathbb{P}^3$ such that the fiber $\pi'^{-1}(q)$ is a finite set for all $q \in \mathbb{P}^3 - \Delta$.

By Lemmas 1.6.13, 1.6.19, and 1.6.20, $\mathrm{Ch}(S_3(1,3))$ and $\mathrm{Ch}(V_C)$ are at most two-dimensional.

Now, take $p \in \mathbb{P}^3 - \Delta - \mathrm{Ch}(S_3(1,3)) - \mathrm{Ch}(V_C)$. Then π_p satisfies the condition. Q.E.D.

Remark 1.6.23.

1. We avoid Δ for π_p to be birational and $\mathrm{Ch}(S_3(1,3))$ (respectively $\mathrm{Ch}(V_C)$)

FIGURE 1.114

for $\pi_p(C)$ not to have a triple point, and so on (respectively a cusp or a tacnode (see Sec. 2.1, and so forth).

2. It is because of this theorem that nonsingular space curves in $\mathbb{P}^3$ and plane nodal curves are important for the study of intrinsic geometry of curves. This theorem will be used to prove the Riemann-Roch Theorem in Sec. 5.1.
3. The projection with the center a general point is called a generic projection.

In the rest of this section, we consider only space curves in $\mathbb{P}^3$ for simplicity. (Many results can be generalized to space curves in $\mathbb{P}^r$ $(r \geq 3)$.)

Let C be a space curve in $\mathbb{P}^3$ and ϕ: M → C be a nonsingular model. Using the notations in Lemma 1.6.1, ϕ can be written locally as

$$\phi: t \to (x_1, x_2, x_3) = (t^{\alpha_1}, t^{\alpha_2} + \cdots, t^{\alpha_3} + \cdots)$$

where $1 \leq \alpha_1 < \alpha_2 < \alpha_3$, $(x_1, x_2, x_3) = (0,0,0)$ at p, $t = 0$ at m and $\phi(m) = p$. The integers α_1, α_2, and α_3 are independent of the choice of t and affine coordinate systems (x_1, x_2, x_3). Put

$$\alpha(m) = \alpha(m, \phi) = (\alpha_1, \alpha_2, \alpha_3)$$

The closure in $\mathbb{P}^3$ of the affine line $\{x_2 = x_3 = 0\}$ is the tangent line T_pC_1 to the irreducible branch C_1 at p corresponding to m.

The closure in $\mathbb{P}^3$ of the affine plane $\{x_3 = 0\}$ is called the osculating plane to the irreducible branch C_1 at p and is denoted by O_pC_1. It is called an osculating plane to C at p. (In a similar way, an osculating k-plane $(k \geq 2)$ to a space curve C in $\mathbb{P}^r$ $(r \geq 3)$ at a point of C can be defined.) If C has a unique irreducible branch at p (in particular, if C is nonsingular at p), then O_pC_1 is written O_pC and is called the osculating plane to C at p.

The proofs of the following lemmas 1.6.24 to 1.6.28 are left to the reader as exercises (Exercise 1).

Under the above notations

Lemma 1.6.24. Let ℓ (respectively P) be a line (respectively plane) in $\mathbb{P}^3$ passing through p. Let

$$\nu m + \nu_2 m_2 + \cdots + \nu_s m_s \qquad (m_j \neq m \quad \text{for} \quad j \geq 2)$$

be the line (respectively plane) cut by ℓ (respectively by P). Then

$$\alpha_2 = \max\{\nu \mid \ell \ni p\} = \nu \quad \text{for} \quad \ell = T_p C_1,$$

$$\alpha_3 = \max\{\nu \mid P \ni p\} = \nu \quad \text{for} \quad P = O_p C_1 .$$

Lemma 1.6.25. There is an open neighborhood U of m in M such that $\alpha(t) = (1,2,3)$ for all $t \in U - \{m\}$. In other words, $\alpha(m) = (1,2,3)$, except finite points $m \in M$.

Lemma 1.6.26. Suppose that $\alpha_1 = 1$ and $\alpha_2 = 2$. Then

$$p \vee \phi(t) \vee \phi(t') \to O_p C_1 \quad \text{as} \quad t, t' \to m$$

(See Fig. 1.115.)

Lemma 1.6.27. Suppose that $\alpha_1 = 1$ and $\alpha_2 = 2$. Put

$$q = O_p C_1 \cap O_{\phi(t)} C_1 \cap O_{\phi(t')} C_1$$

Then $q \to p$ as $t, t' \to m$.

Lemma 1.6.28. Let U be a neighborhood of m in M.

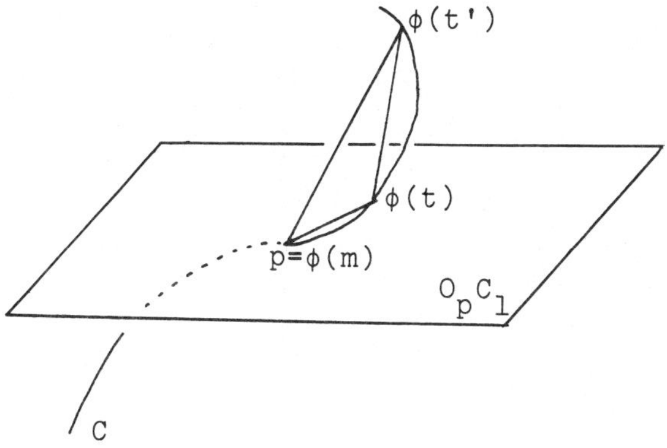

FIGURE 1.115

1. If $q \in \mathbb{P}^3 - O_pC_1$, then the irreducible branch $C_1' = \pi_q(U)$ of the plane curve $C' = \pi_q(M)$ at $r = \pi_q(m)$ has the multiplicity α_1 at r and $I_r(C_1', T_rC_1') = \alpha_2$.
2. If $q \in O_pC_1 - T_pC_1$, then $C_1' = \pi_q(U)$ has the multiplicity α_1 at r and $I_r(C_1', T_rC_1') = \alpha_3$.
3. If $q \in T_pC_1 - C_1$, then $C_1' = \pi_q(U)$ has the multiplicity α_2 at r and $I_r(C_1', T_rC_1') = \alpha_3$.

Using these lemmas, we can generalize Theorem 1.6.22 for not necessarily nonsingular space curves in $\mathbb{P}^3$.

By Lemma 1.6.25, there are only finitely many $m_1, \ldots, m_s \in M$ such that $\phi(m_j)$ is a singular point of C or $\alpha(m_j) = (\alpha_{j1}, \alpha_{j2}, \alpha_{j3}) \neq (1,2,3)$.

In the proof of Theorem 1.6.22, we took p in $\mathbb{P}^3 - \Delta - Ch(S_3(1,3)) - Ch(V_C)$. In the present general case, we take p in

$$\mathbb{P}^3 - \Delta - Ch(S_3(1,3)) - Ch(V_C) - \cup_j O_{\phi(m_j)} C_j$$

where C_j is the irreducible branch of C at $\phi(m_j)$ corresponding to m_j.

Then $\pi_p: C \to C' = \pi_p(C) \subset \mathbb{P}^2$ is birational and C' has only nodes as singular points, except $\pi_p(m_1), \ldots, \pi_p(m_s)$. At $\pi_p(m_j)$, (1) C' has the same number of irreducible branches as C at $\phi(m_j)$ and (2) C' has the same multiplicity as C at $\phi(m_j)$ by Lemma 1.6.28. Thus

<u>Theorem 1.6.22'</u>. Let C be a space curve in $\mathbb{P}^3$ and $p_1, \ldots, p_s$ be the set of all singular points of C. Let s_j and m_j be the number of irreducible branches and the multiplicity at p_j, respectively. Then a generic projection $\pi_p: C \to \mathbb{P}^2$ ($p \in \mathbb{P}^3 - C$) is birational onto its image $C' = \pi_p(C)$ which has only nodes as singular points, except $\pi_p(p_j)$. At each $\pi_p(p_j)$, C' has the number s_j of irreducible branches and the multiplicity m_j.

The proof of the following lemma is similar to that of Lemma 1.6.18.

<u>Lemma 1.6.29</u>. $C^* = \{P \in \mathbb{P}^{3*} \mid P$ is an oculating plane to $C\}$ is a space curve in $\mathbb{P}^{3*}$. Moreover, there is a natural bimeromorphic map

$$\psi: M \to C^*$$

(C^* is called the <u>dual curve of</u> C.)

The following theorem follows from Lemmas 1.6.26 and 1.6.27, (compare Theorem 1.5.3).

Theorem 1.6.30 (Reciprocity of dual curves). Let C be a space curve in $\mathbb{P}^3$. Then the dual curve to the dual curve C^* of C is C itself; $C^{**} = C$.

We denote by c the degree of C^* in $\mathbb{P}^{3*}$ and call it the class of C. We also denote by d the degree of T_C in $G(1,3) \subset \mathbb{P}^5$ (see Sec. 3.3). Note that

c = the number of osculating planes to C passing through a fixed general point p_0 in $\mathbb{P}^3 - C$.
d = the number of tangent lines to C meeting a fixed general line ℓ_0 in $\mathbb{P}^3$ with $\ell_0 \cap C = \phi$.

Put

$$b_1 = \sum_{m \in M} (\alpha_1 - 1)$$

$$b_2 = \sum_{m \in M} (\alpha_2 - \alpha_1 - 1)$$

$$b_3 = \sum_{m \in M} (\alpha_3 - \alpha_2 - 1)$$

where Σ runs over all $m \in M$. By Lemma 1.6.25, they are nonnegative integers.

Let g be the genus of M and n be the degree of C. By the Riemann-Hurwitz formula (Theorem 4.1.4) applied for π_{ℓ_0},

$$d = 2(g - 1 + n) - b_1 \qquad (1)$$

Next, consider a generic projection $\pi_{p_0} : M \to \mathbb{P}^2$. By Theorem 1.6.22' and the discussion preceding the theorem, the class of $C' = \pi_{p_0}(M)$ is d. The class formula (Theorem 1.5.4) applied for C' is nothing but (1). On the other hand, the flex formula (Theorem 1.5.10) applied for C' gives

$$c = 2(g - 1 + d) - n - b_2 \qquad (2)$$

This follows from (2) of Lemma 1.6.28.

The proof of the following lemma is similar to that of Lemma 1.5.7.

Lemma 1.6.31. Using the same notations as in Lemma 1.6.1, the irreducible branch C_I^* of C^* corresponding to m is given by the image of

$$\psi : s \to (u_1, u_2, u_3) = (s^{\alpha_3-\alpha_2}, s^{\alpha_3-\alpha_1} + \cdots, s^{\alpha_3} + \cdots)$$

where s is a local coordinate with s = 0 at m and (u_1, u_2, u_3) is an affine coordinate system with $(u_1, u_2, u_3) = (0, 0, 0)$ at $p^* = O_pC_1 \in C^*$.

Using this lemma and applying the above formulas (1) and (2) to C*, we get

$$d^* = 2(g - 1 + c) - b_3 \qquad (1^*)$$

$$n = 2(g - 1 + d^*) - c - b_2 \qquad (2^*)$$

where d* is the number d for $C^* \subset \mathbb{P}^{3*}$. By (2) and (2*), we get

$$d^* = d$$

Hence (1*) can be written

$$d = 2(g - 1 + c) - b_3 \qquad (3)$$

We rewrite (1), (2), and (3) as follows:

$$-2n + d = 2g - 2 - b_1 \qquad (1)$$

$$n - 2d + c = 2g - 2 - b_2 \qquad (2)$$

$$d - 2c = 2g - 2 - b_3 \qquad (3)$$

Griffiths-Harris [33] called the formulas (1) to (3) the global <u>Plücker's formulas</u>. (They gave these formulas for space curves in any $\mathbb{P}^r$.)

By (1) to (3), the numbers g, c, and d can be solved as follows:

$$g = 1 + \frac{1}{12}(3b_1 + 2b_2 + b_3 - 4n) \qquad (4)$$

$$c = n - \frac{1}{2}(b_1 - b_3) \qquad (5)$$

$$d = \frac{4}{3}n - \frac{1}{6}(3b_1 - 2b_2 - b_3) \qquad (6)$$

Hence, we can compute the genus g by the degree n of C and by the local data b_1, b_2, and b_1. But it should be noted that formula (4) is not a kind of the Genus Formula. The genus formula for plane curves in Sec. 2.1 will be given by using the degree and the local data which are analytic invariants, while α_2 and α_3 in Lemma 1.6.1 are not necessarily analytic invariants.

No genus formula for space curves is known. It seems that there is no such formula, because two nonsingular space curves of the same degree may have different genera. For example, a nonsingular space quartic curve in $\mathbb{P}^3$ is either an elliptic quartic curve (g = 1) or a rational curve (g = 0) which is the image of a generic projection of the rational normal curve in $\mathbb{P}^4$.

However, for space curves in $\mathbb{P}^3$ (or in $\mathbb{P}^r$ in general), an inequality with respect to the genus is known.

Theorem 1.6.32 (Castelnuovo). For a space curve in $\mathbb{P}^3$,

$$g \leq \frac{1}{4}(n-2)^2 \qquad \text{for even } n$$

$$\leq \frac{1}{4}(n-1)(n-3) \qquad \text{for odd } n$$

For the proof of this theorem, see, for example, Griffiths-Harris [33] or Hartshorne [39].

Note 1.6.33. See Griffiths-Harris [33], Arbarello-Cornalba-Griffiths-Harris [6], Harris [37], Hartshorne [39], and so forth, for more about space curves.

Exercises

1. Prove Lemmas 1.6.1, 1.6.2, 1.6.11, and 1.6.24 through 1.6.28.
2. Let C be a nonsingular space curve in $\mathbb{P}^3$. If C is neither a twisted cubic nor an elliptic quartic curve, then dim $S_3(1,3) = 1$. In the exceptional two cases, $S_3(1,3)$ is empty. (See Hartshorne [39, p. 355].)
3. If C is a nonsingular space curve in $\mathbb{P}^r$ ($r \geq 3$) of degree $\geq r + 2$, then dim $S_r(r-2, r) = r - 2$.
4. Under the same notations as in this section, $b_1 = b_2 = b_3 = 0$ if and only if C is a twisted cubic. (In this case $c = 3$ and $d = 4$.)
5. If C is an elliptic quartic curve in $\mathbb{P}^3$, then $b_1 = b_2 = 0$, $b_3 = 16$, $c = 12$, and $d = 8$.
6. Compute b_1, b_2, b_3, c, and d for a canonical sextic curve C_K in $\mathbb{P}^3$ (g = 4).
7. Let C be an irreducible plane nodal curve of degree n (≥ 3) with k-nodes. Let ϕ: M $\to$ C be a nonsingular model. Let Λ be the linear system on M of all line cuts. Then Λ is complete if either (1) n is even and $k < \frac{1}{4}n(n-2)$ or (2) n is odd and $k < \frac{1}{4}(n-1)^2$.

2

SINGULAR CURVES OF LOWER DEGREE

2.1 GENUS FORMULA FOR PLANE CURVES

Let S be a compact complex manifold of dimension 2. Let C be an irreducible curve on S and ϕ: M $\to$ C be a nonsingular model of C. For a point $p \in C$, put

$$\phi^{-1}(p) = \{p_1, \dots, p_s\} \quad (p_j \neq p_k \text{ for } j \neq k)$$

($s = s_p$ is the number of irreducible branches at p.)

The following lemma is easy to prove and is left to the reader as an exercise (Exercise 1).

Lemma 2.1.1. There are (1) a local coordinate system (x, y) around p with $p = (0, 0)$ and (2) for every j ($1 \leq j \leq s$), a local coordinate t_j around p_j with $t_j = 0$ at p_j, such that ϕ can be written as

$$\phi: t_j \to (x, y) = (t_j^{m_j}, h_j(t_j)), \qquad 1 \leq j \leq s$$

where m_j is the multiplicity of the irreducible branch C_j at p corresponding to p_j and $h_j(t_j)$ is a holomorphic function of t_j with $\text{ord}_{t_j=0}\, h_j(t_j) \geq m_j$. [$\text{ord}_{t_j=0}\, h_j(t_j)$ is as in Sec. 1.3, the order of zero of $h_j(t_j)$ at $t_j = 0$.] In particular, if $s = 1$, then we may take

$$\text{ord}_{t=0}\, h(t) \geq m + 1 \quad \text{and} \quad \text{ord}_{t=0}\, h(t) \not\equiv 0 \pmod{m}$$

where $t = t_1$, $m = m_1$, and $h = h_1$. (If p is a nonsingular point, then we may take $h(t) = 0$ identically.) (See Fig. 2.1.)

(s = 3)

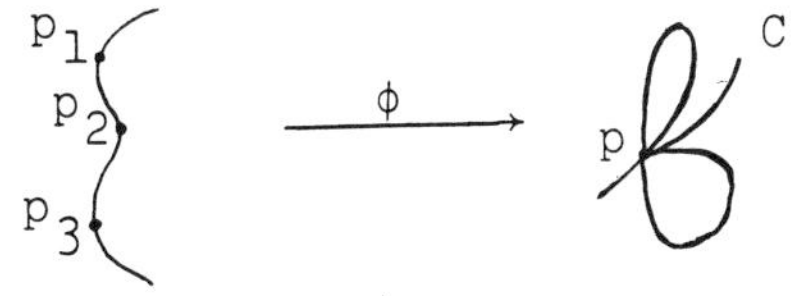

FIGURE 2.1

Now, put

$$\omega_j = \exp\left(\frac{2\pi\sqrt{-1}}{m_j}\right) \quad \text{and} \quad m = m_p = m_1 + \cdots + m_s \text{ (the multiplicity}$$

of C at p)

Put

$$R_j(x,y) = \prod_{\nu=0}^{m_j-1} [y - h_j(\omega_j^{\nu} x^{1/m_j})]$$

$$= y^{m_j} + a_{j1}(x)y^{m_j-1} + \cdots + a_{jm_j}(x)$$

and

$$R(x,y) = \prod_{j=1}^{s} R_j(x,y) = y^m + a_1(x)y^{m-1} + \cdots + a_m(x)$$

Then it is easy to see that $a_{jk}(x)$ and $a_k(x)$ are holomorphic functions of x (compare the proof of Lemma 1.3.6) and

$$\operatorname{ord}_{x=0} a_{jk}(x) \geq k \quad \text{and} \quad \operatorname{ord}_{x=0} a_k(x) \geq k$$

Hence, R(x,y) is a Weierstrass polynomial (see Definition 3.2.1). It is then clear that, around p,

$$C = \{R(x,y) = 0\}$$

Let $\mathcal{O}_p$ be the ring of germs of holomorphic functions on S at p (see Sec. 3.2). It can be identified with the ring $\mathbb{C}\{x,y\}$ of convergent power series. Put

$$I_p = \{f \in \mathcal{O}_p \mid f = 0 \quad \text{on} \quad C\}$$

Then I_p is an ideal of $\mathcal{O}_p$. The ring

$$\mathcal{O}_{C,p} = \frac{\mathcal{O}_p}{I_p}$$

is called the ring of germs of holomorphic functions on C at p.

Lemma 2.1.2. $I_p = R \cdot \mathcal{O}_p$.

Proof. Take a $f \in I_p$. By the Weierstrass Division Theorem (see Theorem 3.2.4),

$$f = gR + h$$

where $g,\ h \in \mathcal{O}_p$ and h is a polynomial with respect to y of degree $\leq m - 1$. Then $h = 0$ on C. But, for a general small x, there are distinct m points $(x, y_1), \ldots, (x, y_m)$ on C. (y_k are the roots of the equation $R(x, y) = 0$.) Hence, $h = 0$ identically. Q.E.D.

A stronger assertion than Lemma 2.1.2 in fact holds as follows. We leave its proof to the reader (Exercise 1).

Lemma 2.1.3. There is a neighborhood U of p in S such that

$$I_q = R \cdot \mathcal{O}_q \quad \text{for all } q \in U$$

In general, if a holomorphic function R on an open set U in S has the property of this lemma, then $R(x, y) = 0$ is called a minimal equation of C on U.

We cover S by a finite number of open sets U_k. Let $f_k = 0$ be a minimal equation of C on U_k. (Put $f_k = 1$ identically, if $U_k \cap C = \phi$.) Put

$$f_{jk} = \frac{f_j}{f_k} \quad \text{on } U_j \cap U_k$$

Then f_{jk} is a never-vanishing holomorphic function on $U_j \cap U_k$ and

$$f_{ik} = f_{ij} f_{jk} \quad \text{on } U_i \cap U_j \cap U_k$$

Hence, $\{f_{jk}\}$ determines a (holomorphic) line bundle on S (see Sec. 3.1) which is denoted by [C] and is called the line bundle on S determined by C.

More generally, let D be a divisor on S. It is, by definition, a formal finite sum

$$D = \alpha_1 C_1 + \cdots \alpha_r C_r$$

where C_j are irreducible curves and α_j are integers (see Sec. 3.4). Put

$$[D] = \alpha_1[C_1] + \cdots + \alpha_r[C_r]$$

and call it the line bundle determined by the divisor D.

Now, let p, $\phi^{-1}(p) = \{p_1, \ldots, p_s\}$, t_j, R, and so forth, be as before. Consider a meromorphic one-form (see Sec. 3.2), around p_j:

$$\sigma_j dt_j = \frac{d(t_j^{m_j})}{\partial R(t_j^{m_j}, h_j(t_j))/\partial y}$$

$$= t_j^{-c_j}(b_{j0} + b_{j1}t_j + \cdots)\, dt_j \qquad (b_{j0} \neq 0)$$

Then, by Lemma 2.1.1, $c_j \geq 0$. Put

$$c_p = c_1 + \cdots + c_s$$

Put

$$\mathcal{O}_j = \mathbb{C}\{t_j\}, \qquad \text{the power series ring, and}$$

$$\hat{\mathcal{O}} = \oplus_j \mathcal{O}_j, \qquad \text{the direct sum}$$

Consider the homomorphisms

$$\phi_j^*: G(x,y) \in \mathcal{O}_p \to G(t_j^{m_j}, h_j(t_j)) \in \mathcal{O}_j$$

$$\phi^*: G(x,y) \in \mathcal{O}_p \to f_1^*G + \cdots + f_s^*G \in \hat{\mathcal{O}}$$

Then the kernel of ϕ^* is clearly I_p. Hence, $\mathcal{O}_{C,p} = \mathcal{O}_p/I_p$ can be regarded as a subring of $\hat{\mathcal{O}}$: $\mathcal{O}_{C,p} \subset \hat{\mathcal{O}}$. $\hat{\mathcal{O}}$ is then regarded as the integral closure of $\mathcal{O}_{C,p}$.

Put

$$\mathcal{O}' = \oplus_j t_j^{c_j} \mathcal{O}_j$$

Then $\mathcal{O}'$ is a subring of $\hat{\mathcal{O}}$.

The next theorem is a deep result. For a proof, see Kodaira [56]. (see also Exercise 5 of Sec. 4.2.)

Theorem 2.1.4 (Gorenstein-Rosenlicht). (1) $\mathcal{O}' \subset \mathcal{O}_{C,p} \subset \hat{\mathcal{O}}$ and (2) $\dim(\hat{\mathcal{O}}/\mathcal{O}_{C,p}) = \dim(\mathcal{O}_{C,p}/\mathcal{O}') = \frac{1}{2}c_p$. In particular, $\delta_p = \frac{1}{2}c_p$ is a nonnegative integer and is an analytic invariant.

The integer δ_p in this theorem is important.

Proposition 2.1.5. $\delta_p \geq \frac{1}{2}m_p(m_p - 1)$. The equality holds if and only if (1) every irreducible branch has p as either a nonsingular point or a simple cusp (of multiplicity m_j) (see Sec. 1.5) and (2) the tangent lines to the irreducible branches at p are mutually distinct.

Proof. We may take (x, y) such that ϕ is given around p_1 by

$$\phi\colon t_1 \to (t_1^{m_1}, h_1(t_1))$$

where $\operatorname{ord} h_1(t_1) \geq m_1 + 1$. This means that T_pC_1 = the x-axis. Then

$$\operatorname{ord}_{x=0} a_{1k}(x) \geq k + 1$$

For $j \geq 2$,

$$\operatorname{ord}_{x=0} a_{jk}(x) \geq k$$

Hence, we get

$$\operatorname{ord}_{x=0} a_k(x) \geq k \qquad \text{for } m - k \geq m_1$$

$$\operatorname{ord}_{x=0} a_k(x) \geq k + 1 \qquad \text{for } m - k < m_1$$

Now,

$$\frac{\partial R}{\partial y}(t_1^{m_1}, h_1(t_1)) = (m-1)h_1(t_1)^{m-1} + (m-2)a_1(t_1^{m_1})h_1(t_1)^{m-2} + \cdots + a_{m-1}(t_1^{m_1})$$

Note that

$$\operatorname{ord}_{t_1=0} a_k(t_1^{m_1})h_1(t_1)^{m-k-1} \geq km_1 + (m-k-1)(m_1+1)$$
$$= mm_1 + m - m_1 - k - 1 \quad \text{for } m-k \geq m_1,$$
$$\operatorname{ord}_{t_1=0} a_k(t_1^{m_1})h_1(t_1)^{m-k-1} \geq (k+1)m_1 + (m-k-1)(m_1+1)$$
$$= mm_1 + m - k - 1 \qquad \text{for } m-k < m_1$$

The minimal number of the right-hand sides is $mm_1 - 1$ when $k = m - m_1$. Hence,

$$c_1 \geq mm_1 - 1 - (m_1 - 1) = mm_1 - m_1$$

In a similar way,

$$c_j \geq mm_j - m_j \quad \text{for } j \geq 2$$

Adding them, we get

$$c_p \geq m(m-1)$$

The equality holds if and only if

$$\operatorname{ord}_{t_j=0} h_j(t_j) = m_j + 1 \quad \text{and} \quad \operatorname{ord}_{x=0} a_{m-m_j}(x) = m - m_j$$

for $1 \leq j \leq s$. The first condition means that p is either a nonsingular point of C_j or a simple cusp of C_j. Consider the second condition. Put $j = 1$ for simplicity. Note that

$$\begin{aligned} a_{m-m_1}(x) &= a_{2m_2}(x) \cdots a_{sm_s}(x) \\ &\quad + a_{11}(x)\{a_{2(m_2-1)}(x)\, a_{3m_3}(x) \cdots a_{sm_s}(x) + \cdots\} \\ &\quad + a_{12}(x)\{\cdots\} \\ &\quad + \cdots \end{aligned}$$

Then

$$\mathrm{ord}_{x=0}\, a_{2m_2}(x) \cdots a_{sm_s}(x) \geq m_2 + \cdots + m_s = m - m_1,$$

$$\mathrm{ord}_{x=0}\, a_{11}(x)\{a_{2(m_2-1)}(x)\, a_{3m_3}(x) \cdots a_{sm_s}(x) + \cdots\} \geq m - m_1 + 1,$$

$$\mathrm{ord}_{x=0}\, a_{12}(x)\{\cdots\} \geq m - m_1 + 1,$$

$$\cdots$$

Hence,

$$\mathrm{ord}_{x=0}\, a_{m-m_1}(x) = m - m_1$$

if and only if

$$\mathrm{ord}_{x=0}\, a_{2m_2}(x) = m_2, \ldots, \mathrm{ord}_{x=0}\, a_{sm_s}(x) = m_s$$

This means that $T_pC_2, \ldots, T_pC_s$ are distinct from T_pC_1 = the x-axis.

Q.E.D.

Corollary 2.1.6.

1. $\delta_p = 0$ if and only if p is a nonsingular point of C.
2. $\delta_p = 1$ if and only if p is either a node or a simple cusp of multiplicity 2.

The proof of the following lemma is left to the reader (see Exercise 1).

Lemma 2.1.7. Let p be a cusp ($s_p = 1$) with multiplicity 2. Let ϕ be locally given by

$$t \to (x,y) = (t^2, b_k t^k + b_{k+1} t^{k+1} + \cdots) \qquad (b_k \neq 0)$$

where $k \geq 3$ and k is odd. Then $\delta_p = \frac{1}{2}(k+1)$.

A cusp p on C with multiplicity 2 is called a double cusp (respectively a ramphoid cusp) if $\delta_p = 2$ (respectively $\delta_p = 3$). A singular point p on C with $s_p = m_p = 2$ and $T_pC_1 = T_pC_2$ is called a tacnode (respectively an osnode) if $\delta_p = 2$ (respectively $\delta_p = 3$). (See Fig. 2.2.)

	s_p	m_p	δ_p	Name
(i)	1	2	1	Simple cusp of multiplicity 2
(ii)	1	2	2	Double cusp
(iii)	1	2	3	Ramphoid cusp
(iv)	1	3	3	Simple cusp of multiplicity 3
(v)	2	2	1	Node
(vi)	2	2	2	Tacnode
(vii)	2	2	3	Osnode
(viii)	3	3	3	Ordinary triple point

Now, put

$$D_C = \sum_{p \in C} \sum_{p_j \in \phi^{-1}(p)} c_j p_j$$

If C is a singular curve (respectively a nonsingular curve), then D_C is a positive divisor (respectively a zero divisor) on M and is called the conductor of C.

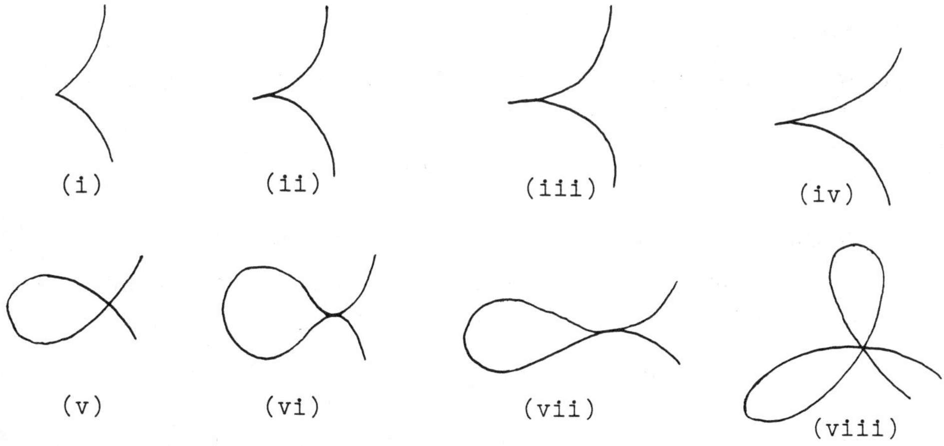

FIGURE 2.2

Theorem 2.1.8 (Adjunction Formula). Let K_S and K_M be the canonical bundle of S and M, respectively (see Sec. 3.1). Then

$$K_M = \phi^*[C] + \phi^*K_S - [D_C]$$

where $[D_C]$ is the line bundle determined by the divisor D_C (see Sec. 3.4).

Proof. Let $S = \bigcup_{j \in A} U_j$ be a finite covering of open sets U_j with a coordinate system (x_j, y_j). K_S is then defined by the Jacobian

$$J_{jk} = \frac{\partial(x_k, y_k)}{\partial(x_j, y_j)} = \begin{vmatrix} \frac{\partial x_k}{\partial x_j} & \frac{\partial y_k}{\partial x_j} \\ \frac{\partial x_k}{\partial y_j} & \frac{\partial y_k}{\partial y_j} \end{vmatrix}$$

Let $M = \bigcup_{\beta \in B} \Delta_\beta$ be a finite covering of open sets Δ_β with a coordinate t_β. K_M is then defined by $\{K_{\beta\gamma}\}$, where

$$K_{\beta\gamma} = \frac{dt_\gamma}{dt_\beta}$$

We may assume that there is a map j: B → A with the following properties:

1. $\phi(\Delta_\beta) \subset U_{j(\beta)}$,
2. If $\phi(\Delta_\beta)$ contains no singular point of C, then $U_{j(\beta)}$ contains no singular point of C. In this case, moreover, ϕ is given by

$$\phi: t_\beta \to (x_j, y_j) = (t_\beta, 0) \qquad (j = j(\beta))$$

(A minimal equation of C on $U_{j(\beta)}$ is in this case given by $y_j = 0$.)

3. If $\phi(\Delta_\beta)$ contains a singular point p of C, then p is a unique singular point of C in U_j $(j = j(\beta))$, $p = (0, 0) = \phi(0)$ and

$$\phi: t_\beta \to (x_j, y_j) = (t_\beta^{m_\beta}, h_\beta(t_\beta))$$

(In this case, a minimal equation $R_j = 0$ of C is given as before.)

In Case 2, $\sigma_\beta dt_\beta = dt_\beta$ and $c_\beta = 0$. In Case 3,

$$\sigma_\beta t_\beta = \frac{dt_\beta^{m_\beta}}{\partial R_j(t_\beta^{m_\beta}, h_\beta(t_\beta))/\partial y_j}$$

Note that

$$[D_C] = \{\sigma_{\beta\gamma}\} = \left\{\frac{\sigma_\gamma}{\sigma_\beta}\right\}$$

$$[C] = \{R_{jk}\} = \left\{\frac{R_j}{R_k}\right\}$$

$$\phi^*[C] = \{R_{\beta\gamma}\} = \{R_{jk} \cdot \phi\} = \left\{\frac{R_j \cdot \phi}{R_k \cdot \phi}\right\}$$

$$\phi^* K_S = \{J_{\beta\gamma}\} = \{J_{jk} \cdot \phi\}$$

where $j = j(\beta)$ and $k = j(\gamma)$.

Now, on C,

$$J_{jk} \frac{dx_j}{\partial R_j/\partial y_j} = \frac{dx_k}{\partial R_j/\partial y_k} + \frac{dx_k}{(\partial R_{jk}/\partial y_k)R_k + R_{jk}(\partial R_k/\partial y_k)}$$

Operating ϕ^* and noting that $\phi^*(R_k) = 0$, we get

$$J_{\beta\gamma}\, \sigma_\beta\, dt_\beta = \frac{1}{R_{\beta\gamma}} \sigma_\gamma\, dt_\gamma$$

Hence,

$$K_{\beta\gamma} = \frac{dt_\gamma}{dt_\beta} = R_{\beta\gamma} J_{\beta\gamma} \sigma^{-1}_{\beta\gamma} \qquad \text{Q.E.D.}$$

Now, we put $S = \mathbb{P}^2$. In this case, we have easily

$$K_{\mathbb{P}^2} = [-3H]$$

where H is a line on $\mathbb{P}^2$. If C is an irreducible plane curve of degree n, then

$$[C] = [nH]$$

Let ϕ: M → C be a nonsingular model of C. Then ϕ determines a linear system Λ on M of all line cuts of C. Note that

$$[D] = \phi^*[H] \quad \text{for } D \in \Lambda$$

By the Adjunction Formula,

$$K_M = \phi^*[nH] + \phi^*K_{\mathbb{P}^2} - [D_C]$$

$$= (n - 3)[D] - [D_C]$$

Comparing the degree of both sides, we get

$$2g - 2 = n(n - 3) - \deg D_C$$

where g is the genus of M (see Proposition 5.1.2). Thus

Theorem 2.1.9 (The Genus Formula). Let C be an irreducible plane curve of degree n, ϕ: M → C be a nonsingular model of C, and g be the genus of M. Then

$$g = \tfrac{1}{2}(n - 1)(n - 2) - \Sigma\delta_p$$

where Σ is extended over all singular points p of C.

Corollary 2.1.10. Let C, ϕ: M → C and g be as in the theorem.

1. $g \leq \frac{1}{2}(n - 1)(n - 2) - \Sigma\frac{1}{2}m_p(m_p - 1)$,
2. If C is nonsingular, then $g = \frac{1}{2}(n - 1)(n - 2)$,
3. If C is a nodal curve with k nodes, then $g = \frac{1}{2}(n - 1)(n - 2) - k$.

Remark 2.1.11. If one knows sheaves and cohomology, then Theorem 2.1.9 can be generalized as follows: Let C be an irreducible curve of degree n in $\mathbb{P}^r$ ($r \geq 2$). Then

$$g = p_a - \Sigma\delta_p$$

where $p_a = \dim H^1(C, \mathcal{O}_C)$, the arithmetic genus of C, $\delta_p = \text{length}(\hat{\mathcal{O}}_{C,p}/\mathcal{O}_{C,p})$ and Σ is extended over all singular points on C (see Hartshorne [39, p. 298]). For example, if C is the complete intersection (see Sec. 3.3) of two surfaces in $\mathbb{P}^3$ of degree a and b, then $p_a = \frac{1}{2}ab(a + b - 4) + 1$. Hence,

$$g = \tfrac{1}{2}ab(a + b - 4) + 1 - \Sigma\delta_p$$

Exercises

1. Prove Lemmas 2.1.1, 2.1.3, and 2.1.7.
2. Prove Corollary 2.1.10 using the Riemann-Hurwitz Formula (see Theorem 4.1.14).
3. Under the notations of this section, note that

$$\frac{dx}{\partial R/\partial y} = \frac{-dy}{\partial R/\partial x} \quad \text{on } C$$

   Using this, prove that every c_j is an analytic invariant.
4. Compute δ_p of the singular points of the following curves and determine their genera: (1) $X_0^2X_1^3 + X_0X_1^4 + X_2^5 = 0$ and (2) $X_0^2X_1^3 + X_0X_2^4 + X_2^5 = 0$.

2.2 SINGULAR PLANE QUARTICS

Now, our problem is naive.

Problem. Given an integer n (≥ 3), what singular irreducible plane curves of degree n exist?

For n = 3 and 4, the solution of this problem is classically well known. But, for $n \geq 5$, it seems that the problem has not been solved completely. This is because the computations become very messy for $n \geq 5$.

Following the idea of Veronese [99], we solve the problem for n = 5. For this purpose, we use the results in Part II freely. Our method is intuitive and projective geometric, and needs little computations. It can be applied to curves of degree ≤ 4 and more or less to curves of higher degree.

We talk about the cases where n = 3 and 4 in this section and the case n = 5 in the next section.

The Case n = 3. The solution of the problem for plane cubic curves is very simple. Let C be an irreducible plane cubic curve and $\phi: M \to C$ be a nonsingular model of C. The genus g of M is given by

$$g = 1 - \Sigma\delta_p$$

by the Genus Formula, where Σ is extended over all singular points of C.

Hence, $g = 1$ or 0. If $g = 1$, then C is a nonsingular cubic curve. If $g = 0$, then C has a unique singular point p such that $\delta_p = 1$. By Corollary 2.1.6, p is either a node or a simple cusp of multiplicity 2.

Proposition 2.2.1. An irreducible plane cubic curve C with (1) a node is projectively equivalent to the (closure in $\mathbb{P}^2$ of the affine) curve

$$y^2 = x^3 + x^2$$

and with (2) a simple cusp of multiplicity 2 is projectively equivalent to the curve

$$y^2 = x^3$$

(See Fig. 2.3.)

Proof. We can prove the proposition by a direct calculation on cubic polynomials. But we prove it by a geometric method as follows.

Let $\phi\colon M = \mathbb{P}^1 \to C$ be a nonsingular model of C. Then the line cuts of C define a linear system $\Lambda = g_3^2$ of dimension 2 and degree 3 on $\mathbb{P}^1$. Hence, ϕ can be obtained by the projection

$$\pi_p : C_0 \to \mathbb{P}^2$$

with the center a point $p \in \mathbb{P}^3 - C_0$, where $C_0 = \Phi_{|3(\infty)|}(\mathbb{P}^1)$ is the twisted cubic. That is, $\phi = \pi_p \cdot \Phi_{|3(\infty)|}$ (see Example 5.3.1). We identify $\mathbb{P}^1$ with C_0 through $\Phi_{|3(\infty)|}$.

To prove the first part, suppose that C has a node. By Plücker's Formula, C has just three flexes: r_1, r_2, and r_3. Put $q_j = \pi_p^{-1}(r_j)$,

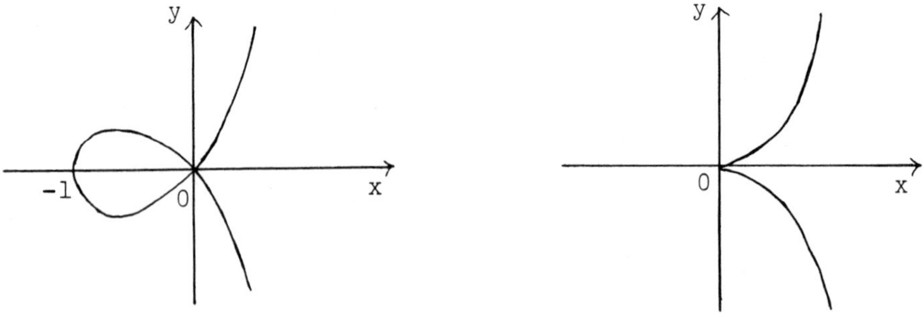

FIGURE 2.3

$1 \leq j \leq 3$. Then p is on the osculating planes to C_0 at q_j, $1 \leq j \leq 3$. These osculating planes determine p as their intersection. (In fact, if the intersection of the planes is a line ℓ, then the projection π_p: $\mathbb{P}^1 = C_0 \to \mathbb{P}^1$ is a rational function of degree 3 with at least three points of ramification with the ramification index 3 (see Sec. 4.1). This contradicts the Riemann-Hurwitz Formula. Now consider $\sigma \in \mathrm{Aut}(\mathbb{P}^1)$ such that

$$\sigma(q_1) = \infty, \quad \sigma(q_2) = \sqrt{-\frac{1}{3}} \quad \text{and} \quad \sigma(q_3) = -\sqrt{-\frac{1}{3}}$$

σ is uniquely extended to $\sigma \in \mathrm{Aut}(\mathbb{P}^3)$ (see Example 5.2.18). The intersection of the osculating planes to C_0 at $\sigma(q_j)$, $1 \leq j \leq 3$, is

$$\sigma(p) = (0 : 1 : 0 : 1)$$

The image of $\pi_{\sigma(p)}$ is clearly the curve $y^2 = x^3 + x^2$.

To prove the second part, suppose that C has a simple cusp r_1 of multiplicity 2. By Plücker's Formula, C has a unique flex r_2. Put $q_j = \pi_p^{-1}(r_j)$, j = 1, 2. p is on the tangent line $\ell = T_{q_1}C_0$ to C_0 at q_1, and on the osculating plane P to C_0 at q_2. Note that $\ell \not\subset P$, so $p = \ell \cap P$. Take a $\sigma \in \mathrm{Aut}(\mathbb{P}^1)$ such that

$$\sigma(q_1) = 0 \quad \text{and} \quad \sigma(q_2) = \infty$$

σ is extended to $\sigma \in \mathrm{Aut}(\mathbb{P}^3)$. Then

$$\sigma(\ell) \cap \sigma(P) = \sigma(p) = (0 : 1 : 0 : 0)$$

The image of $\pi_{\sigma(p)}$ is clearly the curve $y^2 = x^3$. Q.E.D.

Remark 2.2.2. The action of $\mathrm{Aut}(\mathbb{P}^1)$ on $\mathbb{P}^3$ has three orbits. The orbits are

$$C_0, \quad \mathrm{Tan}(C_0) - C_0, \quad \text{and} \quad \mathbb{P}^3 - \mathrm{Tan}(C_0)$$

where $\mathrm{Tan}(C_0)$ is the tangent surface to C_0 (see Lemma 1.6.21). In particular, the quotient space $\mathbb{P}^3/\mathrm{Aut}(\mathbb{P}^1)$ is not Hausdorff.

The Case n = 4. Let C be an irreducible plane quartic curve and ϕ: $M \to C$ be a nonsingular model. The genus g of M is given by

$$g = 3 - \Sigma\, \delta_p$$

Hence, $0 \leq g \leq 3$.

1°. $g = 3$. In this case, C is nonsingular. C is a nonhyperelliptic canonical curve of genus 3 (see Sec. 5.1).

2°. $g = 2$. The line cuts of C define a linear system $\Lambda = g_4^2$ of dimension 2 and degree 4 on M. Take $D \in \Lambda$. Then, by the Riemann-Roch Theorem (see Theorem 5.1.5), $\dim |D| = 2$. Hence, $\Lambda = |D|$ is complete.

Conversely, let M be a compact Riemann surface of genus 2 and D be a positive divisor of degree 4. Then $|D|$ has dimension 2 and no fixed point, for $\dim |D - p| = 1$ for any point $p \in M$. Consider the map

$$\phi = \Phi_{|D|} : M \to \mathbb{P}^2$$

Its image $C = \phi(M)$ is an irreducible plane curve. Two cases occur.

I. $\phi: M \to C$ is a 2—1 map. In this case, $\deg C = 2$, so C is an irreducible conic. Hence, for any q, $r \in C$, $\phi^{-1}(q) \sim \phi^{-1}(r)$ (linearly equivalent). Hence, $\phi^{-1}(q) \in |K|$, the canonical linear system. Hence,

$$D \sim \phi^{-1}(q + r) = \phi^{-1}(q) + \phi^{-1}(r) \sim 2K$$

Conversely, $\Phi_{|2K|}$ is a 2—1 map of M onto an irreducible conic.

II. $\phi: M \to C$ is bimeromorphic. Let q be the uniquely determined singular point of C and put $\phi^{-1}(q) = p + p'$. If $p \neq p'$ (respectively $p = p'$), then q is a node (respectively a simple cusp of multiplicity 2) of C. (See Fig. 2.4.) Let $\pi_q: M \to \mathbb{P}^1$ be the projection with the center q. It is a meromorphic function of degree 2, so $\pi_q = \Phi_{|K|}$. Hence,

$$D \sim p + p' + K$$

Note that $p + p' \not\sim K$ (not linearly equivalent).

Conversely, choose any two points p and p' on M such that $p + p' \not\sim K$. Then $|K + p + p'|$ gives a bimeromorphic map ϕ on M onto an irreducible plane quartic curve with a unique singular point $\phi(p) = \phi(p')$, which is either a node ($p \neq p'$) or a simple cusp of multiplicity 2 ($p = p'$).

Remark 2.2.3. Under the notations in Sec. 5.4, every fiber of

$$\pi: \mathbb{L}_{2,4}^2 \to \mathbb{M}_2$$

FIGURE 2.4

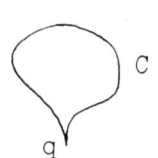

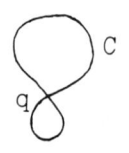

is biholomorphic to $J(M_t)/\mathrm{Aut}(M_t)$, where $J(M_t)$ is the Jacobian variety of M_t (see Section 5.2). Thus, there are a lot of such curves with prescribed moduli.

3°. $g = 1$. Let $M = \mathbb{C}/(\mathbb{Z}\omega_1 + \mathbb{Z}\omega_2)$, $\mathrm{Im}(\omega_2/\omega_1) > 0$, be a complex one-torus. $\phi: M \to C \subset \mathbb{P}^2$ defines a linear system $\Lambda = g_4^2$ without fixed point. By Example 5.3.1, Λ is equivalent to a linear subsystem in $|4(0)|$ under the action of Aut(M). Hence, we may assume that Λ itself is a linear subsystem of $|4(0)|$. This means that C can be obtained as the image of the projection

$$\pi_p: C_0 \to \mathbb{P}^2$$

with a point $p \in \mathbb{P}^3 - C_0$, where C_0 is the image of $\Phi_{|4(0)|}$, that is, the image of

$$\Phi_{|4(0)|}: z \in M \to (1: \wp(z): \wp'(z): \wp(z)^2) \in \mathbb{P}^3$$

(see Sec. 4.3). We identify M with C_0 through $\Phi_{|4(0)|}$.

$\pi_p(C_0)$ and $\pi_q(C_0)$ are projectively equivalent if and only if p and q are equivalent under the action of the finite group B on $\mathbb{P}^3$ (see Example 5.3.1).

There is a one-parameter family $\{Q_\lambda\}_{\lambda \in \mathbb{P}^1}$ of quadric surfaces which contain C_0 (see Example 5.3.15). The finite group B acts on the family $\{Q_\lambda\}$. in fact, it can be easily shown that

$$T_x(Q_{\wp(z)}) = Q_{\wp(z+2x)} \qquad \text{for } z \in M \text{ and } x \in M \text{ with } 4x = 0$$

$$S_{-1}(Q_\lambda) = Q_\lambda \qquad \text{for } \lambda \in \mathbb{P}^1$$

$$S_i(Q_\lambda) = Q_{-\lambda} \qquad \text{for } \lambda \in \mathbb{P}^1, \text{ if } (\omega_1, \omega_2) = (1, i),\ i = \sqrt{-1}$$

$$S_\zeta(Q_\lambda) = Q_{\zeta^4\lambda} \qquad \text{for } \lambda \in \mathbb{P}^1, \text{ if } (\omega_1, \omega_2) = (1, \zeta),\ \zeta = \frac{1 + \sqrt{-3}}{2}$$

(Note that 2x is one of 0, $\omega_1/2$, $(\omega_1 + \omega_2)/2$, $\omega_2/2$. If, say, $2x = \omega_1/2$, then

$$\wp(z + 2x) = e_1 + \frac{(e_1 - e_2)(e_1 - e_3)}{\wp(z) - e_1}$$

where $e_1 = \wp(\omega_1/2)$, $e_3 = \wp(\omega_1 + \omega_2)/2$, and $e_2 = \wp(\omega_2/2)$ (see Exercise 5 of Sec. 4.3). Hence, T_x acts on the parameter space $\mathbb{P}^1$ of $\{Q_\lambda\}$. In particular, B acts transitively on the four quadric cones Q_{e_j}, $0 \leq j \leq 3$, where $e_0 = \infty = \wp(0)$.

Now, every point $p \in \mathbb{P}^3 - C_0$ is on a unique Q_λ in $\{Q_\lambda\}$. Two cases occur.

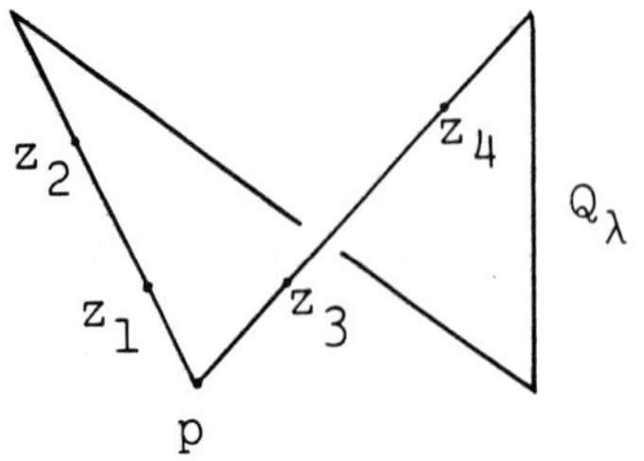

FIGURE 2.5

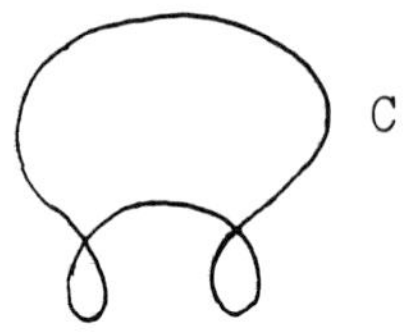

FIGURE 2.6

FIGURE 2.7

FIGURE 2.8

I. <u>Q_λ is nonsingular</u>. In this case, there are two one-parameter families of lines on Q_λ. p is then the intersection point of two lines which belong to different families. These lines cut C_0 in two points z_1, z_2 and z_3, z_4. (See Fig. 2.5.)

If $z_1 \neq z_2$ and $z_3 \neq z_4$, then $C = \pi_p(C_0)$ is an elliptic plane quartic curve with two nodes $\pi_p(z_1) = \pi_p(z_2)$ and $\pi_p(z_3) = \pi_p(z_4)$. (See Fig. 2.6.)

If $z_1 = z_2$ and $z_3 \neq z_4$, then C has a simple cusp $\pi_p(z_1)$ of multiplicity 2 and a node $\pi_p(z_3) = \pi_p(z_4)$. (See Fig. 2.7.)

If $z_1 = z_2$ and $z_3 = z_4$, then C has two simple cusps $\pi_p(z_1)$ and $\pi_p(z_3)$ of multiplicity 2. (See Fig. 2.8.)

Every case actually occurs. For example, the last case occurs when and only when p is the intersection point of the tangent lines $T_{z_1}C_0$ and $T_{z_3}C_0$ to C_0 at z_1 and z_3 such that $2z_1 = -2z_3$ and $\wp(2z_1) = \lambda$. Hence, there are 16 such points p on each Q_λ.

II. <u>Q_λ is a cone</u>. In this case, we may assume that

$$Q_\lambda = Q_\infty = \{X_0 : X_1 : X_2 : X_3) \in \mathbb{P}^3 \mid X_0 X_3 = X_1^2\}$$

The vertex of Q_∞ is $v = (0: 0: 1: 0)$.

Suppose that $p \neq v$. Then there passes a unique line on Q_∞ through p, which cuts C_0 at two points z_1 and z_2 such that $z_1 + z_2 = 0$. (See Fig. 2.9.)

If $z_1 \neq z_2$, that is, $2z_1 \neq 0$, then the elliptic plane quartic curve $C = \pi_p(C_0)$ has a unique singular point $\pi_p(z_1) = \pi_p(z_2)$, a tacnode. (See Fig. 2.10.)

If $z_1 = z_2$, that is, $2z_1 = 0$, then $C = \pi_p(C_0)$ has a unique singular point $\pi_p(z_1)$, a double cusp. (See Fig. 2.11.)

Since $T_{-z_1}(z_2) = \underline{0}$, we may consider only the case $z_1 = 0 = (0: 0: 0: 1)$. Then p is on the line $\overline{vo}$. Put $p = (0: 0: 1: t)$. Then the equation of $C = \pi_p(C_0)$ can be written as

$$(x^2 - y)^2 = t^2 y(4x^3 - g_2 xy^2 - g_3 y^3)$$

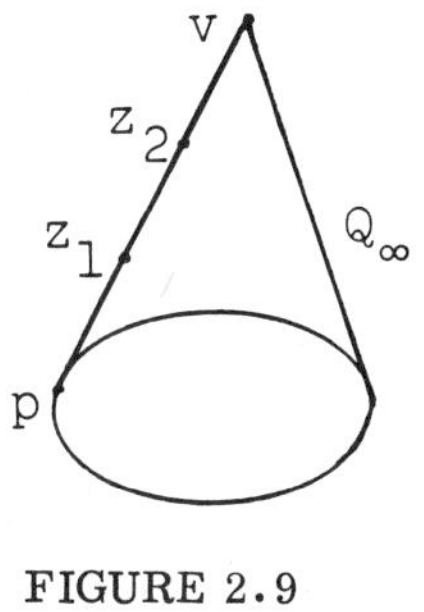

FIGURE 2.9

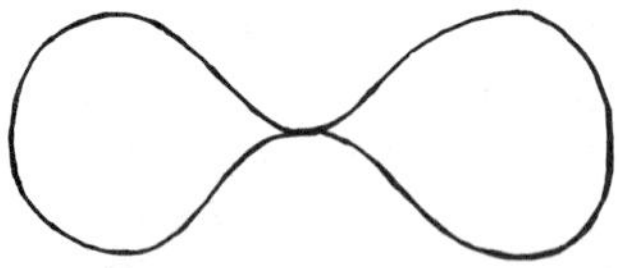

FIGURE 2.10

where $y^2 = 4x^3 - g_2x - g_3$ is the image of $z \in M \to (1:\ \wp(z):\ \wp'(z)) \in \mathbb{P}^2$ (see Sec. 4.3). The origin (0,0) is the double cusp of C.

When p tends to v, t tends to 0. If p = v, then we get a double conic $(x^2 - y)^2 = 0$. In fact, the projection π_V: $C_0 \to \mathbb{P}^2$ is a 2—1 map of C_0 onto an irreducible conic.

4°. g = 0. The line cuts of a rational plane quartic curve C define a linear system $\Lambda = g_4^2$ without fixed point on $\mathbb{P}^1$, so Λ is a linear subsystem of $|4(\infty)|$. This means that C can be obtained as the image of the projection

$$\pi_\ell: C_0 \to \mathbb{P}^2$$

where C_0 is the rational normal curve, that is, the image of

$$\Phi_{|4(\infty)|}: t \in \mathbb{P}^1 \to (1: t: t^2: t^3: t^4) \in \mathbb{P}^4$$

and ℓ is a line in $\mathbb{P}^4$ such that $C_0 \cap \ell = \phi$ (see Example 5.3.1). We identify $\mathbb{P}^1$ with C_0 through $\Phi_{|4(\infty)|}$.

By the Genus Formula, $\Sigma\delta_p = 3$. Hence, there is at least one and at most three singular points on C. By (1) of Corollary 2.1.10, there is no singular point with the multiplicity ≥ 4. Hence, two cases occur.

I. There is a triple point p on C. In this case, p is a unique singular point of C by (1) of Corollary 2.1.10. Put $\pi_\ell^{-1}(p) = p_1 + p_2 + p_3$. Then ℓ is on the plane $P = p_1 \vee p_2 \vee p_3$, spanned by p_1, p_2, and p_3.

FIGURE 2.11

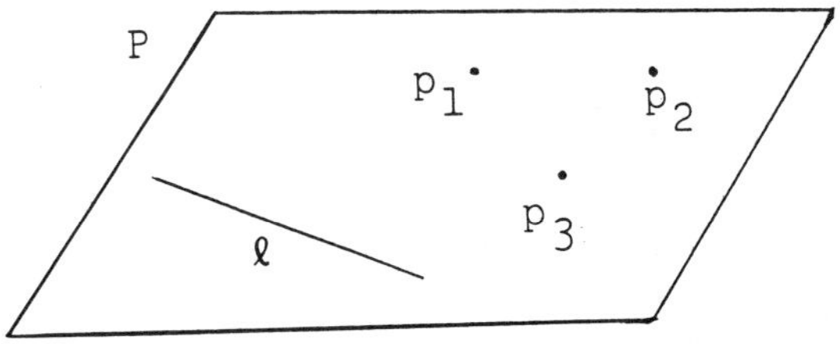

FIGURE 2.12

Conversely, if ℓ is a line on a plane $P = p_1 \vee p_2 \vee p_3$, then $C = \pi_\ell(C_0)$ has a triple point $p = \pi_\ell(p_1) = \pi_\ell(p_2) = \pi_\ell(p_3)$.

a. Suppose that p_1, p_2, and p_3 are mutually distinct. (See Fig. 2.12.) In this case, p is an ordinary triple point of C. (See Fig. 2.13.)

b. Suppose that $p_1 = p_2 \neq p_3$. In this case, ℓ is on the plane P spanned by the tangent line $T_{p_1}C_0$ and p_3. (See Fig. 2.14.) p is then a triple point with two irreducible branches C_1 and C_2 meeting transversally such that C_1 is smooth and C_2 has p as a simple cusp of multiplicity 2. (See Fig. 2.15.)

c. Suppose that $p_1 = p_2 = p_3$. In this case, ℓ is on the osculating 2-plane $O_{p_1}C_0$ to C_0 at p_1. (See Fig. 2.16.) p is then a simple cusp of multiplicity 3. (See Fig. 2.17.)

It can be shown that C is, in this case, projectively equivalent to either

$$x^4 - x^3y + y^3 = 0 \tag{1}$$

(the image of π_ℓ: $t \in \mathbb{P}^1 \to (t-1: t^3: t^4) \in \mathbb{P}^2$ with the center $\ell = \{X_0 - X_1 = 0, X_3 = X_4 = 0\}$), or

$$x^4 - y^3 = 0 \tag{2}$$

(the image of π_ℓ: $t \in \mathbb{P}^1 \to (1: t^3: t^4) \in \mathbb{P}^2$ with the center $\ell = \{X_0 = X_3 = X_4 = 0\}$).

(It can be seen also from Proposition 2.1.5 that only these cases a, b, and c are possible.)

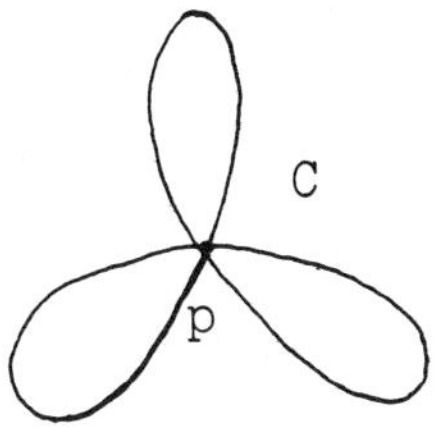

FIGURE 2.13

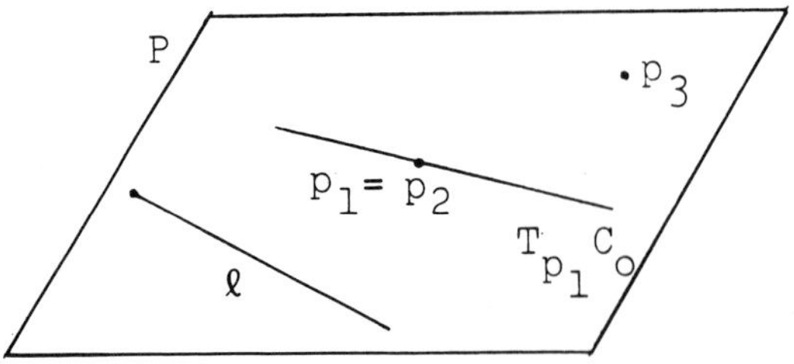

FIGURE 2.14

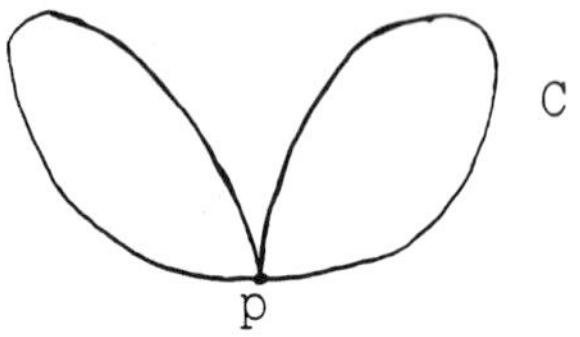

FIGURE 2.15

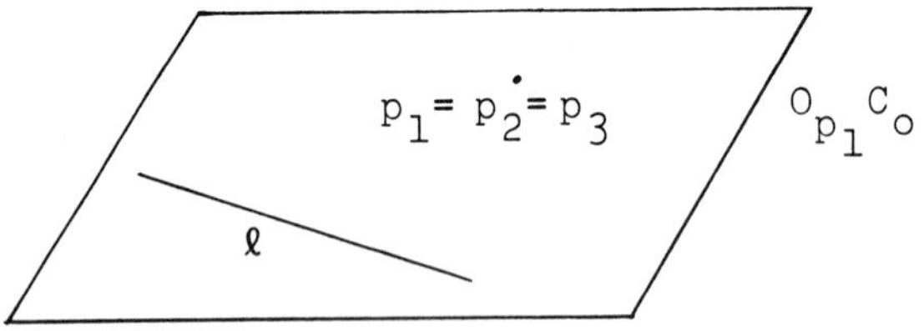

FIGURE 2.16

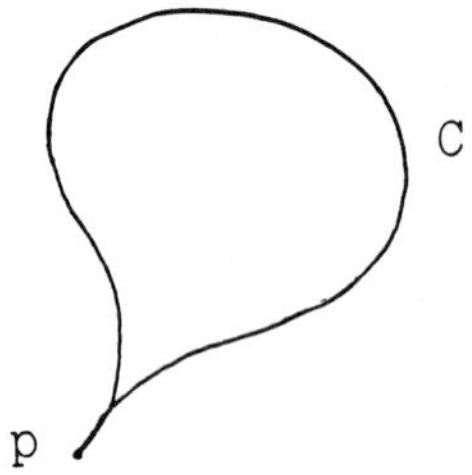

FIGURE 2.17

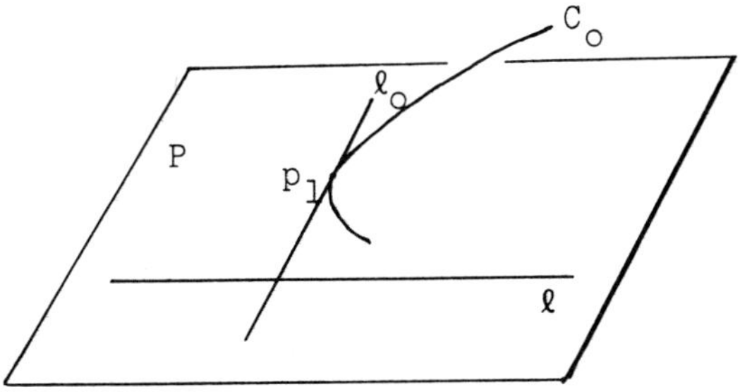

FIGURE 2.18

<u>II. C has only double points</u>. In this case, let p be a double point and put $\pi_\ell(p) = p_1 + p_2$. There are two cases

$$p_1 = p_2 \quad \text{and} \quad p_1 \neq p_2$$

We suppose that $p_1 = p_2$ in the sequel. The case $p_1 \neq p_2$ can be treated in a similar way.

Put $\ell_0 = T_{p_1} C_0$. Then ℓ and ℓ_0 must span a 2-plane P which is not the osculating 2-plane to C_0 at p_1, for p is a double point of C. (See Fig. 2.18.)

What we do is to choose first a 2-plane P with $P \neq O_{p_1} C_0$ which contains ℓ_0 and then a line ℓ on P.

Take a 2-plane P' such that $P' \cap \ell_0 = \phi$ and consider the projection

$$\pi_0 = \pi_{\ell_0} : C_0 \to P'$$

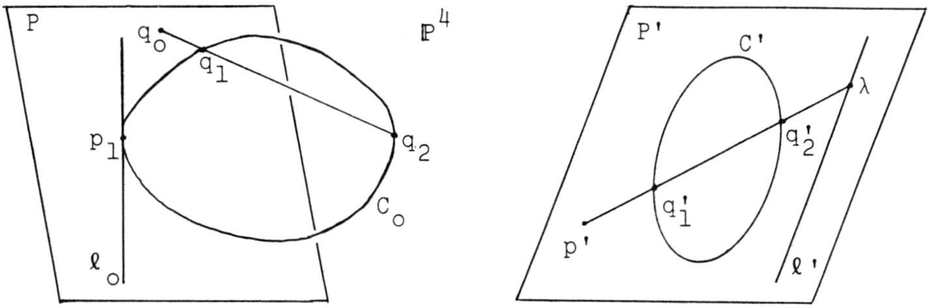

FIGURE 2.19

with the center ℓ_0. Note that

$$\pi_0 = \Phi_{|4(\infty)-2(\infty)|} = \Phi_{|2(\infty)|}$$

so $C' = \pi_0(C_0)$ is an irreducible conic in P'. Two-planes P containing ℓ_0 such that $P \neq O_{p_1} C_0$ are in 1–1 correspondence with the points in $P' - C'$. The correspondence is given by

$$P \to p' = P \cap P', \qquad p' \to P = p' \vee \ell_0$$

Take a point $p' \in P' - C'$ and fix it. Take a line ℓ' on P' such that $p' \notin \ell'$. (See Fig. 2.19.) For a point $\lambda \in \ell'$, the line $\overline{p'\lambda}$ cuts C' at $q_1' = \pi_0(q_1)$ and $q_2' = \pi_0(q_2)$. ($q_1' = q_2'$ if $\overline{p'\lambda}$ is tangent to C'.) Then the line $\overline{q_1 q_2}$ ($T_{q_1} C_0$ if $q_1 = q_2$) intersects with $P = p' \vee \ell_0$ at one point q_0, provided $q_1 \neq p_1$ and $q_2 \neq p_1$. In fact, $H_\lambda = P \vee \overline{p'\lambda}$ is a hyperplane in $\mathbb{P}^4$ such that $C_0 \cap H_\lambda = 2p_1 + q_1 + q_2$.

Now consider the meromorphic map

$$\Psi: \lambda \in \ell' \to q_0 \in P$$

It can be extended to a holomorphic map $\Psi: \ell' \to P$. Ψ is birational onto the image $\Psi(\ell')$, because any four points on C_0 are in general position (see Exercise 2 of Sec. 5.1).

Lemma 2.2.4. The image $\Psi(\ell')$ is either an irreducible conic or a line in P. The latter case occurs if and only if p' is on the tangent line to C' at $\pi_0(p_1)$. Moreover, (1) if $\Psi(\ell')$ is an irreducible conic, then p_1 is on $\Psi(\ell')$ and ℓ_0 is the tangent line to $\Psi(\ell')$ at p_1, and (2) if $\Psi(\ell')$ is a line, then p_1 is not on $\Psi(\ell')$.

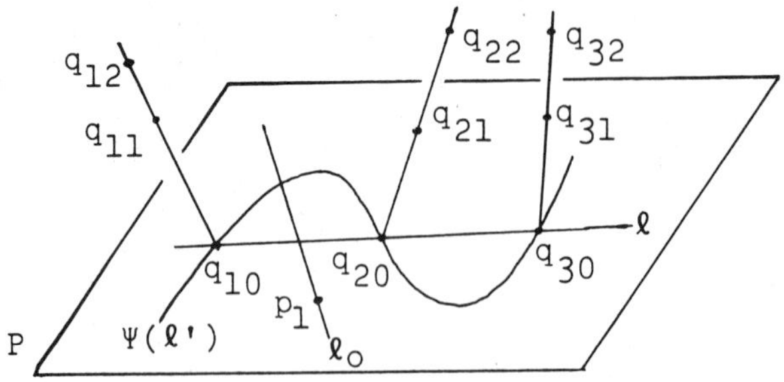

FIGURE 2.20

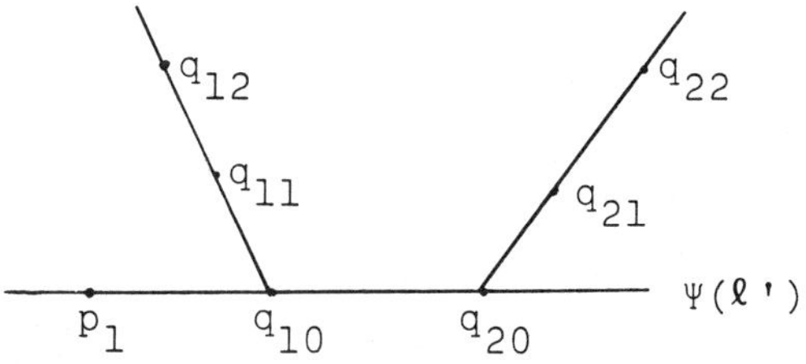

FIGURE 2.21

Proof. Suppose that $d = \deg \Psi(\ell') \geq 3$. Take a general line ℓ on P passing through distinct d points $q_{j0} = \Psi(\lambda_j)$, $1 \leq j \leq d$. Put

$$\overline{p'\lambda_j} \cap C' = \pi_0(q_{j1}) + \pi_0(q_{j2}), \quad 1 \leq j \leq d$$

Then q_{j0}, q_{j1}, and q_{j2} are collinear. (See Fig. 2.20.) The projection $\pi_\ell : C_0 \to \mathbb{P}^2$ is then birational onto its image $\pi_\ell(C_0)$, which has the $d + 1$ singular points (double points)

$$\pi_\ell(p_1),\ \pi_\ell(q_{11}) = \pi_\ell(q_{12}),\ \ldots,\ \pi_\ell(q_{d1}) = \pi_\ell(q_{d2})$$

By Corollary 2.1.10,

$$0 \leq \tfrac{1}{2}(4 - 1)(4 - 2) - (d + 1) = 2 - d < 0$$

a contradiction. Hence, $\deg \Psi(\ell') \leq 2$.

Suppose that p' is not on the tangent line to C' at $\pi_0(p_1)$. This means that there is a hyperplane H with $H \supset P$ such that

$$C_0 \cap H = 3p_1 + q$$

where $q \neq p_1$. Then $\overline{p_1 q}$ meets P at p_1. Hence, p_1 is on $\Psi(\ell')$. Assume that $\Psi(\ell')$ is a line. (See Fig. 2.21.) Take general points q_{10} and q_{20} on $\Psi(\ell')$. Then there are points q_{11} and q_{12} (respectively q_{21} and q_{22}) on C_0 such that q_{10}, q_{11}, q_{12} (respectively q_{20}, q_{21}, q_{22}) are collinear. Then the points p_1, q_{11}, q_{12}, q_{21}, q_{22} on C_0 are on a hyperplane in $\mathbb{P}^4$, which is impossible, for $\deg C_0 = 4$. This shows that $\Psi(\ell')$ is an irreducible conic containing p_1. If $\Psi(\ell') \cap \ell_0 = p_1 + q_{10}$ with $p_1 \neq q_{10}$, then the 2-plane $\ell_0 \vee \overline{q_{11}q_{12}}$ contains four points $2p_1 + q_{11} + q_{12}$, that is, $C_0 \cap (\ell_0 \vee \overline{q_{11}q_{12}}) = 2p_1 + q_{11} + q_{12}$. (Here q_{11} and q_{12} are the points on C_0 such that q_{10}, q_{11}, and q_{12} are collinear.) But this is impossible (see Exercise 2 of Sec. 5.1). Hence, ℓ_0 is tangent to $\Psi(\ell')$.

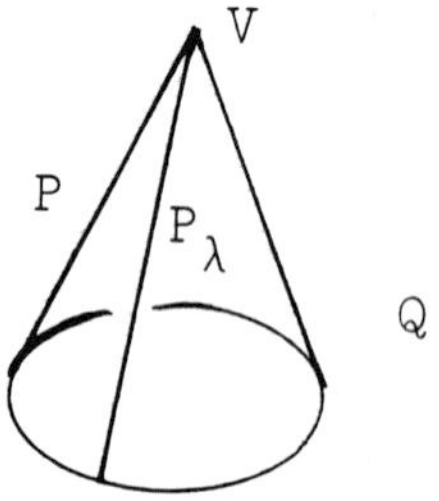

FIGURE 2.22

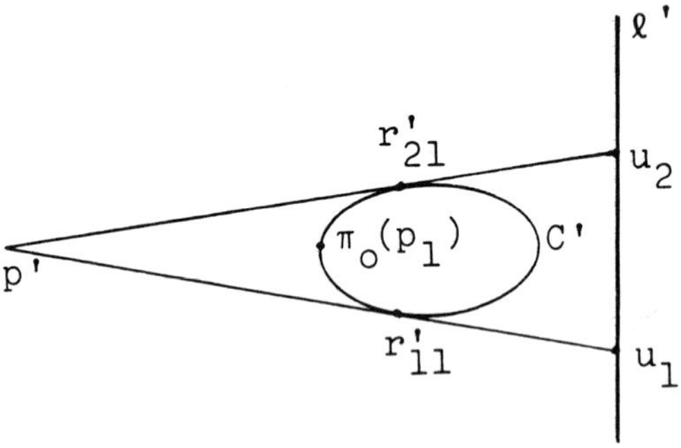

FIGURE 2.23

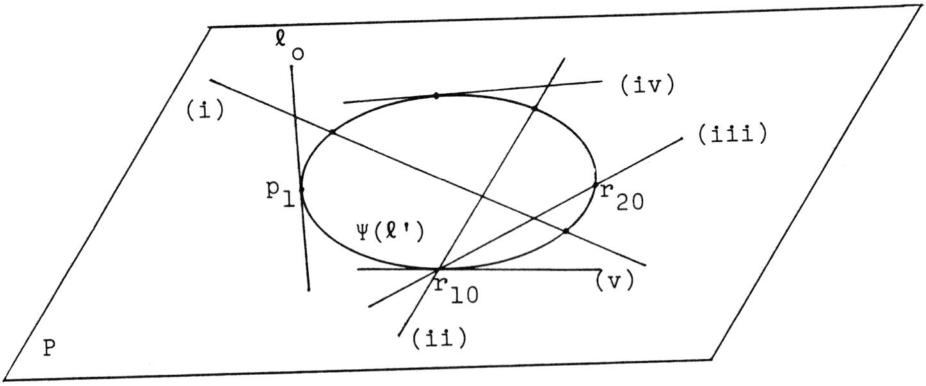

FIGURE 2.24

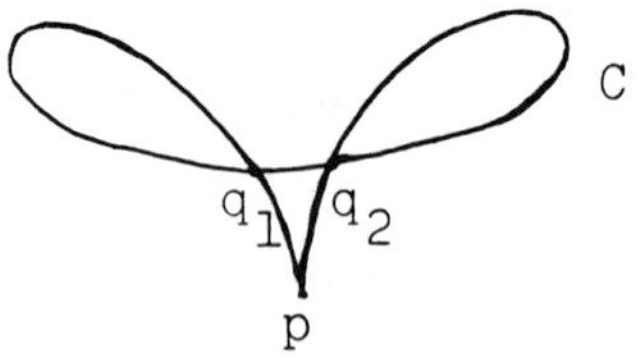

FIGURE 2.25

Next, suppose that p' is on the tangent line to C' at $\pi_0(p_1)$. For $\lambda \in \ell'$, put $H_\lambda = P \vee \overline{p'\lambda}$ as above. Then $\{H_\lambda\}$ is a one-parameter family of hyperplanes containing P. Put

$$C_0 \cap H_\lambda = 2p_1 + q_1(\lambda) + q_2(\lambda)$$

Then $\Lambda = \{q_1(\lambda) + q_2(\lambda) \mid \lambda \in \ell'\}$ is a linear pencil on $\mathbb{P}^1$. By the assumption, Λ contains $2p_1$. Hence, by Lemma 5.3.22, a quadric hypersurface $Q = Q(\Lambda, \Lambda)$ of rank 3 in $\mathbb{P}^4$ with $C_0 \subset Q$ can be constructed. (See Fig. 2.22.) Note that the vertex $V = V_Q$ of Q is a line. Note also that

$$V \subset P \subset Q$$

Take another 2-plane P_λ in Q which contains V and $q_1(\lambda)$ and $q_2(\lambda)$. Since $P \cap P_\lambda = V$, the line $\overline{q_1(\lambda)q_2(\lambda)}$ meets P at a point on V. Hence, $\Psi(\ell') = V$ is a line. Since Λ has no fixed point, $V = \Psi(\ell')$ has no point on C_0 (see Lemma 5.3.22). This completes the proof of the lemma. Q.E.D.

a. Suppose that $\Psi(\ell')$ is an irreducible conic. Let μ_1 and μ_2 ($\mu_1 \neq \mu_2$) be points on ℓ' such that $\overline{p'\mu_1}$ and $\overline{p'\mu_2}$ are tangent to C'. (See Fig. 2.23.) Put $2r'_{11} = 2\pi_0(r_{11}) = C' \cap \overline{p'\mu_1}$ and $2r'_{21} = 2\pi_0(r_{21}) = C' \cap \overline{p'\mu_2}$. Also put $r_{10} = \Psi(\mu_1)$ and $r_{20} = \Psi(\mu_2)$.

Now, let ℓ be a line on P. Then $p = \pi_\ell(p_1)$ is a simple cusp of multiplicity 2 on $C = \pi_\ell(C_0)$. (See Fig. 2.24.)

i. Suppose that ℓ is a general line on P. In this case ℓ cuts $\Psi(\ell')$ at two points ($\neq r_{10}, \neq r_{20}$). Hence, $C = \pi_\ell(C_0)$ has three singular points p, q_1, and q_2, where q_1 and q_2 are nodes. (See Fig. 2.25.)

ii. Suppose that ℓ is a general line on P passing through r_{10}. In this case $C = \pi_\ell(C_0)$ has three singular points p, q, and $r = \pi_\ell(r_{11})$, where q is a node and r is a simple cusp of multiplicity 2. (See Fig. 2.26.)

iii. Suppose that $\ell = \overline{r_{10}r_{20}}$. In this case, $C = \pi_\ell(C_0)$ has three simple cusps p, $r_1 = \pi_\ell(r_{11})$, and $r_2 = \pi_\ell(r_{21})$ of multiplicity 2. (See Fig. 2.27.)

iv. Suppose that ℓ is a general tangent line to $\Psi(\ell')$. In this case, $C = \pi_\ell(C_0)$ has two singular points p and q, where q is a tacnode. (See Fig. 2.28.)

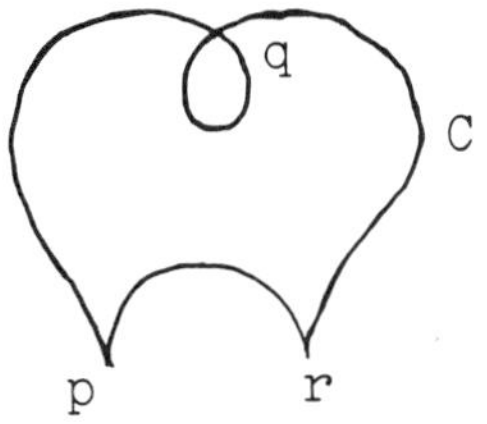

FIGURE 2.26

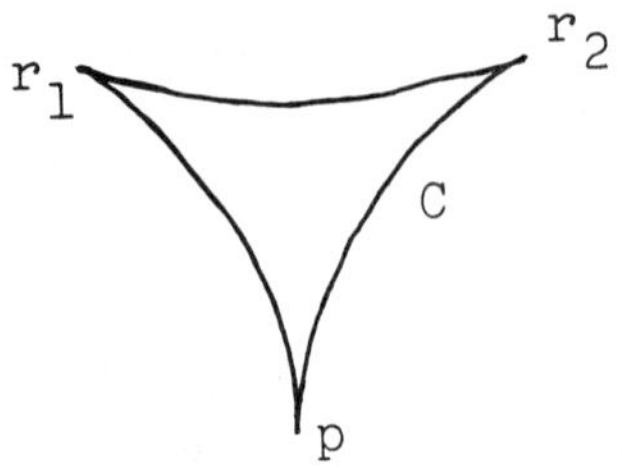

FIGURE 2.27

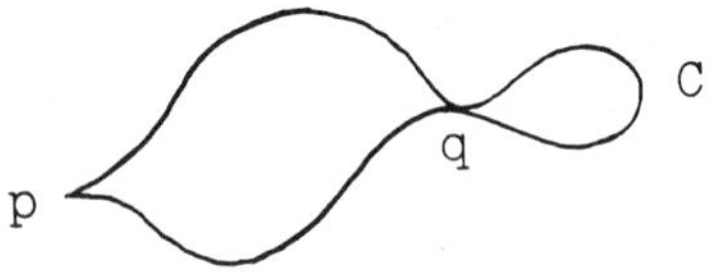

FIGURE 2.28

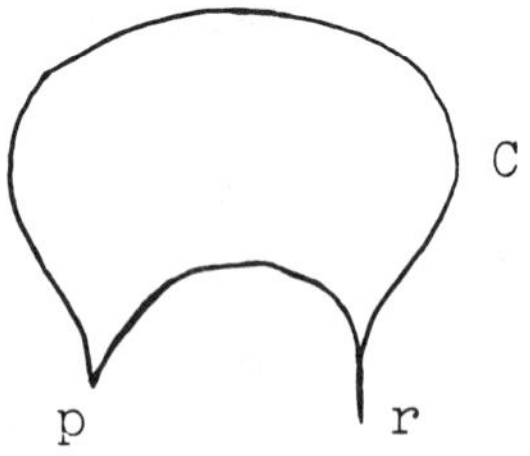

FIGURE 2.29

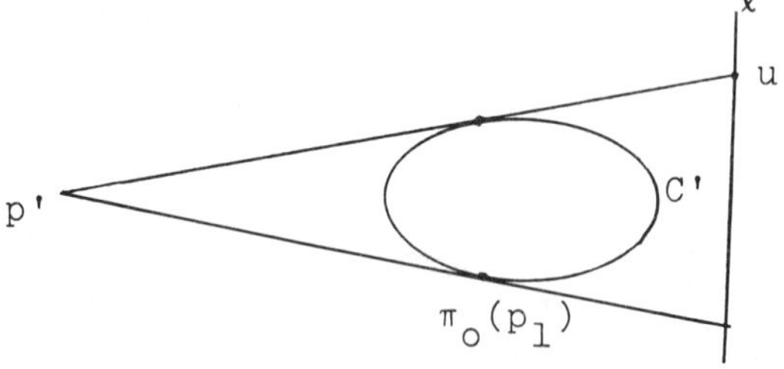

FIGURE 2.30

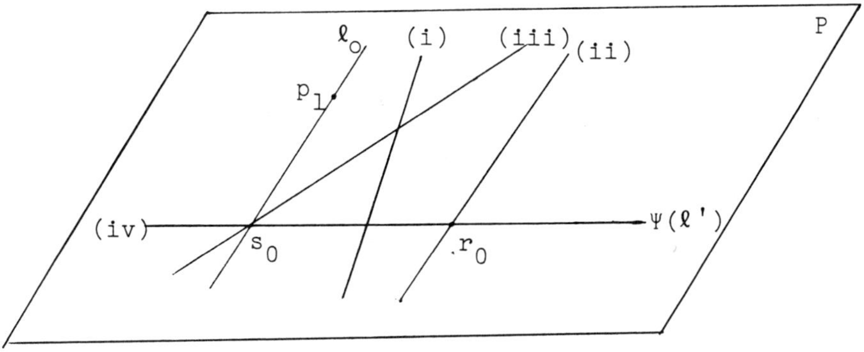

FIGURE 2.31

v. Suppose that $\ell = T_{r_{10}}\Psi(\ell')$. In this case, $C = \pi_\ell(C_0)$ has two singular points p and $r = \pi_\ell(r_{11})$, which is a double cusp. (See Fig. 2.29.)

b. Suppose that $\Psi(\ell')$ is a line. Let μ be the point on ℓ' such that $\overline{p'\mu}$ is another tangent line than $T_{\pi_0(p_1)}C'$ passing through p'. (See Fig. 2.30.) Put $2r_1' = 2\pi_0(r_1) = C' \cap \overline{p'\mu}$ and $r_0 = \Psi(\mu)$. Also put $s_0 = \ell_0 \cap \Psi(\ell')$.

Let ℓ be a line on P. (See Fig. 2.31.)

i. Suppose that ℓ is a general line on P. In this case, $C = \pi_\ell(C_0)$ has two singular points: a double cusp p and a node q. (See Fig. 2.32.)

ii. Suppose that ℓ passes through r_0 and $\ell \neq \Psi(\ell')$. In this case, $C = \pi_\ell(C_0)$ has two singular points: a double cusp p and a simple cusp $r = \pi_\ell(r_1)$ of multiplicity 2. (See Fig. 2.33.)

iii. Suppose that ℓ passes through s_0 and $\ell \neq \Psi(\ell')$. In this case, $C = \pi_\ell(C_0)$ has a unique singular point p which is a ramphoid cusp. (See Fig. 2.34.)

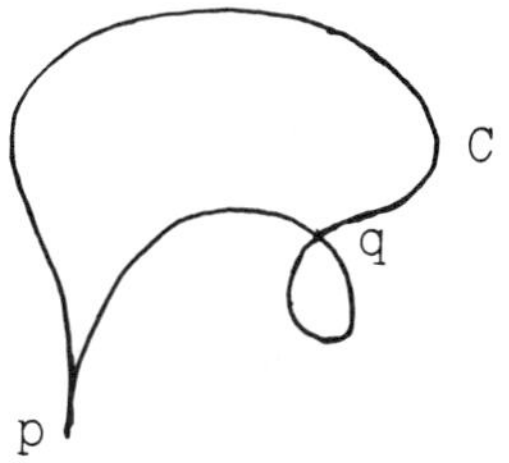

FIGURE 2.32

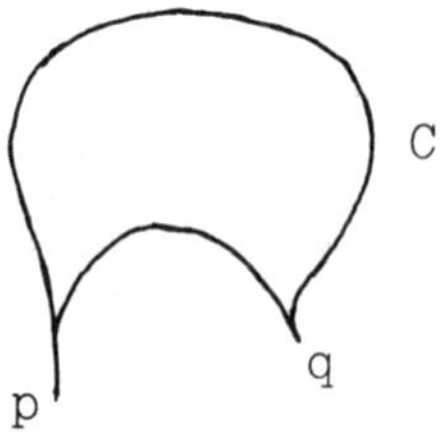

FIGURE 2.33

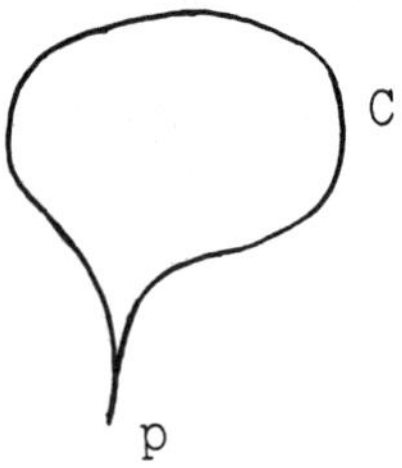

FIGURE 2.34

3 nodes   a node and a tacnode   an osnode

FIGURE 2.35

(1)

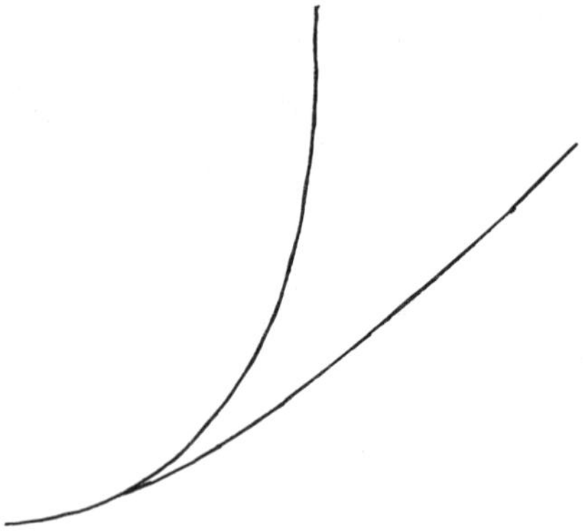

FIGURE 2.36

(2)

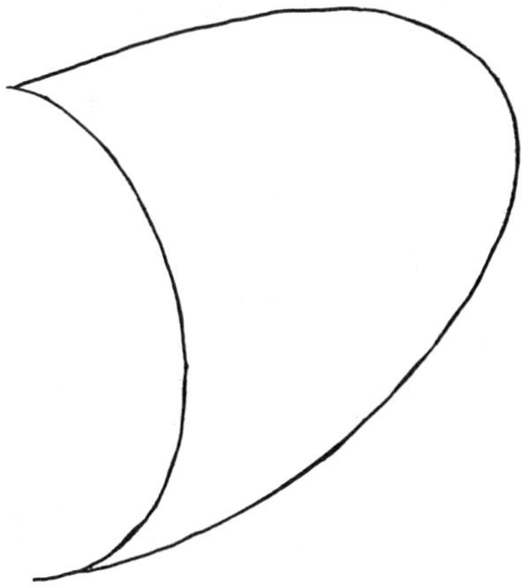

FIGURE 2.37

iv. Suppose that $\ell = \Psi(\ell')$. In this case, π_ℓ: $C_0 \rightarrow \mathbb{P}^2$ is a double covering onto an irreducible conic.

So far, we have assumed that $p_1 = p_2$. A similar argument can be applied for the case $p_1 \neq p_2$ and we can obtain various curves $C = \pi_\ell(C_0)$. They have already appeared above, except the three types shown in Fig. 2.35.

It is a little complicated to classify rational plane quartic curves up to projective equivalence. We only state the following theorem.

(3)

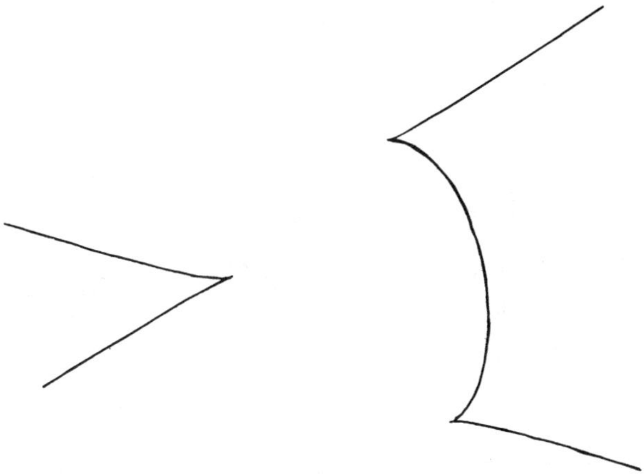

FIGURE 2.38 (Drawings courtesy of Dr. Hut)

Theorem 2.2.5. Let C be a rational plane quartic curve.

1. If C has a ramphoid cusp, then C is projectively equivalent to

$$(y - x^2)^2 = xy^3$$

(the image of $t \in \mathbb{P}^1 \to (1 + t^3: t^2: t^4) \in \mathbb{P}^2$).

2. If C has a double cusp and a simple cusp of multiplicity 2, then C is projectively equivalent to

$$(y - x^2) = x^3y$$

(the image of $t \in \mathbb{P}^1 \to (1 + t: t^2: t^4) \in \mathbb{P}^2$).

3. If C has three simple cusps of multiplicity 2, then C is projectively equivalent to

$$(2y + x^2)^2 = 4x^2(x - 2)(x + y)$$

(the image of $t \in \mathbb{P}^1 \to (t - \frac{1}{2}: t^2: t^4 - 2t^3) \in \mathbb{P}^2$), which is dual to

$$y^2 = x^3 + x^2$$

(see Example 1.5.14). (See Figs. 2.36, 2.37, and 2.38.)

Finally, let us give a few interesting examples of rational plane quartic curves.

1. C: $(x^2 + y^2)^2 = x^2 - y^2$ (Lemniscate) (See Fig. 2.39.)

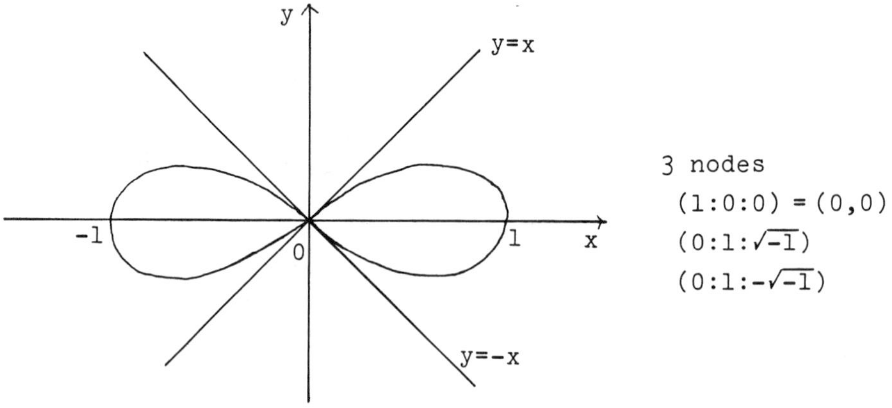

FIGURE 2.39

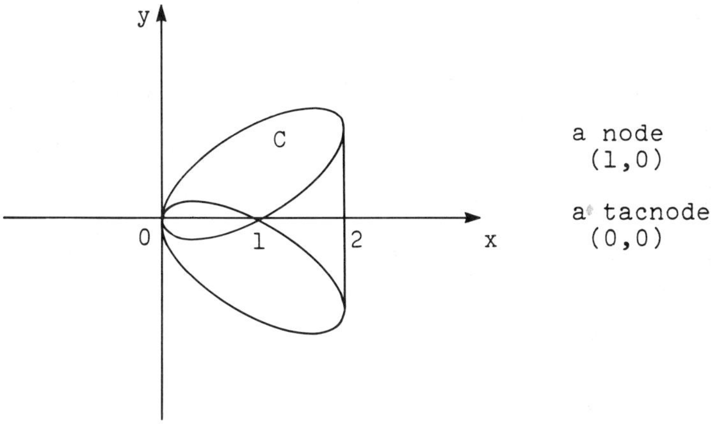

FIGURE 2.40

2. C: $(x^2 + y^2 - 3x)^2 = 4x^2(2 - x)$ (See Fig. 2.40.)

3. C: $(x^2 + y^2 - 2x)^2 = x^2 + y^2$ (Limaçon) (See Fig. 2.41.)

4. C: $(x^2 + y^2 - x)^2 = x^2 + y^2$ (Cardioid) (See Fig. 2.42.)

Notes 2.2.6. The above method is for nothing but to analyze the chordal variety $Ch(C_0)$ of the rational normal curve C_0 (see Veronese [99]).

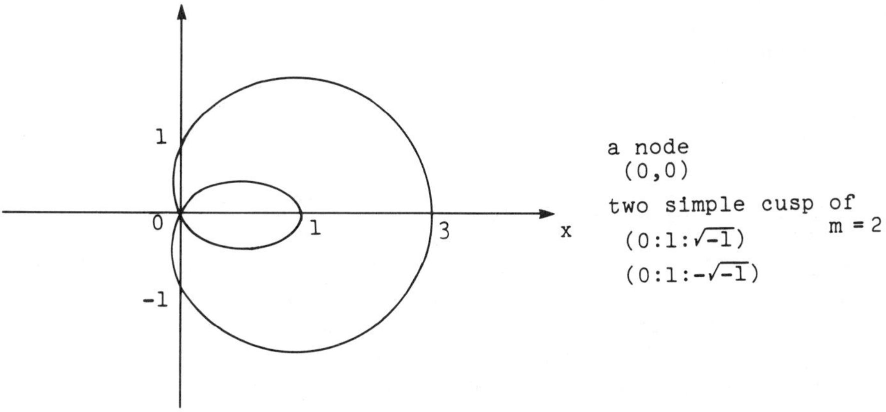

FIGURE 2.41

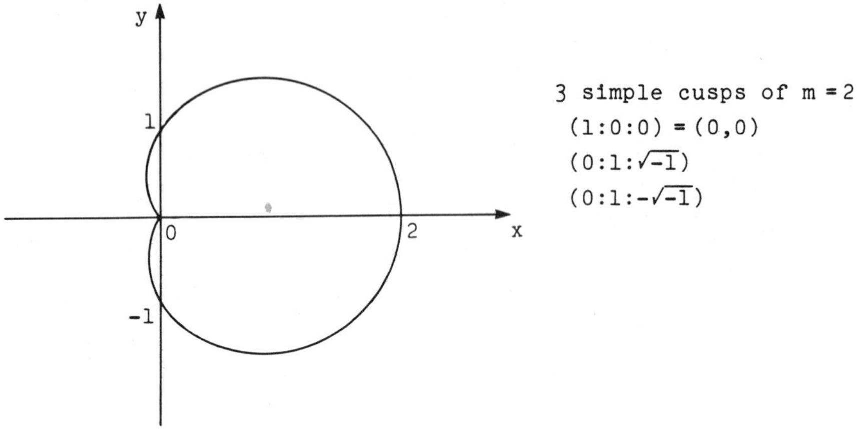

FIGURE 2.42

Exercises

1. Prove Theorem 2.2.5.
2. Classify rational plane quartics up to projective equivalence.

2.3 SINGULAR PLANE QUINTICS

Now, we classify singular irreducible plane quintic curves. Let C be an irreducible plane quintic curve and ϕ: M $\rightarrow$ C be a nonsingular model of C. The line cuts of C define a linear system $\Lambda = g_5^2$ of dimension 2 and of degree 5 on M without fixed point. Conversely, any such a linear system Λ on M gives an irreducible plane quintic curve $C = \Phi_\Lambda(M)$, because 5 is a prime number.

By the Genus Formula, the genus of M is given by

$$g = 6 - \Sigma\, \delta_p$$

where Σ is extended over all singular points $p \in C$. Hence, $g \leq 6$.

1°. g = 6. In this case, C is a nonsingular plane quintic curve. Λ is complete and is a unique g_5^2 on M = C (see Theorem 5.3.17). Hence, for nonsingular plane quintic curves C_1 and C_2, C_1 and C_2 are biholomorphic if and only if they are projectively equivalent (see Corollary 5.3.19).

2°. g = 5. In this case, C has a unique singular point p which is either a node or a simple cusp of multiplicity 2. Put $\phi^{-1}(p) = p_1 + p_2$. The projection

$$\pi_p : M \rightarrow \mathbb{P}^1$$

with the center p gives a meromorphic function on M of degree 3. (See Fig. 2.43.) Hence, M is a trigonal compact Riemann surface (see Definition 5.3.10). In particular, M is nonhyperelliptic.

Conversely, let M be a trigonal compact Riemann surface of genus 5. Then a linear system $\Lambda_0 = g_3^1$ on M of dimension 1 and of degree 3 is complete and uniquely determined (see Theorem 5.3.12). Let $\Lambda = g_5^2$ be a linear system on M of dimension 2 and of degree 5. By the Riemann-Roch Theorem (see Theorem 5.1.5),

$$\dim |D| - \dim |K - D| = 1 \quad \text{for} \quad D \in \Lambda$$

Since $\dim |D| \geq \dim \Lambda = 2$, we have $\dim |K - D| \geq 1$. Hence, D is a special divisor (see Sec. 5.3). By Clifford's Theorem (Theorem 5.3.7)

$$\dim |D| \leq \frac{5}{2}$$

Hence, $\dim |D| = 2$ and $\dim |K - D| = 1$. In particular, $\Lambda = |D|$ is complete. $|K - D|$ is a g_3^1, so $|K - D| = \Lambda_0$ is uniquely determined. Hence

$$\Lambda = |K - D_0| \quad \text{for } D_0 \in \Lambda_0$$

so Λ is uniquely determined. A similar argument shows that $\Lambda = |K - D_0|$ has no fixed point. Thus we get

Proposition 2.3.1. (1) Let M be a trigonal compact Riemann surface of genus 5. Then a linear system $\Lambda = g_5^2$ on M is uniquely determined and has no fixed point; $\Lambda = |K - D_0|$, where $D_0 \in \Lambda_0 = g_3^1$. The singular point of $C = \Phi_\Lambda(M)$ is $p = \Phi_\Lambda(p_1) = \Phi_\Lambda(p_2)$, where the divisor $p_1 + p_2$ satisfies $2D_0 + p_1 + p_2 \sim K$. (2) Two irreducible plan quintic curves of genus 5 are birational if and only if they are projectively equivalent.

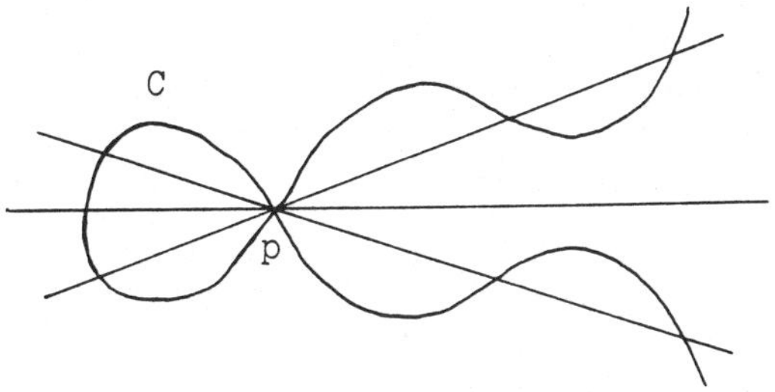

FIGURE 2.43

Remark 2.3.2. Using Λ_0 and a linear pencil Λ_t in Λ, a quadric hypersurface $Q_t = Q(\Lambda_0, \Lambda_t)$ of rank ≤ 4 in $\mathbb{P}^4$ containing the canonical curve C_K of M can be constructed (see Lemma 5.3.22). Since dim $\Lambda = 2$, the linear pencils in Λ form a 2-parameter family $\{\Lambda_t\}_{t\in\mathbb{P}^2}$, so the quadric hypersurfaces Q_t form also a 2-parameter family $\{Q_t\}_{t\in\mathbb{P}^2}$. The intersection $\bigcap Q_t$ is not C_K itself but a rational ruled surface F_1 imbedded in $\mathbb{P}^4$ (see Saint-Donat [84] and Andreotti-Mayer [5]). In this family $\{Q_t\}$, there is a unique Q_0 of rank 3. This is in fact $Q(\Lambda_0, \Lambda_0 + p_1 + p_2)$. The vertex line of Q_0 cuts C_K at just two points $p_1 + p_2$.

Example 2.3.3. (1) C: $x^5 + xy + y^5 = 0$. ($p = (0,0)$ is a node.) (See Fig. 2.44.) (2) C: $x^5 + x^3 + y^5 + y^2 = 0$. ($p = (0,0)$ is a simple cusp of multiplicity 2.) (See Fig. 2.45.)

3°. $g = 4$. In this case, $\Sigma\delta_p = 2$, so C cannot have a triple point. Let p be a double point and put $\phi^{-1}(p) = p_1 + p_2$. The projection

$$\pi_p : M \to \mathbb{P}^1$$

with the center p gives a meromorphic function on M of degree 3. In particular, M is nonhyperelliptic.

Conversely, let M be a nonhyperelliptic compact Riemann surface of genus 4 and $\Lambda = g_5^2$ be a linear system on M of dimension 2 and of degree 5. By the Riemann-Roch Theorem,

$$\dim |D| - \dim |K - D| = 2 \quad \text{for} \quad D \in \Lambda$$

Since dim $|D| \geq \dim \Lambda = 2$, we get dim $|K - D| \geq 0$. Hence, D is a special

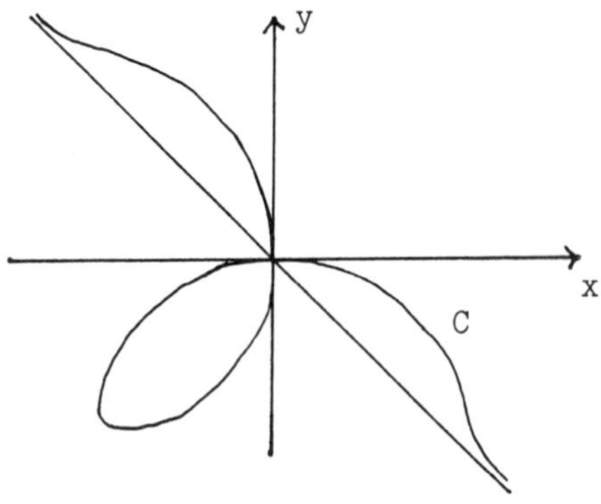

FIGURE 2.44

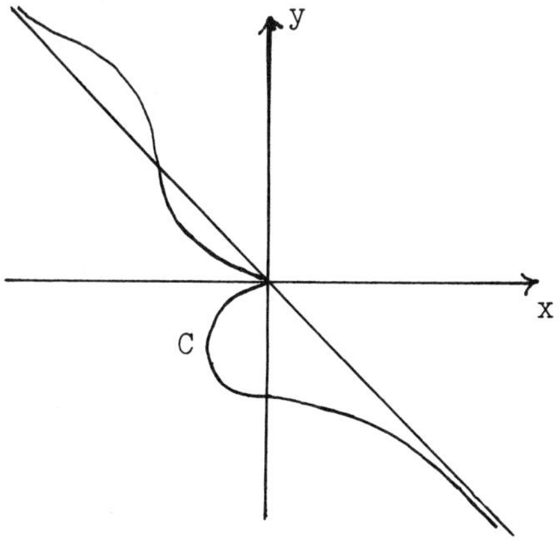

FIGURE 2.45

divisor. By Clifford's Theorem

$$\dim |D| \leq \frac{5}{2}$$

Hence, $\dim |D| = 2$ and $\dim |K - D| = 0$. In particular, $\Lambda = |D|$ is complete. $\dim |K - D| = 0$ means that there is a unique point $p_0 \in M$ such that $D \sim K - p_0$. Hence, we have

$$\Lambda = |K - p_0|$$

This implies that Λ has no fixed point and

$$\Phi_\Lambda = \pi_{p_0} \circ \Phi_K$$

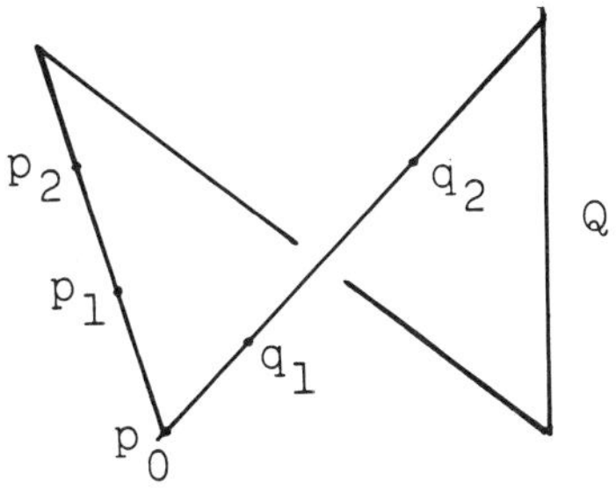

FIGURE 2.46

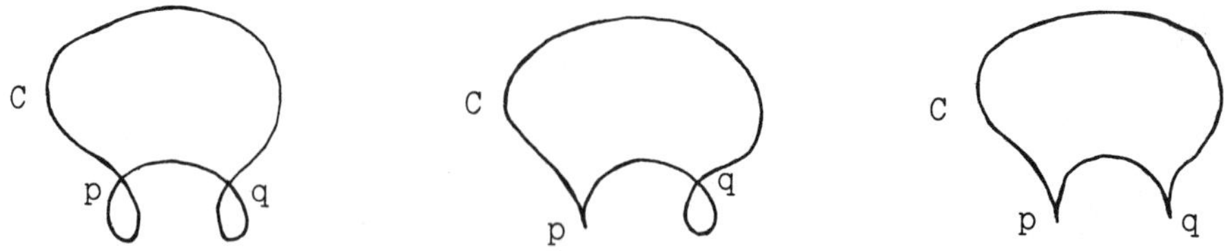

FIGURE 2.47

where π_{p_0} is the projection

$$\pi_{p_0} : C_K \to \mathbb{P}^2 \qquad (C_K = \Phi_K(M) \subset \mathbb{P}^3)$$

with the center $\Phi(p_0)$. Thus, $C = \Phi_\Lambda(M)$ can be obtained as the image of the projection π_{p_0} with the center $p_0 \in C_K$. (We identify M with C_K through Φ_K.)

The canonical curve C_K ($\subset \mathbb{P}^3$) is a complete intersection of a quadric surface Q and a cubic surface R: $C_K = Q \cap R$ (see Example 5.1.21). Two cases occur.

I. Q is nonsingular. In this case, there are two lines ℓ and ℓ' ($\ell \neq \ell'$) on Q passing through p_0. Put

$$C_K \cap \ell = p_0 + p_1 + p_2 \quad \text{and} \quad C_K \cap \ell' = p_0 + q_1 + q_2$$

(See Fig. 2.46.) Then $C = \pi_{p_0}(C_K)$ has just two singular points

$$p = \pi_{p_0}(p_1) = \pi_{p_0}(p_2) \quad \text{and} \quad q = \pi_{p_0}(q_1) = \pi_{p_0}(q_2)$$

If $p_1 \neq p_2$ (respectively $p_1 = p_2$), then p is a node (respectively a simple

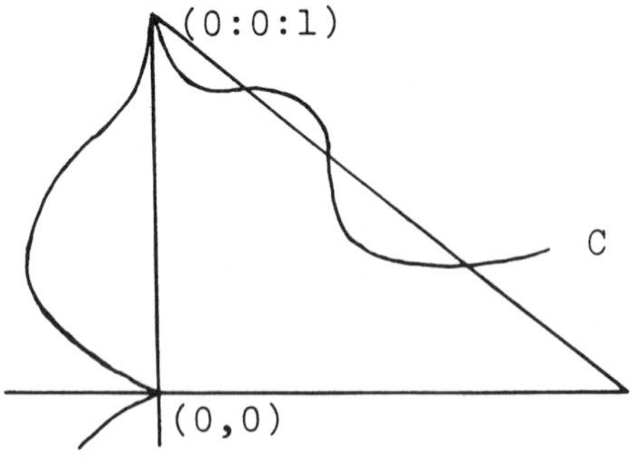

FIGURE 2.48

cusp of multiplicity 2). Hence, three types of C may be possible. (See Fig. 2.47.)

The first two types of C are actually possible for any M, that is, there are points $p_0 \in M$ such that $C = \pi_{p_0}(M)$ are the first two types.

But the last type of C is not possible for some M.

Example 2.3.4. (1) If M is a nonsingular model of the plane curve $C_0: y^3(x^3 + 1) = x^3 - 1$, then the last type of C is not possible for M (Exercise 1). (2) The plane quintic curve C: $x^5 + x^3 + x^2y^3 + y^2 = 0$ has genus 4 and just two simple cusps (0,0) and (0: 0: 1) of multiplicity 2. (See Fig. 2.48.)

II. Q is a cone. In this case, there is a unique line ℓ on Q passing through p_0. Put $C_K \cap \ell = p_0 + p_1 + p_2$. (See Fig. 2.49.) Then $C = \pi_{p_0}(C_K)$ has a unique singular point

$$p = \pi_{p_0}(p_1) = \pi_{p_0}(p_2)$$

If $p_1 \neq p_2$ (respectively $p_1 = p_2$), then p is a tacnode (respectively a double cusp). (See Fig. 2.50.)

More precisely,

1. if p_0, p_1, and p_2 are mutually distinct, then p is a tacnode with $\lambda_p = (2,2)$ (see Sec. 1.5),
2. if $p_0 = p_1 \neq p_2$, then p is a tacnode with $\lambda_p = (3,2)$,
3. if $p_0 \neq p_1 = p_2$, then p is a double cusp with $\lambda_p = I_p(C, T_pC) = 4$,
4. if $p_0 = p_1 = p_2$, then p is a double cusp with $\lambda_p = 5$

The first three cases of C occur for any $C_K = Q \cap R$ with a cone Q, while the last case of C occurs for a special C_K. In fact, if Q is fixed and R is taken general, then Case 4 cannot occur.

The curve

$$C: x^5 - y^5 - y^2 = 0$$

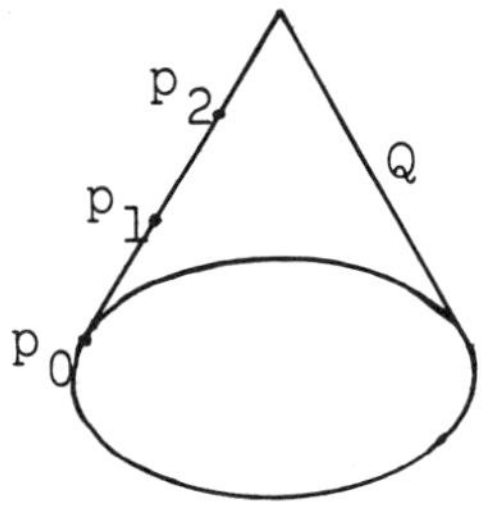

FIGURE 2.49

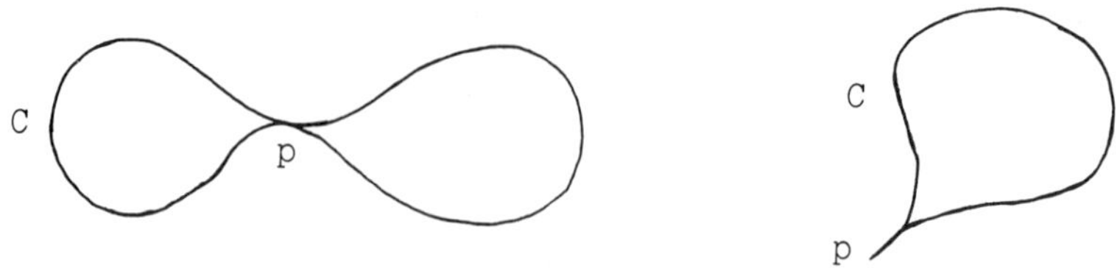

FIGURE 2.50

is an example of irreducible plane quintic curve of genus 4 with a double cusp $p = (0,0)$ with $\lambda_p = 5$.

4°. $g = 3$. Let $\Lambda = g_5^2$ be as before the linear system on M of all line cuts of C. Then, by the Riemann-Roch Theorem,

$$\dim |D| = 2 \quad \text{for } D \in \Lambda$$

($\deg D = 5 > 4 = 2g - 2$.) Hence, $\Lambda = |D|$ is complete.

Conversely, let D be a positive divisor on M of degree 5. Then $\dim |D| = 2$, so $|D|$ is a g_5^2. $|D|$ has a fixed point $p_0 \in M$ if and only if $\dim |D - p_0| = 2$. Since $\deg(D - p_0) = 4$, this means that $|D - p_0| = |K|$, that is, $|D| = |K + p_0|$. Conversely, $|K + p_0|$ is a g_5^2 with a fixed point p_0.

Thus, the curve C is the image of $\Phi_{|D|}$, where $|D|$ is a g_5^2 on M such that

$$\dim |D - K| = -1 \tag{*}$$

Now, $\Sigma\delta_p = 3$ implies that two cases occur.

I. C has a triple point. In this case, p is a unique singular point on C. The projection

$$\pi_p : M \to \mathbb{P}^1$$

with the center p is a meromorphic function of degree 2. Hence, M is hyper-

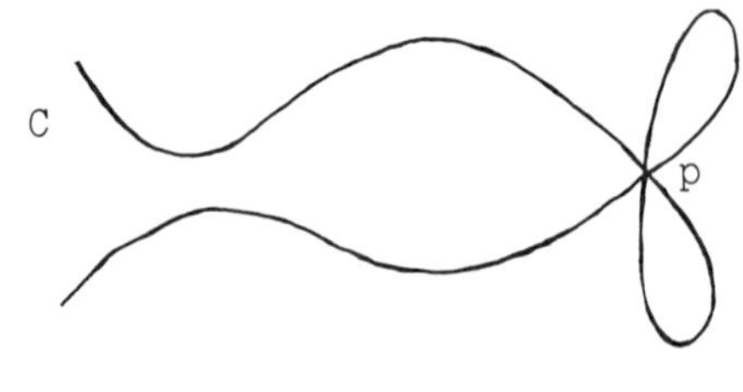

FIGURE 2.51

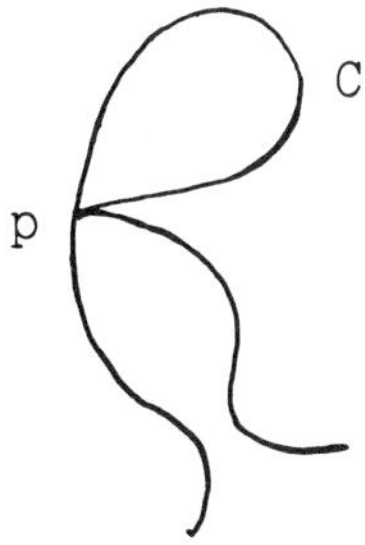

FIGURE 2.52

elliptic. π_p defines a linear system $|D_0| = g_2^1$ on M. Note that $|D_0|$ is a unique g_2^1 on M. Put $\phi^{-1}(p) = p_1 + p_2 + p_3$. Then,

$$\Lambda = |D_0 + p_1 + p_2 + p_3|$$

Note that $|2D_0| = |K|$. Hence, by (*), we have

$$p_1 + p_2 \notin |D_0|, \quad p_2 + p_3 \notin |D_0|, \text{ and } p_3 + p_1 \notin |D_0| \qquad (**)$$

Conversely, let M be hyperelliptic and $p_1 + p_2 + p_3$ be any positive divisor M which satisfies (**). Then $\Lambda = |D_0 + p_1 + p_2 + p_3|$ is a g_5^2 such that $C = \Phi_\Lambda(M)$ is an irreducible plane quintic curve with a triple point $p = \Phi_\Lambda(p_1) = \Phi_\Lambda(p_2) = \Phi_\Lambda(p_3)$. p is either

1. an ordinary triple point ($p_1 \neq p_2 \neq p_3 \neq p_1$) (see Fig. 2.51)
2. or the intersection point of a smooth (that is, nonsingular) branch C_1 and a branch C_2 having p as a simple cusp of multiplicity 2 meeting transversally ($p_1 \neq p_2 = p_3$) (see Fig. 2.52)
3. or a simple cusp of multiplicity 3 ($p_1 = p_2 = p_3$) (see Fig. 2.53).

(See Proposition 2.1.5.)

FIGURE 2.53

II. C has no triple point. Take a double point $p \in C$ and put $\phi^{-1}(p) = p_1 + p_2$. The projection

$$\pi_p : M \to \mathbb{P}^1$$

with the center p is a meromorphic function on M of degree 3. Hence, M is nonhyperelliptic. Its canonical curve C_K is a nonsingular plane quartic curve. We identify M with C_K through Φ_K.

Then there is a unique point $p_0 \in C_K$ such that $\pi_p = \pi_{p_0}$, where

$$\pi_{p_0} : C_K \to \mathbb{P}^1$$

is the projection of the canonical curve C_K with the center p_0 (see 1 of Theorem 5.3.17). Hence,

$$D - (p_1 + p_2) \sim K - p_0 \qquad \text{for } D \in \Lambda$$

Hence,

$$\Lambda = |D| = |K - p_0 + p_1 + p_2|$$

By (*) above, $p_0 \neq p_1$ and $p_0 \neq p_2$.

Conversely, if p_0, p_1, and p_2 are points on $M = C_K$ such that $p_0 \neq p_1$ and $p_0 \neq p_2$, then $\Lambda = |K - p_0 + p_1 + p_2|$ is a g_5^2 without fixed point such that $C = \Phi_\Lambda(M)$ has a double point $p = \Phi_\Lambda(p_1) = \Phi_\Lambda(p_2)$.

Suppose that C has another double point q. Put $\phi^{-1}(q) = q_1 + q_2$. Then, by the same reason as above, there is a unique point q_0 such that

$$\Lambda = |D| = |K - q_0 + q_1 + q_2|$$

Hence, $-p_0 + p_1 + p_2 \sim -q_0 + q_1 + q_2$, so

$$q_0 + p_1 + p_2 \sim p_0 + q_1 + q_2$$

This means that there is $r_0 \in C_K$ such that

$$q_0 + p_1 + p_2 + r_0 \sim p_0 + q_1 + q_2 + r_0 \sim K$$

Hence, q_0, p_1, p_2, and r_0 (respectively p_0, q_1, q_2, and r_0) are collinear on C_K.

Suppose that C has another double point r. Put $\phi^{-1}(r) = r_1 + r_2$. Then, by the same reason as above, there is a unique point r_0' such that

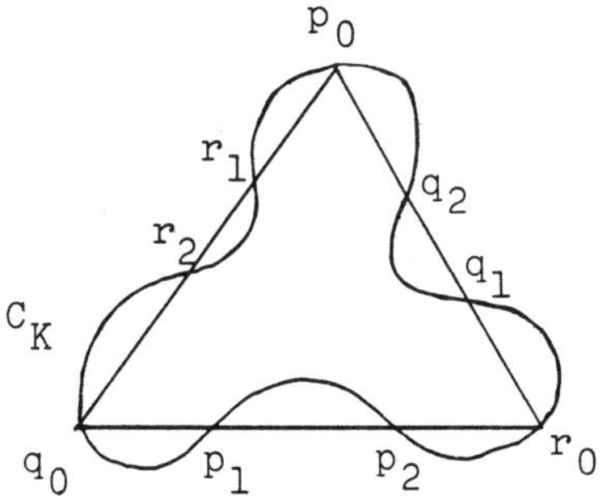

FIGURE 2.54

$$r_0' + p_1 + p_2 \sim p_0 + r_1 + r_2$$

and

$$r_0' + q_1 + q_2 \sim q_0 + r_1 + r_2$$

Hence, we get $r_0' = r_0$ and the configuration around C_K in Fig. 2.54.

Note that

$$\Lambda = |D| = |2K - p_0 - q_0 - r_0|$$

Hence, p_0, q_0, and r_0 are not collinear on C_K, for Λ has no fixed point. If $p_0 = q_0 = r_0$, then this means that p_0 is not a flex of C_K. In other words, C can be obtained as the image of the projection

$$\pi_P\colon C_{2K} \to \mathbb{P}^2$$

where $C_{2K} = \Phi_{2K}(M)$ ($\subset \mathbb{P}^5$) and P is the 2-plane spanned by $\Phi_{2K}(p_0)$, $\Phi_{2K}(q_0)$, and $\Phi_{2K}(r_0)$.

a. Suppose that p_0, q_0, and r_0 are mutually distinct. In this case, there are just three singular points p, q, and r on C, which are either nodes ($p_1 \neq p_2$, say) or simple cusps of multiplicity 2 ($p_1 = p_2$, say). Hence, we get the correspondences in Figs. 2.55-2.58. (Here, "$\Longleftrightarrow$" means that, from the configuration around C_K of the left-hand side, the plane quintic curve C of the right-hand side can be obtained, and vice versa.)

It is clear that these configurations, except the last one, can be drawn around any C_K. Hence, every type of C above, except the last one, is possible for any M.

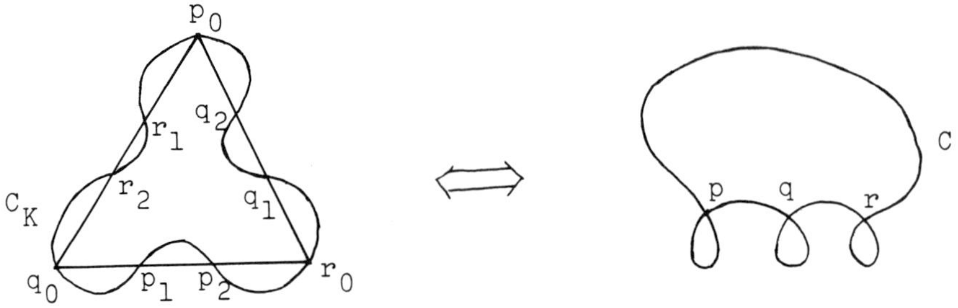

FIGURE 2.55

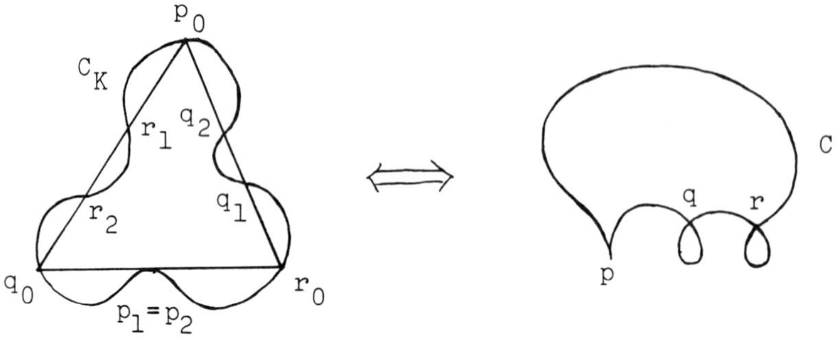

FIGURE 2.56

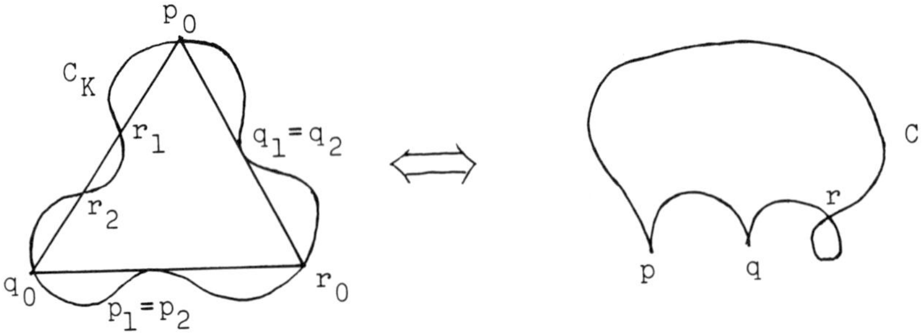

FIGURE 2.57

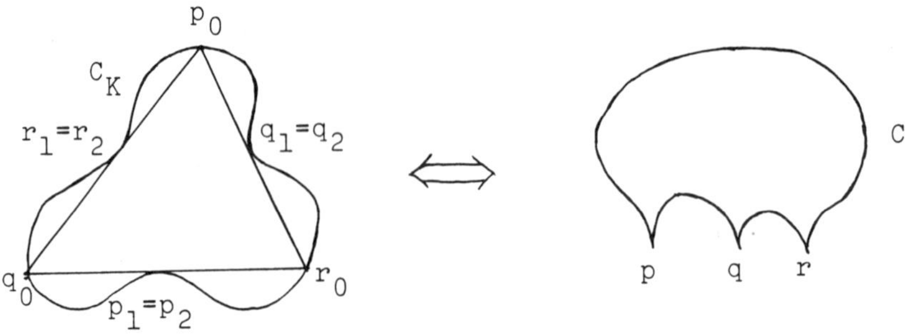

FIGURE 2.58

As for the last configuration, it is possible to construct a nonsingular plane quintic C_K with the last configuration as follows.

Example 2.3.5. Consider a linear system $\{C_\xi\}_{\xi \in \mathbb{P}^2}$ of plane quartic curves

$$C_\xi: aX_0X_1(X_0 + X_1)^2 + bX_1X_2(X_1 + X_2)^2 + cX_2X_0(X_2 + X_0)^2 = 0$$

where $\xi = (a: b: c) \in \mathbb{P}^2$. $\{C_\xi\}$ has the base points

$$p_0 = (1: 0: 0), \quad q_0 = (0: 1: 0), \quad r_0 = (0: 0: 1)$$

$$p_1 = (0: 1: -1), \quad q_1 = (1: 0: -1), \quad r_1 = (1: -1: 0)$$

If a, b, and c are mutually distinct and $abc \neq 0$, then these six points are nonsingular points of C_ξ ($\xi = (a: b: c)$). Hence, by Bertini's Theorem (Theorem 3.4.14), C_ξ is nonsingular for a general $\xi \in \mathbb{P}^2$. Clearly, such a C_ξ has the last configuration. (See Fig. 2.59.)

Remark 2.3.6. Recently, Prof. Mukai informed us that, using the theory of correspondence of algebraic curves (see Griffiths-Harris [33]), the last configuration always exists around any C_K. In fact, he showed that 288 such configurations exist.

b. Suppose that $p_0 \neq q_0 = r_0$. This case occurs if and only if the line $\overline{p_1p_2}$ is tangent to C_K at $q_0 = r_0$. (See Fig. 2.60.) In this case, there are just two singular points p and q on C corresponding to C_K. Here, p is either a node ($p_1 \neq p_2$) or a simple cusp of multiplicity 2 ($p_1 = p_2$), while q is either a tacnode ($q_1 \neq q_2$) or a double cusp ($q_1 = q_2$). Every combination is possible for any M. (See Figs. 2.61-2.63.)

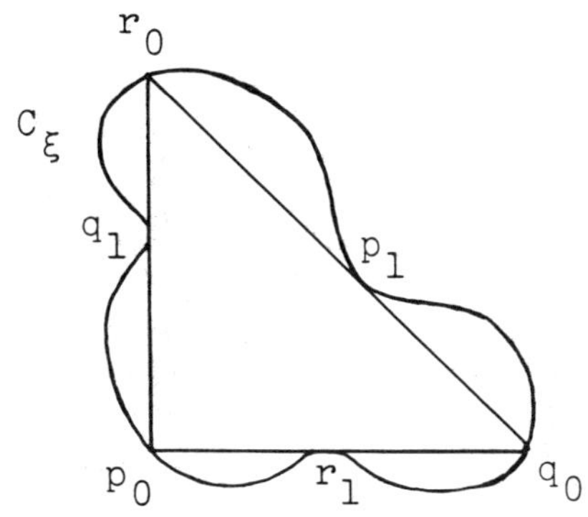

FIGURE 2.59

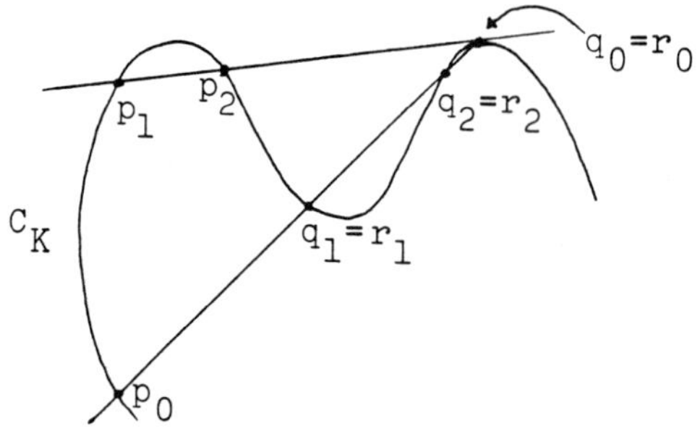

FIGURE 2.60

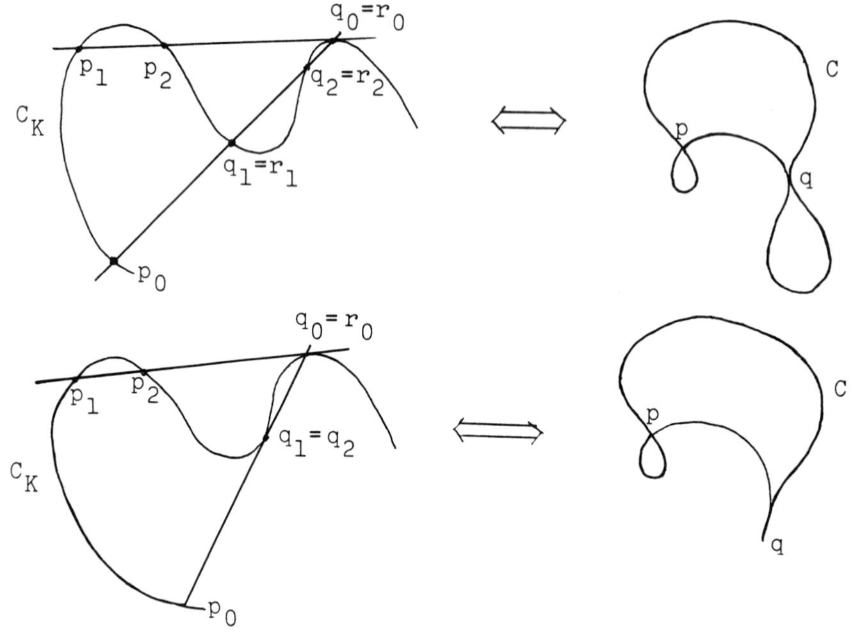

FIGURE 2.61

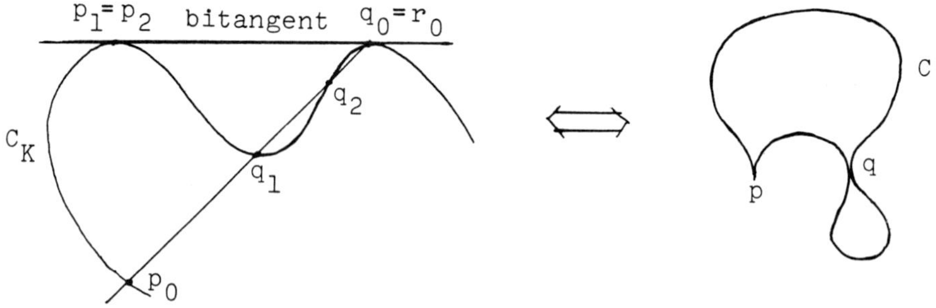

FIGURE 2.62

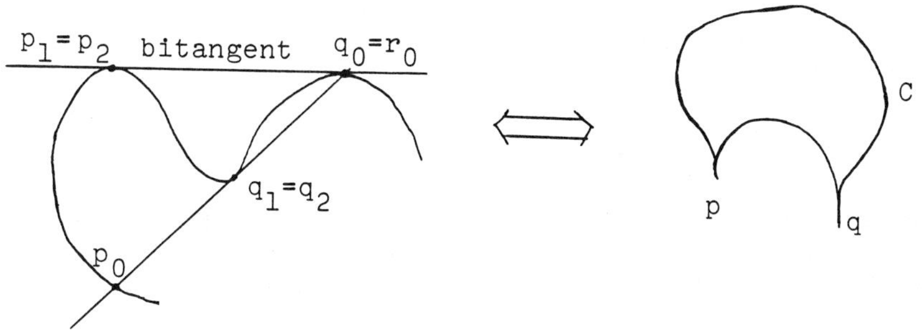

FIGURE 2.63

In the last configuration, if $p_0 \neq q_1 = q_2$, then $\lambda_q = I_q(C, T_qC) = 4$. If $p_0 = q_1 = q_2$ (a flex), then $\lambda_q = 5$.

c. Suppose that $p_0 = q_0 = r_0$. In this case, p is a unique singular point of C corresponding to C_K. Here, p is either an osnode ($p_1 \neq p_2$) or a ramphoid cusp ($p_1 = p_2$). Both are possible for any M. (See Fig. 2.64.)

5°. g = 2. We omit this case, because the method and the results in this case are more or less similar to those in the next case g = 1. (See also Namba [73].)

6°. g = 1. Let $M = \mathbb{C}/(\mathbb{Z}\omega_1 + \mathbb{Z}\omega_2)$, $\omega_2/\omega_1 \in \mathbb{H}$, be a complex one-torus. Let C_0 be the image curve in $\mathbb{P}^4$ of the holomorphic imbedding

$$\Phi = \Phi_{|5(0)|} : z \in M \to (1: \wp(z): \wp'(z): \wp(z)^2: \wp(z)\,\wp'(z) \in \mathbb{P}^4$$

(0 is the zero of the additive group M.) We identify M with C_0 through Φ. Then, as in the case of plane elliptic quartic curves (see Sec. 2.2), we may

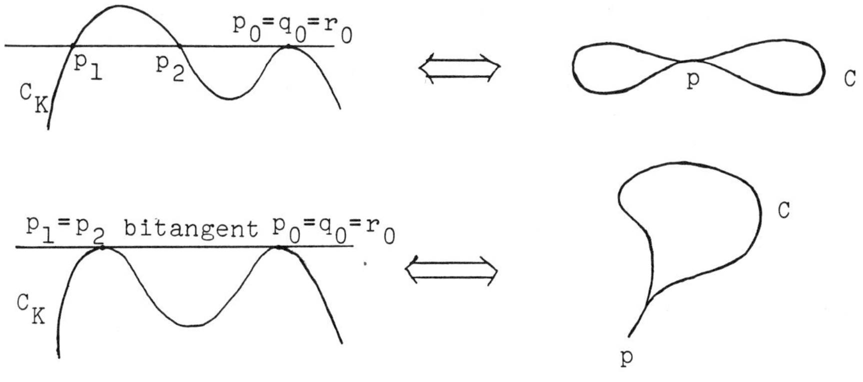

FIGURE 2.64

assume that a given plane elliptic quintic curve C can be obtained as the image of the projection

$$\pi_\ell: C_0 \to \mathbb{P}^2$$

with the center a line ℓ in $\mathbb{P}^4$ such that $\ell \cap C_0 = \phi$.

$C = \pi_\ell(C_0)$ and $C' = \pi_{\ell'}(C_0)$ are projectively equivalent if and only if ℓ and ℓ' are in the same orbit of the finite subgroup B of Aut (M) acting on $\mathbb{P}^4$ (see Sec. 5.3).

Now C cannot have a singular point of multiplicity 4 by Proposition 2.1.5, so two cases occur.

I. C has a triple point. This case can be treated in a similar way to the case of plane rational quartic curves with only double points (see Sec. 2.2). We leave this case to the reader as an exercise (Exercise 2).

II. C has no triple point. Take a double point $p \in C$ and put $\pi_\ell^{-1}(p) = p_1 + p_2$. There are two cases

$$p_1 = p_2 \quad \text{and} \quad p_1 \neq p_2$$

Henceforth, we treat the case $p_1 = p_2$. The case $p_1 \neq p_2$ can be treated in a similar way.

Put $\ell_0 = T_{p_1} C_0$. Then the lines ℓ_0 and ℓ must span a 2-plane P in $\mathbb{P}^4$ which is not the osculating 2-plane $O_{p_1} C_0$ to C_0 at p_1. (See Fig. 2.65.)

What we do is to choose first a 2-plane P containing ℓ_0 such that $P \neq O_{p_1} C_0$ and then choose a line ℓ on P.

Take a 2-plane P' such that $P' \cap \ell_0 = \phi$ and consider the projection

$$\pi_0 = \pi_{\ell_0}: C_0 \to P'$$

with the center ℓ_0. Then π_0 is a holomorphic imbedding. In fact,

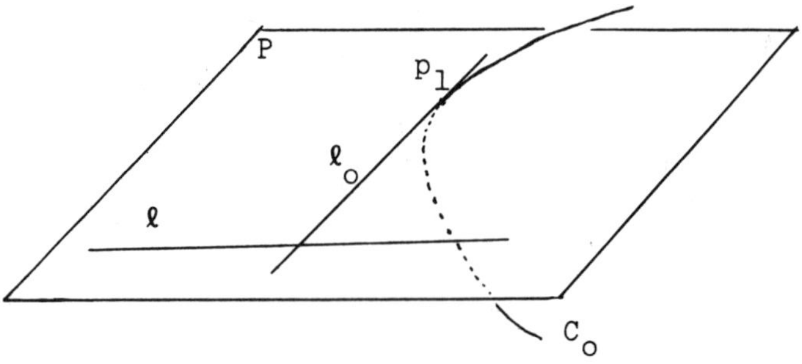

FIGURE 2.65

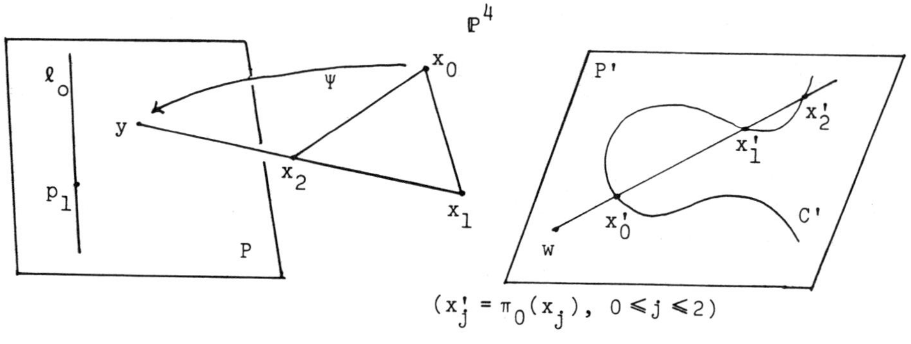

FIGURE 2.66

$$\pi_0 = \Phi_{|5(0)-2(p_1)|} : M \to \mathbb{P}^2$$

The image $C' = \pi_0(C_0)$ is a nonsingular plane cubic curve.

The 2-planes P are in 1-1 correspondence to the points in $P' - C'$. In fact the correspondence is given by

$$P \to P \cap P' = \{w\}, \quad w \in P' - C' \to P = w \vee \ell_0$$

The linear pencil of 3-planes containing P determines a linear pencil $\Lambda = g_3^1$ on C_0. Λ corresponds to the projection

$$\pi_w : C' \to \mathbb{P}^1$$

with the center w. From Λ and $\Lambda_0 = |2(p_1)|$, we can construct a quadric hypersurface $Q = Q(\Lambda, \Lambda_0)$ in $\mathbb{P}^4$ of rank 4 containing C_0 (see Lemma 5.3.22). The vertex point v of Q is on P.

For a point $x_0 \in C_0$, let x_1 and x_2 be the points in C_0 such that $x_0 + x_1 + x_2 \in \Lambda$. Let $y \in P$ be the intersection point of $\overline{x_1 x_2}$ and P. Consider the map

$$\Psi : x_0 \in C_0 \to y \in P$$

Ψ can be extended to a holomorphic map Ψ. (See Fig. 2.66.)

In a similar way to the proof of Lemma 2.2.4, we can prove the following lemma (see Exercise 1).

<u>Lemma 2.3.7.</u> (1) Ψ is a birational map of C_0 onto the image curve $\Psi(C_0)$ in P.

(2) The degree of $\Psi(C_0)$ is either 3 or 4. It is 3 if and only if w is on the tangent line $T_{p_1'}C'$ to C' at $p_1' = \pi_0(p_1)$. Moreover, this occurs if and only if the vertex v of Q is on ℓ_0.
(3) p_1 is on $\Psi(C_0)$ and ℓ_0 is tangent to $\Psi(C_0)$ at p_1.
(4) If deg $\Psi(C_0) = 3$, then $\Psi = \Phi_{|5(0)-2(p_1)|}$, and there is a projective transformation $\sigma: P' \to P$ such that $\sigma(C') = \Psi(C_0)$ and $\sigma(w) = v$.
(5) If deg $\Psi(C_0) = 4$, then $\Psi(C_0)$ is an elliptic quartic curve in P having p_1 as a tacnode or a double cusp.

a. Suppose that deg $\Psi(C_0) = 3$. In this case, p is a cusp with $\delta_p \geq 2$. By 2 of the above lemma, $\overline{wp_1'} = T_{p_1'}C'$, $(p' \quad \pi_0(p_1))$. This line passes through another point $p_0' \in C'$. Then Λ contains $2(p_1) + (p_0)$, where $\pi_0(p_0) = p_0'$.

i. Suppose that $p_0 \neq p_1$. In this case, $\Psi(p_0) = p_1$ and $\Psi(p_1) = \sigma(p_0') \neq p_1$. Moreover, p_1, $\Psi(p_1)$, and v (the vertex of Q) are on the line ℓ_0 and are mutually distinct. (See Fig. 2.67.)

Now we choose a line ℓ on P. ℓ should pass through neither p_1 nor v. (If ℓ passes through v, then $C = \pi_\ell(C_0)$ must have a triple point.)

i-1. Suppose that $\Psi(p_1) \notin \ell$. In this case, p is a double cusp of $C = \pi_\ell(C_0)$ with $\lambda_p = I_p(C, T_pC) = 4$. ℓ cuts $\Psi(C_0)$ at three pointx x, y, and z. The singular points on C other than p correspond to these points x, y, and z. By 4 of the above lemma, Λ can be obtained by the projection

$$\pi_v: \Psi(C_0) \to \mathbb{P}^1$$

with the center v. Hence, putting

$$\Psi(C_0) \cdot \overline{xv} = x + \Psi(q_1) + \Psi(q_2)$$

the singular point corresponding to x is

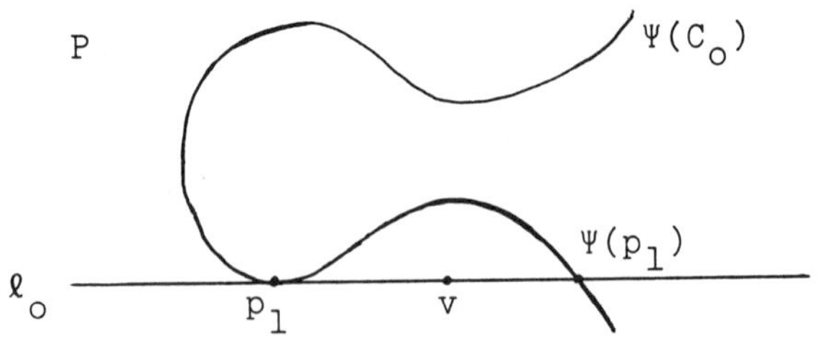

FIGURE 2.67

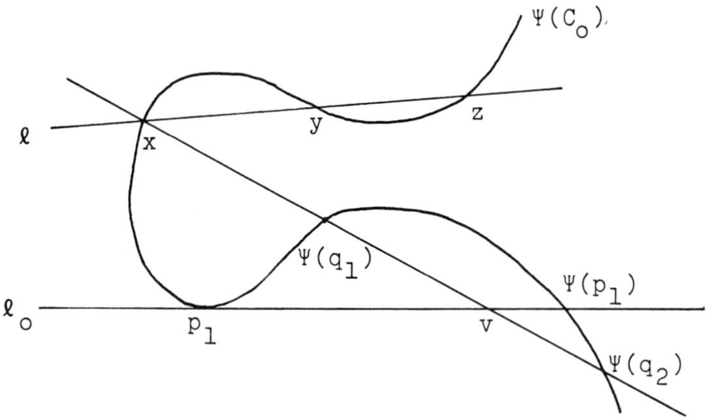

FIGURE 2.68

$$q = \pi_\ell(q_1) = \pi_\ell(q_2)$$

(See Fig. 2.68.)

If x, y, and z are mutually distinct, then q is either a node ($q_1 \neq q_2$) or a simple cusp of multiplicity 2 ($q_1 = q_2$). The latter case occurs if and only if the line $\overline{xv}$ is tangent to $\Psi(C_0)$ at $\Psi(q_1)$.

If $x = y \neq z$, then q is either a tacnode ($q_1 \neq q_2$) or a double cusp ($q_1 = q_2$). In the latter case, $\lambda_q = 4$ in general. (It can be shown that $\lambda_q = 5$ if and only if $\Psi(p_1)$, z, and $\Psi(q_1)$ are collinear.)

If $x = y \neq z$, then q is either an osnode ($q_1 \neq q_2$) or a ramphoid cusp ($q_1 = q_2$).

Note that both $\Psi(C_0)$ and C' are projectively equivalent to the standard nonsingular plane cubic curve

$$E: y^2 = 4x^3 - g_2x - g_3$$

which is the image of

$$\Phi_{|3(0)|}: z \in M \to (1: \wp(z): \wp'(z)) \in \mathbb{P}^2$$

(see Example 5.3.1). Hence, a configuration around $\Psi(C_0)$ can be obtained from that around E by changing the names of the points.

Thus we get the correspondences in Figs. 2.69-2.78. (Here, the symbol "$\Longleftrightarrow$" has the same meaning as in the case g = 3.)

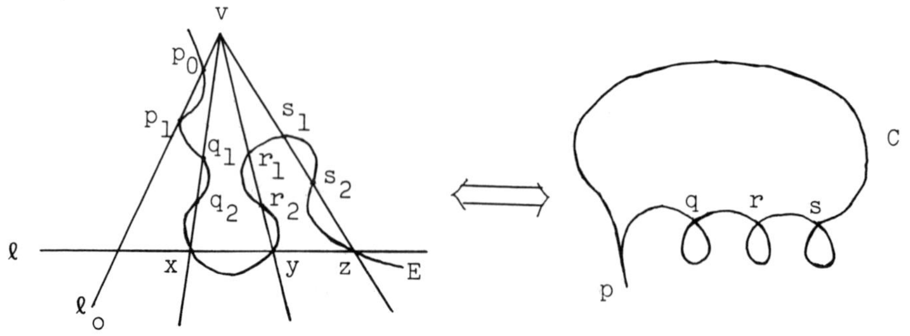

FIGURE 2.69

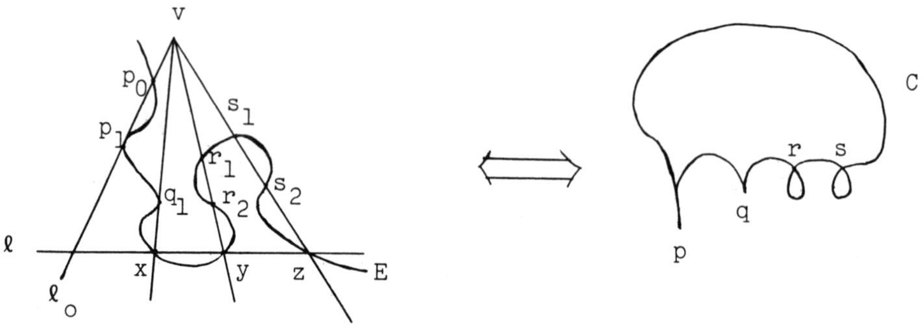

FIGURE 2.70

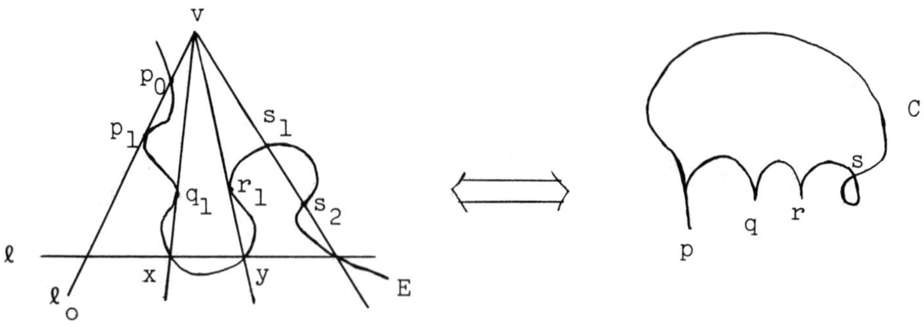

FIGURE 2.71

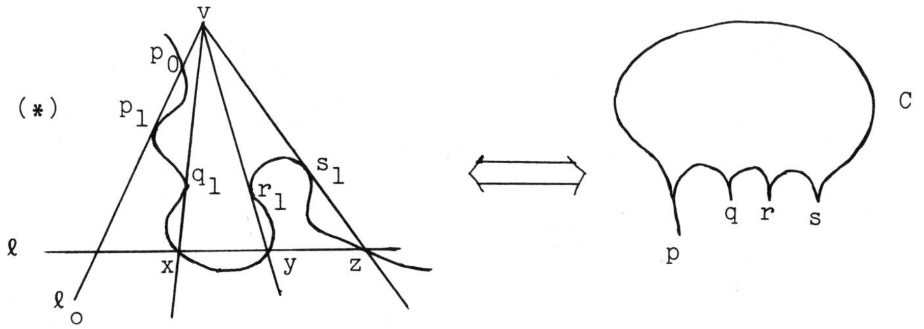

FIGURE 2.72

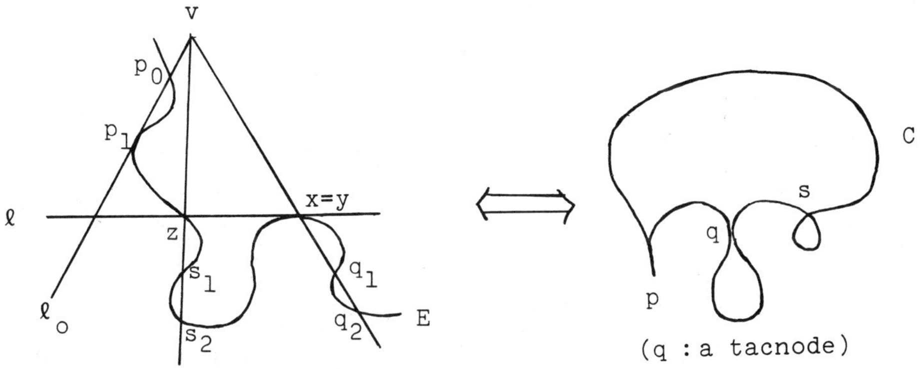

FIGURE 2.73

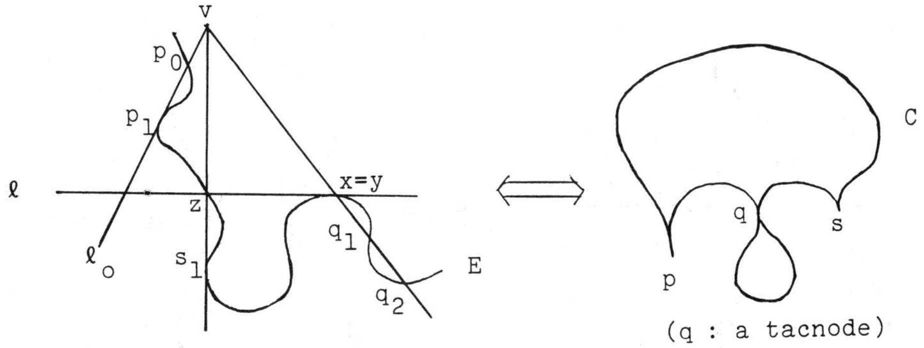

FIGURE 2.74

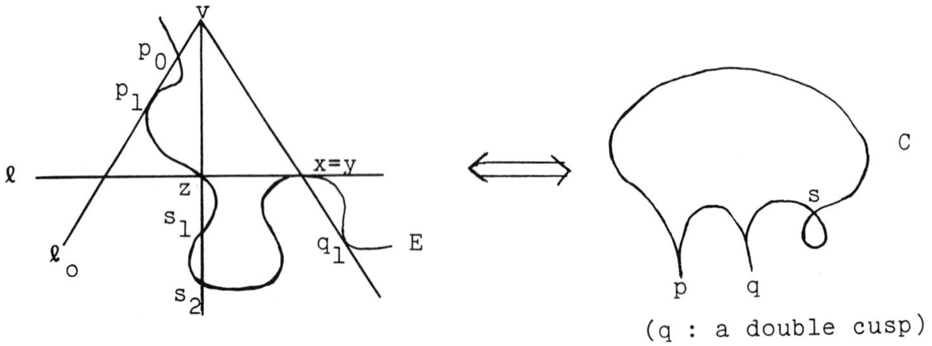

FIGURE 2.75

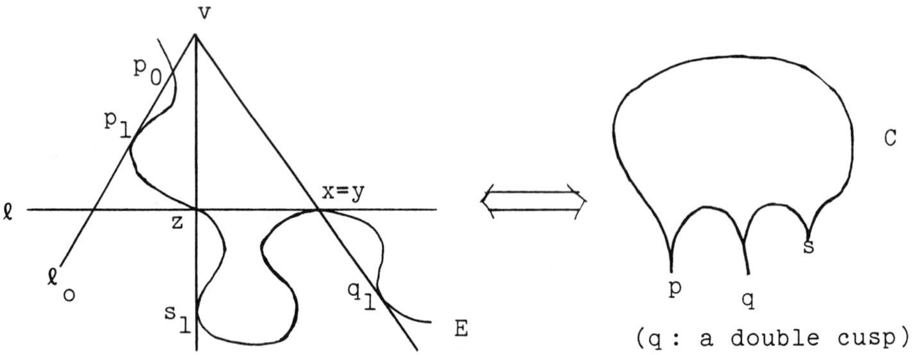

FIGURE 2.76

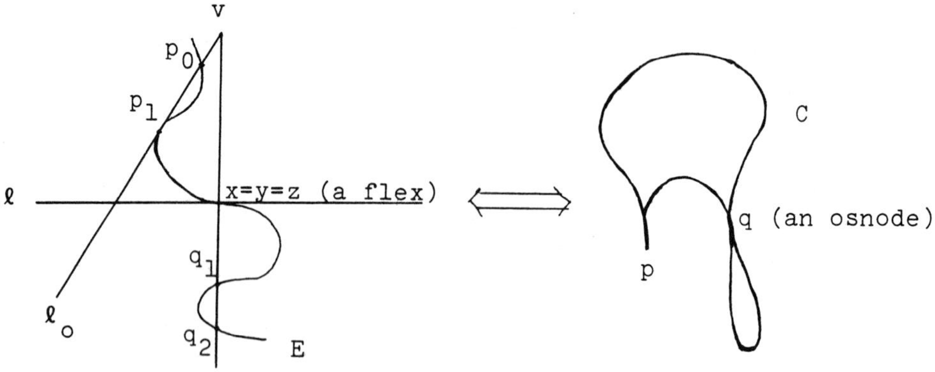

FIGURE 2.77

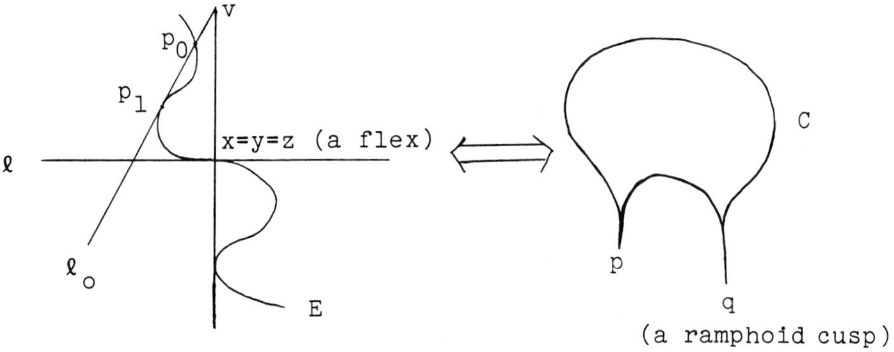

FIGURE 2.78

It is clear that there exist these configurations around any E, except (*). Moreover, it can be shown that the configuration (*) also exists around any E. In fact, for an E with $g_2 \neq 0$, we take $v = (0: 1: 0)$ and the configuration in Fig. 2.79. (If $g_2 \neq 0$, then, by the Riemann-Hurwitz Formula applied for $\pi_v = \mathfrak{p}'$, a point p_1 as in the picture exists.)

If $g_2 = 0$, then E is clearly projectively equivalent to

$$E_0 : y^2 = 4x^3 - 1$$

so we may use E_0 instead of E. Take $v = (0, \lambda)$, where $\lambda \neq 0$, $\lambda \neq \pm\sqrt{-1}$ and $\lambda \neq \pm\sqrt{3}$. Since $\lambda \neq \pm\sqrt{-1}$, v is not on E_0. The projection π_v with the center

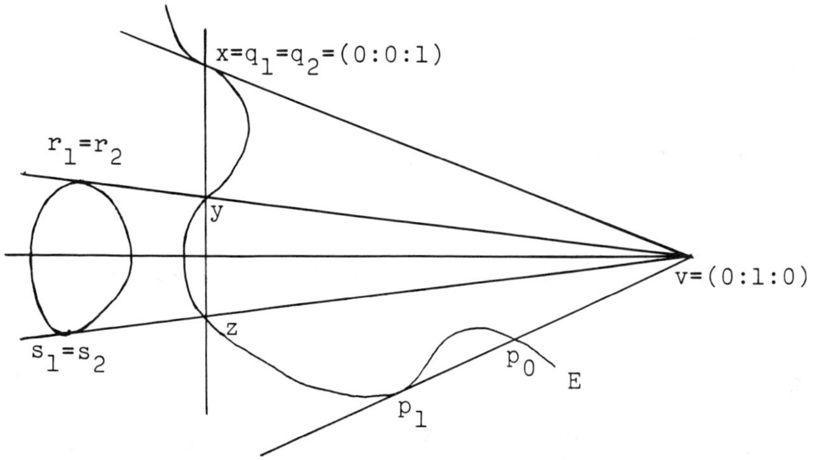

FIGURE 2.79

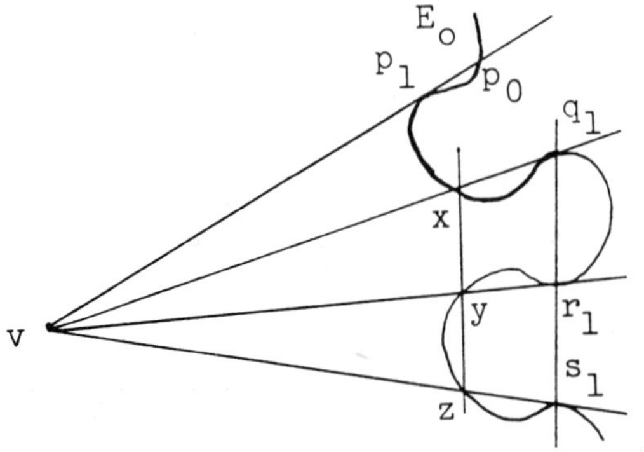

FIGURE 2.80

v is given by

$$\pi_v = \frac{\wp' - \lambda}{\wp}$$

Then

$$(\pi_v)' = \frac{2\wp^3 + 1 + \lambda \wp'}{\wp^2}$$

Hence, the points $(x, y) \in E_0$ such that $T_{(x,y)}E_0$ passes through v are the intersection points of the curves

$$\begin{cases} y^2 = 4x^3 - 1 \\ 2x^3 + 1 + \lambda y = 0 \end{cases}$$

Let α and β be the roots of the equation

$$(2X + 1)^2 = \lambda^2(4X - 1)$$

Since $\lambda \neq 0$ and $\lambda \neq \pm\sqrt{3}$, we have $\alpha = \beta$. The intersection points are then given by

$$q_1 = \left(\sqrt[3]{\alpha}, \frac{-(2\alpha + 1)}{\lambda}\right), \quad r_1 = \left(\sqrt[3]{\alpha\omega}, \frac{-(2\alpha + 1)}{\lambda}\right), \quad s_1 = \left(\sqrt[3]{\alpha\omega^2}, \frac{-(2\alpha + 1)}{\lambda}\right),$$

$$p_1 = \left(\sqrt[3]{\beta}, \frac{-(2\beta + 1)}{\lambda}\right), \quad \left(\sqrt[3]{\beta\omega}, \frac{-(2\beta + 1)}{\lambda}\right), \quad \left(\sqrt[3]{\beta\omega^2}, \frac{-(2\beta + 1)}{\lambda}\right)$$

($\omega = (-1 + \sqrt{-3})/2$). Note that the points q_1, r_1, and s_1 are collinear, that is,

$$q_1 + r_1 + s_1 = 0$$

on the additive group $M = E_0$. Hence,

$$x + y + z = 0$$

where $x = -2q_1$, $y = -2r_1$, and $z = -2s_1$ on the additive group $M = E_0$. (See Fig. 2.80.)

Thus, for any given M, there is a plane quintic curve C of every type above which is birational to M.

i-2. Suppose that $\Psi(p_1) \in \ell$. In this case, p is such that $\delta_p \geq 3$.

If ℓ is not tangent to $\Psi(C_0)$ at $\Psi(p_1)$, then $\delta_p = 3$, so p is a ramphoid cusp.

If ℓ is tangent to $\Psi(C_0)$ at $\Psi(p_1)$ and $\Psi(p_1)$ is not a flex of $\Psi(C_0)$, then p is a cusp with $\delta_p = 4$.

If ℓ is tangent to $\Psi(C_0)$ at a flex $\Psi(p_1)$, then p is a cusp with $\delta_p = 5$. In this case, p is a unique singular point of C.

As in Case i-1, we have the correspondences in Figs. 2.81-2.88.

It is clear that these configurations exist around any E. (As for the configuration (*), take $v = (0: 1: 0)$ and $p_0 = p_1 = (0: 0: 1)$.) Thus, for any given M, there is a plane quintic curve C of every type above which is birational to M.

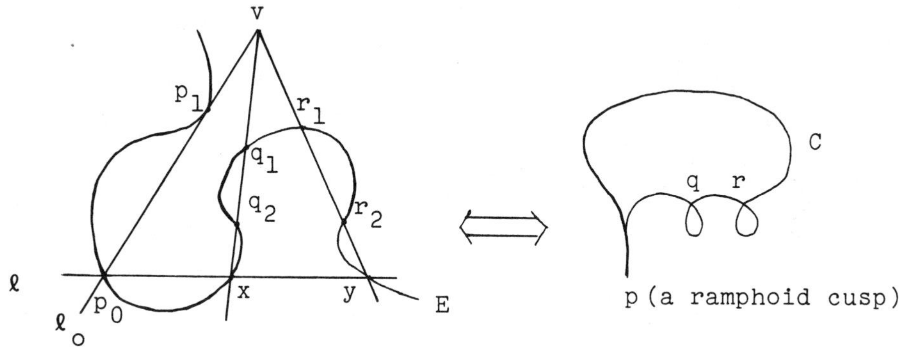

FIGURE 2.81

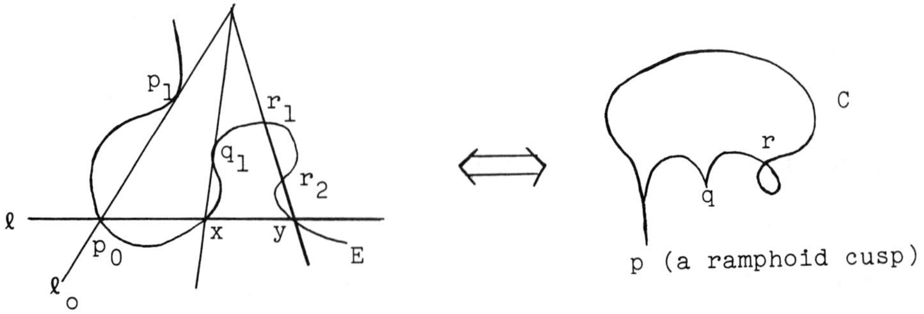

FIGURE 2.82

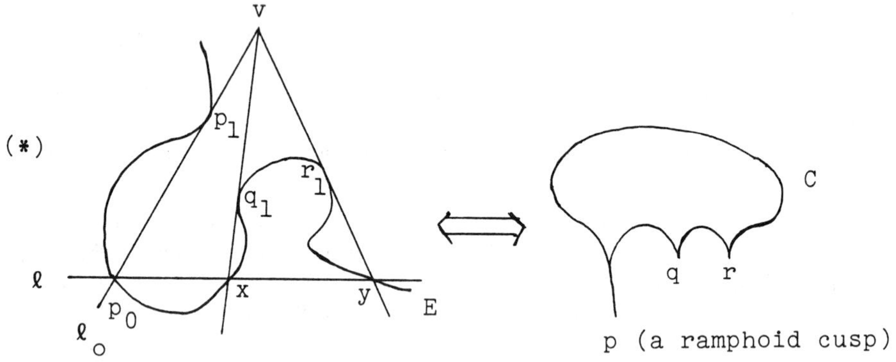

FIGURE 2.83

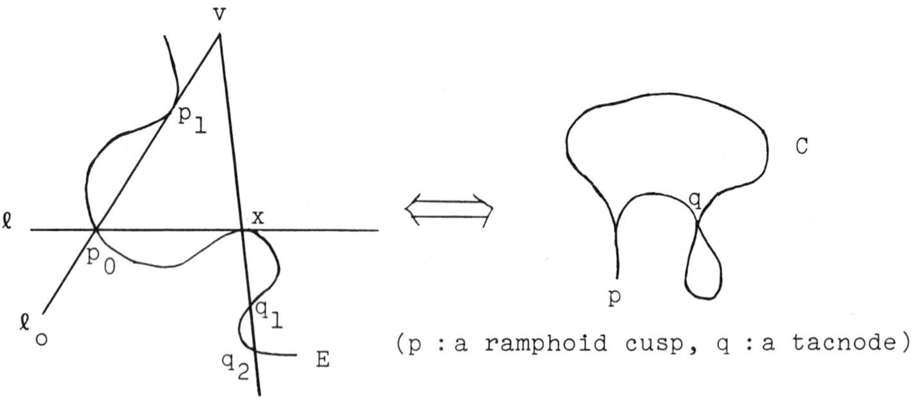

FIGURE 2.84

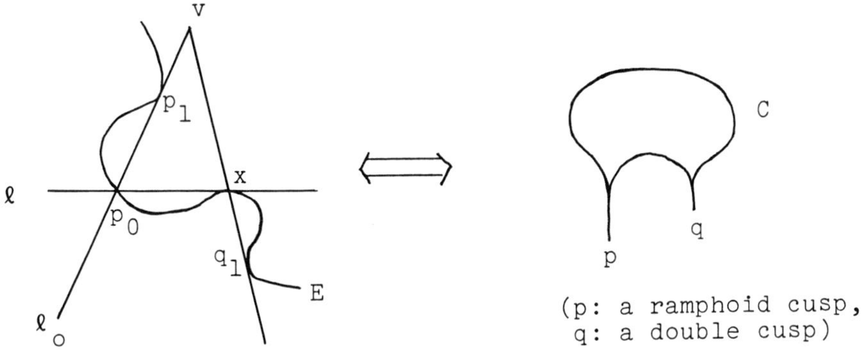

FIGURE 2.85

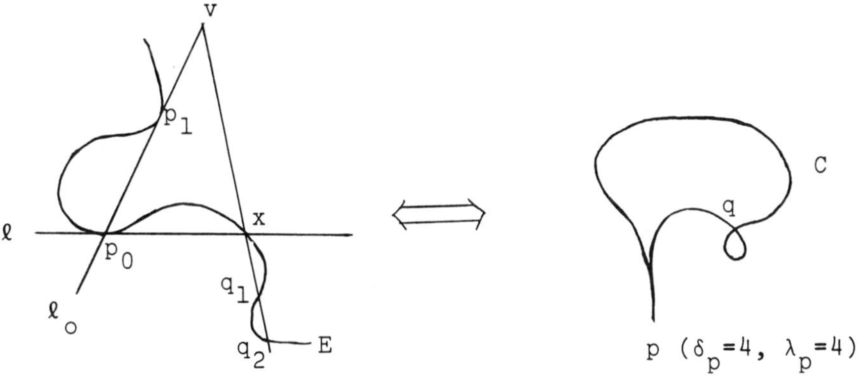

FIGURE 2.86

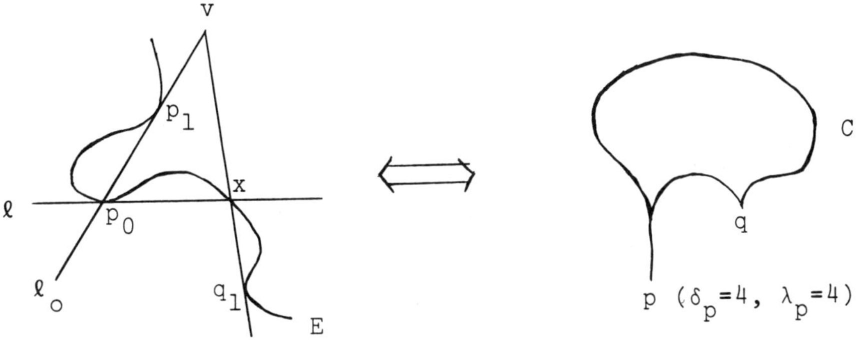

FIGURE 2.87

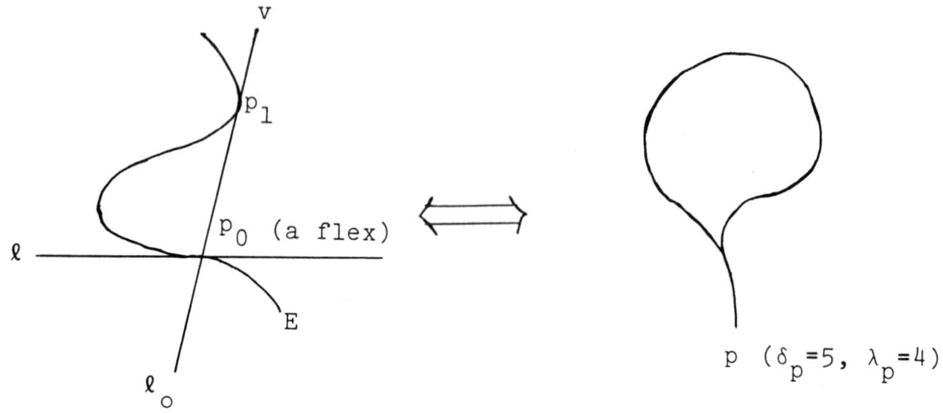

FIGURE 2.88

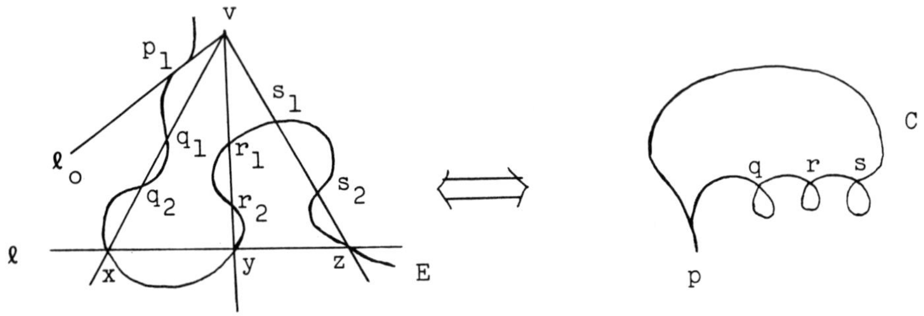

FIGURE 2.89

(*)

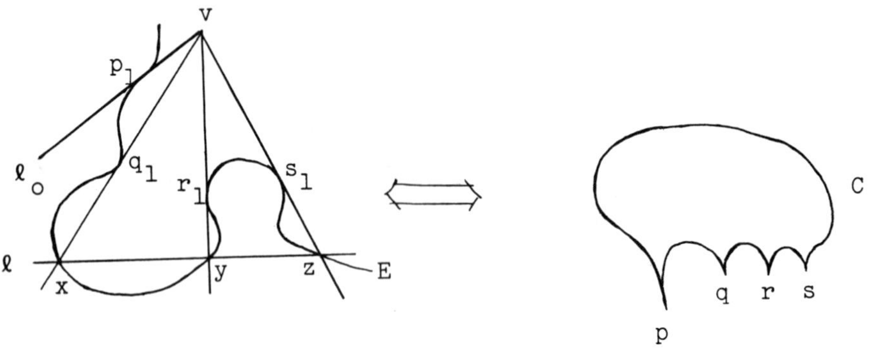

FIGURE 2.90

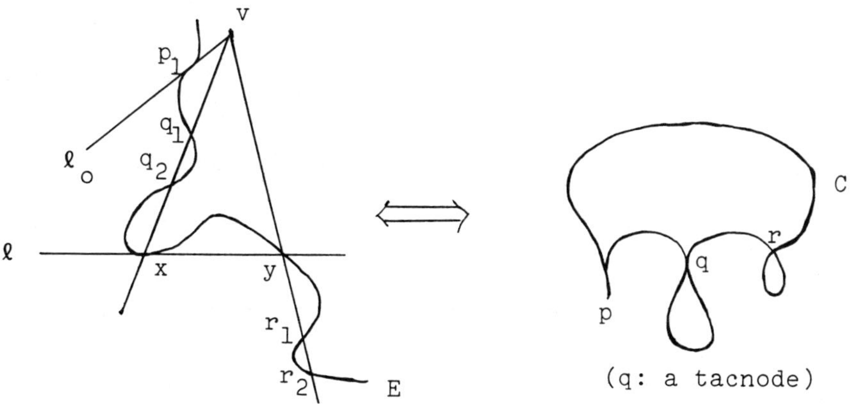

FIGURE 2.91

ii. Suppose that $p_0 = p_1$. In this case, $5p_1 = 0$ on the additive group M. moreover, $\Psi(p_1) = p_1$, so p_1 is a flex of $\Psi(C_0)$. p is in this case a double cusp with $\lambda_p = 5$.

Similar correspondences to Case i exist in this case (see Figs. 2.89-2.91). However, in contrast to Case i,

Lemma 2.3.8. There is no configuration (*) around any E. Hence, there is no curve C corresponding to (*).

Proof. Taking a suitable homogeneous coordinate system $(X : X_1 : X_2)$, we may consider the following equation for E and the configuration in Fig. 2.92.

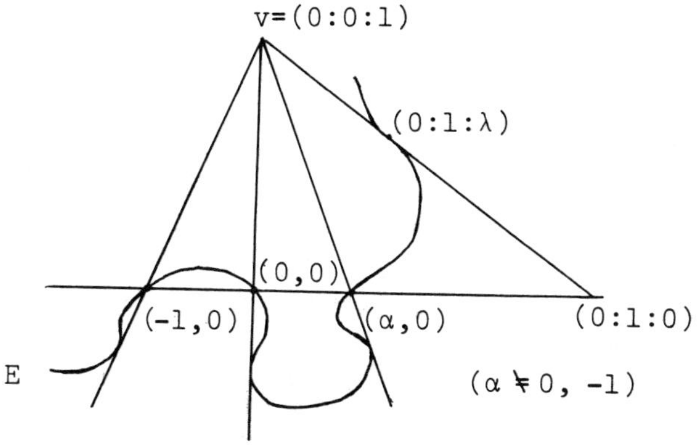

FIGURE 2.92

$$E: ax + by + cx^2 + dxy + ey^2 + (\lambda x - y)^3 = 0, \qquad (x = X_1/X_0,\ y = X_2/X_0)$$

Note that $\lambda \neq 0$, $a \neq 0$, $a \neq \lambda^3$, $c = a + \lambda^3$, $\lambda = -a/\lambda^3$. The conditions that the lines $\{x = 0\}$, $\{x = -1\}$, $\{x = \alpha\}$ are tangent to E can be written as

$$e^2 + 4b = 0, \quad 3\lambda^2 + 6\lambda e + 4d - e^2 - 4b = 0, \quad 3\lambda^2\alpha^2 - 6e\lambda\alpha - 4d\alpha - e^2 - 4b = 0$$

respectively. From these equations, we get $\alpha(\alpha + 1) = 0$, a contradiction. Hence, such a configuration cannot exist. Q.E.D.

All other configurations in this case exist, so the corresponding curves C exist.

b. Suppose that $\deg \Psi(C_0) = 4$. In this case, p is a simple cusp of multiplicity 2. Let p_0 and p_0' be the points on C_0 such that

$$(p_1) + (p_0) + (p_0') \in \Lambda$$

Then, by 2 of Lemma 2.3.7, $p_0 \neq p_1$ and $p_0' \neq p_1$. By 5 of the lemma, the quartic curve $C'' = \Psi(C_0)$ in P has $p_1 = \Psi(p_0) = \Psi(p_0')$ as a tacnode ($p_0 \neq p_0'$) or a double cusp ($p_0 = p_0'$). The vertex v of Q is on C'' in this case. The projection π_v with the center v gives the linear pencil Λ again.

Take a line ℓ on P such that $p_1 \notin \ell$ and $v \notin \ell$. As in Case a, we have the configuration around C'' in Fig. 2.93. From this configuration, we have the plane elliptic quintic curve $C = \pi_\ell(C_0)$. (See Fig. 2.94.)

On the other hand, from a given plane elliptic quartic curve C'' with a tacnode or a double cusp and a configuration as above, we can construct a plane elliptic quintic curve C as above (and vice versa). In fact, C'' is given by the projection

$$\pi_a : C_1 = \Phi_{|2(p_0)+2(p_0')|}(M) \to \mathbb{P}^2$$

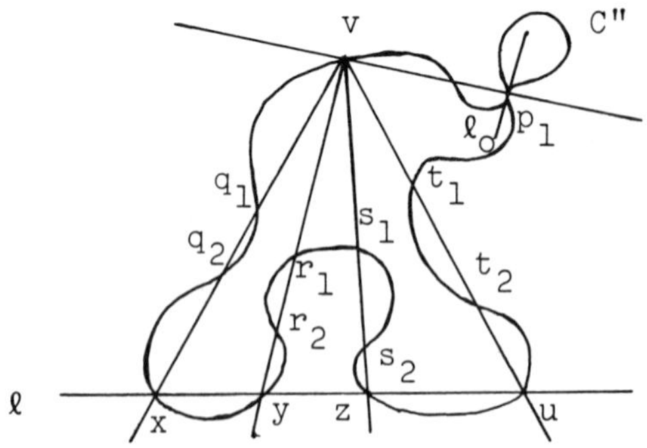

FIGURE 2.93

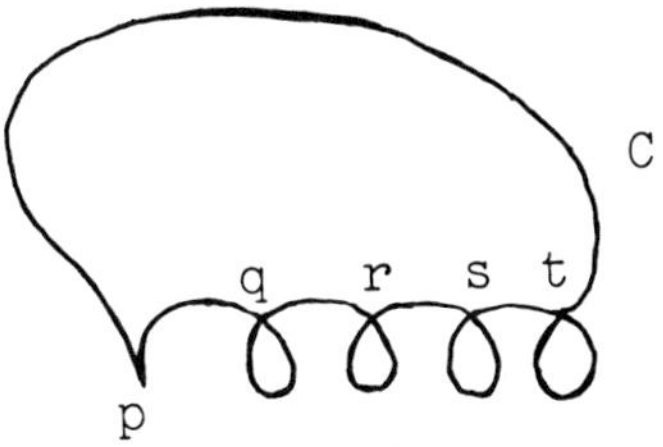

FIGURE 2.94

with the center a point a on the line $\overline{p_0 p_0'}$. (See Fig. 2.95.)

The tacnode (or double cusp) on C'' is $\pi_a(p_0) = \pi_a(p_0')$ (see Sec. 2.2). In $\mathbb{P}^3$, the points v, x, q_1, q_2 are on a plane S, so the point a is uniquely determined by $\{a\} = \overline{p_0 p_0'} \cap S$.

Take $p_1 \in M$ such that $3p_1 + p_0 + p_0' = 0$. Put $C_0 = \Phi_{|5(0)|}(M)$ again. Consider again $C' = \pi_0(C_0)$ in P', where $\pi_0 = \pi_{\ell_0}$ and $\ell_0 = T_{p_1} C_0$. The linear pencil $L = g_3^1$ given by π_v is also given by

$$\pi_w: C' \to \mathbb{P}^1$$

with the uniquely determined point w on $\mathbb{P}^2 - C'$. ℓ_0 and w span a 2-plane P in $\mathbb{P}^4$.

Now we construct the map Ψ as before. Then it is clear that $\Psi(C_0)$ and C'' are projectively equivalent and have the same configurations around them. The line ℓ' in P corresponding to ℓ gives the plane quintic curve $C = \pi_{\ell'}(C_0)$ with the prescribed singularities.

Thus, as in Case a, we have the correspondences in Figs. 2.96 and 2.97. We do not know if there is the last configuration around any C''. But there are C'' with the last configuration. For example

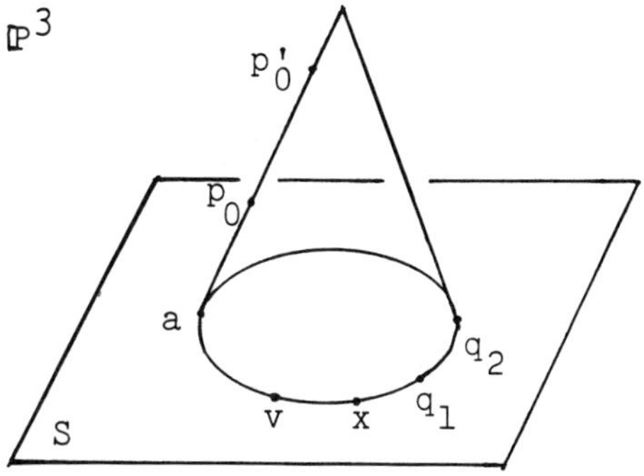

FIGURE 2.95

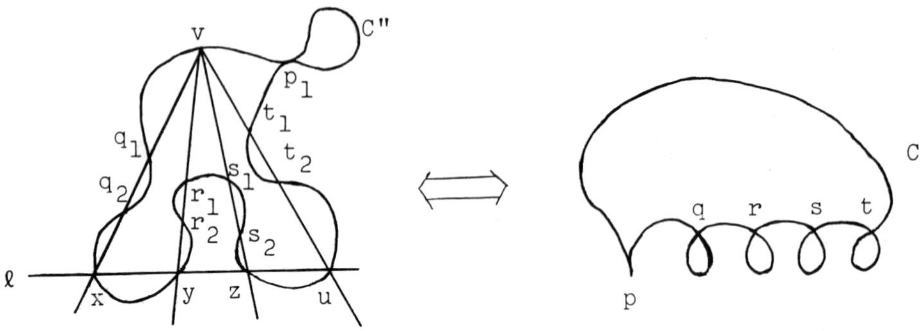

FIGURE 2.96

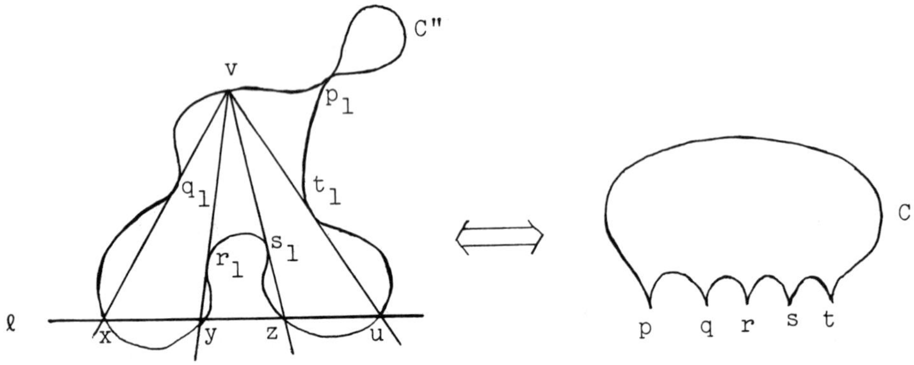

FIGURE 2.97

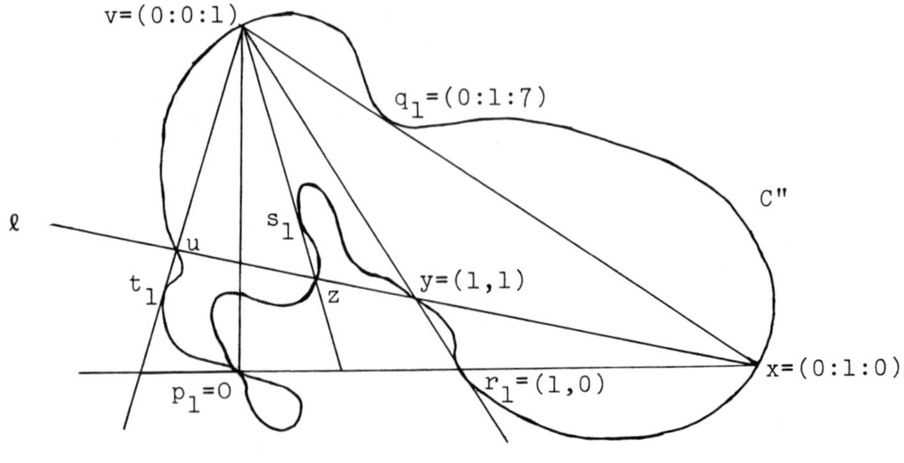

FIGURE 2.98

$$C'': 7(x+y)^2 - (x+y)(7x^2 - 42xy + 31y^2) - xy(7x-y)^2 = 0$$

See Fig. 2.98, where $z = (\sqrt{3/7}, 1)$, $u = (-\sqrt{3/7}, 1)$, $s_1 = (\sqrt{3/7}, (-9 + 5\sqrt{3/7})/(31 + \sqrt{3/7}))$, and $t_1 = (-\sqrt{3/7}, (-9 - 5\sqrt{3/7})/(31 - \sqrt{3/7}))$.

By tracing the above discussion, it can be shown that the curve $C = \pi_\ell(C_0)$ corresponding to this configuration is given by the image of the map

$$(x,y) \in C'' \rightarrow \left(1 : \frac{1}{x} : h(x,y)\right) \in \mathbb{P}^2$$

where

$$h(x,y) = \frac{64x}{x+y} + \frac{31(x+y)+xy}{x^2} + \frac{31(x+y)+xy}{x^3}$$

See Fig. 2.99, where $k = \sqrt{3/7}$, $\alpha = h(1/k, (-9+5k)/(31+k))$, $\beta = h(-1/k, (-9-5k)/(31-k))$.

Note 2.3.9. A plane elliptic quintic curve with five simple cusps of multiplicity 2 is called a Del Pezzo quintic. It was found by Del Pezzo [25]. See also Lefschetz [60] and Zariski [103].

7°. $g = 0$. The method in this case is more or less similar to those in the previous cases. But the result is messy, so we only state the following theorem.

Theorem 2.3.10. Plane rational quintic curves with only cusps as singular points are the following and no others, up to projective equivalence.

1. $t \in \mathbb{P}^1 \rightarrow (1 : t^4 : t^5) \in \mathbb{P}^2$

   $t \in \mathbb{P}^1 \rightarrow (1 + at - (1+a)t^2 : t^4 : t^5) \in \mathbb{P}^2$, $a \in \mathbb{C}$

   (See Fig. 2.100.)

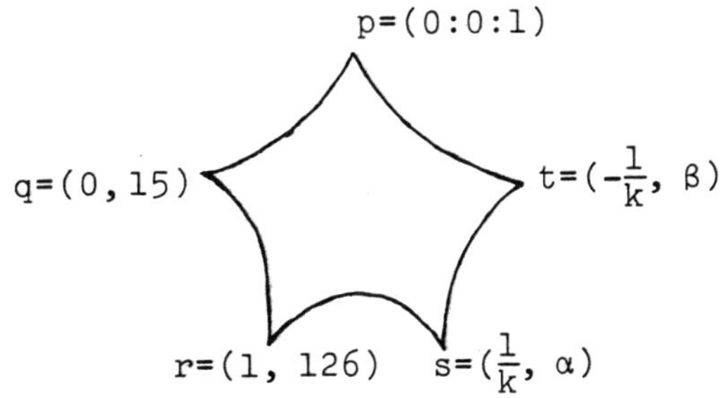

FIGURE 2.99

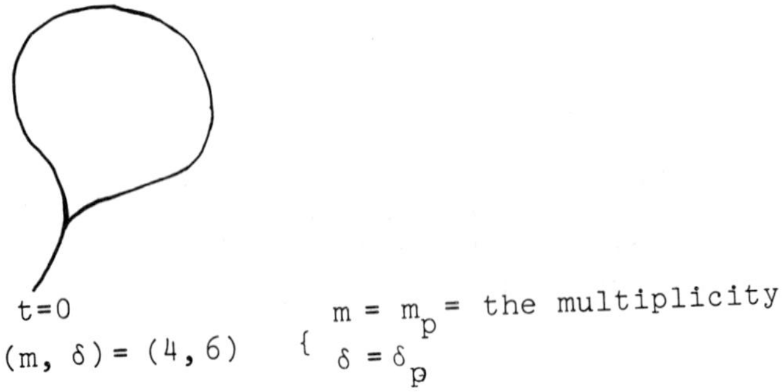

FIGURE 2.100

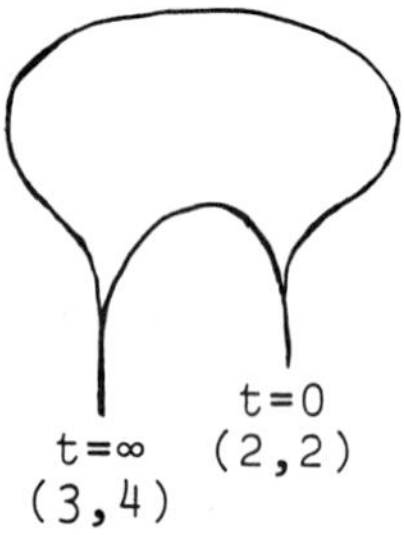

FIGURE 2.101

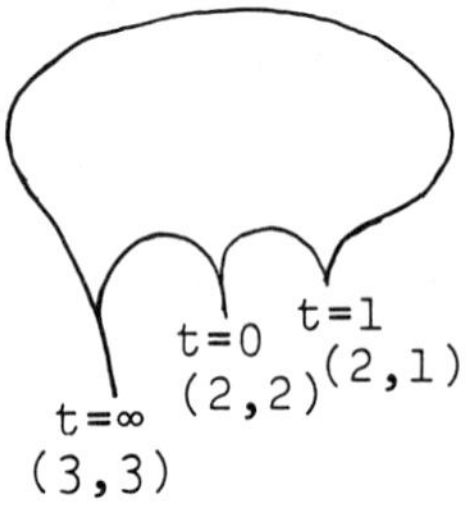

FIGURE 2.102

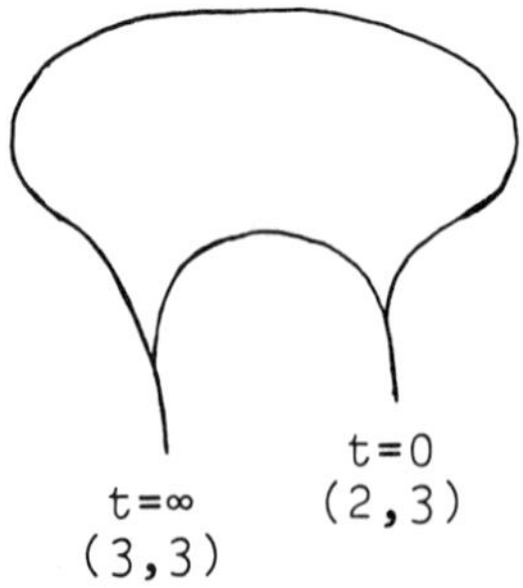

FIGURE 2.103

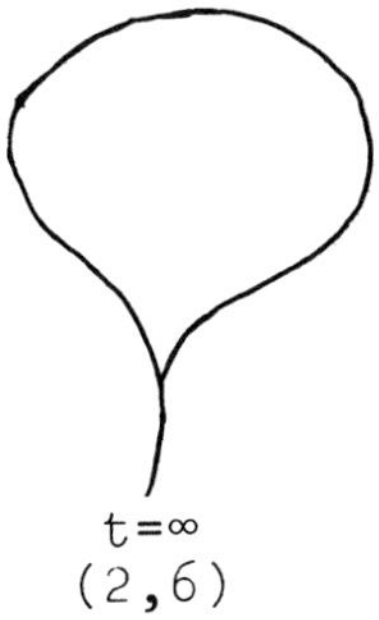

FIGURE 2.104

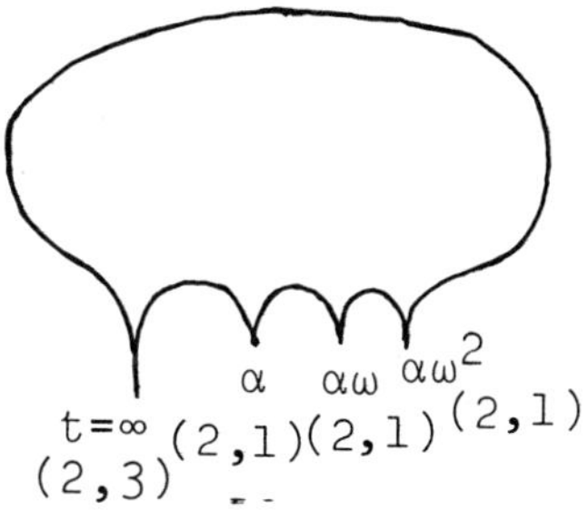

$$\begin{cases} \alpha = \sqrt[3]{-1/2} \\ \omega = \dfrac{-1+\sqrt{-3}}{2} \end{cases}$$

FIGURE 2.105

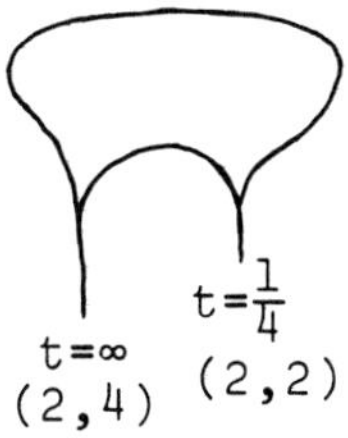

FIGURE 2.106

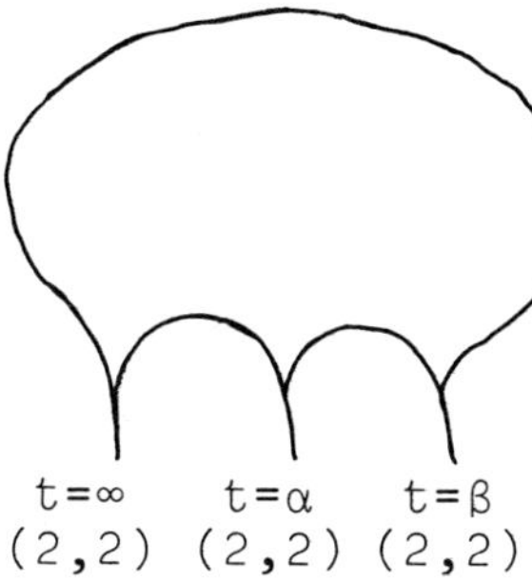

$$\left(\alpha,\ \beta = \frac{5 \pm 6\sqrt{5}}{16}\right)$$

FIGURE 2.107

2. $t \to (1: t^2: t^4 + t^5)$

$t \to (1: t^2: t^5)$

(See Fig. 2.101.)

3. $t \to \left(t - \frac{1}{2}: t^2: -\frac{3}{2}t^4 + t^5\right)$ (See Fig. 2.102.)

4. $t \to \left(t - \frac{1}{2}: t^2: \frac{1}{2}t^4 + t^5\right)$ (See Fig. 2.103.)

5. $t \to (t: t^3 - 1: t^5 - 2t^2)$ (See Fig. 2.104.)

6. $t \to (t: t^3 - 1: t^5 + 2t^2)$ (See Fig. 2.105.)
(This is the dual curve of the plane rational quartic curve with a ramphoid cusp: $t \to (1 + t^3: t^2: t^4)$ (see Sec. 2.2).)

7. $t \to \left(t - 1: t^3 - \frac{5}{32}: -\frac{47}{128} + \frac{11}{16}t^2 + t^4 + t^5\right)$ (See Fig. 2.106.)

8. $t \to \left(t - 1: t^3 - \frac{5}{32}: -\frac{125}{128} - \frac{25}{16}t^2 - 5t^4 + t^5\right)$ (See Fig. 2.107.)

Note 2.3.11. Curves 2, 4, and 7 were found by Yoshihara [102] by a different method. He proved that the plane rational quintic curves with just two cusps as singular points are curves 2, 4, and 7, up to projective equivalence. See also Hwang [47].

Exercises

1. Prove the assertion in Example 2.3.4 and Lemma 2.3.7.
2. Classify plane elliptic quintic curves with a triple point.
3. Complete the discussion in the case $g = 0$.

2.4 SINGULAR SPACE CURVES

In this section, we consider the following problem.

Problem. Given an integer n (≥ 4), what singular space curves in $\mathbb{P}^3$ of degree n exist?

By a similar method to that in the previous section, we can solve the problem for $n \leq 7$, but the result is very messy.

So we content ourselves only to treat the case of elliptic space sextic curves in $\mathbb{P}^3$, for this case is neither so trivial nor so messy.

In the case of plane curves, the analytic invariant δ_p at a singular point p was fundamental. On the other hand, in the case of space curves, no genus formula is known. So, in this case, we use only the following invariants: Let

C be a space curve in $\mathbb{P}^3$ and ϕ: M $\to$ C be a nonsingular model. For a singular point $p \in C$, we use the analytic invariants

$$m_p = \text{the multiplicity of C at p,}$$

$$s_p = \text{the number of irreducible branches of C at p,}$$

and the projective invariant

$$\alpha_z = (\alpha_1, \alpha_2, \alpha_3) \quad \text{for each } z \in \phi^{-1}(p),$$

where

$\alpha_1 = m'_p$ = the multiplicity at p of the irreducible branch C' corresponding to z,

$$\alpha_2 = I_p(C', T_pC'),$$

$$\alpha_3 = I_p(C', O_pC')$$

(O_pC' = the osculating plane at p to C') (see Sec. 1.6).

Now, let C be a space elliptic sextic curve in $\mathbb{P}^3$. Then we may assume that C can be obtained as the image of the projection

$$\pi_\ell: C_0 \to \mathbb{P}^3$$

with the center a line in $\mathbb{P}^5$ such that $\ell \cap C_0 = \phi$ (see Example 5.3.1). Here C_0 is the image of the holomorphic imbedding

$$\Phi_{|6(0)|}: z \in M \to (1: \wp(z): \wp'(z): \wp(z)^2: \wp'(z)\,\wp(z): \wp(z)^3) \in \mathbb{P}^5$$

($M = \mathbb{C}/(\mathbb{Z}\omega_1 + \mathbb{Z}\omega_2)$). We identify M with C_0 through $\Phi_{|6(0)|}$.

If ℓ is taken general, then π_ℓ is a holomorphic imbedding, so C is nonsingular (see Theorem 1.6.22).

Suppose that $C = \pi_\ell(C_0)$ has a singular point p. Put

$$\pi_\ell^{-1}(p) = (p_1) + \cdots + (p_m), \quad (m = m_p \geq 2)$$

Then $p_1, \ldots, p_m$ and ℓ must be on a 2-plane in $\mathbb{P}^5$. Hence,

$$m \leq 3$$

by the following lemma (see Exercise 1).

Lemma 2.4.1. Any n - 1 points on $C_0 = \Phi_{|n(0)|}(M)$, $(M = \mathbb{C}/\mathbb{Z}\omega_1 + \mathbb{Z}\omega_2))$, are in general position in $\mathbb{P}^{n-1}$.

Hence, there are two cases.

I. $m = m_p = 3$. In this case, ℓ must be on the 2-plane P spanned by p_1, p_2, and p_3. Conversely, if ℓ is on P, then $C = \pi_\ell(C_0)$ has $p = \pi_\ell(p_1) = \pi_\ell(p_2) = \pi_\ell(p_3)$ as a triple point.

There is no other singular point on C in this case. In fact, if q is a double point, say, on C, then putting $\pi_\ell^{-1}(q) = (q_1) + (q_2)$,

$$p_1, p_2, p_3, q_1, q_2$$

must be on a 3-plane in $\mathbb{P}^5$, which contradicts the above lemma.

a. Suppose that p_1, p_2, and p_3 are mutually distinct. In this case, p is an ordinary triple point, in the sense that three irreducible branches at p have mutually distinct tangent lines and osculating planes. (See Fig. 2.108.)

b. Suppose that $p_1 \neq p_2 = p_3$. In this case, P is spanned by p_1 and $T_{p_2}C_0$, and $\ell \subset P$. C has then two irreducible branches C' and C'' at p such that (1) C' is smooth, (2) C'' has p as a cusp with $\alpha_Z = (\alpha_1, \alpha_2, \alpha_3) = (2, 3, *)$, $4 \leq * \leq 6$, and (3) C' and C'' meet transversally, in the sense that they have distinct tangent lines and osculating planes. (Here * indicates a not necessarily uniquely determined positive integer.) (See Fig. 2.109.)

c. Suppose that $p_1 = p_2 = p_3$. In this case, $\ell \subset O_{p_1}C_0$, the osculating 2-plane. p is then a cusp with $(\alpha_1, \alpha_2, \alpha_3) = (3, 4, *)$, $(* = 5 \text{ or } 6)$. (See Fig. 2.110.)

II. $m = m_p = 2$. In this case, even if there are other singular points on C, they must be double points, as was shown in Case I.

Put $\pi_\ell^{-1}(p) = (p_1) + (p_2)$. We assume that $p_1 = p_2$. The case $p_1 \neq p_2$ can be treated in a similar way. Put $\ell_0 = T_{p_1}C_0$. The lines ℓ and ℓ_0 must span a 2-plane P. (See Fig. 2.111.)

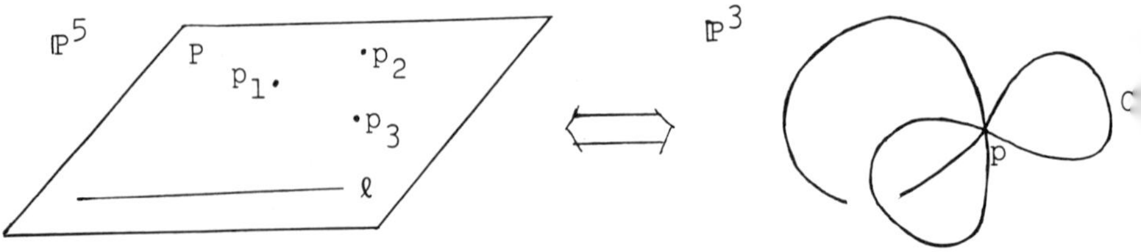

FIGURE 2.108

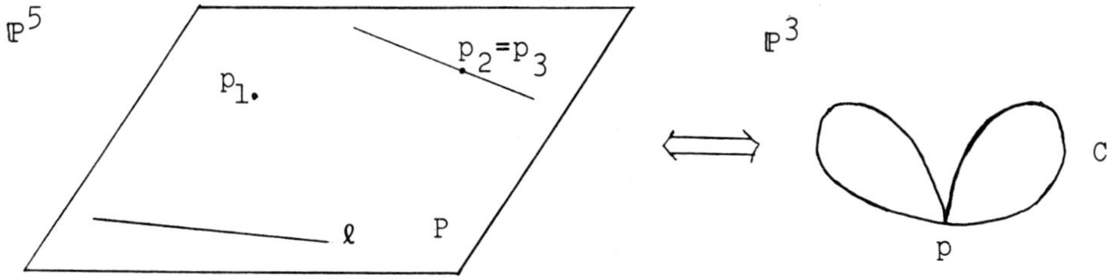

FIGURE 2.109

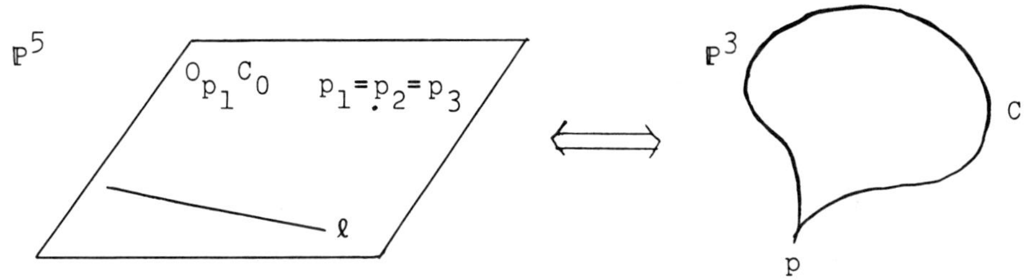

FIGURE 2.110

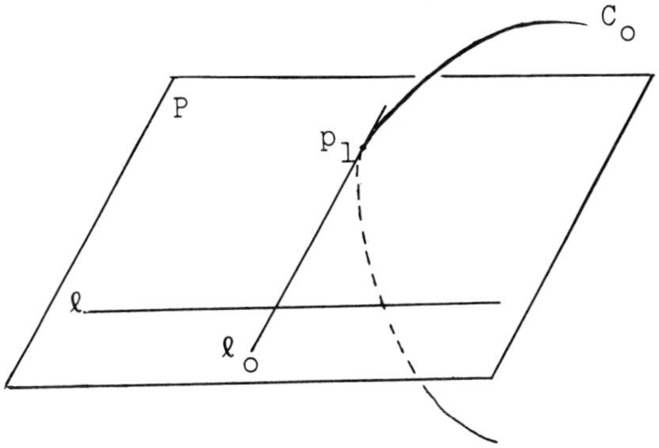

FIGURE 2.111

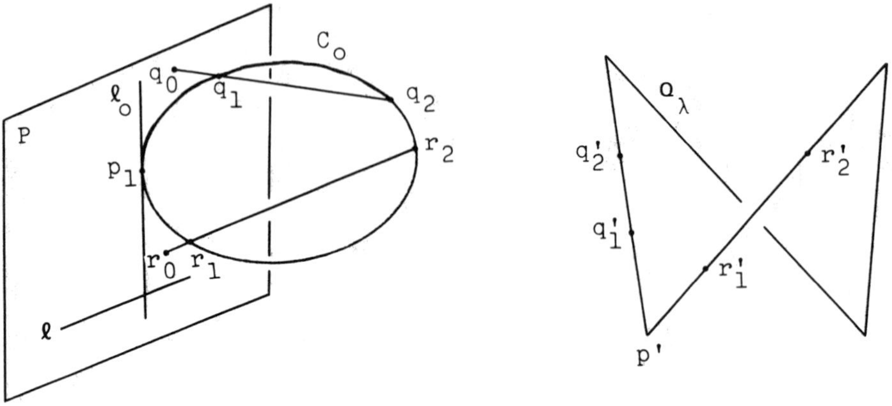

FIGURE 2.112

What we do is to choose first a 2-plane P with $\ell_0 \subset P$ such that $P \neq O_{p_1} C_0$ and then to choose a line $\ell \subset P$. Let

$$\pi_0 = \pi_{\ell_0} : C_0 \to \mathbb{P}^3$$

be the projection with the center ℓ_0. Then π_0 is a biholomorphic map onto an elliptic quartic curve C' in $\mathbb{P}^3$. (In fact, $C' = \Phi_{|6(0)-2(p_1)|}(M)$.) Then there is a 1-1 correspondence between 2-planes P as above and points $p' \in \mathbb{P}^3 - C'$

$$P \to p' = P \cap \mathbb{P}^3, \quad p' \to P = p' \vee \ell_0$$

Let $\{Q_\lambda\}_{\lambda \in \mathbb{P}^1}$ be the one-parameter family of quadric surfaces Q_λ such that $C' \subset Q_\lambda$. p' is then contained in a unique Q_λ.

a. Suppose that $p' \in Q_\lambda$ for a nonsingular Q_λ. Let ℓ' and ℓ'' be lines on Q_λ passing through p', and put

$$\ell' \cap C' = (q_1') + (q_2')$$

$$\ell'' \cap C' = (r_1') + (r_2')$$

Put $\pi_0^{-1}(q_1') = q_1$, and so forth. Put also

$$q_0 = \overline{q_1 q_2} \cap P \quad \text{and} \quad r_0 = \overline{r_1 r_2} \cap P$$

(See Fig. 2.112.)

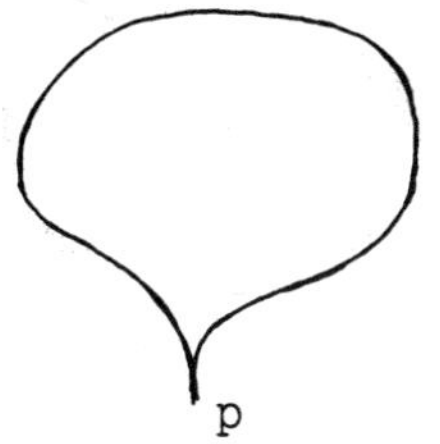

FIGURE 2.113

i. Suppose that p_1 is equal to none of q_1, q_2, r_1, r_2. In this case, p is a cusp with

$$\alpha_p = (\alpha_1, \alpha_2, \alpha_3) = (2, 3, *), \quad 4 \leq * \leq 6$$

i-1. Suppose that ℓ passes through neither q_0 nor r_0. In this case, p is a unique singular point of $C = \pi_\ell(C_0)$. (See Fig. 2.113.)

i-2. Suppose that ℓ passes through q_0, but does not pass through r_0. In this case, C has just two singular points p and $q = \pi_\ell(q_1) = \pi_\ell(q_2)$. Here q is either a node ($q_1 \neq q_2$), in the sense that the two smooth irreducible branches at q have distinct tangent lines, or a cusp ($q_1 = q_2$) with

$$\alpha_q = (\alpha_1, \alpha_2, \alpha_3) = (2, 3, *), \quad 4 \leq * \leq 6$$

Both cases are possible. (See Fig. 2.114.)

i-3. Suppose that $\ell = \overline{q_0 r_0}$. In this case, C has just three singular points p, $q = \pi_\ell(q_1) = \pi_\ell(q_2)$ and $r = \pi_\ell(r_1) = \pi_\ell(r_2)$. Here q and r are either nodes or cusps with $(\alpha_1, \alpha_2, \alpha_3) = (2, 3, *)$, $4 \leq * \leq 6$, depending on $q_1 \neq q_2$ or $q_1 = q_2$, and so forth. Every combination is possible if p' varies. (See Fig. 2.115.)

FIGURE 2.114

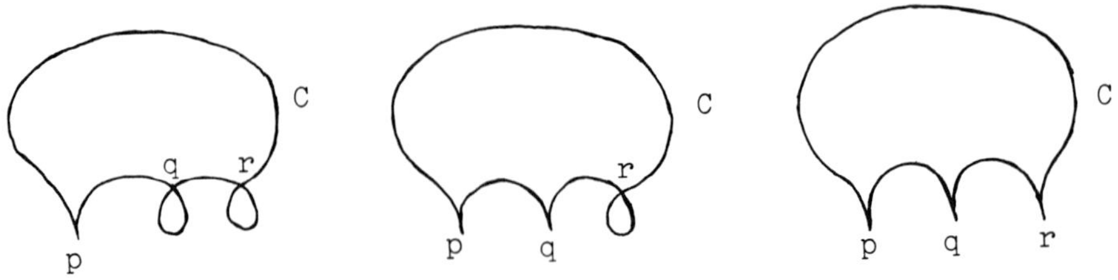

FIGURE 2.115

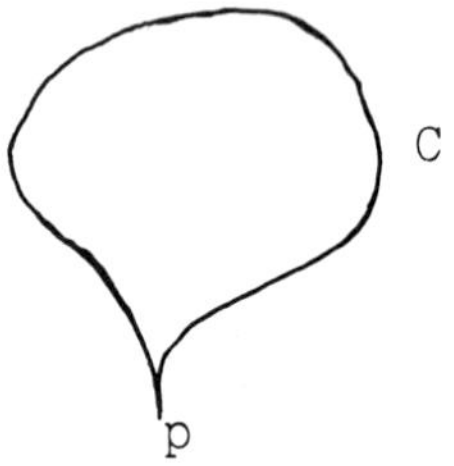

FIGURE 2.116

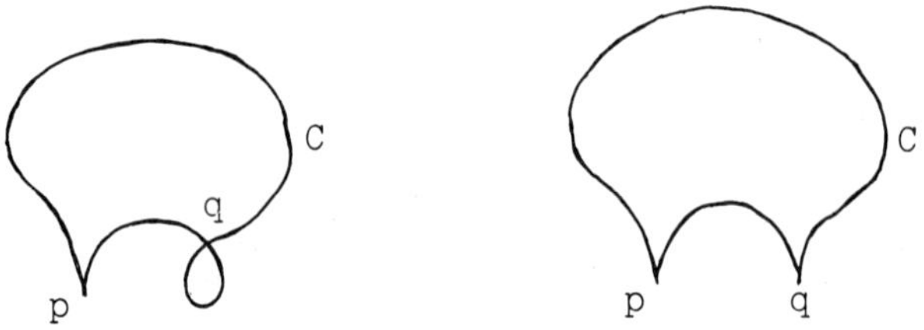

FIGURE 2.117

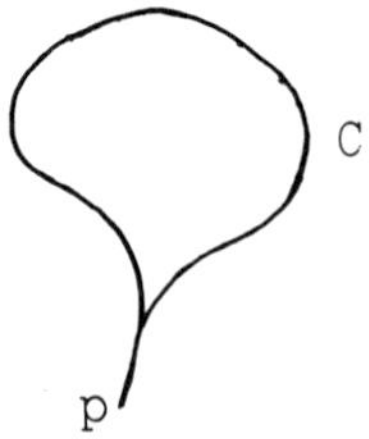

FIGURE 2.118

ii. Suppose that $p_1 = r_1 \neq r_2$. In this case $r_0 = p_1$. p is, in this case, a cusp with

$$\alpha_p = (\alpha_1, \alpha_2, \alpha_3) = (2, 3, *), \quad 4 \leq * \leq 6$$

ii-1. Suppose that ℓ does not pass through q_0. In this case, p is a unique singular point of C. (See Fig. 2.116.)

ii-2. Suppose that $q_0 \in \ell$. In this case, C has just two singular points p and q. Here q is either a node ($q_1 \neq q_2$) or a cusp ($q_1 = q_2$) with

$$\alpha_q = (\alpha_1, \alpha_2, \alpha_3) = (2, 3, *), \quad 4 \leq * \leq 6$$

Both cases are possible. (See Fig. 2.117.)

iii. Suppose that $p_1 = r_1 = r_2$. In this case, r_0 is not determined. p is in this case a cusp with

$$\alpha_p = (\alpha_1, \alpha_2, \alpha_3) = (2, 4, 5)$$

iii-1. Suppose that $q_0 \notin \ell$. In this case, p is a unique single point of C. (See Fig. 2.118.)

iii-2. Suppose that $q_0 \in \ell$. In this case, C has just two singular points p and q, which is either a node ($q_1 \neq q_2$) or a cusp ($q_1 = q_2$) with

$$\alpha_q = (\alpha_1, \alpha_2, \alpha_3) = (2, 3, *), \quad 4 \leq * \leq 6$$

(See Fig. 2.119.)

b. Suppose that $p' \in Q_\lambda$ for a cone Q_λ and $p' \neq v$, where v is the vertex of Q_λ. Let ℓ' be the line on Q_λ passing through p'. Put

$$C' \cap \ell' = (q_1') + (q_2')$$

FIGURE 2.119

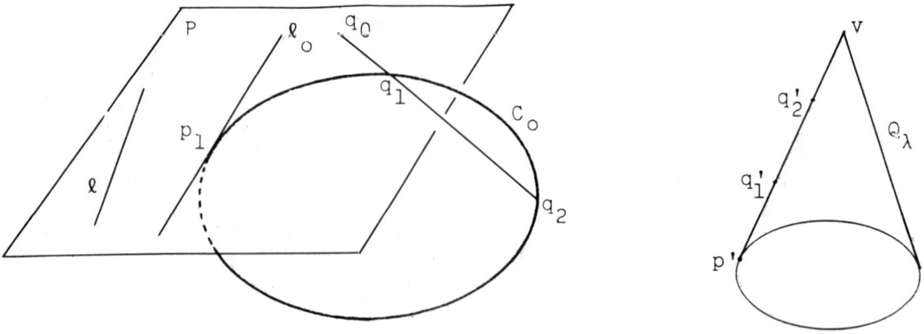

FIGURE 2.120

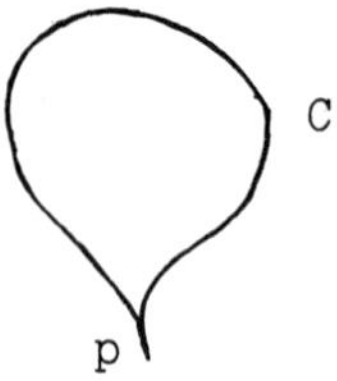

FIGURE 2.121

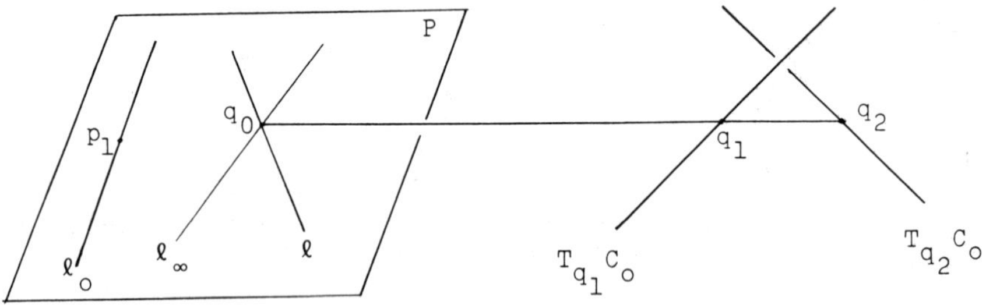

FIGURE 2.122

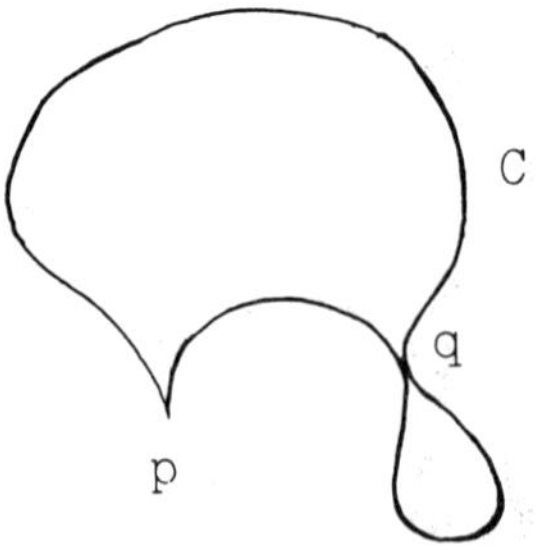

FIGURE 2.123

$$\pi_0^{-1}(q_1') = q_1, \quad \pi_0^{-1}(q_2') = q_2$$

$$q_0 = P \cap \overline{q_1 q_2}$$

(See Fig. 2.120.)

i. Suppose that $p_1 \neq q_1$ and $p_1 \neq q_2$. In this case, p is a cusp with

$$\alpha_p = (\alpha_1, \alpha_2, \alpha_3) = (2, 3, *), \qquad 4 \leq * \leq 6$$

i-1. Suppose that $q_0 \notin \ell$. In this case, p is a unique singular point of C. (See Fig. 2.121.)

i-2. Suppose that $q_0 \in \ell$ and $q_1 \neq q_2$. In this case, put

$$\ell_\infty = P \cap (T_{q_1} C_0 \vee T_{q_2} C_0)$$

where $T_{q_1} C_0 \vee T_{q_2} C_0$ is the 3-space in $\mathbb{P}^5$ spanned by the tangent lines $T_{q_1} C_0$ and $T_{q_2} C_0$. Then ℓ_∞ is a line on P passing through q_0. (See Fig. 2.122.)

If $\ell = \ell_\infty$, then C has just two singular points p and $q = \pi_\ell(q_1) = \pi_\ell(q_2)$, which is of tacnode type, in the sense that the irreducible branches at q have the same tangent lines. (See Fig. 2.123.)

If $\ell \neq \ell_\infty$, then C has just two singular points p and q, which is a node. In this case, a new geometric phenomenon occurs. Let ℓ_1 and ℓ_2 be the tangent lines at q to the irreducible branches of C. Then $p \in \ell_1 \vee \ell_2$. (See Fig. 2.124.)

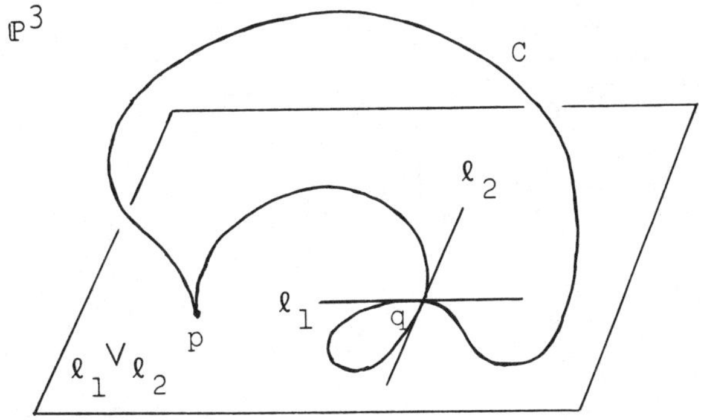

FIGURE 2.124

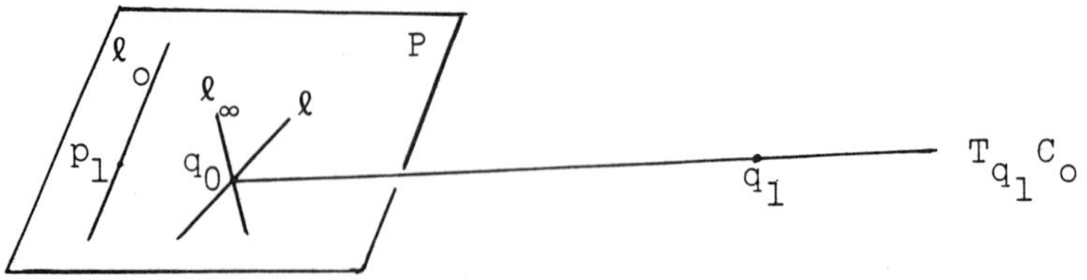

FIGURE 2.125

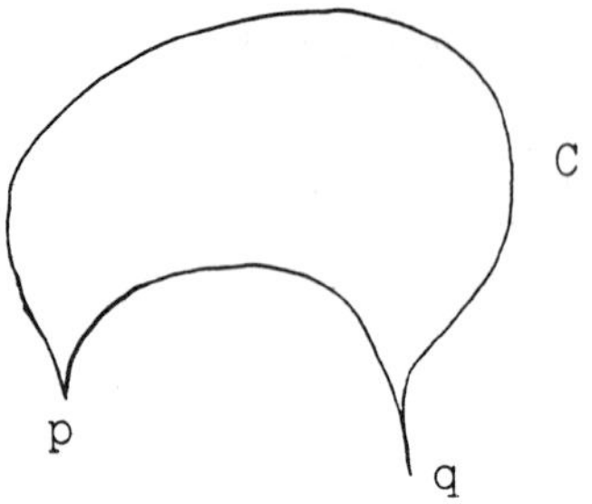

FIGURE 2.126

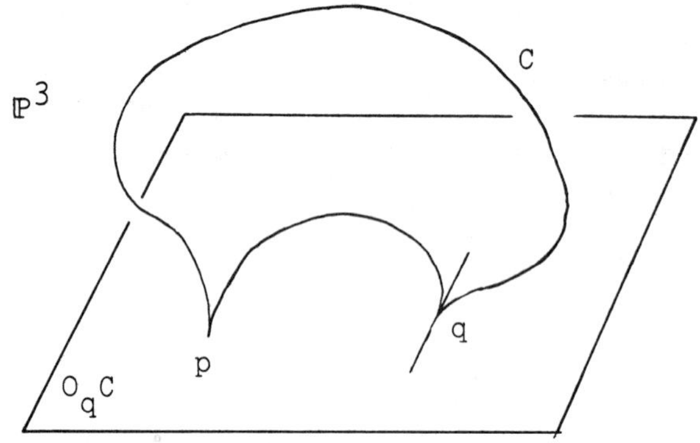

FIGURE 2.127

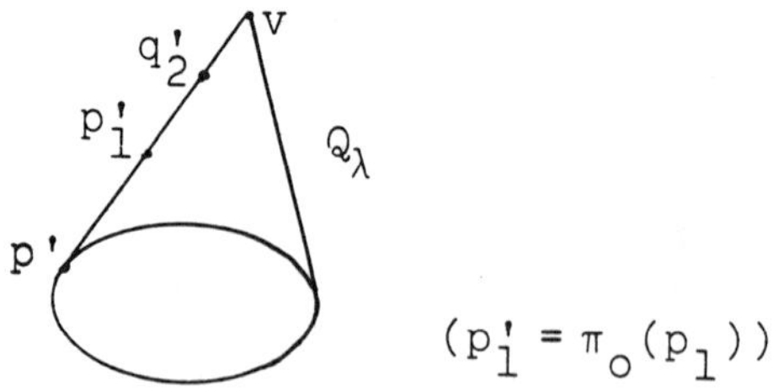

FIGURE 2.128

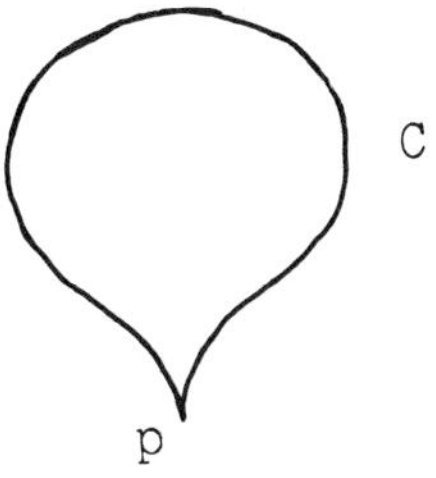

FIGURE 2.129

i-3. Suppose that $q_0 \in \ell$ and $q_1 = q_2$. In this case, put

$$\ell_\infty = P \cap O^{(3)}_{q_1} C_0$$

where $O^{(3)}_{q_1} C_0$ is the osculating 3-plane to C_0 at q_1. Then ℓ_∞ is a line in P passing through q_0. (See Fig. 2.125.)

If $\ell = \ell_\infty$, then C has just two singular points p and q, which is a cusp with

$$\alpha_q = (\alpha_1, \alpha_2, \alpha_3) = (2, 4, *), \quad 5 \leq * \leq 6$$

(See Fig. 2.126.)

If $\ell \neq \ell_\infty$, then C has just two singular points p and q, which is a cusp with

$$\alpha_q = (\alpha_1, \alpha_2, \alpha_3) = (2, 3, *), \quad 4 \leq * \leq 6$$

such that $p \in O_q C$. (See Fig. 2.127.)

ii. Suppose that $p_1 = q_1 \neq q_2$. In this case,

1. $4p_1 + 2q_2 = 0$ in the additive group M (see Fig. 2.128), and
2. $q_0 = p_1$

p is, in this case, a unique singular point of C, a cusp with

$$\alpha_p = (\alpha_1, \alpha_2, \alpha_3) = (2, 3, 4)$$

(See Fig. 2.129.)

iii. Suppose that $p_1 = q_1 = q_2$. In this case, (1) $6p_1 = 0$ in the additive group M (see Fig. 2.130), and (2) q_0 is not determined.

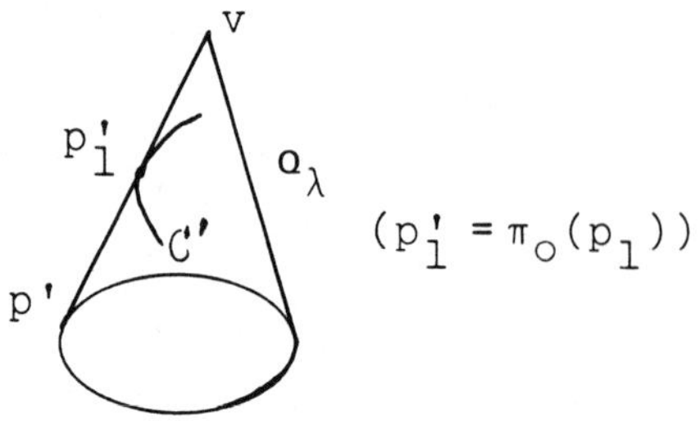

FIGURE 1.30

p is, in this case, a unique singular point of C, a cusp with

$$\alpha_p = (\alpha_1, \alpha_2, \alpha_3) = (2, 4, 6)$$

(See Fig. 2.131.)

c. Suppose that $p' = v$, the vertex of a cone Q_λ. Let

$$\pi_v : Q_\lambda - \{v\} \to \mathbb{P}^2$$

be the projection with the center v. Then the image B is an irreducible conic. For $\xi \in B$, put

$$C' \cap \overline{\xi v} = (q_1') + (q_2')$$

$$\pi_0^{-1}(q_1') = q_1, \quad \pi_0^{-1}(q_2') = q_2$$

$$P \cap \overline{q_1 q_2} = q_0$$

(See Fig. 2.132.)

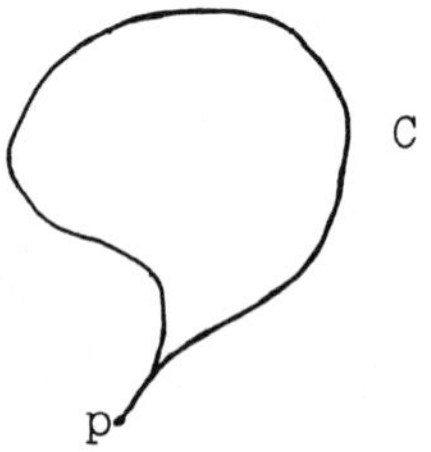

FIGURE 2.131

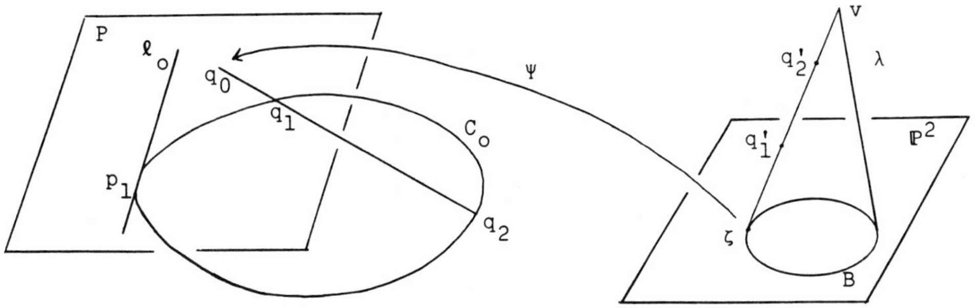

FIGURE 2.132

Define a holomorphic map

$$\Psi: \zeta \in B \to q_0 \in P$$

In a similar way to the proof of Lemma 2.2.4, we can prove the following lemma (see Exercise 1).

Lemma 2.4.2. (1) Ψ is birational onto the image curve $\Psi(B)$. (2) $\Psi(B)$ is either an irreducible conic or a line. $\Psi(B)$ is a line if and only if $6p_1 = 0$ in the additive group M. (3) If $\Psi(B)$ is a conic, then $p_1 \in \Psi(B)$ and ℓ_0 is tangent to $\Psi(B)$ at p_1. (4) If $\Psi(B)$ is a line, then $p_1 \notin \Psi(B)$.

Since π_v: C' → B is a double covering, there are just four branch points ζ_j, $1 \le j \le 4$. Put

$$s_j = \Psi(\zeta_j), \quad 1 \le j \le 4$$

i. Suppose that $\Psi(B)$ is an irreducible conic. In this case, p is a cusp of C with

$$\alpha_p = (\alpha_1, \alpha_2, \alpha_3) = (2, 3, 4)$$

(See Fig. 2.133.)

i-1. Suppose that ℓ is a general line on P. In this case, C has just three singular points p, q, and r (see line 1 in Fig. 2.133). Here q and r are nodes such that

$$p \in \ell_1 \vee \ell_2 \quad \text{and} \quad p \in \ell_3 \vee \ell_4$$

where ℓ_1 and ℓ_2 (respectively ℓ_3 and ℓ_4) are the tangent lines at q (respectively r) of irreducible branches of C. (See Fig. 2.134.)

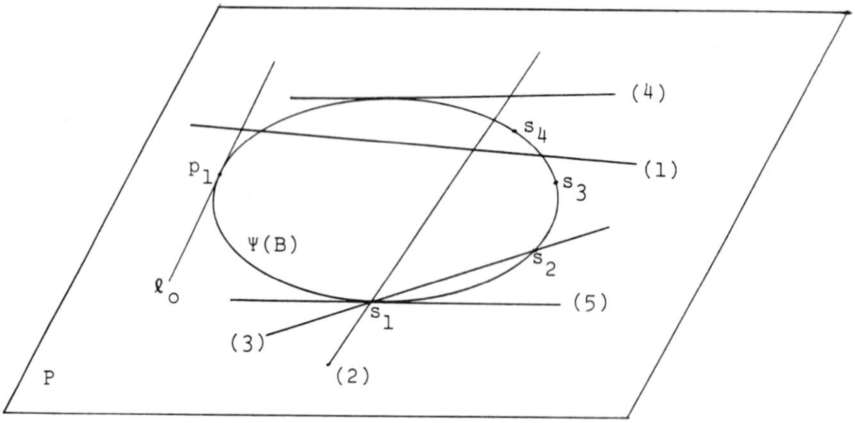

FIGURE 2.133

i-2. Suppose that ℓ cuts $\Psi(B)$ at two points s_1 and q_0 such that $q_0 \neq s_j$, $2 \leq j \leq 4$. In this case, C has just three singular points p, q, and r. Here q is a node and t is a cusp with

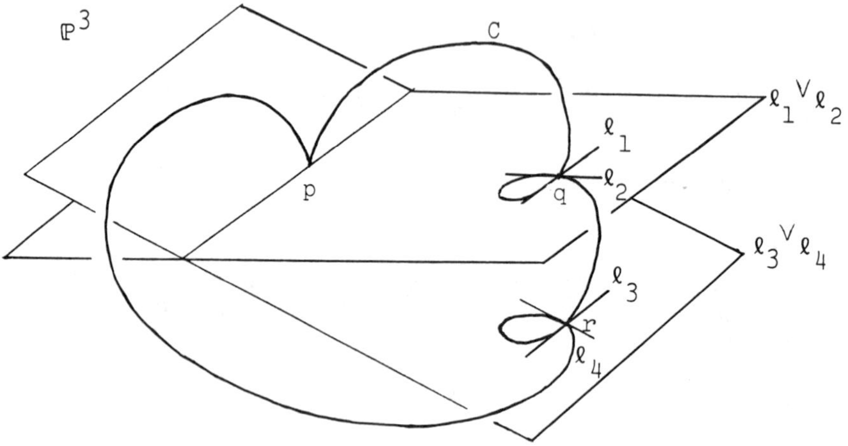

FIGURE 2.134

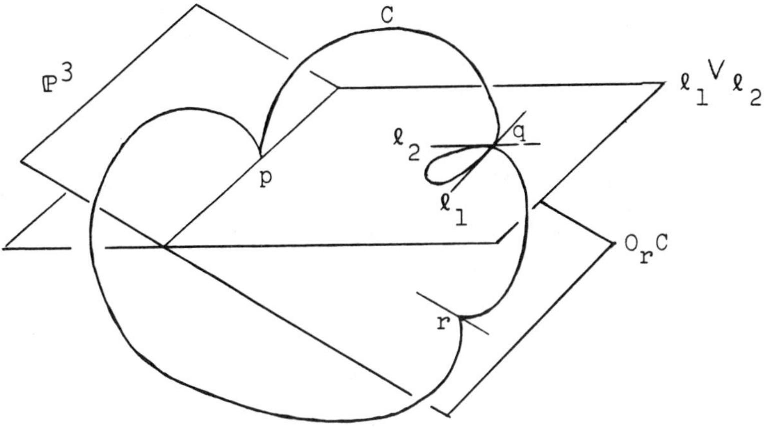

FIGURE 2.135

$$\alpha_r = (\alpha_1, \alpha_2, \alpha_3) = (2, 3, *), \quad 4 \leq * \leq 6$$

such that

$$p \in \ell_1 \vee \ell_2 \quad \text{and} \quad p \in O_rC$$

where ℓ_1 and ℓ_2 are the tangent lines at q to the irreducible branches of C. (See Fig. 2.135.)

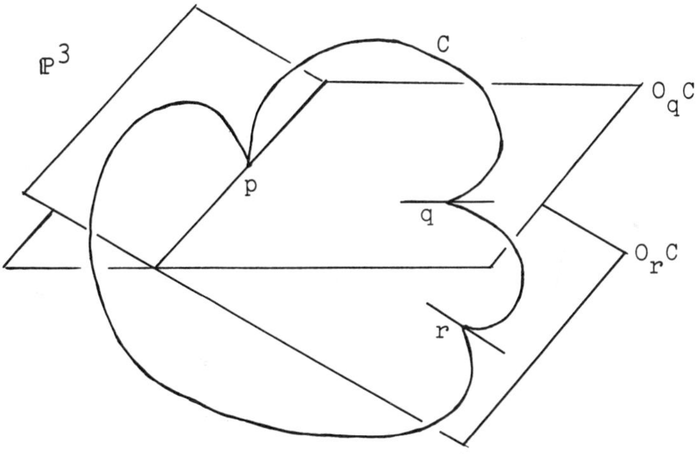

FIGURE 2.136

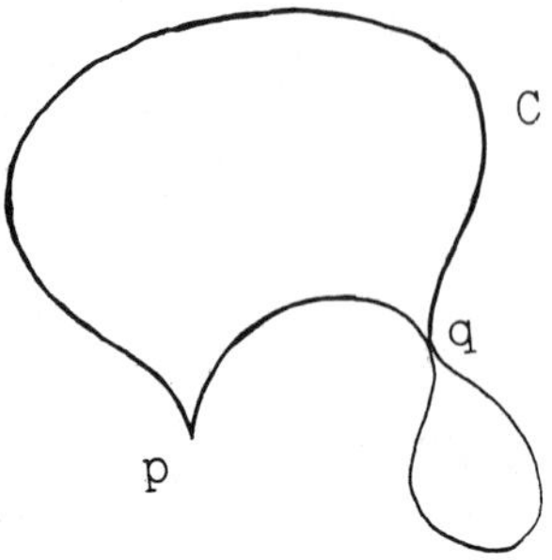

FIGURE 2.137

i-3. Suppose that $\ell = \overline{s_1 s_2}$. In this case, C has just three singular points p, q, and r. Here q and r are cusps with

$$(\alpha_1, \alpha_2, \alpha_3) = (2, 3, *), \quad 4 \leq * \leq 6$$

such that $p \in O_q C$ and $p \in O_r C$. (See Fig. 2.136.)

i-4. Suppose that ℓ is tangent to Ψ(B) at q_0 such that $q_0 \neq s_j$, $1 \leq j \leq 4$. In this case, C has just two singular points p and q. Here q is of tacnode type. (See Fig. 2.137.)

i-5. Suppose that ℓ is tangent to Ψ(B) at s_1. In this case, C has just two singular points p and q. Here q is a cusp with

$$\alpha_q = (\alpha_1, \alpha_2, \alpha_3) = (2, 4, *), \; 5 \leq * \leq 6$$

(See Fig. 2.138.)

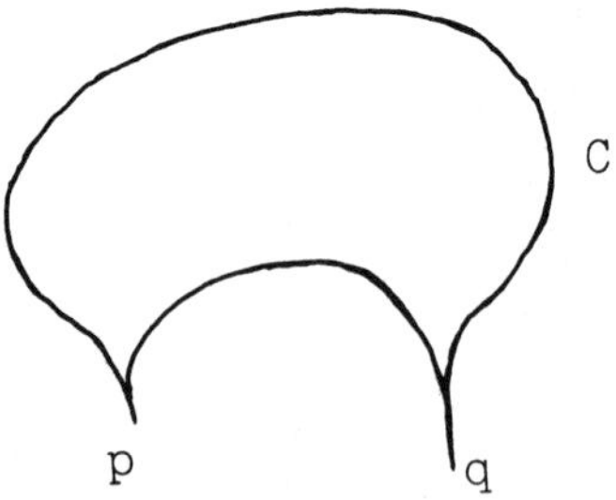

FIGURE 2.138

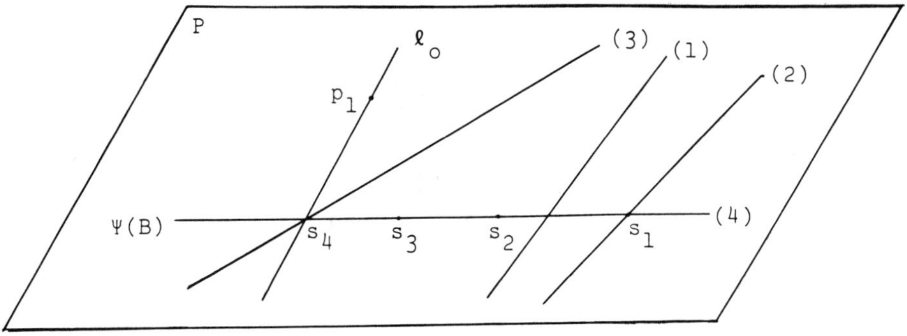

FIGURE 2.139

ii. Suppose that Ψ(B) is a line. In this case, if ℓ is a line on P such that $\ell \neq \Psi(B)$, then p is a cusp of C with

$$\alpha_p = (\alpha_1, \alpha_2, \alpha_3) = (2, 4, 6)$$

(See Fig. 2.139.)

ii-1. Suppose that ℓ is a general line on P. In this case, C has just two singular points p and q (see line 1 in the above picture). Here q is a node such that

$$p \in \ell_1 \vee \ell_2$$

where ℓ_1 and ℓ_2 are tangent lines at q of the irreducible branches of C. (See Fig. 2.140.)

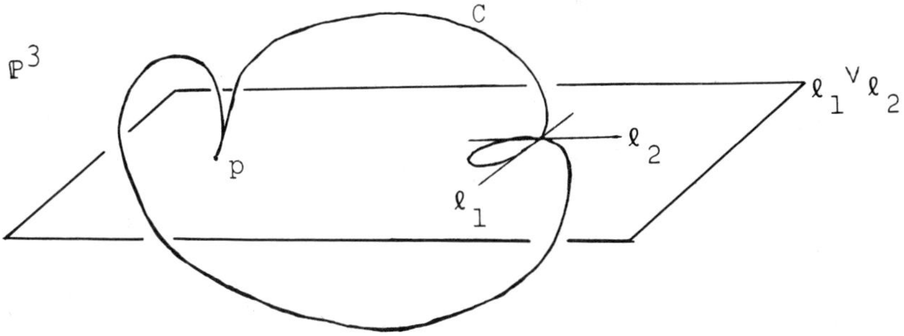

FIGURE 2.140

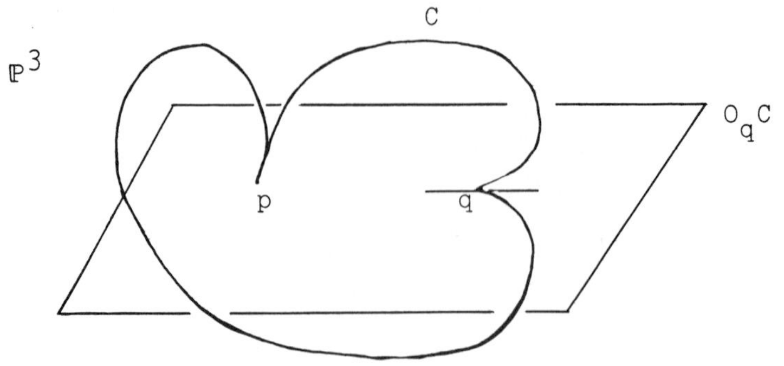

FIGURE 2.141

ii-2. Suppose that $s_1 \in \ell$ and $\ell \neq \Psi(B)$. In this case, C has just two singular points p and q. Here, q is a cusp with

$$\alpha_q = (\alpha_1, \alpha_2, \alpha_3) = (2, 3, 6)$$

such that $p \in O_qC$. (See Fig. 2.141.)

ii-3. Suppose that ℓ passes through $s_4 = \ell_0 \cap \Psi(B)$ and $\ell \neq \Psi(B)$. In this case C has a unique singular point p. (See Fig. 2.142.)

ii-4. If $\ell = \Psi(B)$, then

$$\pi_\ell: C_0 \rightarrow \mathbb{P}^3$$

is a double covering over a rational normal curve.

Exercises

1. Prove Lemmas 2.4.1 and 2.4.2.
2. Classify singular space rational sextic curves in $\mathbb{P}^3$.

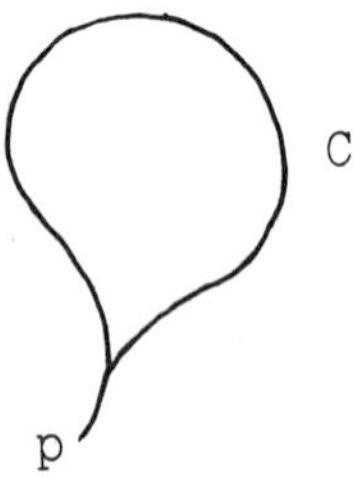

FIGURE 2.142

PART II

INTRINSIC GEOMETRY OF CURVES

3

COMPLEX MANIFOLDS AND PROJECTIVE VARIETIES

3.1 COMPLEX MANIFOLDS

Let Ω be an open set in $\mathbb{C}^n$. A complex valued function f on Ω is said to be holomorphic if it is (1) continuous and (2) holomorphic in each variable, that is, $f(z) = f(z_1, \ldots, z_n)$ is holomorphic in z_1 if $z_2, \ldots, z_n$ are fixed, and so forth.

Many fundamental properties of holomorphic functions of one variable can be easily generalized to those of several variables. We list some of them.

1. (Cauchy's Integral Formula) If f is holomorphic in a neighborhood of the closure of the polydisc

$$\Delta = \Delta(z_1^\circ, r_1) \times \cdots \times \Delta(z_n^\circ, r_n), \qquad (\Delta(z_j^\circ, r_j) = \{z \in \mathbb{C} \mid |z - z_j^\circ| < r_j\})$$

then

$$f(z_1, \ldots, z_n) = \frac{1}{(2\pi\sqrt{-1})^n} \int_\Gamma \frac{f(\zeta_1, \ldots, \zeta_n)}{(\zeta_1 - z_1) \cdots (\zeta_n - z_n)} d\zeta_1 \cdots d\zeta_n$$

for any point $(z_1, \ldots, z_n) \in \Delta$, where

$$\Gamma = \partial\Delta(z_1^\circ, r_1) \times \cdots \times \partial\Delta(z_n^\circ, r_n), \qquad (\partial\Delta(z_j^\circ, r_j) = \{z \in \mathbb{C} \mid |z - z_j^\circ| = r_j\})$$

(Note that Γ is not the topological boundary of Δ but a part of it.)

2. A holomorphic function is complex analytic, that is, at every point $p \in \Omega$, it can be expanded into a convergent power series

$$f(z_1, \ldots, z_n) = \sum_{\alpha_j \geq 0} a_{\alpha_1 \cdots \alpha_n} (z_1 - z_1^\circ)^{\alpha_1} \cdots (z_n - z_n^\circ)^{\alpha_n}$$

in a neighborhood of $p = (z_1^\circ, \ldots, z_n^\circ)$, and vice versa.

3. A holomorphic function f is a C^∞-function which satisfies the Cauchy-Riemann's equations:

$$\frac{\partial f}{\partial \bar{z}_j} = 0 \quad 1 \leq j \leq n$$

where

$$\frac{\partial}{\partial \bar{z}_j} = \frac{1}{2}\left(\frac{\partial}{\partial x_j} + \sqrt{-1}\,\frac{\partial}{\partial y_j}\right), \quad (z_j = x_j + \sqrt{-1}\,y_j)$$

and vice versa.

4. (Maximum Principle) For a holomorphic function f on a domain (that is, a connected open set) Ω of $\mathbb{C}^n$, if the absolute value $|f(z)|$ takes its maximum value at a point in Ω, then f is necessarily a constant.
5. (Principle of Analytic Continuation) For holomorphic functions f and g on a domain Ω in $\mathbb{C}^n$, if $f = g$ on a (nonempty) open set in Ω, then $f = g$ on Ω.
6. (Riemann's Extension Theorem) Let Ω be a domain in $\mathbb{C}^n$ and S be a thin subset of Ω. Let f be holomorphic on $\Omega - S$ such that, for any $p \in S$, there is a neighborhood U of p in Ω such that f is bounded on $U-S$. Then there is a unique holomorphic function $\tilde{f}$ on Ω such that $\tilde{f} = f$ on $\Omega - S$. (Here, a subset S is said to be thin if, for every point $p \in S$, there are a neighborhood U of p in Ω and a nonconstant holomorphic function h on U such that $h = 0$ on $U \cap S$. S is then nowhere dense by Property 5.)

For the proofs of these properties, see, for example, R. Narasimhan [75], a book which is strongly recommended for a beginner.

A map

$$f : z = (z_1, \ldots, z_n) \to f(z) = (f_1(z), \ldots, f_m(z))$$

of an open set Ω in $\mathbb{C}^n$ into $\mathbb{C}^m$ is said to be holomorphic if all functions $f_k(z)$, $1 \leq k \leq m$, are holomorphic.

A homeomorphism f of Ω onto an open set Ω' in $\mathbb{C}^n$ is said to be biholomorphic (or holomorphically isomorphic) if f and f^{-1} are holomorphic.

For a holomorphic map $f = (f_1, \ldots, f_m) : \Omega \to \mathbb{C}^m$, we define the Jacobian matrix $J_f(p)$ (respectively the Jacobian $\det J_f(p)$, if $m = n$) of f at $p \in \Omega$ by the matrix

$$J_f(p) = \frac{\partial(f_1, \ldots, f_m)}{\partial(z_1, \ldots, z_n)} = \begin{pmatrix} \frac{\partial f_1}{\partial z_1}(p) & \cdots & \frac{\partial f_m}{\partial z_1}(p) \\ \cdots & & \\ \frac{\partial f_1}{\partial z_n}(p) & \cdots & \frac{\partial f_m}{\partial z_n}(p) \end{pmatrix}$$

(respectively by its determinant).

7. (Inverse Mapping Theorem) Let $f: \Omega\ (\subset \mathbb{C}^n) \to \mathbb{C}^n$ be a holomorphic map. Suppose that, for a point $p \in \Omega$, $\det J_f(p) \neq 0$. Then there are neighborhoods U of p and U' of f(p) such that f maps U biholomorphically onto U'.
8. (Implicit Mapping Theorem) Let $f = (f_1, \ldots, f_m): \Omega\ (\subset \mathbb{C}^n) \to \mathbb{C}^m$ be a holomorphic map. For a point $p_0 \in \Omega$, suppose that (1) the rank k of $J_f(p)$ is constant in a neighborhood of p_0, (2) $f_1(p_0) = \cdots = f_m(p_0) = 0$, and (3) $\det\left(\frac{\partial(f_1, \ldots, f_k)}{\partial(z_1, \ldots, z_k)}(p_0)\right) \neq 0$. Then there are a neighborhood U of p_0 in Ω and uniquely determined holomorphic functions $\phi_1, \ldots, \phi_k$ of $(z_{k+1}, \ldots, z_n)$ such that the simultaneous equations on U

$$f_1(z_1, \ldots, z_n) = \cdots = f_m(z_1, \ldots, z_n) = 0$$

have unique solutions

$$z_1 = \phi_1(z_{k+1}, \ldots, z_n), \ \ldots, \ z_k = \phi_k(z_{k+1}, \ldots, z_n)$$

Theorem 7 can be proved by the usual Inverse Mapping Theorem together with the Cauchy-Riemann's equations. Theorem 8 follows from Theorem 7 (compare Exercise 1).

Now, we are ready to define a complex manifold. A complex manifold is, by definition, a connected Hausdorff space M with an open covering $\{U_\alpha\}$ such that, for every α, there is a homeomorphism ϕ_α of U_α onto an open set Ω_α in $\mathbb{C}^n$ with the following property: if $U_\alpha \cap U_\beta \neq \phi$, then $\phi_\alpha \cdot \phi_\beta^{-1}: \phi_\beta(U_\alpha \cap U_\beta) \to \phi_\alpha(U_\alpha \cap U_\beta)$ is biholomorphic. (See Fig. 3.1.)

The pair (U_α, ϕ_α) is called a chart of M. A coordinate system $(z_1^\alpha, \ldots, z_n^\alpha)$ in Ω_α is called a local coordinate system on U_α.

We have assumed the connectedness of M in the above definition. Some authors do not assume it. We call such a mathematical object a not necessarily connected complex manifold.

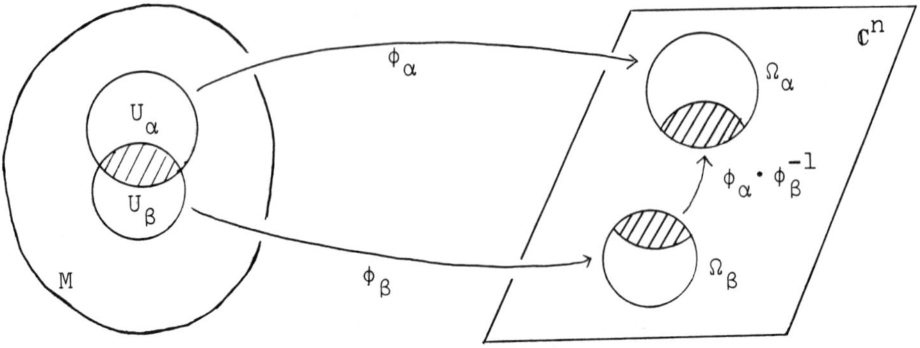

FIGURE 3.1

The integer n in the above definition is called the dimension of M and is denoted by dim M. It should be noted that the topological dimension of M is 2n. In fact, as is easily shown,

Proposition 3.1.1. An n-dimensional complex manifold M can be regarded as a 2n-dimensional orientable differentiable manifold. (Its differentiable structure is called the underlying differentiable structure of M.)

However, the converse is not true. For example, it is known that the even dimensional spheres

$$S^4, S^8, S^{10}, S^{12}, \ldots$$

cannot be complex manifolds. ($S^2 = \mathbb{P}^1(\mathbb{C})$ is a complex manifold. It is an unsolved problem whether S^6 can be a complex manifold.)

If U is a connected open set in a complex manifold M, then U is naturally a complex manifold.

If M and N are complex manifolds, then the product $M \times N$ is naturally a complex manifold.

$\mathbb{C}^n$ and its connected open sets are primary examples of complex manifolds.

One of the most important examples of compact complex manifolds is the complex projective space $\mathbb{P}^n = \mathbb{P}^n(\mathbb{C})$ (see Sec. 1.1).

Now, let M be a complex manifold and U be an open set of M. A complex valued function f on U is said to be holomorphic if, for any chart (U_α, ϕ_α) with $U_\alpha \cap U \neq \phi$, the function $f \cdot \phi_\alpha^{-1}$ on $\phi_\alpha(U_\alpha \cap U)$ is holomorphic.

By the Maximum Principle,

Proposition 3.1.2. A holomorphic function on a compact complex manifold is necessarily a constant.

As for $M = \mathbb{P}^1$, this proposition is known as Liouville's Theorem.

A continuous map f of a complex manifold M into a complex manifold N is said to be holomorphic if, for any chart (U_α, ϕ_α) on M and (W_ν, ψ_ν) on N such that $U_\alpha \cap f^{-1}(W_\nu) \neq \phi$, the map $\psi_\nu f \phi_\alpha^{-1}$ on $\phi_\alpha(U_\alpha \cap f^{-1}(W_\nu))$ is holomorphic. A bijective map f of M onto N is said to be biholomorphic (or holomorphically isomorphic) if both f and f^{-1} are holomorphic. M and N are said to be biholomorphic (or holomorphically equivalent if there is a biholomorphic map f: M → N. An equivalence class is called a complex structure. If M and N are holomorphically equivalent, they are said to have the same complex structure.

If M and N have the same complex structure, then they have the same underlying differentiable structure. But the converse is not true, as will be shown later by an example.

A biholomorphic map of M onto itself is called an automorphism. The set of all automorphisms of M forms a group Aut(M), called the automorphism group of M, under the composition of maps. Roughly speaking, this group measures how M is symmetric. In the extreme case, if Aut(M) acts on M transitively, then M is said to be homogeneous. For example, $\mathbb{P}^1 = \hat{\mathbb{C}}$ is homogeneous; $\mathrm{Aut}(\hat{\mathbb{C}})$ is the group of all linear fractional transformations

$$\phi: z \in \hat{\mathbb{C}} \to \frac{az+b}{cz+d} \in \hat{\mathbb{C}}, \qquad ad - bc \neq 0$$

Note that $\mathrm{Aut}(\hat{\mathbb{C}})$ itself can be considered as a complex manifold

$$\mathrm{Aut}(\hat{\mathbb{C}}) = \{(a: b: c: d) \in \mathbb{P}^3 \mid ad - bc \neq 0\}$$

an open set in $\mathbb{P}^3$. This is a special case of

Theorem 3.1.3 (Bochner-Montgomery [13]). For a compact complex manifold M, Aut(M) is a complex Lie group such that the action

$$(\phi, p) \in \mathrm{Aut}(M) \times M \to \phi(p) \in M$$

is holomorphic.

Here, a complex Lie group G is a group which is a not necessarily connected complex manifold such that the group action

$$(x, y) \in G \times G \to xy^{-1} \in G$$

is holomorphic.

$$GL(n, \mathbb{C}) = \{n \times n \text{ nonsingular matrices}\}$$

$$SL(n, \mathbb{C}) = \{n \times n \text{ matrices A with } \det A = 1\}$$

called the general linear group and the special linear group, respectively, are important examples of complex Lie groups. For a n-dimensional complex vector space V, the set GL(V) of all linear transformations of V onto itself forms a complex Lie group which is isomorphic to $GL(n, \mathbb{C})$.

In the above theorem, the compactness of M is an essential assumption. For example, $\mathrm{Aut}(\mathbb{C}^2)$ is too big to be a Lie group. However,

Theorem 3.1.4 (Cartan [16]). If a complex manifold M is biholomorphic to a bounded domain of $\mathbb{C}^n$, then Aut(M) is a (real) Lie group.

For example, the upper half plane

$$\mathbb{H} = \{z = x + \sqrt{-1}\, y \in \mathbb{C} \mid y > 0\}$$

is biholomorphic to the unit disc

$$\mathbb{D} = \{z \in \mathbb{C} \mid |z| < 1\}$$

In fact, a biholomorphic map $f: \mathbb{H} \to \mathbb{D}$ is given by

$$f: z \in \mathbb{H} \to \frac{z - \sqrt{-1}}{z + \sqrt{-1}} \in \mathbb{D}$$

(See Fig. 3.2.) It is well known that

$$\mathrm{Aut}(\mathbb{H}) = \{z \to \frac{az + b}{cz + d} \mid a, b, c, d \in R,\ ad - bc \neq 0\}$$

which can be identified with an open set of the real projective space $\mathbb{P}^3(\mathbb{R})$.

Now, let p be a point of a n-dimensional complex manifold M and $(z_1, \ldots, z_n)$ be a local coordinate system around p. If we regard M as a 2n-dimensional differentiable manifold, then the (real) tangent space $T_p^{\mathbb{R}}M$ is defined. It has the following basis

$$\left(\frac{\partial}{\partial x_1}\right)_p, \left(\frac{\partial}{\partial y_1}\right)_p, \ldots, \left(\frac{\partial}{\partial x_n}\right)_p, \left(\frac{\partial}{\partial y_n}\right)_p$$

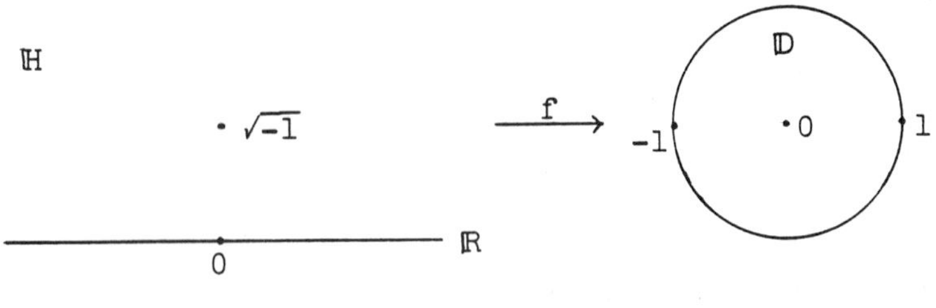

FIGURE 3.2

($z_k = x_k + \sqrt{-1}\, y_k$, $1 \leq k \leq n$). Consider the complex vector space $T_p^{\mathbb{R}} M \otimes \mathbb{C}$. We take the following basis of it.

$$\left(\frac{\partial}{\partial z_k}\right)_p = \frac{1}{2}\left(\left(\frac{\partial}{\partial x_k}\right)_p - \sqrt{-1}\left(\frac{\partial}{\partial y_k}\right)_p\right), \quad \left(\frac{\partial}{\partial \bar{z}_k}\right)_p = \frac{1}{2}\left(\left(\frac{\partial}{\partial x_p}\right)_p + \sqrt{-1}\left(\frac{\partial}{\partial y_k}\right)_p\right)$$

($1 \leq k \leq n$). We denote by T_pM the complex vector subspace of $T_p^{\mathbb{R}} M \otimes \mathbb{C}$ spanned by $\left(\frac{\partial}{\partial z_k}\right)_p$, $1 \leq k \leq n$, and call it the tangent space at p to the complex manifold M (It can easily be shown that T_pM does not depend on the choice of $(z_1, \ldots, z_n)$.) Its dual space T_p^*M, called the cotangent space at p to M, has the dual basis $\{(dz_1)_p, \ldots, (dz_n)_p\}$ to $\{(\partial/\partial z_1)_p, \ldots, (\partial/\partial z_n)_p\}$, where $(dz_k)_p = (dx_k)_p + \sqrt{-1}\,(dy_k)_p$, $1 \leq k \leq n$.

A holomorphic vector field X on M is a correspondence

$$X\colon p \in M \to X_p \in T_pM$$

which can be locally written as

$$X_p = a_1(p)\left(\frac{\partial}{\partial z_1}\right)_p + \cdots + a_n(p)\left(\frac{\partial}{\partial z_n}\right)_p$$

where $a_k(p)$, $1 \leq k \leq n$, are holomorphic functions. We write X locally as

$$X = a_1 \frac{\partial}{\partial z_1} + \cdots + a_n \frac{\partial}{\partial z_n}$$

A holomorphic m-form ($1 \leq m \leq n$) ω on M is a correspondence

$$\omega\colon p \in M \to \omega_p \in \Lambda^m T_p^*M$$

($\Lambda^m T_p^*M$ = the m-th exterior product of T_p^*M) which can be locally written as

$$\omega_p = \sum_{1 \leq \alpha_1 < \cdots < \alpha_m \leq n} a_{\alpha_1 \cdots \alpha_m}(p)(dz_{\alpha_1})_p \wedge \cdots \wedge (dz_{\alpha_m})_p$$

where $a_{\alpha_1 \cdots \alpha_m}(p)$ are holomorphic functions. We write ω locally as

$$\omega = \sum a_{\alpha_1 \cdots \alpha_m} dz_{\alpha_1} \wedge \cdots \wedge dz_{\alpha_m}$$

For a holomorphic map $f: M \to N$ and $p \quad M$, <u>the differential</u>

$$(df)_p : T_pM \to T_qN \quad (q = f(p))$$

at p is defined by

$$(df)_p\left(\left(\frac{\partial}{\partial z_k}\right)_p\right) = \sum_j \left(\frac{\partial f_j}{\partial z_k}\right)_p \left(\frac{\partial}{\partial w_j}\right)_q$$

where

$$f: z = (z_1, \ldots, z_n) \to w = (w_1, \ldots, w_m) = (f_1(z), \ldots, f_m(z))$$

If f is biholomorphic, then $(df)_p$ is a linear isomorphism. The converse is true <u>locally</u> by the inverse mapping theorem.

f is called a <u>holomorphic immersion</u> if $(df)_p$ is injective for all $p \in M$. f is called a <u>holomorphic imbedding</u> if it is injective and is a holomorphic immersion.

A closed subset S of a complex manifold M is a <u>complex submanifold</u> of M if (1) S is a complex manifold and (2) the inclusion map $\iota: S \hookrightarrow M$ is a holomorphic imbedding. By the implicit mapping theorem,

<u>Proposition 3.1.5</u>. If $f: M \to N$ is a holomorphic imbedding such that f(M) is closed in N, then (1) f(M) is a complex submanifold of N and (2) $f: M \to f(M)$ is a holomorphic isomorphism.

Next, let B, M, and F be complex manifolds. Let G be a subgroup of Aut(F) and suppose that G is a complex Lie group acting holomorphically on F. A surjective holomorphic map

$$\pi: B \to M$$

(or simply B itself) is called a <u>(holomorphic) fiber bundle on</u> M <u>with the standard fiber</u> F <u>and the structure group</u> G, if there are an open covering $\{U_\alpha\}$ of M and holomorphic isomorphisms

$$\psi_\alpha: \pi^{-1}(U_\alpha) \to U_\alpha \times F$$

such that (1) the diagram

$$\begin{array}{ccc} \pi^{-1}(U_\alpha) & \longrightarrow & U_\alpha \times G \\ & {\scriptstyle \pi} \searrow \quad \swarrow {\scriptstyle \pi_0} & \\ & U_\alpha & \end{array}$$

is commutative, where π_0 is the natural projection, (2) the automorphism

$$g_{\alpha\beta}(p): \zeta \in F \to g_{\alpha\beta}(p, \zeta) \in F$$

is an element of G for every $p \in U_\alpha \cap U_\beta$, where

$$\psi_\alpha \psi_\beta^{-1}(p, \zeta) = (p, g_{\alpha\beta}(p, \zeta)) \quad \text{for} \quad (p, \zeta) \in (U_\alpha \cap U_\beta) \times F$$

and (3) the map

$$g_{\alpha\beta}: p \in U_\alpha \cap U_\beta \to g_{\alpha\beta}(p) \in G$$

is holomorphic.

The maps $g_{\alpha\beta}$ are called the transition functions of the fiber bundle. They satisfy the following relations:

$$g_{\alpha\alpha}(p) = 1 \ (\in G) \quad \text{for } p \in U_\alpha, \text{ and}$$

$$g_{\alpha\beta}(p)\, g_{\beta\gamma}(p) = g_{\alpha\gamma}(p) \quad \text{for } p \in U_\alpha \cap U_\beta \cap U_\gamma$$

Conversely, given M, F, G, and a collection of holomorphic maps $g_{\alpha\beta}$: $U_\alpha \cap U_\beta \to G$ satisfying these relations, we can construct a complex manifold B and a fiber bundle $\pi: B \to M$ with the transition function $g_{\alpha\beta}$ as follows

$$B = \bigcup_\alpha (U_\alpha \times F)/\sim$$

where $\bigcup$ is a disjoint union and the equivalence relation is given by

$$(p, \zeta) \in U_\alpha \times F \sim (q, \xi) \in U_\beta \times F$$

if and only if (1) $p = q$ and (2) $\zeta = g_{\alpha\beta}(p)\xi$. Then B and the natural projection $\pi: B \to M$ satisfy the condition.

If F is compact and $G = \mathrm{Aut}(F)$, then $\pi: B \to M$ is simply called a F-bundle. A $\mathbb{P}^r$-bundle is sometimes called a projective bundle.

A holomorphic section of a fiber bundle $\pi: B \to M$ is, by definition, a holomorphic map $\sigma: M \to B$ such that

$$\pi\sigma = \text{the identity map on } M$$

σ is naturally identified with a collection of holomorphic maps

$$\xi_\alpha : U_\alpha \to F$$

such that

$$\xi_\alpha(p) = g_{\alpha\beta}(p)\xi_\beta(p) \quad \text{for } p \in U_\alpha \cap U_\beta$$

Note that the above definition of a fiber bundle is parallel to that of differentiable (or topological) fiber bundle. Hence, we can define morphisms (isomorphisms) of fiber bundles in a similar way.

In the above definition of a fiber bundle, if $F = G$ is a complex Lie group and G acts on itself as the right (or left) translations

$$T_h : g \in G \to gh \in G \quad \text{for } h \in G$$

then $\pi: B \to M$ is called a principal bundle with the structure group G. (In this case, G acts on B and $M = B/G$ is the quotient space.)

In the above definition of a fiber bundle again, if F is a complex vector space and $G = GL(F)$, then $\pi: B \to M$ is called a (holomorphic) vector bundle on M.

The disjoint union

$$TM = \bigcup_p T_pM \quad (\text{respectively } \Lambda^m T^*M = \bigcup_p \Lambda^m T_p^*M)$$

is an important example of vector bundles on M, called the tangent bundle of M (respectively the m-th cotangent bundle of M). Its holomorphic sections are nothing but holomorphic vector fields (respective holomorphic m-forms).

A vector bundle with $\dim F = 1$ is called a (holomorphic) line bundle on M. $K_M = \Lambda^n T^*M$ ($n = \dim M$) is a very important example of line bundles on M and is called the canonical bundle of M.

The transition functions of a line bundle on M are nonvanishing holomorphic functions $g_{\alpha\beta}$ on $U_\alpha \cap U_\beta$ such that

$$g_{\alpha\alpha} = 1 \quad \text{on } U_\alpha$$

$$g_{\alpha\beta}g_{\beta\gamma} = g_{\alpha\gamma} \quad \text{on } U_\alpha \cap U_\beta \cap U_\gamma$$

Conversely, such a collection $\{g_{\alpha\beta}\}$ of nonvanishing holomorphic functions determines an (isomorphism class of) line bundle on M. We identify the collection $\{g_{\alpha\beta}\}$ with the line bundle determined by it.

Note that the transition functions of K_M are given by the Jacobians

$$g_{\alpha\beta} = \det \frac{\partial(z_1^\beta, \ldots, z_n^\beta)}{\partial(z_1^\alpha, \ldots, z_n^\alpha)}$$

where $(z_1^\alpha, \ldots, z_n^\alpha)$ (respectively $(z_1^\beta, \ldots, z_n^\beta)$) is a local coordinate system on U_α (respectively on U_β).

Let $L_1 = \{g_{\alpha\beta}\}$ and $L_2 = \{h_{\alpha\beta}\}$ be line bundles on M. The tensor product $L_1 \otimes L_2$ is, by definition, the line bundle $\{g_{\alpha\beta} h_{\alpha\beta}\}$ on M. $L_1 \otimes L_2$ is sometimes written as $L_1 + L_2$, if no confusion is probable. (Hence, $L_1^{\otimes k}$ is sometimes written as kL_1.)

The set Pic(M) of all (isomorphic classes of) line bundles on M forms a group under the tensor product, called the Picard group of M.

Let M and N be complex manifolds and $f: M \to N$ be a holomorphic map. Let $L = \{g_{\alpha\beta}\}$ be a line bundle on N. Then $f^*L = \{g_{\alpha\beta} \cdot f\}$ is a line bundle on M, called the pull back of L by f. The map

$$f^*: L \in \mathrm{Pic}(N) \to f^*L \in \mathrm{Pic}(M)$$

is a homomorphism. If $\xi = \{\xi_\alpha\}$ is a holomorphic section of L, then $f^*\xi = \{\xi_\alpha \cdot f\}$ is a holomorphic section of f^*L.

Finally, we give an important example of compact complex manifolds, a complex torus. Let ω_1 and ω_2 be nonzero complex numbers such that the imaginary part $\mathrm{Im}(\omega_2/\omega_1)$ of ω_2/ω_1 is positive, that is, $\omega_2/\omega_1 \in \mathbb{H}$. We denote by $\Gamma = \mathbb{Z}\omega_1 + \mathbb{Z}\omega_2$, the free additive group generated by ω_1 and ω_2. ($\mathbb{Z}$ is the additive group of all integers.) Γ acts on $\mathbb{C}$ as follows

$$(a\omega_1 + b\omega_2, z) \in \Gamma \times \mathbb{C} \to z + a\omega_1 + b\omega_2 \in \mathbb{C} \quad (a, b \in \mathbb{Z})$$

The quotient space $\mathbb{C}/\Gamma$ is called a complex 1-torus. It is easy to see that $\mathbb{C}/\Gamma$ is a compact complex manifold of dimension 1; its complex structure is uniquely determined by the requirement that the natural projection

$$\pi: \mathbb{C} \to \mathbb{C}/\Gamma$$

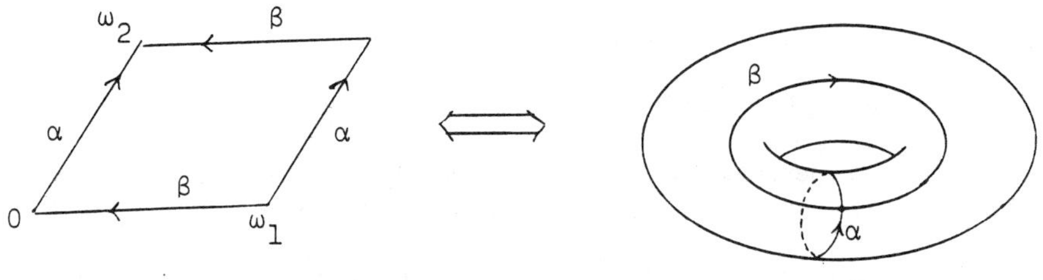

FIGURE 3.3

is holomorphic. Topologically, $\mathbb{C}/\Gamma$ is $S^1 \times S^1$. (See Fig. 3.3.)

In a similar way, let $\omega_1, \ldots, \omega_{2n}$ be nonzero vectors in $\mathbb{C}^n$ which are linearly independent over $\mathbb{R}$, the real numbers. Put $\Gamma = \mathbb{Z}\omega_1 + \cdots + \mathbb{Z}\omega_{2n}$. Γ then acts on $\mathbb{C}^n$ as above and $\mathbb{C}^n/\Gamma$ is a n-dimensional compact complex manifold, called a complex n-torus. Topologically, $\mathbb{C}^n/\Gamma$ is $S^1 \times \cdots \times S^1$ (2n-times). The $(2n \times n)$-matrix

$$\Omega = \begin{pmatrix} \omega_1 \\ \vdots \\ \omega_{2n} \end{pmatrix}$$

is called a period matrix. We sometimes write $\mathbb{C}^n/\Omega$ instead of $\mathbb{C}^n/\Gamma$.

An interesting phenomenon occurs for complex tori; two complex tori may have different complex structures, while they have the same underlying differentiable structure $S^1 \times \cdots \times S^1$. We explain this for complex 1-tori.

Let $M = \mathbb{C}/\Gamma$ and $M' = \mathbb{C}/\Gamma'$ be complex 1-tori, where

$$\Gamma = \mathbb{Z}\omega_1 + \mathbb{Z}\omega_2 \quad \text{and} \quad \Gamma' = \mathbb{Z}\omega_1' + \mathbb{Z}\omega_2'$$

Assume that there is a nonconstant holomorphic map

$$f : M \to M'$$

Since the natural projections $\pi : \mathbb{C} \to M$ and $\pi' : \mathbb{C} \to M'$ are the universal covering spaces, there is a continuous map $\tilde{f} : \mathbb{C} \to \mathbb{C}$ such that the diagram

$$\begin{array}{ccc} \mathbb{C} & \xrightarrow{\tilde{f}} & \mathbb{C} \\ \pi \downarrow & & \downarrow \pi' \\ M & \xrightarrow{f} & M' \end{array}$$

is commutative. Locally, $\tilde{f} = \pi'^{-1} f \pi$, so $\tilde{f}$ is a holomorphic function on $\mathbb{C}$. By the diagram, for any $z \in \mathbb{C}$, there are integers a, b, c, d such that

$$\begin{aligned} \tilde{f}(z + \omega_1) &= \tilde{f}(z) + a\omega_1' + b\omega_2' \\ \tilde{f}(z + \omega_2) &= \tilde{f}(z) + c\omega_1' + b\omega_2' \end{aligned} \tag{1}$$

Then a, b, c, and d clearly depend continuously on z, so they must be constants. Hence,

$$\tilde{f}'(z + \omega_1) = \tilde{f}'(z) \quad \text{and} \quad \tilde{f}'(z + \omega_2) = \tilde{f}'(z) \quad \text{for } z \in \mathbb{C}$$

where $\tilde{f}' = d\tilde{f}/dz$. This implies that $\tilde{f}'$ is a holomorphic function on M, so is a constant by Proposition 3.1.2. Hence, $\tilde{f}$ can be written as

$$\tilde{f}(z) = \lambda z + \mu, \qquad (\lambda, \mu \in \mathbb{C}, \lambda \neq 0)$$

In (1), put z = 0. Then we get

$$\begin{aligned} \lambda\omega_1 &= a\omega_1' + b\omega_2' \\ \lambda\omega_2 &= c\omega_1' + b\omega_2' \end{aligned} \tag{2}$$

Hence,

$$\frac{\omega_2}{\omega_1} = \frac{c + d(\omega_2'/\omega_1')}{a + b(\omega_2'/\omega_1')}$$

Since $\mathrm{Im}(\omega_2/\omega_1) > 0$ and $\mathrm{Im}(\omega_2'/\omega_1') > 0$,

$$ad - bc > 0$$

Conversely, if λ satisfies (2) with $ad - bc > 0$ and μ is arbitrary, then $\tilde{f}: z \in \mathbb{C} \to \lambda z + \mu \in \mathbb{C}$ induces a holomorphic map

$$f: M \to M'$$

It is an (unramified) covering map. The covering degree is

$$[\Gamma': \tilde{f}(\Gamma)] = ad - bc$$

In particular, $f: M \to M'$ is biholomorphic if and only if

$$ad - bc = 1, \text{ that is, } \begin{pmatrix} a & b \\ c & d \end{pmatrix} \in SL(2, \mathbb{Z})$$

Thus, we conclude

<u>Theorem 3.1.6.</u> (1) $\mathbb{C}/(\mathbb{Z}\omega_1 + \mathbb{Z}\omega_2)$ is biholomorphic to $\mathbb{C}/(\mathbb{Z} + \mathbb{Z}\tau)$, where $\tau = \omega_2/\omega_1 \in \mathbb{H}$. (2) $\mathbb{C}/(\mathbb{Z} + \mathbb{Z}\tau_1)$ and $\mathbb{C}/(\mathbb{Z} + \mathbb{Z}\tau_2)$ are biholomorphic if and only if τ_1 and τ_2 are equivalent under the following action of SL(2, $\mathbb{Z}$) on $\mathbb{H}$

$$\left(\begin{pmatrix} a & b \\ c & d \end{pmatrix}, \tau\right) \in SL(2, \mathbb{Z}) \in \mathbb{H} \to \frac{c + d\tau}{a + b\tau} \in \mathbb{H}$$

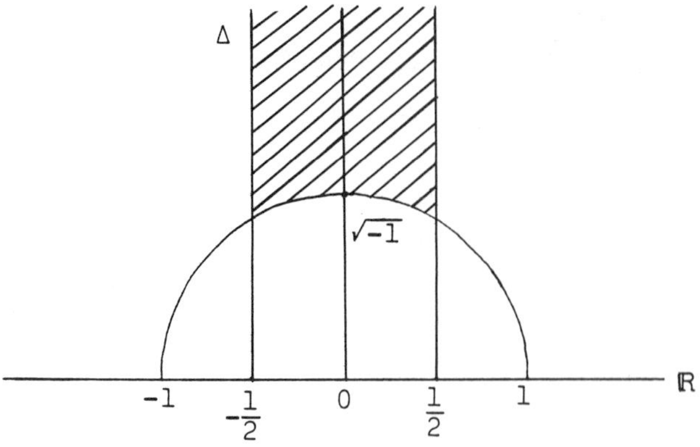

FIGURE 3.4

Remark 3.1.7. Note that SL(2, $\mathbb{Z}$) itself is not (regarded as) a subgroup of Aut($\mathbb{H}$), but SL(2, $\mathbb{Z}$)/$\{\pm E\}$ $\left(E = \begin{pmatrix} 1 & 0 \\ 0 & 1 \end{pmatrix}\right)$ is. This group is called the elliptic modular group.

A fundamental domain Δ of the action of SL(2, $\mathbb{Z}$) on $\mathbb{H}$ is given by Fig. 3.4. (Here, a fundamental domain Δ means a domain such that (1) every point of H is equivalent under the action of SL(2, $\mathbb{Z}$) to some point of $\bar{\Delta}$ and (2) any two points of Δ are not equivalent.) Note that two points of the boundary of Δ are equivalent if and only if they are symmetric with respect to the imaginary axis.

For the proof of the above facts, see, for example, Serre [88].

Figure 3.4 (more precisely, a figure almost equivalent to this) was drawn by Gauss in his notebook in 1827. Nobody could understand its meaning for more than 50 years!

The quotient space $\mathbb{H}$/SL(2, $\mathbb{Z}$) is called the moduli space of complex 1-tori. It is the set of all complex structures on $S^1 \times S^1$, and is a complex manifold biholomorphic to $\mathbb{C}$. See Corollary 5.2.20.

In the above argument, if we put $\Gamma' = \Gamma$, then we get information about holomorph maps of $\mathbb{C}/\Gamma$ onto itself (see Exercise 11). In particular (compare Exercise 10)

Theorem 3.1.8.

1. If ω_2/ω_1 is equivalent under SL(2, $\mathbb{Z}$) to neither $i = \sqrt{-1}$ nor $\zeta = (1 + \sqrt{-3})/2$, then

$$\mathrm{Aut}(\mathbb{C}/(\mathbb{Z}\omega_1 + \mathbb{Z}\omega_2)) = \langle \mathrm{Aut}_0,\ S_{-1} \rangle$$

2. $\mathrm{Aut}(\mathbb{C}/(\mathbb{Z} + \mathbb{Z}i)) = \langle \mathrm{Aut}_0, S_i \rangle$

3. $\mathrm{Aut}(\mathbb{C}/(\mathbb{Z} + \mathbb{Z}\xi)) = \langle \mathrm{Aut}_0, S_\xi \rangle$

($G = \langle A, B \rangle$ means that the group G is generated by A and B.) Here, Aut_0 is the set of all translations $T_y: x \to x + y$ and S_λ is defined by $S_\lambda: x \to \lambda x$.

Note 3.1.9. As for the function theory of several complex variables, see, for example, R. Narasimhan [75] and Hörmander [45]. As for complex manifolds, see, for example, Chern [19], Kodaira-Morrow [57], and Griffiths-Harris [33, Chapter 0].

Exercises

1. Prove (1) Cauchy's Integral Formula—(8) Implicit Mapping Theorem.

2. (Open Mapping Theorem) Let Ω be a domain in $\mathbb{C}^n$ and f be a nonconstant holomorphic function on Ω. Then $f: \Omega \to \mathbb{C}$ is an open map, that is, f(U) is open in $\mathbb{C}$ if U is open in Ω.

3. (Hartog's Theorem) For $r > 0$, put

$$\Delta(0,r) = \left\{ z = (z_1, \ldots, z_n) \in \mathbb{C}^n \;\middle|\; |z_k| < r,\ 1 \le k \le n \right\}$$

If $n \ge 2$, then any holomorphic function on $\Delta(0,r) - \overline{\Delta(0,r')}$ with $0 < r' < r$ can be uniquely extended to a holomorphic function on $\Delta(0,r)$.

4. Let M be a complex manifold. A subgroup G of Aut(M) is said to act on M (1) properly discontinuously, if, for any compact sets K_1 and K_2 in M, $\{\sigma \in G \mid (\sigma K_1) \cap K_2 \neq \phi\}$ is a finite subset of G and (2) without fixed point, if every $\sigma \in G - \{1\}$ has no fixed point, that is, $\sigma(p) \neq p$ for all $p \in M$. Prove that, if a subgroup G of Aut(M) acts on M properly discontinuously without fixed point, then the quotient space M/G becomes a complex manifold such that the canonical projection $\pi: M \to M/G$ is a holomorphic (unramified) covering map.

5. (Primary Hopf Manifolds) Put $W = \mathbb{C}^n - \{0\}$ and take $\alpha \in \mathbb{C}$ such that $0 < |\alpha| < 1$. Put $G = G_\alpha = \{\alpha^m \mid m \in \mathbb{Z}\}$. G acts on W by $\alpha^m(z_1, \ldots, z_n) = (\alpha^m z_1, \ldots, \alpha^m z_n)$. Show that the quotient space $M = W/G$ is a complex manifold which is diffeomorphic to $S^1 \times S^{2n-1}$. M is called a primary Hopf manifold. When are W/G_α and W/G_β holomorphically isomorphic? (See Fig. 3.5.)

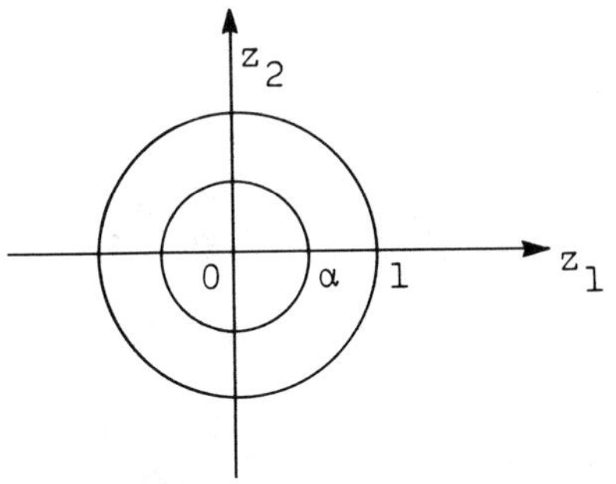

FIGURE 3.5

6. (Iwasawa Manifold) Identify $\mathbb{C}^3$ with the set of all (3×3)-matrices of the form

$$A = \begin{pmatrix} 1 & z_1 & z_2 \\ 0 & 1 & z_3 \\ 0 & 0 & 1 \end{pmatrix}$$

Put

$$G = \left\{ A = \begin{pmatrix} 1 & z_1 & z_2 \\ 0 & 1 & z_3 \\ 0 & 0 & 1 \end{pmatrix} \;\middle|\; z_k \in \mathbb{Z} + \sqrt{-1}\,\mathbb{Z},\ 1 \leq k \leq 3 \right\}$$

G acts on $\mathbb{C}^3$ by the matrix multiplication from the right. Show that the quotient space $\mathbb{C}^3/G$ is a compact complex manifold. (This is called the Iwasawa manifold.)

7. Put $P = \{(z_1, z_2) \in \mathbb{C}^2 \mid |z_1| < 1 \text{ and } |z_2| < 1\}$ (a polydisc) and $B = \{(z_1, z_2) \in \mathbb{C}^2 \mid |z_1|^2 + |z_2|^2 < 1\}$ (the unit ball). Compute the Lie groups Aut(P) and Aut(B) and conclude that P and B cannot be biholomorphic.

8. (1) Every holomorphic vector field on $\mathbb{P}^1$ can be written as

$$X = (a_0 + a_1 z + a_2 z^2) \frac{\partial}{\partial z}, \qquad (a_0, a_1, a_2 \in \mathbb{C})$$

where z is an affine coordinate on $\mathbb{P}^1 - \{\infty\}$. (2) Any holomorphic one-form on $\mathbb{P}^1$ is identically zero.

9. Let f: $M \to N$ be a holomorphic map of a complex manifold M into a complex manifold N. Then the graph $\Gamma_f = \{(p, f(p)) \in M \times N\}$ is a complex submanifold of $M \times N$.

10. Prove Theorem 3.1.8.

11. Let $M = \mathbb{C}/(\mathbb{Z} + \mathbb{Z}\tau)$ $(\tau \in \mathbb{H})$ be a complex 1-torus and m be a positive integer. (1) If τ is not a quadratic imaginary, then there is an (unramified) covering map $M \to M$ of degree m if and only if $m = k^2$ for an integer k. (2) If $a\tau^2 + b\tau + c = 0$, where $a > 0$, $a, b, c \in \mathbb{Z}$, $(a, b, c) = 1$ (coprime), then there is an (unramified) covering map $M \to M$ of degree m if an only if there is an integral solution (X, Y) for the quadratic Diophantus equation

$$X^2 - bXY + acY^2 = m$$

12. Let G be a complex Lie group and H be a complex Lie subgroup of G, that is, a subgroup of G which is a complex submanifold. Then the quotient space G/H is a complex manifold and the natural projection $G \to G/H$ is a principal bundle on G/H with the structure group H.

13. Complex n-tori $\mathbb{C}^n/\Omega$ and $\mathbb{C}^n/\Omega'$ are biholomorphic if and only if there are $A \in SL(2n, \mathbb{Z})$ and $B \in GL(n, \mathbb{C})$ such that

$$\Omega' = A\Omega B$$

14. A holomorphic map of a projective space into a complex torus is necessarily a constant map.

3.2 COMPLEX ANALYTIC SETS

Let M be an n-dimensional complex manifold and p be a point of M. Holomorphic functions f and g defined on neighborhoods U and V, respectively, of p are said to be equivalent at p if $f = g$ on a neighborhood $W \subset U \cap V$ of p. The equivalence class is called a germ of holomorphic functions at p and is denoted by $[f]_p$. However, by abuse of notation, we write simply f instead of $[f]_p$, when p is fixed.

The set of all germs at p forms a commutative ring $\mathcal{O}_p = \mathcal{O}_{M,p}$, called the ring of germs of holomorphic functions at p. The definition being local, $\mathcal{O}_p$ is $\mathbb{C}$-isomorphic to ${}_n\mathcal{O} = \mathcal{O}_{\mathbb{C}^n,0}$ (0 = the origin of $\mathbb{C}^n$), which is $\mathbb{C}$-isomorphic to the ring $\mathbb{C}\{z_1, \ldots, z_n\}$ of all convergent power series

$$\mathcal{O}_p \cong {}_n\mathcal{O} \cong \mathbb{C}\{z_1, \ldots, z_n\}$$

$f \to$ the power series expansion of f at $p = 0$

We identify these three rings through the isomorphisms.

Definition 3.2.1.

1. A germ $f \in {}_n\mathcal{O}$ is said to be regular of order k (>0) in z_n if

$$f(0, \ldots, 0, z_n) = c_k z_n^k + c_{k+1} z_n^{k+1} + \cdots, \qquad c_k \neq 0$$

2. A Weierstrass polynomial h of degree k (>0) in z_n is a (germ of) holomorphic function in $(z_1, \ldots, z_n)$ of the form

$$h(z_1, \ldots, z_n) = z_n^k + a_1(z_1, \ldots, z_{n-1}) z_n^{k-1} + \cdots + a_k(z_1, \ldots, z_{n-1})$$

where $a_j(z_1, \ldots, z_{n-1})$, $1 \leq j \leq k$, are (germs of) holomorphic functions in $(z_1, \ldots, z_{n-1})$ such that $a_j(0, \ldots, 0) = 0$.

Remark 3.2.2. Any nonconstant f such that $f(0) = 0$ can be regular (of order k) in z_n, after a suitable linear change of coordinate systems.

Theorem 3.2.3 (Weierstrass Preparation Theorem). If $f \in {}_n\mathcal{O}$ is regular of order k in z_n, then there are a Weierstrass polynomial h of order k in z_n and a unit $u \in {}_n\mathcal{O}$ such that $f = uh$. Moreover, h and u are uniquely determined. (Note that $u \in {}_n\mathcal{O}$ is a unit if and only if $u(0) \neq 0$.)

Proof. For $r = (r_1, \ldots, r_n)$ with $r_j > 0$, $1 \leq j \leq n$, put

$$\Delta(0, r) = \left\{ z = (z_1, \ldots, z_n) \in \mathbb{C}^n \,\middle|\, |z_1| < r_1, \ \ldots, \ |z_n| < r_n \right\}$$

a polydisc. Taking $r_1, \ldots, r_n$ sufficiently small, we may assume that f is holomorphic in a neighborhood of the closure $\overline{\Delta(0, r)}$ of $\Delta(0, r)$.

By the assumption, $f(0, \ldots, 0, z_n)$ has the zero of order k at $z_n = 0$. Taking r_n sufficiently small, we may assume that

$$f(0, \ldots, 0, z_n) \neq 0 \quad \text{for } 0 < |z_n| \leq r_n$$

Put

$$\epsilon = \inf \left\{ |f(0, \ldots, 0, z_n)| \,\middle|\, |z_n| = r_n \right\} \ (>0)$$

By the compactness of the set $\{(0, \ldots, 0, z_n) \mid |z_n| = r_n\}$, there are sufficiently small $r_1, \ldots, r_n > 0$ such that

$$|f(z_1, \ldots, z_{n-1}, z_n) - f(0, \ldots, 0, z_n)| < \epsilon$$

for

$$|z_j| < r_j, \quad 1 \leq j \leq n-1 \quad \text{and} \quad |z_n| = r_n$$

By Rouche's Theorem, for a fixed $(z_1, \ldots, z_{n-1})$ with $|z_j| < r_j$, $1 \leq j \leq n-1$, the equation

$$f(z_1, \ldots, z_{n-1}, z_n) = 0$$

for z_n has just k zeros (counting multiplicities) in $|z_n| < r_n$, which we denote by

$$\alpha_1 = \alpha_1(z'), \ldots, \alpha_k = \alpha_k(z'), \qquad (z' = (z_1, \ldots, z_{n-1}))$$

Note that $\alpha_j(0) = 0$, $1 \leq j \leq k$. Put

$$h(z_1, \ldots, z_n) = \prod_{\nu=1}^{k} (z_n - \alpha_\nu) = z_n^k + a_1(z')z_n^{k-1} + \cdots + a_k(z')$$

where $a_j(z')$ are the elementary symmetric functions in $\alpha_1, \ldots, \alpha_k$.

On the other hand, we have

$$\sum_{\nu=1}^{k} \alpha_\nu^p = \frac{1}{2\pi\sqrt{-1}} \int_{|\zeta|=r_n} \frac{\partial f(z_1, \ldots, z_{n-1}, \zeta)}{\partial \zeta} \cdot \frac{\zeta^p d\zeta}{f(z_1, \ldots, z_{n-1}, \zeta)}$$

for $p = 1, 2, \ldots$, a well-known formula in one variable. Hence, $S_p = \Sigma \alpha_\nu^p$ is holomorphic in $z' = (z_1, \ldots, z_{n-1})$ with $|z_j| < r_j$, $1 \leq j \leq n-1$. Note that $a_1, \ldots, a_k$ are polynomials (over the rational numbers $\mathbb{Q}$) in $S_1, \ldots, S_k$. Hence, $a_1, \ldots, a_k$ are holomorphic in z'. Note that $a_j(0, \ldots, 0) = 0$. Thus h is a Weierstrass polynomial.

We show that $u = f/h$ is holomorphic and nonzero at 0. For any fixed $z' = (z_1, \ldots, z_{n-1})$, with $|z_j| < r_j$, $1 \leq j \leq n-1$, $u(z) = u(z', z_n)$ is clearly holomorphic and nonzero in $|z_n| < r_n$. Put

$$A = \sup \left\{ |f(z)| \,\middle|\, z \in \overline{\Delta(0, r)} \right\}$$

$$B = \inf \left\{ |h(z)| \,\middle|\, |z_j| < r_j/2, \ 1 \leq j \leq n-1, \ \text{and} \ |z_n| = r_n \right\}$$

Note that $B > 0$. By the Maximum Principle in one variable,

$$|u(z)| \leq A/B \quad \text{for } z = (z', z_n) \quad \text{with} \quad |z_j| < r_j/2, \ 1 \leq j \leq n-1,$$

$$\text{and} \quad |z_n| \leq r_n$$

By Riemann's Extension Theorem (see Sec. 3.1), u(z) is holomorphic in z with $|z_1| < r_1/2, \ldots, |z_{n-1}| < r_{n-1}/2$ and $|z_n| < r_n$.

The uniqueness of h and u is clear, for h is, by the construction, a unique Weierstrass polynomial which has the same zeros as f in a neighborhood of 0. Q.E.D.

Theorem 3.2.4 (Weierstrass Division Theorem). Let h be a Weierstrass polynomial in z_n of degree k. Then, for any $f \in {}_n\mathcal{O}$, there are uniquely determined $g \in {}_n\mathcal{O}$ and a polynomial $r \in {}_{n-1}\mathcal{O}[z_n]$ over ${}_{n-1}\mathcal{O}$ of degree $< k$ such that $f = gh + r$. Moreover, if $f \in {}_{n-1}\mathcal{O}[z_n]$, then $g \in {}_{n-1}\mathcal{O}[z_n]$.

Proof. Taking $r_1, \ldots, r_n$ sufficiently small, we may assume that (1) f and h are holomorphic in a neighborhood of $\overline{\Delta(0,r)}$ and (2) $h(z_1, \ldots, z_n) \neq 0$ for $|z_j| < r_j$, $1 \leq j \leq n-1$, and $|z_n| = r_n$. Then

$$g(z) = \frac{1}{2\pi\sqrt{-1}} \int_{|\zeta|=r_n} \frac{f(z_1, \ldots, z_{n-1}, \zeta)}{h(z_1, \ldots, z_{n-1}, \zeta)} \cdot \frac{d\zeta}{(\zeta - z_n)}$$

is holomorphic in $\Delta(0,r)$. Put $r = f - gh$. Then

$$r(z) = \frac{1}{2\pi\sqrt{-1}} \int_{|\zeta|=r_n} \frac{f(z', \zeta)}{h(z', \zeta)} \left[\frac{h(z', \zeta) - h(z', z_n)}{\zeta - z_n} \right] d\zeta$$

($z' = (z_1, \ldots, z_{n-1})$). Note that $(h(z', \zeta) - h(z', z_n))/(\zeta - z_n)$ is a polynomial in z_n of degree $< k$. Moreover, by the construction, if $f \in {}_{n-1}\mathcal{O}[z_n]$, then $g \in {}_{n-1}\mathcal{O}[z_n]$.

Next, suppose that

$$f = gh + r = g_*h + r_*$$

Then

$$r - r_* = h(g_* - g)$$

An argument similar to the proof of Theorem 3.2.3 shows that h has just k zeros (counting multiplicities) in $|z_n| < r_n$ for a fixed $z' = (z_1, \ldots, z_{n-1})$ with $|z_j| < r_j$, $1 \leq j \leq n-1$. On the other hand, $r - r_*$ has at most $k - 1$ zeros, unless $r - r_* = 0$ identically. Hence, $r = r_*$ and $g = g_*$. Q.E.D.

The above two theorems are fundamental. Using these theorems,

Proposition 3.2.5. For a Weierstrass polynomial h in z_n, h is reducible in ${}_n\mathcal{O}$ if and only if it is so in ${}_{n-1}\mathcal{O}[z_n]$. Moreover, if h is reducible, then all of its factors are Weierstrass polynomials, modulo units of ${}_{n-1}\mathcal{O}[z_n]$.

Proof. Suppose that h is reducible in ${}_n\mathcal{O}$; $h = f_1 f_2$, where $f_1(0) = f_2(0) = 0$. Since h is regular in z_n, so are f_1 and f_2. By the Weierstrass Preparation Theorem,

$$f_1 = u_1 h_1 \quad \text{and} \quad f_2 = u_2 h_2$$

Then

$$h = (u_1 u_2)(h_1 h_2)$$

Note that $h_1 h_2$ is a Weierstrass polynomial and $u_1 u_2$ is a unit. By the uniqueness,

$$u_1 u_2 = 1 \quad \text{and} \quad h = h_1 h_2$$

Hence, h is reducible in ${}_{n-1}\mathcal{O}[z_n]$ and its factors are Weierstrass polynomials.

Conversely, suppose that h is reducible in ${}_{n-1}\mathcal{O}[z_n]$: $h = f_1 f_2$, where $f_1, f_2 \in {}_{n-1}\mathcal{O}[z_n]$ are nonunits. We show that f_1 and f_2 are nonunits in $\mathcal{O}_n$. In fact, if f_1 is a unit in $\mathcal{O}_n$, then, applying the Weierstrass Division Theorem to $f_2 = (1/f_1) \cdot h$, we conclude that $1/f_1 \in {}_{n-1}\mathcal{O}[z_n]$. This means that f_1 is a unit in ${}_{n-1}\mathcal{O}[z_n]$, a contradiction. Q.E.D.

Now, we prove

Theorem 3.2.6. ${}_n\mathcal{O}$ is a unique factorization domain (UFD).

Proof (By induction on n). If n = 0, then ${}_0\mathcal{O} = \mathbb{C}$ is a field. Suppose that ${}_{n-1}\mathcal{O}$ is a UFD. By Gauss' Theorem (see Van der Waerden [98]), ${}_{n-1}\mathcal{O}[z_n]$ is also a UFD. Take $f \in {}_n\mathcal{O}$. After a suitable linear change of coordinate systems, we may assume that f is regular in z_n. By the Weierstrass Preparation Theorem, we may write f = uh. Since ${}_{n-1}\mathcal{O}[z_n]$ is a UFD,

$$h = h_1 \cdots h_s$$

where $h_1, \ldots, h_s$ are irreducible in ${}_{n-1}\mathcal{O}[z_n]$. They are unique up to order and units in ${}_{n-1}\mathcal{O}[z_n]$. Hence,

$$f = uh_1 \cdots h_s$$

By Proposition 3.2.5, this is a factorization in ${}_n\mathcal{O}$ of f into irreducible factors. If

$$f = u'f_1 \cdots f_t$$

is another factorization in ${}_n\mathcal{O}$ into irreducible factors, then, by writing

$$f_j = u_j h'_j, \quad 1 \le j \le t$$

we have

$$f = u''h'_1 \cdots h'_t$$

Hence,

$$u = u'' \quad \text{and} \quad h_1 \cdots h_s = h'_1 \cdots h'_t$$

Since ${}_{n-1}\mathcal{O}[z_n]$ is a UFD and h'_j are irreducible in ${}_{n-1}\mathcal{O}[z_n]$ by Proposition 3.2.5, we have

$$t = s \quad \text{and} \quad h'_j = h_j, \quad 1 \le j \le s$$

after a suitable change of order. Q.E.D.

The following Theorem 3.2.7–Proposition 3.2.9 can be shown also by using theorems 3.2.3 and 3.2.4 (compare Exercise 1).

Theorem 3.2.7. ${}_n\mathcal{O}$ is a Noetherian ring.

Proposition 3.2.8. For f, g $\in {}_n\mathcal{O}$, suppose that (1) f has no multiple factor in ${}_n\mathcal{O}$ and (2) g vanishes on $\{f = 0\}$ (in a neighborhood of 0). Then g can be divided by f in ${}_n\mathcal{O}$.

Proposition 3.2.9. Let f and g be holomorphic functions on an open set W of a complex manifold M. For a point $p \in W$, assume that $[f]_p$ and $[g]_p$ are coprime in $\mathcal{O}_p$. Then there is a neighborhood U of p in W such that $[f]_q$ and $[g]_q$ are coprime in $\mathcal{O}_q$ for all $q \in U$.

Now, a subset S of a complex manifold M is called a complex analytic set in M, or simply an analytic set in M, if, for any point $p \in M$, there are a neighborhood U of p in M and finitely many holomorphic functions $f_1, \ldots, f_m$ on U such that

$$S \cap U = \{q \in U \mid f_1(q) = \cdots = f_m(q) = 0\}$$

We call $f_1 = \cdots = f_m = 0$ local equations of S around p.

For a point $p \in M$, the set of all germs of holomorphic functions which vanish on S around p forms an ideal $I_{S,p}$ of $\mathcal{O}_p$.

Local equations $f_1 = \cdots = f_m = 0$ of S around p are said to be minimal if (1) $f_1, \ldots, f_m$ generate $I_{S,q}$ for all $q \in U$ and (2) $f_1, \ldots, f_{m-1}$ (and so forth) do not generate $I_{S,p}$. It can be shown that, for any point $p \in M$, there are minimal local equations of S around p (see Gunning-Rossi [36, p. 141]).

If we can take m = 1 for any point $p \in M$, then S is called a hypersurface of M.

If W is a (connected) open set in M and S an analytic set in M, then $S \cap W$ is an analytic set in W.

If S_1 and S_2 are analytic sets in M, then $S_1 \cap S_2$ and $S_1 \cup S_2$ are also analytic sets in M. This is clear from the definition.

If M and N are complex manifolds, $f: M \to N$ a holomorphic map and S an analytic set in N, then $f^{-1}(S)$ is an analytic set in M. In particular, every fiber $f^{-1}(q)$, $q \in N$, of f is an analytic set in M. This is a very important example of analytic sets.

The proof of the following proposition is left to the reader (Exercise 1).

Proposition 3.2.10. Let S be an analytic set in a complex manifold M such that $S \neq M$. Then (1) S is closed in M, (2) M - S is dense in M, and (3) M - S is connected.

Let S be an analytic set in M. For a point $p \in S$, let $f_1 = \cdots = f_m = 0$ be minimal local equations of S around p. Suppose that

$$\operatorname{rank} \frac{\partial(f_1, \ldots, f_m)}{\partial(z_1, \ldots, z_n)} = k$$

is constant in a neighborhood of p in M, where $(z_1, \ldots, z_n)$ is a local coordinate system. It can be shown that this condition depends only on $I_{S,p}$ and does not depend on the choice of $f_1, \ldots, f_m$. In this case, p is called a nonsingular point of S. Otherwise, p is called a singular point of S.

If p is a nonsingular point of S, then, by the implicit mapping theorem (see Sec. 3.1), there is a connected neighborhood U of p in M such that $S \cap U$ is a complex submanifold of U.

If every point of S is nonsingular, then S is said to be nonsingular. A nonsingular connected analytic set is a complex submanifold of M, and vice versa.

The set Sing(S) of all singular points of S is called the singular locus of S. It can be shown that Sing(s) is also an analytic set in M, which is nowhere dense in S. (Roughly speaking, Sing(S) is locally defined by

$$f_1 = \cdots = f_m = 0 \quad \text{and} \quad \Delta_\nu = 0, \quad 1 \leq \nu \leq t,$$

where Δ_ν are some minors of $\dfrac{\partial(f_1, \ldots, f_m)}{\partial(z_1, \ldots, z_n)}$. For a rigorous proof, see Narasimhan [74, p. 58].)

An analytic set S of a complex manifold M is said to be reducible if $S = S_1 \cup S_2$, where S_1 and S_2 are analytic sets in M such that $S_1 \neq S$ and $S_2 \neq S$. Otherwise, S is said to be irreducible.

For the proof of the following proposition, see Narasimhan [74, p. 58].

Proposition 3.2.11. An analytic set S is a complex manifold M is irreducible if and only if S–Sing(s) is connected.

For an irreducible analytic set S in M, the dimension, dim S, of S is, by definition, the dimension of S–Sing(S) as a complex manifold.

Any analytic set S in M can be uniquely written as a union of irreducible analytic sets S_α

$$S = \bigcup S_\alpha, \quad (S_\alpha \not\subset S_\beta \quad \text{if } \alpha \neq \beta)$$

This expression is called the irreducible decomposition of S. Each S_α is called an irreducible component of S. If S is compact, then the number of irreducible components is finite. Otherwise, it may be infinite.

The dimension, dim S, of S is, by definition,

$$\dim S = \max_\alpha \dim S_\alpha$$

If every component has the same dimension k, then S is said to be pure k-dimensional. A pure one-dimensional S is called an analytic curve in M. For a n-dimensional M, a pure (n − 1)-dimensional S is a hypersurface and vice versa (compare Exercise 4).

Next, let p be a point of a complex manifold M. Two analytic sets S and T in some (connected) neighborhoods of p in M are said to be equivalent at p if there is a neighborhood U of p in M such that $S \cap U = T \cap U$. This is an equivalence relation. The equivalence class in which S belongs is denoted by $[S]_p$ and is called a germ of analytic sets at p. By abuse of notation, however, we write simply S instead of $[S]_p$, unless any confusion is probable.

Let $S = [S]_p$ be a germ of analytic sets at p. A germ $[f]_p$ of holomorphic functions at p is said to vanish on S if there are $f \in [f]_p$ and $S \in [S]_p$ such that f vanishes on S.

The set I_S of all germs of holomorphic functions at p which vanish on $S = [S]_p$ is an ideal of $\mathcal{O}_p$.

For the proof of the following theorem, see Narasimhan [74, p. 43].

Theorem 3.2.12 (Hilbert Nullstellensatz). I_S is equal to its radical: $I_S = \sqrt{I_S}$. Conversely, if an ideal I of $\mathcal{O}_p$ satisfies $I = \sqrt{I}$, then there is a unique germ $S = [S]_p$ of analytic sets at p such that $I = I_S$.

Here, the radical $\sqrt{I}$ of an ideal I of a commutative ring R is, by definition, the ideal

$$\sqrt{I} = \{x \in R \mid x^m \in I \quad \text{for some } m \geq 1\}$$

This theorem says that germs of analytic sets at p are in 1-1 correspondence with the set of all ideals I of $\mathcal{O}_p$ such that $I = \sqrt{I}$.

$S = [S]_p$ is said to be irreducible if I_S is a prime ideal of $\mathcal{O}_p$. Otherwise, it is said to be reducible.

Since $I_S = \sqrt{I_S}$, I_S can be written as the intersection of a finite number of prime ideals:

$$I_S = P_1 \cap \cdots \cap P_t \qquad (P_j \not\subset P_k \quad \text{for } j \neq k)$$

(see Zariski-Samuel [104] or Northcott [76]). This expression is unique up to order. By the Hilbert Nullstellensatz, there is $S_j = [S_j]_p$ such that $P_j = I_{S_j}$. Each S_j is called an irreducible branch of $S = [S]_p$ (at p). We write

$$S = \bigcup S_j$$

and call this the irreducible decomposition of S (at p).

The dimension, dim S, of $S = [S]_p$ is, by definition,

$$\dim S = \inf \{\dim T \mid T \in [S]_p\}$$

Note that

$$\dim S = \max \dim S_j$$

where S_j are the irreducible branches of S at p.

For an analytic set S in M and $p \in S$, we put

$$\dim_p S = \dim [S]_p$$

and call it the dimension of S at p.

Now, we state a beautiful theorem by Remmert-Stein. For the proof, see Narasimhan [74, p. 123].

Theorem 3.2.13 (Remmert-Stein Continuation Theorem). Let S be an analytic set in a complex manifold M and T be an analytic set in M - S. Suppose that

$$\dim S < \dim_p T \quad \text{for all } p \in T$$

Then the closure $\bar{T}$ of T in M is an analytic set in M.

We also state here the Proper Mapping Theorem by Remmert. This is a very important theorem. For the proof, see, again, Narasimhan [74, p. 129]. A continuous map f: X → Y between Hausdorff spaces X and Y is said to be proper if $f^{-1}(K)$ is compact for any compact set K in Y. (If X is compact, then f is always proper.)

Theorem 3.2.14 (Proper Mapping Theorem). (1) Let $\tilde{f}: M \to N$ be a holomorphic map of a complex manifold M into a complex manifold N. Let S be an analytic set in M. Suppose that the restriction $f: S \to N$ of $\tilde{f}$ to S is proper. Then the image f(S) is an analytic set in N. (If S is irreducible, then f(S) is irreducible (compare Exercise 6).) (2) The dimension of f(S) is given as follows: For a point p, $f^{-1}f(p) = S \cap \tilde{f}^{-1}\tilde{f}(p)$ is an analytic set in M. Put

$$r_p(f) = \dim_p S - \dim_p f^{-1}f(p) \quad \text{and} \quad r(f) = \max\{r_p(f) \mid p \in S\}$$

Then $\dim f(S) = r(f)$.

Now we define a meromorphic map following Remmert [83]. Let M and N be complex manifolds and X be an irreducible analytic set in M. Let $\tilde{f}: M \to N$ be a holomorphic map. The restriction $f: X \to N$ of $\tilde{f}$ to X is called a proper modification if (1) f is surjective and proper and (2) there are analytic sets S in M with $\text{Sing}(X) \subset S \subset X$ and T in N with $T \neq N$ such that f induces a biholomorphic map of X - S onto N - T.

Definition 3.2.15. Let M and N be complex manifolds. A map f of M into the power set of N is called a meromorphic map of M into N if (1) the graph

$$\Gamma_f = \{(p, q) \in M \times N \mid q \in f(p)\}$$

of f is an irreducible analytic set in M × N and (2) the natural projection $\pi_1: \Gamma_f \to M$ is a proper modification. Moreover, f is called a bimeromorphic map (respectively a surjective meromorphic map) of M onto N if (3) the natural projection $\pi_2: \Gamma_f \to N$ is also a proper modification (respectively a surjective map).

Note that a meromorphic map f of M into N is not necessarily a map in the usual sense. But we denote $f: M \to N$ by abuse of notation.

By the definition, for a meromorphic map $f: M \to N$, there is the smallest subset S_f of M such that f induces a holomorphic map of $M - S_f$ into N. S_f is called the set of points of indeterminacy of f. It is known that

Proposition 3.2.16 (Remmert [83]). S_f is an analytic set in M such that $\dim S_f \leq \dim M - 2$.

Definition 3.2.17. A meromorphic function f on a complex manifold M is a holomorphic function on M - S for some analytic set S in M with $S \neq M$ such that, for any point $p \in M$, there are a neighborhood U of p and holomorphic functions g and h on U such that $f(q) = g(q)/h(q)$ for all $q \in U - S$.

Note again that a meromorphic function on M is not necessarily a function in the usual sense. In the definition of a meromorphic function f on M, we may assume that g and h are coprime in $\mathcal{O}_q$ for all $q \in U$ (see Proposition 3.2.9). Let Γ be an analytic set in $M \times \mathbb{P}^1$ defined in $U \times \mathbb{P}^1$ by

$$\Gamma \cap (U \times \mathbb{P}^1) = \{(p, (\lambda_0 : \lambda_1)) \in U \times \mathbb{P}^1 \mid \lambda_0 g(p) - \lambda_1 h(p) = 0\}$$

Then Γ defines a meromorphic map $\Phi : M \to \mathbb{P}^1$ such that $\Gamma_\Phi = \Gamma$. Φ can be written as $\Phi = (1: f)$. In general,

Proposition 3.2.18 (Remmert [83]). Let $f_1, \ldots, f_r$ be meromorphic functions on a complex manifold M. Then

$$\Phi: p \in M \to (1: f_1(p): \cdots f_r(p)) \in \mathbb{P}^r$$

is a meromorphic map.

The set of all meromorphic functions on M forms a field $\mathbb{C}(M)$ in a natural way. The transcendence degree of $\mathbb{C}(M)$ over $\mathbb{C}$ is called the algebraic dimension of M and is denoted by alg. dim M.

Theorem 3.2.19 (Siegel [92]). If M is compact, then

$$\text{alg. dim } M \leq \dim M$$

There are examples of compact complex manifolds M such that a meromorphic function on M is necessarily a constant, that is, $\mathbb{C}(M) = \mathbb{C}$. The complex 2-torus $M = \mathbb{C}^2/\Gamma$, where

$$\Gamma = \mathbb{Z}\begin{pmatrix} 1 \\ 0 \end{pmatrix} + \mathbb{Z}\begin{pmatrix} 0 \\ 1 \end{pmatrix} + \mathbb{Z}\begin{pmatrix} \sqrt{-2} \\ \sqrt{-3} \end{pmatrix} + \mathbb{Z}\begin{pmatrix} \sqrt{-5} \\ \sqrt{-7} \end{pmatrix}$$

is such an example (see Siegel [93, p. 104]).

Let $\pi: B \to M$ be a vector bundle with the standard fiber $\mathbb{C}^n$. Let $\{U_\alpha\}$ be an open covering of M and $g_{\alpha\beta}$ be the transition functions of the vector bundle (see Sec. 3.1). A collection $\{\xi_\alpha\}$ of vector valued meromorphic functions

$$\xi_\alpha = (\xi_{\alpha 1}, \ldots, \xi_{\alpha n}): U_\alpha \to \mathbb{C}^n$$

(that is, every $\xi_{\alpha j}$ is a meromorphic function on U_α) is called a meromorphic section of the vector bundle if

$$\xi_\alpha = g_{\alpha\beta}\xi_\beta \quad \text{on } U_\alpha \cap U_\beta$$

A meromorphic section of a m-cotangent bundle $\Lambda^m T^*M$ is called a meromorphic m-form on M.

Note 3.2.20. The content of this section is mainly taken from Gunning-Rossi [36] and R. Narasimhan [74]. The latter is particularly recommended to the beginner. In these books, many results in this section are generalized to complex analytic spaces.

Exercises

1. Prove Theorem 3.2.7–Proposition 3.2.10.
2. A compact analytic set in $\mathbb{C}^n$ is a finite point set.
3. Let S be an analytic set in an n-dimensional complex manifold M such that $\dim S \leq n - 2$. Then every holomorphic function on $M - S$ can be extended to a holomorphic function on M (see Narasimhan [74, p. 59]).
4. Let f be a holomorphic function on an n-dimensional complex manifold M. Then $V(f) = \{p \in M \mid f(p) = 0\}$ is an analytic set in M such that $\dim_p V(f) = n - 1$ for all $p \in V(f)$. The converse does not necessarily hold. But if $[S]_p$ is a pure $(n - 1)$-dimensional germ of analytic sets at $p \in M$, then there is a holomorphic function f on a neighborhood of p such that $[S]_p = [V(f)]_p$ (see Gunning-Rossi [36, p. 113]).
5. (Zariski tangent space) Let S be an analytic set of a complex manifold M. For a point $p \in S$, the Zariski tangent space T_pS to S at p is, by definition, the linear subspace

$$T_pS = \{X \in T_pM \mid Xf = 0 \quad \text{for all} \quad f \in I_{S,p}\}$$

of T_pM. Prove that (1) $\dim_p S \leq \dim T_pS$ and (2) $\dim_p S = \dim T_pS$ if and only if p is a nonsingular point of S.

6. Let $f: M \to N$ be a proper holomorphic map. If S is an irreducible analytic set of M, then f(S) is an irreducible analytic set in N.

7. (Semi-Continuity Theorem) Let $f: M \to N$ be a holomorphic map. Then $\dim_p f^{-1}f(p)$ is an upper semi-continuous function of $p \in M$, that is, for any point $p \in M$, there is a neighborhood U of p in M such that $\dim_q f^{-1}f(q) \leq \dim_p f^{-1}f(p)$ for all $q \in U$.

8. Let $f: M \to N$ be a holomorphic map and k be a positive integer. Then $D_f^k = \{p \in M \mid \operatorname{rank}(df)_p < k\}$ is an analytic set in M.

9. Let M and N be compact complex manifolds and $f: M \to N$ be a holomorphic map. Then f is surjective if and only if there is a point $p \in M$ such that $\operatorname{rank}(df)_p = \dim N$.

10. Let M and N be complex manifolds and $f: M \to N$ be a surjective proper holomorphic map. Put $\dim M = n$ and $\dim N = r$. Then (1) $f(D_f^r) \neq N$ (use Sard's Theorem (see Matsushima [65])), (2) for any point $p \in M - D_f^r$, there are neighborhoods U of p in M, W of f(p) in N with $f(U) = W$ and a biholomorphic map $\alpha: U \to W \times \Omega$, where Ω is a domain in $\mathbb{C}^{n-r}$ such that $\pi \cdot \alpha = f$, where $\pi: W \times \Omega \to W$ is a natural projection, (3) for any point $p \in M - D_f^r$, there is a holomorphic local section $\sigma: W \to U$ passing through p, that is, a holomorphic map σ such that $f \cdot \sigma = 1$ on W and $\sigma(f(p)) = p$, (4) for any point $q \in N - f(D_f^r)$, the fiber $f^{-1}(q)$ is a not necessarily connected $(n - r)$-dimensional complex submanifold of M, (5) the number of connected components of $f^{-1}(q)$ is finite and constant for $q \in N - f(D_f^r)$, and (6) for any $q_1, q_2 \in N - f(D_f^r)$, $f^{-1}(q_1)$ and $f^{-1}(q_2)$ are diffeomorphic (see Kuranishi [59]).

11. Let $f: M \to N$ be a holomorphic map. Then f is an open map if and only if $\dim_p f^{-1}f(p) = \dim M - \dim N$ for all $p \in M$.

12. Let $f: M \to N$ be a holomorphic map. Then (1)

$$\dim_p f^{-1}f(p) + \operatorname{rank}(df)_p \leq \dim M \quad \text{for any point } p \in M$$

and (2) if the equality in (1) holds at $p \in M$, then p is a nonsingular point of $f^{-1}f(p)$ (use Exercise 5).

13. A bijective holomorphic map $f: M \to N$ of a complex manifold M onto a complex manifold N is biholomorphic (see Griffiths-Harris [33, p. 19]).

3.3 PROJECTIVE VARIETIES

Let $F_1, \dots, F_m$ be homogeneous polynomials in the variables $X_0, \dots, X_N$. Then the set

$$V(F_1,\dots,F_m) = \{(X_0 : \cdots : X_N) \in \mathbb{P}^N \mid F_j(X_0,\dots,X_N) = 0,\ 1 \le j \le m\}$$

is well defined and is called a projective algebraic set in $\mathbb{P}^N$ or simply an algebraic set in $\mathbb{P}^N$. In particular, for a single $F = F_1$, $V(F)$ is called a hypersurface in $\mathbb{P}^N$. The Fermat variety

$$\{(X_0 : \cdots : X_N) \in \mathbb{P}^N \mid X_0^d + \cdots + X_N^d = 0\}$$

is such an example. Note that

$$V(F_1,\dots,F_m) = V(F_1) \cap \cdots \cap V(F_m)$$

A projective algebraic set is an analytic set in $\mathbb{P}^N$. In fact, putting $U_0 = \{X_0 \ne 0\}$, say,

$$V(F_1,\dots,F_m) \cap U_0 = \{(x_1,\dots,x_N) \in U_0 \mid F_j(1,x_1,\dots,x_N) = 0,\ 1 \le j \le m\}$$

where $x_k = X_k/X_0$, $1 \le k \le N$.

A surprising theorem by Chow asserts the converse:

Theorem 3.3.1 (Chow). An analytic set in $\mathbb{P}^N$ is a projective algebraic set.

Proof (Cartan). Let $\pi: \mathbb{C}^{N+1} - \{0\} \to \mathbb{P}^N$ be the canonical projection. Let V be an analytic set in $\mathbb{P}^N$. Then

$$A = \pi^{-1}(V)$$

is an analytic set in $\mathbb{C}^{N+1} - \{0\}$. For any point $z \in A$, the set A contains the line passing through z and the origin 0 (except 0). Hence, $\dim_z A \ge 1$, so, by the Remmert-Stein Continuation Theorem (Theorem 3.2.13), the closure $\bar{A} = A \cup \{0\}$ of A in $\mathbb{C}^{N+1}$ is an analytic set in $\mathbb{C}^{N+1}$.

Let $\Delta = \Delta(0,r)$ $(r > 0)$ be a small polydisc and $f_1, \dots, f_s$ be holomorphic functions on Δ such that

$$\bar{A} \cap \Delta = \{z \in \Delta \mid f_1(z) = \cdots = f_s(z) = 0\}$$

We expand each f_j into a convergent series of homogeneous polynomials in Δ

$$f_j(z) = \sum_{\nu=1}^{\infty} f_{j\nu}(z), \qquad (\deg f_{j\nu} = \nu)$$

Put

$$B = \{z \in \mathbb{C}^{N+1} \mid f_{j\nu}(z) = 0, \quad 1 \leq j \leq s, \ \nu \geq 1\}$$

We claim that $\bar{A} = B$. It is clear that $B \cap \Delta \subset \bar{A} \cap \Delta$. Note that B and $\bar{A}$ are invariant under the multiplication by a complex number; $\alpha(B) \subset B$ and $\alpha(\bar{A}) \subset \bar{A}$ for $\alpha \in \mathbb{C}$. Hence, $B \subset \bar{A}$. In order to show that $\bar{A} \subset B$, it suffices to show that $\bar{A} \cap \Delta \subset B \cap \Delta$. For any point $z \in \bar{A} \cap \Delta$ and $t \in \mathbb{C}$ with $|t| \leq 1$, we have $tz \in \bar{A} \cap \Delta$. Hence,

$$\sum_{\nu=1}^{\infty} f_{j\nu}(tz) = \sum_{\nu=1}^{\infty} t^{\nu} f_{j\nu}(z) = 0 \quad \text{for } 1 \leq j \leq s$$

This holds for all $t \in \mathbb{C}$ with $|t| \leq 1$. Hence,

$$f_{j\nu}(z) = 0 \quad \text{for } 1 \leq j \leq s \quad \text{and} \quad \nu \geq 1$$

Hence, $z \in B \cap \Delta$. Thus we conclude that $B = \bar{A}$.

Now $\mathbb{C}[z_0, \ldots, z_n]$ is Noetherian (see Van der Waerden [98]), so there is a finite subset $\{F_1, \ldots, F_m\}$ of $\{f_{j\nu}\}$ such that

$$\bar{A} = B = \{z \in \mathbb{C}^{N+1} \mid F_1(z) = \cdots = F_m(z) = 0\}$$

This means that $V = V(F_1, \ldots, F_m)$. Q.E.D.

A projective algebraic set $V = V(F_1, \ldots, F_m)$ in $\mathbb{P}^N$ is said to be <u>reducible</u> (respectively <u>irreducible</u>) if it is reducible (respectively irreducible) as an analytic set in $\mathbb{P}^N$. V can be uniquely decomposed into a finite number of <u>irreducible components</u>

$$V = V_1 \cup \cdots \cup V_s \quad (V_j \not\subset V_k \ \text{for } j \neq k)$$

called the <u>irreducible decomposition</u> of V. The <u>dimension</u>, dim V, of V is the dimension of V as an analytic set in $\mathbb{P}^N$.

If every irreducible component of V has dimension 1 (respectively 2), then V is called a <u>projective algebraic curve</u> (respectively <u>surface</u>) in $\mathbb{P}^N$.

An ideal I of the polynomial ring $\mathbb{C}[X_0, \ldots, X_N]$ is said to be <u>homogeneous</u> if it is generated by a finite number of homogeneous polynomials. For a homogeneous ideal I,

$$V(I) = \{(X_0 : \cdots : X_N) \in \mathbb{P}^N \mid F(X_0, \ldots, X_N) = 0 \quad \text{for all } F \in I\}$$

is clearly a well-defined projective algebraic set. Conversely, any projective algebraic set $V = V(F_1, \ldots, F_m)$ in $\mathbb{P}^N$ can be written as $V = V(I)$, where $I = (F_1, \ldots, F_m)$ is the homogeneous ideal generated by $F_1, \ldots, F_m$.

If I is homogeneous, then its radical $\sqrt{I}$ (see Sec. 3.2) is also homogeneous (see Exercise 2). It is clear that

$$V(I) = V(\sqrt{I})$$

Note that, if $V(I)$ is nonempty, then $\sqrt{I} \neq (X_0, \ldots, X_N)$, the ideal generated by $X_0, \ldots, X_N$.

For the proof of the following theorem, see Mumford [70, p. 22].

Theorem 3.3.2 (Projective Nullstellensatz). Let I be a homogeneous ideal of $\mathbb{C}[X_0, \ldots, X_N]$. If a homogeneous polynomial F vanishes on $V(I)$, then $F \in \sqrt{I}$.

This theorem implies that there is a 1-1 correspondence between (nonempty) projective algebraic sets in $\mathbb{P}^N$ and homogeneous ideals I of $\mathbb{C}[X_0, \ldots, X_N]$ such that $\sqrt{I} = I$, $I \neq (X_0, \ldots, X_N)$ and $I \neq \mathbb{C}[X_0, \ldots, X_N]$.

The irreducible decomposition

$$V = V_1 \cup \cdots \cup V_s \quad (V_j \not\subset V_k \quad \text{for } j \neq k)$$

of $V = V(I)$ corresponds to the prime decomposition

$$\sqrt{I} = P_1 \cap \cdots \cap P_s \quad (P_k \not\subset P_j \quad \text{for } j \neq k)$$

of $\sqrt{I}$, where each P_j is a homogeneous prime ideal such that

$$V_j = V(P_j) \quad 1 \leq j \leq s$$

In particular,

Proposition 3.3.3. $V = V(I)$ is irreducible if and only if $\sqrt{I}$ is a prime ideal.

Here is an interesting example. The projective curve

$$V = V(X_0 X_2 - X_1^2,\ X_1 X_3 - X_2^2)$$

in $\mathbb{P}^3$ is reducible. The irreducible decomposition of V is

$$V = V_1 \cup V_2$$

where

$$V_1 = V(X_1, X_2), \quad \text{a line in } \mathbb{P}^3, \quad \text{and}$$

$$V_2 = V(X_0X_2 - X_1^2, X_1X_3 - X_2^2, X_0X_3 - X_1X_2)$$

the twisted cubic curve in $\mathbb{P}^3$ (see Sec. 1.6). The ideal

$$P = (X_0X_2 - X_1^2, X_1X_3 - X_2^2, X_0X_3 - X_1X_2)$$

is a prime ideal, which cannot be generated by just two homogeneous polynomials (see Exercise 4). But, as a set,

$$V_2 = V(X_1X_3 - X_2^2) \cap V(X_0^2X_3 - 2X_0X_1X_2 + X_1^3)$$

Hence, V_2 is, set theoretically, the intersection of two surfaces in $\mathbb{P}^3$.

It is an unsolved problem if any irreducible curve in $\mathbb{P}^3$ is set theoretically an intersection of two surfaces.

In general, an irreducible algebraic set V of dimension n in $\mathbb{P}^N$ is called a complete intersection if the homogeneous prime ideal P such that V = V(P) is generated by N - n homogeneous polynomials. V is called a set-theoretic complete intersection if V is the intersection of N - n hypersurfaces.

Let V be a projective algebraic set in $\mathbb{P}^N$. A point $p \in V$ is said to be nonsingular (respectively singular) if it is a nonsingular (respectively singular) point of V as an analytic set in $\mathbb{P}^N$. V is said to be nonsingular if every point of V is nonsingular.

An irreducible projective algebraic set (in $\mathbb{P}^N$) is called a projective algebraic variety, or simply a projective variety.

A nonsingular projective variety is called a projective algebraic manifold, or simply a projective manifold. It is nothing but a complex submanifold of $\mathbb{P}^N$. The Fermat variety is such an example.

Under the induced topology from $\mathbb{P}^N$, a projective variety V in $\mathbb{P}^N$ is a compact Hausdorff space. V has another topology, called the Zariski topology, whose closed sets are projective algebraic sets contained in V. The Zariski topology is weaker than the usual topology. It is not even Hausdorff. A Zariski closed set in V is sometimes called a (projective) algebraic set in V.

Let P be a homogeneous prime ideal of $\mathbb{C}[X_0, \ldots, X_N]$ such that $P \neq (X_0, \ldots, X_N)$ and $P \neq \mathbb{C}[X_0, \ldots, X_N]$. Let V = V(P) be the projective variety corresponding to P. The integral domain

$$\Gamma_h(V) = \frac{\mathbb{C}[X_0, \ldots, X_N]}{P}$$

is called the homogeneous coordinate ring of V.

A nonzero element of $\Gamma_h(V)$ is a coset $F + P$, where $F \in \mathbb{C}[X_0, \ldots, X_N]$ and $F \notin P$. If F is homogeneous, then the degree, deg (F + P), of F + P is, by definition, deg F. This is well defined, for P is a homogeneous ideal (compare Exercise 1). Hence, $\Gamma_h(V)$ is a graded ring.

Elements of the quotient field of $\Gamma_h(V)$ cannot be necessarily regarded as "functions" on V. But if F and G are homogeneous polynomials with deg F = deg G and $G \notin P$, then

$$\left(\frac{F + P}{G + P}\right)(p) = \frac{F(p)}{G(p)}$$

is well defined for $p \in V$ with $G(p) \neq 0$. Here, we may assume that F and G are coprime, that is, have no component in common. We call an element $(F + P)/(G + P)$ of the quotient field of $\Gamma_h(V)$ with deg F = deg G a rational function on V. Note that it is not a function on V but a function on the Zariski open set

$$\{p \in V \mid G(p) \neq 0\}$$

of V. $f = (F + P)/(G + P)$ is written as F/G. The Zariski closed sets

$$D_\infty(f) = \{p \in V \mid G(p) = 0\}$$

$$D_0(f) = \{p \in V \mid F(p) = 0\}$$

$$D_\infty(f) \cap D_0(f)$$

are called the set of poles, the set of zeros, and the set of points of indeterminacy, respectively. (Note that we have assumed that F and G are coprime.)

If V is a projective manifold, then a rational function is clearly a meromorphic function. The converse is also true:

Theorem 3.3.4. A meromorphic function on a projective manifold is a rational function.

For the proof, see, for example, Griffiths-Harris [33, p. 168].

The set $\mathbb{C}(V)$ of all rational functions on a projective variety V forms a subfield of the quotient field of $\Gamma_h(V)$. $\mathbb{C}(V)$ is called the field of rational functions on V.

For the proof of the following proposition, see Mumford [70, p. 36].

Proposition 3.3.5. For a projective variety V, the transcendence degree of $\mathbb{C}(V)$ over $\mathbb{C}$ is dim V. More precisely, there are algebraically independent $f_1, \ldots, f_n$ (n = dim V) in $\mathbb{C}(V)$ such that $\mathbb{C}(V)$ is a finite extension of

$\mathbb{C}(f_1, \ldots, f_n)$. (Such a field is called an algebraic function field of n variables.)

The converse to the proposition also holds (see Exercise 5). The functions $f_1, \ldots, f_n$ in the proposition can be, in fact, chosen from

$$X_1/X_0, \ldots, X_N/X_0$$

In particular, if $V = \mathbb{P}^N$, then

$$\mathbb{C}(\mathbb{P}^N) = \mathbb{C}(X_1/X_0, \ldots, X_N/X_0)$$

a purely transcendental extension over $\mathbb{C}$.

For a point $p \in V$, put

$$\mathcal{O}_p(V) = \{f \in \mathbb{C}(V) \mid f(p) \quad \text{is defined}\}$$

Then $\mathcal{O}_p(V)$ is a local ring with the maximal ideal

$$m_p(V) = \{f \in \mathcal{O}_p(V) \mid f(p) = 0\}$$

For a Zariski open set U of V, put

$$\Gamma(U) = \bigcap_{p \in U} \mathcal{O}_p(V)$$

It is a subring of $\mathbb{C}(V)$. Rational functions in $\Gamma(U)$ are defined at every point of U, and are called regular functions on U.

Let V ($\subset \mathbb{P}^N$) and W ($\subset \mathbb{P}^r$) be projective varieties. Let U be a Zariski open set in V and

$$f: U \to W$$

be a map. f is called a regular map of U into W if (1) f is continuous in Zariski topology and (2) for any Zariski open set W' of W and for any $h \in \Gamma(W')$, the composition $h \cdot f$ belongs to $\Gamma(f^{-1}(W'))$.

If U and W are nonsingular, then a regular map of U into W is a holomorphic map.

Let f_1 and f_2 be regular maps of Zariski open sets U_1 and U_2 in V, respectively, into W. We say that f_1 and f_2 are equivalent, $f_1 \sim f_2$, if

$$f_1(p) = f_2(p) \quad \text{for all } p \in U_1 \cap U_2$$

This is an equivalence relation, for f_j are continuous and every Zariski open set is dense in V. An equivalence class $f = \{f_j\}$ is called a rational map of V into W. Note that it is not necessarily a map on the whole V. The domain of definition of a rational map $f = \{f_\alpha\}$ is the union $U = \cup U_\alpha$, where

$$f_\alpha : U_\alpha \to W$$

moves over all elements of the equivalence class. Note that f is a regular map on U, but is not defined outside of U. The Zariski closed set V - U is called the set of points of indeterminacy. We write, however, $f : V \to W$, not specifying the domain of definition.

The proofs of the following Theorem 3.3.6–Corollary 3.3.8 are left to the reader (see Exercise 6).

Theorem 3.3.6. For projective manifolds V and W, a meromorphic map $f : V \to W$ is a rational map, and vice versa.

Proposition 3.3.7. Let V be a projective variety in $\mathbb{P}^N$ and $F_0, \ldots, F_r$ be homogeneous polynomials in $\mathbb{C}[X_0, \ldots, X_N]$ of the same degree such that (1) $F_0, \ldots, F_r$ are coprime and (2) $V \not\subset V(F_0, \ldots, F_r)$. Then

$$f : p \in V \to (F_0(p) : \cdots : F_r(p)) \in \mathbb{P}^r$$

is a rational map of V into $\mathbb{P}^r$ such that the set of points of indeterminacy is contained in $V \cap V(F_0, \ldots, F_r)$. Conversely, any rational map of V into $\mathbb{P}^r$ can be obtained in this way.

Corollary 3.3.8. Let $f_1, \ldots, f_r$ be rational functions on a projective variety V. Then

$$f : p \in V \to (1 : f_1(p) : \cdots : f_r(p)) \in \mathbb{P}^r$$

is a rational map of V into $\mathbb{P}^r$. Conversely, any rational map of V into $\mathbb{P}^r$ can be obtained in this way. In particular, any rational function on V can be regarded as a rational map of V into $\mathbb{P}^1$, and vice versa.

In Proposition 3.3.7, if

$$\deg F_0 = \cdots = \deg F_r = 1$$

and $F_0, \ldots, F_r$ are linearly independent, then the rational map

$$f = \pi_S : p \in V \to (F_0(p) : \cdots : F_r(p)) \in \mathbb{P}^r$$

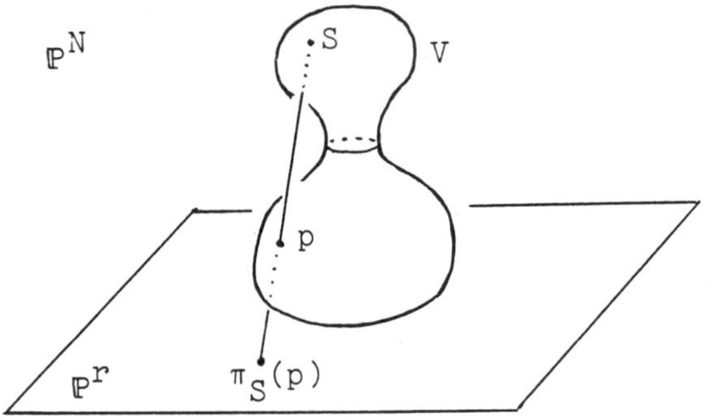

FIGURE 3.6

is called the <u>projection with the center</u>

$$S = V(F_0, \ldots, F_r), \text{ a } (N - r - 1)\text{-plane in } \mathbb{P}^N$$

(See Fig. 3.6.)

A rational map $f: V \to W$, where $W \subset \mathbb{P}^r$, can be regarded as a rational map $f: V \to \mathbb{P}^r$.

The <u>image</u> $f(V)$ <u>of a rational map</u> $f: V \to W$ is, by definition, the Zariski closure in W of $f(U)$, where U is the domain of definition of f. We say that f is <u>surjective</u> or <u>dominating</u> if $f(V) = W$.

A rational map $f: V \to W$ is regular if f is defined everywhere. For example, the projection $\pi_S: V \to \mathbb{P}^r$ with the center S is regular if $S \cap V = \phi$.

If V and W are nonsingular, then a regular map $f: V \to W$ is a holomorphic map. Conversely, by Theorem 3.3.6,

<u>Theorem 3.3.9</u>. For projective manifolds V and W, a holomorphic map $f: V \to W$ is regular.

Theorems 3.3.4, 3.3.6, and 3.3.9 are examples of the <u>GAGA Principle</u>: "A global analytic object on a projective manifold (variety) is algebraic." See Serre [86].

Thus, for projective manifolds, we may use the following terminologies interchangeably:

rational functions	$\longleftrightarrow$	meromorphic functions
rational maps	$\longleftrightarrow$	meromorphic maps
regular maps	$\longleftrightarrow$	holomorphic maps

For projective varieties V and W, a bijective regular map of V onto W is said to be biregular if f^{-1} is also regular. V and W are said to be biregular or isomorphic if there is a biregular map $f: V \to W$. Thus we have the category of projective varieties.

More generally, let U and U' be Zariski open sets in projective varieties V and W, respectively. A bijective map $f: U \to U'$ is said to be biregular if $f: U \to W$ and $f^{-1}: U' \to V$ are regular maps.

A rational map $f: V \to W$ of a projective variety V into a projective variety W is said to be birational if there are Zariski open sets U of V and U' of W such that f is defined on U and is a biregular map of U onto U'. V and W are said to be birational if there is a birational map $f: V \to W$. This is clearly an equivalence relation, which is coarser than the biregular relation.

This is a very important notion. One of the main objects to study in algebraic geometry is the property of projective varieties invariant under birational equivalence (so called the birational geometry). Italian geometers at the beginning of this century constructed a beautiful birational geometry of surfaces.

The proof of the following proposition is left to the reader (Exercise 6).

Proposition 3.3.10. Projective varieties V and W are birational if and only if the fields $\mathbb{C}(V)$ and $\mathbb{C}(W)$ are $\mathbb{C}$-isomorphic.

A deep and fundamental theorem by Hironaka says that

Theorem 3.3.11 (Hironaka [40]). For any projective variety V, there are a projective manifold M and a birational regular map $f: M \to V$ such that f induces a biregular map $f: M - f^{-1}(S) \to V - S$, where $S = \operatorname{Sing} V$ is the singular locus of V.

The map $f: M \to V$ (or M itself) in this theorem is called a nonsingular model of V (or a resolution of singularity of V). The theorem is true even in analytic category (see Hironaka [41]).

A projective variety V of dimension n is called a rational variety if it is birational to $\mathbb{P}^n$, that is, $\mathbb{C}(V)$ is a purely transcendental extension over $\mathbb{C}$.

Example 3.3.12. $\mathbb{P}^1 \times \mathbb{P}^1$ and $\mathbb{P}^2$ are not biregular but birational. In fact, first of all, $\mathbb{P}^1 \times \mathbb{P}^1$ can be considered as a projective manifold; the map

$$((X_0 : X_1), (Y_0 : Y_1)) \in \mathbb{P}^1 \times \mathbb{P}^1 \to (X_0 Y_0 : X_0 Y_1 : X_1 Y_0 : X_1 Y_1) \in \mathbb{P}^3$$

gives a holomorphic imbedding of $\mathbb{P}^1 \times \mathbb{P}^1$ into $\mathbb{P}^3$. Its image is the nonsingular quadric surface

$$Q = \{(Z_0 : Z_1 : Z_2 : Z_3) \in \mathbb{P}^3 \mid Z_0 Z_3 = Z_1 Z_2\}$$

We identify $\mathbb{P}^1 \times \mathbb{P}^1$ with Q. The projection

$$\pi_0: (Z_0: Z_1: Z_2: Z_3) \in Q \to (Z_1: Z_2: Z_3) \in \mathbb{P}^2$$

with the center 0 = (1: 0: 0: 0) gives a birational map of $\mathbb{P}^1 \times \mathbb{P}^1 = Q$ onto $\mathbb{P}^2$. Note that the first (or second) projection $\mathbb{P}^1 \times \mathbb{P}^1 \to \mathbb{P}^1$ is a regular map. On the other hand, there is no regular map of $\mathbb{P}^2$ onto $\mathbb{P}^1$. In fact, if $f: \mathbb{P}^2 \to \mathbb{P}^1$ were a surjective regular map, then the plane curves $f^{-1}(a)$ and $f^{-1}(b)$ would be disjoint for $a \neq b$. This contradicts Bezout's Theorem (see Sec. 1.3). Hence, $\mathbb{P}^1 \times \mathbb{P}^1$ and $\mathbb{P}^2$ cannot be biregular.

In general, $\mathbb{P}^m \times \mathbb{P}^n$ is holomorphically imbedded in $\mathbb{P}^N$, where $N = (m + 1)(n + 1) - 1$:

$$((X_0: \cdots : X_m), (Y_0: \cdots : Y_n)) \in \mathbb{P}^m \times \mathbb{P}^n \to (\cdots : X_jY_k: \cdots) \in \mathbb{P}^N$$

This is called the Segre imbedding. $\mathbb{P}^m \times \mathbb{P}^n$ is a rational variety.

If V ($\subset \mathbb{P}^m$) and W ($\subset \mathbb{P}^n$) are projective varieties, then $V \times W (\subset \mathbb{P}^m \times \mathbb{P}^n \subset \mathbb{P}^N)$ is also a projective variety.

The Proper Mapping Theorem (Theorem 3.2.14) can be considered as a generalization of the classical

Theorem 3.3.13 (Main Theorem of the Elimination Theory). Let $\pi: \mathbb{P}^m \times \mathbb{P}^n \to \mathbb{P}^m$ be the natural projection. If V is a projective algebraic set in $\mathbb{P}^m \times \mathbb{P}^n$, then $\pi(V)$ is a projective algebraic set in $\mathbb{P}^m$.

See Mumford [70, p. 33] for the direct proof of this theorem.

Suggested by the above example $\mathbb{P}^m \times \mathbb{P}^n$, we generalize the notion of projective manifolds as follows:

Definition 3.3.14. A compact complex manifold M is called a projective manifold if there is a holomorphic imbedding $f: M \to \mathbb{P}^N$ of M into a projective space $\mathbb{P}^N$.

Note that we do not specify the imbedding $f: M \to \mathbb{P}^N$. There exist a lot of f's. Note also that many notions, as Zariski closed sets, rational functions, and so on, on a projective manifold M are independent of the choice of the imbedding $M \to \mathbb{P}^N$. That is, they are intrinsic notions for M.

A projective manifold is one of the most fruitful natural mathematical objects while C^∞-ones are artificial. (A prejudice of Kronecker type!)

A complex n-torus is called an Abelian variety if it is a projective manifold. Not every complex n-torus can be an Abelian variety for $n \geq 2$ (see the example after Theorem 3.2.19).

Another important example of projective manifolds is the Grassmann variety. Let r and N be integers such that $0 \leq r \leq N$. The Grassman variety

$G(r, N)$ is, by definition, the set of all r-planes in $\mathbb{P}^N$. (Some authors denote it by $G(r + 1, N + 1)$ or $G(N + 1, r + 1)$.) Note that $G(0, N) = \mathbb{P}^N$ and $G(N - 1, N) = \mathbb{P}^{N*}$, the dual projective space. The proof of the following lemma is left to the reader (Exercise 6).

Lemma 3.3.15. $G(r, N)$ is a homogeneous compact complex manifold of dimension $(r + 1)(N - r)$. Moreover, there is a principal bundle $\pi: B \to G(r, N)$ with the structure group $GL(r + 1, \mathbb{C})$, where

$$B = \{A \mid A \text{ is a } (r + 1) \times (N + 1)\text{-matrix with rank } A = r + 1\}$$

(Hence, $G(r, N) = B/G(r + 1, \mathbb{C})$ is the quotient space.)

Now, $G(r, N)$ is a projective manifold. We explain this in the case of $G(1, 3)$.

A line ℓ in $\mathbb{P}^3$ is determined by two distinct points

$$p_0 = (X_0^0 : X_1^0 : X_2^0 : X_3^0) \quad \text{and} \quad p_1 = (X_0^1 : X_1^1 : X_2^1 : X_3^1)$$

on it. If we choose another pair of distinct points

$$q_0 = (Y_0^0 : Y_1^0 : Y_2^0 : Y_3^0) \quad \text{and} \quad q_1 = (Y_0^1 : Y_1^1 : Y_2^1 : Y_3^1)$$

on ℓ, then there is a nonsingular (2×2)-matrix $A = (a_{jk})$ such that

$$\begin{aligned} Y_0 &= a_{00} X_0 + a_{01} X_1 \\ Y_1 &= a_{10} X_0 + a_{11} X_1 \end{aligned} \tag{*}$$

where $X_0 = (X_0^0, X_1^0, X_2^0, X_3^0)$, and so forth, are vectors in $\mathbb{C}^4$. Conversely, if Y_0 and Y_1 are vectors in $\mathbb{C}^4$ given by (*) for a nonsingular (2×2)-matrix $A = (a_{jk})$, then q_0 and q_1 are on ℓ. Put

$$W_{jk} = W_{jk}(p_0, p_1) = X_j^0 X_k^1 - X_k^0 X_j^1, \qquad 0 \le j < k \le 3$$

the (2×2)-minors of the matrix

$$\begin{pmatrix} X_0 \\ X_1 \end{pmatrix} = \begin{pmatrix} X_0^0 X_1^0 X_2^0 X_3^0 \\ X_0^1 X_1^1 X_2^1 X_3^1 \end{pmatrix}$$

Note that some $W_{jk} \neq 0$. Note also that

$$W_{jk}(q_0, q_1) = (\det A) W_{jk}(p_0, p_1) \quad \text{for } 0 \le j < k \le 3$$

Hence,

$$(W_{01}: W_{23}: W_{02}: W_{13}: W_{03}: W_{12}) \in \mathbb{P}^5$$

depends only on the line $\ell \in G(1, 3)$. The map

$$\Phi: \ell \in G(1, 3) \to (W_{01}: W_{23}: W_{02}: W_{13}: W_{03}: W_{12}) \in \mathbb{P}^5$$

is injective. In fact, if $W_{01} \neq 0$, say, then by putting

$$A = \begin{pmatrix} X_0^0 & X_1^0 \\ X_0^1 & X_1^1 \end{pmatrix}^{-1}$$

in (*), we may assume that

$$\begin{pmatrix} X_0^0 & X_1^0 \\ X_0^1 & X_1^1 \end{pmatrix} = \begin{pmatrix} 1 & 0 \\ 0 & 1 \end{pmatrix}$$

Then

$$(W_{01}: W_{23}: W_{02}: W_{13}: W_{03}: W_{12}) = (1: X_2^0 X_3^1 - X_2^1 X_3^0: X_2^1: -X_3^0: X_3^1: -X_2^0)$$

Hence, the line ℓ through $p_0 = (1: 0: X_2^0: X_3^0)$ and $p_1 = (0: 1: X_2^1: X_3^1)$ is uniquely determined by $(W_{01}: \cdots : W_{12}) \in \mathbb{P}^5$.

It can be easily shown that Φ is a holomorphic imbedding. Hence, $G(1, 3)$ is a projective manifold.

The minors W_{jk} clearly satisfy the relation

$$W_{01}W_{23} - W_{02}W_{13} + W_{03}W_{12} = 0$$

Hence, the image $\Phi(G(1, 3))$ is contained in the irreducible nonsingular quadric hypersurface

$$S = \{(Z_0: Z_1: Z_2: Z_3: Z_4: Z_5) \in \mathbb{P}^5 \mid Z_0 Z_1 - Z_2 Z_3 + Z_4 Z_5 = 0\}$$

in $\mathbb{P}^5$. Note that $\Phi(G(1, 3)) = S$. In fact, the projective algebraic set $\Phi(G(1,3))$ has dimension 4 and is contained in the four-dimensional irreducible S. Hence, $\Phi(G(1, 3)) = S$. Thus $G(1, 3)$ can be identified with S through Φ.

This argument can be generalized to $G(r, N)$; take an $(r + 1)$-ple of points

$$p_0 = (X_0^0: \cdots : X_N^0), \ldots, p_r = (X_0^r: \cdots : X_N^r)$$

in general position on a r-plane H in $\mathbb{P}^N$. The rank of the matrix

$$X = \begin{pmatrix} X_0^0 & \cdots & X_N^0 \\ \cdots & & \\ X_0^r & \cdots & X_N^r \end{pmatrix}$$

is r + 1. The number of (r + 1) × (r + 1)-minors $W_{i_0 \cdots i_r}$ of X is $\binom{N+1}{r+1}$. Put $s = \binom{N+1}{r+1} - 1$. Then it can be shown that the map

$$\Phi: H \in G(r, N) \to (\cdots : W_{i_0 \cdots i_r} : \cdots) \in \mathbb{P}^s$$

is a well-defined holomorphic imbedding. Hence, G(r, N) is a projective manifold (see Hodge-Pedoe [43] for details):

Theorem 3.3.16. The Grassmann variety G(r, N) is a projective manifold of dimension (r + 1)(N - r).

The holomorphic imbedding Φ is called the Plücker imbedding and $(\cdots : W_{i_0 \cdots i_r} : \cdots)$ is called the Plücker coordinate of H ∈ G(r, N).

Note 3.3.17. (1) One of the best references for this section is Mumford [70]. (2) Iitaka introduced the fundamental birational invariant κ(V), called the Kodaira dimension of a projective variety V, and has developed the birational geometry (see Iitaka [49] and Ueno [97]).

Exercises

1. An ideal I of $\mathbb{C}[X_0, \ldots, X_N]$ is a homogeneous ideal if and only if, for any f ∈ I, every homogeneous part of f belongs to I.

2. If I is a homogeneous ideal of $\mathbb{C}[X_0, \ldots, X_N]$, then $\sqrt{I}$ is also homogeneous.

3. For a projective algebraic set V in $\mathbb{P}^N$,

$$\dim V = \sup \{k \mid \text{there is a chain of irreducible algebraic sets } V_0 \subsetneqq V_1 \subsetneqq \cdots \subsetneqq V_k \subsetneqq V\}$$

4. The ideal $P = (X_0X_2 - X_1^2, X_1X_3 - X_2^2, X_0X_3 - X_1X_2)$ of $\mathbb{C}[X_0, X_1, X_2, X_3]$ is a prime ideal which cannot be generated by two homogeneous polynomials.

5. Let K be an algebraic function field of n variables. Then there is a projective variety V (in some $\mathbb{P}^N$) such that K and $\mathbb{C}(V)$ is $\mathbb{C}$-isomorphic.

6. Prove Theorem 3.3.6–Corollary 3.3.8, Proposition 3.3.10, and Lemma 3.3.15.

7. Let $V = V(I)$ be a projective algebraic set in $\mathbb{P}^N$, where $I = (F_1, \dots, F_m)$ be a homogeneous ideal of $\mathbb{C}[X_0, \dots, X_N]$ such that $\sqrt{I} = I$. Then, (1) the Zariski tangent space to V at p can be identified with the affine linear subspace

$$\{(x_1, \dots, x_N) \mid \Sigma_j (\partial F_\nu(1, x_1, \dots, x_N)/\partial x_j)_p (x_j - a_j) = 0, \ 1 \le \nu \le m\}$$

of $\mathbb{C}^N$, where $x_j = X_j/X_0$ and $p = (1: a_1: \cdots : a_N)$, and (2) its closure in $\mathbb{P}^N$ is the linear subspace

$$\left\{(X_0: \cdots : X_N) \in \mathbb{P}^N \,\middle|\, \sum_j \left(\frac{\partial F_\nu}{\partial X_j}\right)_p X_j = 0, \quad 1 \le j \le m\right\}$$

of $\mathbb{P}^N$, which is called the Zariski tangent space to V at p again and is denoted by T_pV.

8. A nonsingular hypersurface in $\mathbb{P}^N$ is necessarily irreducible.

9. Let V be a projective algebraic set in $\mathbb{P}^N$. If every irreducible component of V is $(N - 1)$-dimensional, then V is a hypersurface, that is, $V = V(F)$ for a homogeneous polynomial F.

10. An n-dimensional projective manifold can be holomorphically imbedded in $\mathbb{P}^{2n+1}$.

11. Let V be a projective algebraic set in $\mathbb{P}^N$ and $p_1, \dots, p_m$ be points in $\mathbb{P}^N - V$. Then there is a hypersurface S such that (1) $V \subset S$ and (2) $p_j \notin S$ for $1 \le j \le m$.

12. Let V and W be projective algebraic sets in $\mathbb{P}^N$ such that $V \subset W$. Put $r = \dim W$. Then there are $r + 1$ homogeneous polynomials $F_1, \dots, F_{r+1}$ such that $V = W \cap V(F_1, \dots, F_{r+1})$. In particular (taking $W = \mathbb{P}^N$), for any projective algebraic set V in $\mathbb{P}^N$, there are $N + 1$ homogeneous polynomials $F_1, \dots, F_{N+1}$ such that $V = V(F_1, \dots, F_{N+1})$.

13. Let V be a projective variety of dimension n in $\mathbb{P}^N$. Then, there is a uniquely determined positive integer d such that any general $(N - n)$-plane meets with V at just d distinct points. (d is called the degree, deg V, of V in $\mathbb{P}^N$, which is not a biregular invariant.) Prove that (1) deg V = 1 if and only if V is an n-plane in $\mathbb{P}^N$ and (2) if V is nondegenerate, that is, $V \not\subset H$ for any hyperplane H in $\mathbb{P}^N$, then $\deg V \ge N - n + 1$.

14. (1) Let $F_1, \dots, F_m$ be homogeneous polynomials in $\mathbb{C}[X_0, \dots, X_N]$. If $m \le N$, then $V(F_1, \dots, F_m)$ is nonempty. (2) There is no nonconstant holomorphic map $\mathbb{P}^n \to \mathbb{P}^r$, where $n > r$.

3.4 DIVISORS AND LINEAR SYSTEMS

Let M be a compact complex manifold of dimension n. (Later on, we will assume that M is a projective manifold.) A formal (finite) sum

$$D = a_1D_1 + \cdots + a_mD_m \qquad (a_j \in \mathbb{Z} \text{ for } 1 \leq j \leq m)$$

of irreducible hypersurfaces D_j of M is called a (Weil) divisor on M. It is said to be positive (or effective) and is denoted by $D > 0$ if $a_j \geq 0$ for $1 \leq j \leq m$ and $a_k > 0$ for some k. We write $D \geq 0$ if $D > 0$ or $D = 0$. The set Div(M) of all divisors on M naturally forms an additive group, called the divisor group on M. For D, E $\in$ Div(M), we write $D \geq E$ if $D - E \geq 0$. An irreducible hypersurface on M can be regarded as a positive divisor, called a prime divisor. The support of $D = a_1D_1 + \cdots + a_mD_m$, where $a_j \neq 0$ for $1 \leq j \leq m$, is, by definition, the hypersurface $D_1 \cup \cdots \cup D_m$ of M.

Let D be a divisor on M as above. Let $\{U_\alpha\}$ be a (finite) open covering of M such that every D_j is given by the minimal equation $h_{\alpha j} = 0$ in U_α, where $h_{\alpha j}$ is a holomorphic function on U_α. Put

$$f_\alpha = h_{\alpha 1}^{a_1} \cdots h_{\alpha m}^{a_m}$$

a meromorphic function on U_α. Then

$$f_{\alpha\beta} = \frac{f_\alpha}{f_\beta} \tag{*}$$

is a nonvanishing holomorphic function on $U_\alpha \cap U_\beta$.

Conversely, consider a (finite) open covering $\{U_\alpha\}$ of M and a collection $\{f_\alpha\}$ of meromorphic functions f_α on U_α such that $f_{\alpha\beta} = f_\alpha/f_\beta$ is a nonvanishing holomorphic function on $U_\alpha \cap U_\beta$.

Take another open covering $\{W_\nu\}$ of M and such a collection $\{g_\nu\}$ as $\{f_\alpha\}$. $\{f_\alpha\}$ and $\{g_\nu\}$ are said to be equivalent if

$$h_{\alpha\nu} = \frac{f_\alpha}{g_\nu}$$

is a nonvanishing holomorphic function on $U_\alpha \cap W_\nu$ for any (α, ν) with nonempty $U_\alpha \cap W_\nu$. This is clearly an equivalence relation. An equivalence class is called a Cartier divisor on M, which is denoted by $\{f_\alpha\}$ by abuse of notation.

Hence, a divisor determines uniquely a Cartier divisor on M.

Conversely, a Cartier divisor $\{f_\alpha\}$ determines uniquely a divisor on M. In fact, we may write

$$f_\alpha = \frac{g_\alpha}{h_\alpha} \quad \text{on} \quad U_\alpha$$

where g_α and h_α are holomorphic functions on U_α which are coprime in $\mathcal{O}_q$ for every $q \in U_\alpha$ (see Proposition 3.2.9). By Proposition 3.2.8, the equations $g_\alpha = 0$ (respectively $h_\alpha = 0$) in U_α define a global hypersurface S (respectively T) of M. Let

$$S = S_1 \cup \cdots \cup S_s \text{ (respectively } T = T_1 \cup \cdots \cup T_t)$$

be the irreducible decomposition of S (respectively T). Let $g_{\alpha\nu} = 0$ (respectively $h_{\alpha\nu} = 0$) be the minimal equation of S_ν (respectively T_ν) on U_α. Then, for every point $p \in U_\alpha$, there are positive integers a_ν (respectively b_ν) such that

$$g_\alpha = u_\alpha g_{\alpha 1}^{a_1} \cdots g_{\alpha s}^{a_s} \quad \text{(respectively } h_{\alpha 1}^{b_1} = v_\alpha h_{\alpha 1}^{b_1} \cdots h_{\alpha t}^{b_t}) \quad \text{in } \mathcal{O}_p$$

where u_α (respectively v_α) is a unit. It is easy to see that $a_1, \ldots, a_s$ (respectively $b_1, \ldots, b_t$) are locally constants. Hence, we may assume that they are constants on U_α, and so, on M. Now, we associate with $\{f_\alpha\}$ a divisor

$$D = a_1 S_1 + \cdots + a_s S_s - b_1 T_1 - \cdots - b_t T_t$$

Thus there is a 1-1 correspondence between (Weil) divisors and Cartier divisors on M. We identify them through this correspondence. Note that $D = \{f_\alpha\} \geq 0$ if and only if every f_α is holomorphic on U_α.

Now, the nonvanishing holomorphic functions $f_{\alpha\beta}$ on $U_\alpha \cap U_\beta$ in (*) satisfy

$$f_{\alpha\alpha} = 1 \quad \text{on } U_\alpha$$

$$f_{\alpha\beta} f_{\beta\gamma} = f_{\alpha\gamma} \quad \text{on } U_\alpha \cap U_\beta \cap U_\gamma$$

Hence, $\{f_{\alpha\beta}\}$ determines an (isomorphism class of) line bundle on M, which is denoted by [D] and is called the <u>(isomorphism class of) line bundle determined by the divisor</u> D.

A divisor D on M is called a canonical divisor if $[D] \simeq K_M$, the canonical bundle.

Definition 3.4.1. Divisors D and E on M are said to be linearly equivalent, $D \sim E$, if [D] and [E] are isomorphic. An equivalence class is called a divisor class.

A meromorphic function f on M clearly gives a (Cartier) divisor, called the principal divisor determined by f, and denoted by (f). Put

$$(f) = D_0(f) - D_\infty(f) = a_1D_1 + \cdots + a_sD_s - b_1E_1 - \cdots - b_tE_t$$

$$D_0(f) = a_1D_1 + \cdots + a_sD_s, \quad D_\infty(f) = b_1E_1 + \cdots + b_tE_t$$

where a_ν and b_μ are positive integers. The positive divisor $D_0(f)$ (respectively $D_\infty(f)$) is called the zero divisor (respectively the polar divisor) of f. Note that [(f)] is the trivial bundle, that is, the identity element of the Picard group Pic(M) of M.

Lemma 3.4.2. (1) $D \sim E$ if and only if there is a meromorphic function f on M such that D - E = (f). (2) There is the exact sequence

$$1 \to \mathbb{C}^* \to \mathbb{C}(M)^* \to \mathrm{Div}(M) \to \mathrm{Pic}(M)$$

where $\mathbb{C}(M)^* = \mathbb{C}(M) - \{0\}$ (respectively $\mathbb{C}^* = \mathbb{C} - \{0\}$) is the multiplicative group. (The image of $\mathrm{Div}(M) \to \mathrm{Pic}(M)$ can thus be identified with the group $\mathrm{Div}(M)/\sim$, called the divisor class group of M.)

Proof. Let $D = \{f_\alpha\}$ and $E = \{g_\alpha\}$ be Cartier divisors. If [D] is isomorphic to [E], we may assume that

$$\frac{f_\alpha}{f_\beta} = \frac{g_\alpha}{g_\beta} \quad \text{on} \quad U_\alpha \cap U_\beta$$

Hence,

$$f = \frac{f_\alpha}{g_\alpha} = \frac{f_\beta}{g_\beta}$$

is a meromorphic function on M. This means that D = E + (f).

Conversely, if D - E = (f), then [D] = [E + (f)] = [E] + [(f)] = [E].
(2) follows from (1). Q.E.D.

For a divisor D on M, put

$$L(D) = \{f \mid f \text{ is a meromorphic function with } (f) + D \geq 0 \text{ or is the zero function}\}$$

It is easy to see that L(D) is a vector space over $\mathbb{C}$. If D is positive, then L(D) is called the set of all meromorphic functions on M with poles at most on D.

For a line bundle L on M, we denote by $\Gamma(M, L)$, or simply by $\Gamma(L)$, the vector space over $\mathbb{C}$ of all holomorphic sections of L.

Lemma 3.4.3. L(D) and $\Gamma([D])$ are canonically isomorphic.

Proof. We regard $D = \{f_\alpha\}$ as a Cartier divisor. We associate $\xi = \{\xi_\alpha\} \in \Gamma([D])$ with $f = \xi_\alpha / f_\alpha \in L(D)$. Q.E.D.

A deep theorem says that

Theorem 3.4.4. L(D) ($\simeq \Gamma([D])$) is finite dimensional.

For the proof of this theorem, see, for example, Mumford [70, p. 103]. In the case dim M = 1, see Theorem 4.1.13.

Now, for a divisor D on M, we put

$$|D| = \{E \mid E \text{ is a positive divisor such that } E \sim D\}$$

and call it the complete linear system determined by D. Also, for a line bundle L on M, we put

$$|L| = \{E \mid E \text{ is a positive divisor such that } [E] \simeq L\}$$

and call it the complete linear system determined by L. Note that $|[D]| = |D|$.

For the canonical bundle K_M of M, $|K_M|$ is called the canonical linear system.

The proof of the following lemma is left to the reader (Exercise 1).

Lemma 3.4.5. (1) There is a canonical bijection of |D| onto $\mathbb{P}(L(D))$, the projective space of all one-dimensional vector subspaces of L(D). (2) For a line bundle L on M, there is a canonical bijection of |L| onto $\mathbb{P}(\Gamma(L))$.

We identify |D| (respectively |L|) with the projective space $\mathbb{P}(L(D))$ (respectively $\mathbb{P}(\Gamma(L))$ through this bijection. We also identify |D| with $\mathbb{P}(\Gamma([D]))$.

A linear subspace Λ (of dimension r) of |D| (or |L|) is called a linear system on M (of dimension r). We write $r = \dim \Lambda$. A one-dimensional linear system is called a linear pencil. $\dim \Lambda = 0$ (respectively -1) if $\Lambda = \{D\}$ (respectively is empty).

Let Λ be a linear system on M. A positive divisor F is called a fixed component of Λ if

$$D - F \geq 0 \quad \text{for all } D \in \Lambda$$

If there is no fixed component, then Λ is said to be fixed component free, or without fixed component.

If there is a fixed component, then there is the maximal fixed component F_0 of Λ in the sense that

$$F_0 \geq F \quad \text{for all fixed component } F$$

F_0 is called the fixed part of Λ. Put

$$\Lambda - F_0 = \{D - F_0 \mid D \in \Lambda\}$$

where F_0 is the fixed part of Λ. Then $\Lambda - F_0$ is clearly a fixed component free linear system on V such that $\dim(\Lambda - F_0) = \dim \Lambda$. $\Lambda - F_0$ is called the variable part of Λ.

Let Λ be a fixed component free linear system on M. A point $p \in M$ is called a base point of Λ if p is contained in the support of every $D \in \Lambda$. The set $Bs(\Lambda)$ of all base points of Λ is called the base locus of Λ. Λ is said to be base point free if there is no base point.

Let Λ be an r-dimensional linear system on M. Since Λ is an r-plane in $\mathbb{P}(\Gamma([D]))$, there corresponds to Λ, a unique $(r + 1)$-dimensional vector subspace $\tilde{\Lambda}$ of $\Gamma([D])$. Take a basis $\{\xi_0, \dots, \xi_r\}$ of $\tilde{\Lambda}$. Then it can be easily seen that

$$\Phi_\Lambda : p \in M \to (\xi_0(p) : \cdots : \xi_r(p)) \in \mathbb{P}^r$$

is a well-defined meromorphic map of M into $\mathbb{P}^r$, which is called the meromorphic map associated with the linear system Λ.

For a divisor D on M, we sometimes write Φ_D instead of $\Phi_{|D|}$. For a line bundle L on M, we write Φ_L instead of $\Phi_{|L|}$. For the canonical bundle $K = K_M$ of M, Φ_K is called the canonical map.

The proof of the following lemma is left to the reader (Exercise 1).

Lemma 3.4.6. Let Λ be an r-dimensional linear system on M and F_0 be the fixed part of Λ.

1. $\Phi_\Lambda = \Phi_{\Lambda - F_0}$
2. The set of all points of indeterminacy of $\Phi_\Lambda = \Phi_{\Lambda - F_0}$ is the base locus

$Bs(\Lambda - F_0)$ of $\Lambda - F_0$. In particular $\Phi_\Lambda: M \to \mathbb{P}^r$ is a holomorphic map if and only if $\Lambda - F_0$ is base point free.

3. The base locus $Bs(\Lambda - F_0)$ of $\Lambda - F_0$ is an analytic set in M of dimension at most n - 2 (n = dim M).

4. The meromorphic map Φ_Λ is nondegenerate, that is, the image $\Phi_\Lambda(M)$ is not contained in any hyperplane. (The image $\Phi_\Lambda(M)$ of Φ_Λ is, by definition, the Zariski closure in $\mathbb{P}^r$ of $\Phi_\Lambda(M - Bs(\Lambda - F_0))$.)

Note that Φ_Λ is determined up to the group of all projective transformations of $\mathbb{P}^r$; if we change the basis $\{\xi_0, \ldots, \xi_r\}$ to another basis $\{\zeta_0, \ldots, \zeta_r\}$ of $\tilde{\Lambda}$, then Φ_Λ is changed to the composition $\sigma \cdot \Phi_\Lambda$, where σ is the projective transformation of $\mathbb{P}^r$ such that $\sigma(\xi_j) = \zeta_j$ for $0 \leq j \leq r$.

Henceforth, we assume that M is a projective manifold. Then any nondegenerate (see 4 of Lemma 3.4.6) meromorphic (that is, rational) map $f: M \to \mathbb{P}^r$ determines a fixed component free linear system Λ of dimension r such that $f = \Phi_\Lambda$. In fact, let M be a projective manifold in $\mathbb{P}^N$ and f be given by

$$f: p \in M \to (F_0(p): \cdots : F_r(p)) \in \mathbb{P}^r$$

where $F_0, \ldots, F_r$ are homogeneous polynomials of the same degree m such that they are coprime and $M \not\subset V(F_0, \ldots, F_r)$ (see Proposition 3.3.7). Put

$$U_j = \{(X_0: \cdots : X_N) \in M \mid X_j \neq 0\}, \quad 0 \leq j \leq N$$

Then $\{U_j\}$ is an open covering of M. A hyperplane H in $\mathbb{P}^r$ given by

$$H = \{(Y_0: \cdots : Y_r) \in \mathbb{P}^r \mid a_0Y_0 + \cdots + a_rY_r = 0\}$$

for $(a_0: \cdots : a_r) \in \mathbb{P}^{r*}$ gives a (nonzero) homogeneous polynomial

$$F_H = a_0F_0 + \cdots + a_rF_r$$

of degree m, up to nonzero constants. Then

$$g_j\left(\frac{X_0}{X_j}, \ldots, \frac{X_N}{X_j}\right) = F_H\left(\frac{X_0}{X_j}, \ldots, \frac{X_N}{X_j}\right)$$

is a holomorphic function on U_j. Note that

$$\frac{g_j}{g_k} = \left(\frac{X_k}{X_j}\right)^m$$

is a nonvanishing holomorphic function on $U_j \cap U_k$. Hence, $\{g_j\}$ defines a positive (Cartier) divisor D_H on M, called (the pull back by f of) the hyperplane cut (or section) by H. It can be shown that D_H does not depend on the imbedding $M \subset \mathbb{P}^N$, but depends only on $f: M \to \mathbb{P}^r$ and H. Now

$$\Lambda = \{D_H \mid H \text{ is a hyperplane}\}$$

is clearly a fixed component free linear system of dimension r on M such that $\Phi_\Lambda = f$ (up to the group of all projective transformations of $\mathbb{P}^r$). Λ can be identified with the dual projective space $\mathbb{P}^{r*}$ to $\mathbb{P}^r$. The identification is given by

$$H \in \mathbb{P}^{r^*} \to D_H \in \Lambda$$

The meromorphic map $f = \Phi_\Lambda$ is then given by

$$f(p) = \Phi_\Lambda(p) = \bigcap H$$

where the intersection runs over all hyperplanes H in $\mathbb{P}^r$ such that the support of D_H contains p. The base locus $Bs(\Lambda)$ is nothing but the set of all points of indeterminacy of f. We sometimes call it the base locus of f.

Nondegenerate meromorphic (that is, rational) maps f and g on M into $\mathbb{P}^r$ are said to be projectively equivalent if there is a projective transformation σ of $\mathbb{P}^r$ such that $g = \sigma \cdot f$. In this case, we write

$$f \sim g \ (\text{mod Aut}(\mathbb{P}^r))$$

We conclude that

Theorem 3.4.7. Let M be a projective manifold. Then there is a natural 1-1 correspondence between r-dimensional fixed component free linear systems on M and projective equivalence classes of nondegenerate meromorphic (that is, rational) maps of M into $\mathbb{P}^r$.

A linear subsystem Λ' of a linear system Λ is, by definition, a linear system Λ' such that $\Lambda' \subset \Lambda$. (Hence, a linear system is a linear subsystem of a complete linear system.) The proof of the following proposition is left to the reader (Exercise 1).

Proposition 3.4.8. Let Λ be an r-dimensional linear system on M. For any linear subsystem Λ' of Λ of dimension k ($k < r$) there is a unique $(r - k - 1)$-plane S in $\mathbb{P}^r$ such that the diagram

$$\begin{array}{ccc} M & \xrightarrow{\Phi_\Lambda} & \mathbb{P}^r \\ & {\scriptstyle \Phi_{\Lambda'}} \searrow \quad \swarrow {\scriptstyle \pi_S} & \\ & \mathbb{P}^k & \end{array}$$

is commutative, where π_S is the projection with the center S. The correspondence $\Lambda' \to S$ is bijective.

Next, let M and N be projective (or compact complex) manifolds and

$$f: M \to N$$

be a holomorphic map. Let Λ be a fixed component free linear system of dimension r on N. For $D \in \Lambda$, we regard $D = \{h_\alpha\}$ as a Cartier divisor. Then $f^*D = \{h_\alpha \cdot f\}$ is a (Cartier) divisor on M such that

$$[f^*D] = f^*[D]$$

Moreover, we have clearly

Lemma 3.4.9. $f^*\Lambda = \{f^*D \mid D \in \Lambda\}$ is a linear system of dimension $\leq r$ such that $\Phi_{f^*\Lambda} = \Phi_\Lambda \cdot f$. If f is surjective, then $\dim f^*\Lambda = r$.

Remark 3.4.10. $f^*\Lambda$ may have a fixed component. Note also that $f^*\Lambda$ may not be complete, even if Λ is complete. Let Φ_Λ be given by

$$\Phi_\Lambda: p \in N \to (\xi_0(p): \cdots : \xi_r(p)) \in \mathbb{P}^r$$

where $\xi_j \in \Gamma(N, [D])$ for a $D \in \Lambda$. Then $\Phi_{f^*\Lambda}$ is given by

$$\Phi_{f^*\Lambda}: q \in M \to (\tilde{\xi}_0(q): \cdots : \tilde{\xi}_r(q)) \in \mathbb{P}^r$$

where $\tilde{\xi}_j = f^*\xi_j = \xi_j \cdot f \in \Gamma(M, [f^*D])$, provided f is surjective.

Now, let M be a projective manifold and Λ be a fixed component free linear system on M of dimension r. Put

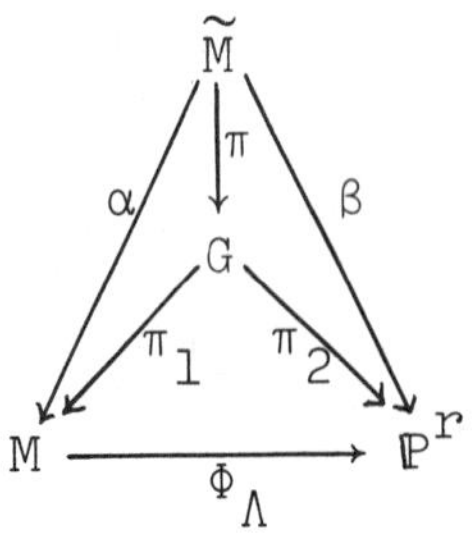

FIGURE 3.7

$B = \mathrm{Bs}(\Lambda)$,

G = the graph of $\Phi_\Lambda : M \to \mathbb{P}^r$,

$\left.\begin{array}{l}\pi_1 : G \to M \\ \pi_2 : G \to \mathbb{P}^r\end{array}\right\}$ the natural projections,

$\pi : \tilde{M} \to G$ a nonsingular model of G (see Theorem 3.3.11),

$\alpha = \pi_1\pi : \tilde{M} \to M$,

$\beta = \pi_2\pi : \tilde{M} \to \mathbb{P}^r$.

(See Fig. 3.7.) Then we have clearly

Lemma 3.4.11.

1. α and β are holomorphic maps.
2. α is a surjective birational map.
3. $\pi_1 : G - \pi_1^{-1}(B) \simeq M - B$. (In particular, Sing $G \subset \pi_1^{-1}(B)$.)
4. $\pi : \tilde{M} - \alpha^{-1}(B) \simeq G - \pi_1^{-1}(B)$.
5. $\alpha : \tilde{M} - \alpha^{-1}(B) \simeq M - B$.

Consider the linear system $\alpha^*\Lambda$ on $\tilde{M}$. It has the dimension r (see Lemma 3.4.9). $\alpha^*\Lambda$ may have a fixed component. Put

$\tilde{F}_0$ = the fixed part of $\alpha^*\Lambda$

Then we have clearly

Lemma 3.4.12.

1. $\beta = \Phi_{\alpha^*\Lambda} = \Phi_{\alpha^*\Lambda - \tilde{F}_0}$. In particular, $\alpha^*\Lambda - \tilde{F}_0$ is base point free.

2. $D \in \Lambda \to \alpha^*D \in \alpha^*\Lambda \to \tilde{D} = \alpha^*D - \tilde{F}_0 \in \alpha^*\Lambda - F_0$ are projective transformations of projective spaces.

Consider the linear system

$$\Lambda_0 = \{H \mid H \text{ is a hyperplane in } \mathbb{P}^r\}$$

on $\mathbb{P}^r$. Then, by 1 of Lemma 3.4.12,

$$\alpha^*\Lambda - \tilde{F}_0 = \beta^*\Lambda_0$$

Hence, by 2 of the same lemma, every $\tilde{D} = \alpha^*D - \tilde{F}_0$ can be written as β^*H for $H \in \Lambda_0$. H is uniquely determined by D. Thus

Lemma 3.4.13. There is a projective transformation

$$D \in \Lambda \to H \in \Lambda_0$$

where $\tilde{D} = \beta^*H$. (D is nothing but D_H defined before.)

Using these considerations, we can now prove the important theorem by Bertini.

Theorem 3.4.14 (Bertini). Let M be a projective manifold and Λ be a fixed component free linear system of dimension r (≥ 1) on M.

1. First theorem by Bertini. If $\dim \Phi_\Lambda(M) \geq 2$, then a general member of Λ is irreducible, that is, a prime divisor. If $\dim \Phi_\Lambda(M) = 1$, then a general member of Λ is the sum of d distinct prime divisors, where d is a constant such that $d \geq r$.
2. Second theorem by Bertini. Let D be a general member of Λ. Then $\mathrm{Sing}(D) \subset \mathrm{Bs}(\Lambda)$. (Here, by (1), D and its support are identified.)

Proof. (1) The proof of the assertion in the case $\dim \Phi_\Lambda(M) \geq 2$ is a little complicated, and so is omitted. See Ueno [97, p. 46] for the proof in this case. Suppose that $\dim \Phi_\Lambda(M) = 1$. Then

$$W = \Phi_\Lambda(M)$$

is a nondegenerate irreducible curve in $\mathbb{P}^r$. Hence,

$$e = \deg W \geq r \quad \text{(see Proposition 1.6.8)}$$

Note that Sing W is a finite set and

$$\beta : \tilde{M} - \beta^{-1}(\mathrm{Sing}\, W) \to W - \mathrm{Sing}\, W$$

is a proper surjective holomorphic map. By Exercise 10 of Sec. 3.2, there is a finite set Z in W - Sing W such that, for every $q \in W$ - Sing W - Z, (1) $\beta^{-1}(q)$ is a nonsingular hypersurface in $\tilde{M} - \beta^{-1}(\text{Sing } W)$ (hence, in $\tilde{M}$), and (2) the number m of connected components of $\beta^{-1}(q)$ is constant.

Let

$$\beta^{-1}(q) = \tilde{D}_1(q) \cup \cdots \cup \tilde{D}_m(q) \qquad \text{(disjoint union)}$$

be the decomposition of $\beta^{-1}(q)$ into the connected (that is, irreducible) components. Consider the divisors

$$\tilde{D}(q) = \tilde{D}_1(q) + \cdots + \tilde{D}_m(q) \qquad \text{on } \tilde{M}$$

and

$$D(q) = D_1(q) + \cdots + D_m(q) \qquad \text{on } M$$

where $D_\nu(q) = \alpha(\tilde{D}_\nu(q))$, $1 \le \nu \le m$, are hypersurfaces in M. Note that $D_\nu(q)$ (respectively $\tilde{D}_\nu(q)$), $1 \le \nu \le m$, are mutually distinct prime divisors.

Now, take a hyperplane H in $\mathbb{P}^r$ such that $H \cap W$ consists of e distinct points $q_1, \ldots, q_e$ of W - Sing W - Z. Then we have clearly

$$\beta^* H = \tilde{D}(q_1) + \cdots + \tilde{D}(q_e)$$

where the supports of $\tilde{D}(q_j)$, $1 \le j \le e$, are mutually disjoint. Hence,

$$\begin{aligned} D_H &= D(q_1) + \cdots + D(q_e) \\ &= D_1(q_1) + \cdots + D_m(q_1) + \cdots + D_1(q_e) + \cdots + D_m(q_e) \end{aligned}$$

is a divisor on M consisting of em distinct prime divisors $D_\nu(q_j)$. Note that

$$d = em \ge e \ge r$$

(2) Let $(Y_0 : \cdots : Y_r)$ be a homogeneous coordinate system in $\mathbb{P}^r$. Put

$$U_j = \{(Y_0 : \cdots : Y_r) \in \mathbb{P}^r \mid Y_j \ne 0\}$$

Then the functions $x_j^k = Y_k/Y_j$ $(k \ne j)$ on U_j form an affine coordinate system on U_j. The holomorphic map β is given by

$$x_j^k = \beta_j^k(z) \quad (k \neq j)$$

where $\beta_j^k(z)$ are holomorphic functions on $\beta^{-1}(U_j)$.

Let H_a be a hyperplane in $\mathbb{P}^r$ given by

$$H_a = \{(Y_0 : \cdots : Y_r) \in \mathbb{P}^r \mid a_0 Y_0 + \cdots + a_r Y_r = 0\}$$

where $a = (a_0 : \cdots : a_r) \in \mathbb{P}^{r*}$. Put

$$U_j^* = \{(a_0 : \cdots : a_r) \in \mathbb{P}^{r*} \mid a_j \neq 0\}$$

and

$$a_j^k = \frac{a_k}{a_j} \quad (k \neq j) \quad \text{on } U_j^*$$

Now, consider the holomorphic functions

$$\gamma_j(z, a) = \sum_{k \neq j} a_j^k \beta_j^k(z) + 1$$

on $\beta^{-1}(U_j) \times U_j^*$ for $0 \leq j \leq r$. Then it is clear that they define a (global) hypersurface

$$\tilde{E} = \{\gamma_j = 0\}$$

in $\tilde{M} \times \mathbb{P}^{r*}$. $\tilde{E}$ is nonsingular. In fact

$$\left(\frac{\partial \gamma_j}{\partial a_j^k}\right)_{(z, a)} = \beta_j^k(z)$$

and one of $\beta_j^k(z)$, for k with $k \neq j$, is not zero for every point $(z, a) \in \beta^{-1}(U_j) \times U_j^*$ such that $\gamma_j(z, a) = 0$.

Let

$$\phi : \tilde{E} \to \mathbb{P}^{r*}$$

be the map induced by the projection $\tilde{M} \times \mathbb{P}^{r*} \to \mathbb{P}^{r*}$. Then, by the construction, it is clear that (1) for any fixed point $a \in \mathbb{P}^{r*}$, $\phi^{-1}(a)$ is a positive divisor on $\tilde{M}$ and is equal to $\beta^* H_a$ and (2)

$$\phi\colon \tilde{E}_1 \to \mathbb{P}^{r*}$$

is surjective for every connected component $\tilde{E}_1$ of $\tilde{E}$. Hence, for a general $a \in \mathbb{P}^{r*}$, β^*H_a is a nonsingular prime divisor on $\tilde{M}$ or a sum of mutually disjoint nonsingular prime divisors on $\tilde{M}$ (see Exercise 10 of Sec. 3.2).

Since $\alpha\colon \tilde{M} - \alpha^{-1}(B) \simeq M - B$ (see (5) of Lemma 3.4.11), a general member D_{H_a} of Λ is nonsingular outside of $B = Bs(\Lambda)$. Q.E.D.

As an application of Bertini's Theorem, we have the following theorem, which will be used to prove Clifford's Theorem (see Sec. 5.3).

Theorem 3.4.15. Let D and D' be positive divisors on a projective manifold M. Then

$$\dim |D| + \dim |D'| \leq \dim |D + D'| \tag{*}$$

Suppose that $|D + D'|$ has no fixed component. Then the equality in (*) holds if and only if there is a positive divisor D_0 on M such that (1) $|D_0|$ has no fixed component, (2) $\dim |jD_0| = j$ for $1 \leq j \leq \dim |D + D'|$, and (3) $D \sim kD_0$ and $D' \sim \ell D_0$ (linearly equivalent) for some $k \geq 1$ and $\ell \geq 1$.

It seems that this theorem is not known, so we give here a proof, whose idea originally comes from Griffiths-Harris [33, p. 250].

The proof is separated into several steps. First of all, note that, putting $E = D + D'$, the map

$$\phi\colon (D_1, D_2) \in |D| \times |D'| \to D_1 + D_2 \in |E|$$

is a finite-to-one holomorphic map of the projective manifold $|D| \times |D'|$ into the projective space $|E|$. In fact, for a given positive divisor E' on M, there are only finitely many ways of expressing E' as a sum $E' = D_1 + D_2$ of two positive divisors D_1 and D_2. Hence,

$$\dim |D| + \dim |D'| = \dim (|D| \times |D'|) = \dim \phi(|D| \times |D'|) \leq \dim |E|$$

by the Proper Mapping Theorem (Theorem 3.2.14).

Next, suppose that

$$\Lambda = |E| = |D + D'|$$

has no fixed component. Consider the meromorphic map

$$\Phi_\Lambda\colon M \to \mathbb{P}^r \quad (r = \dim \Lambda)$$

For a positive divisor D_1 in M, put

$$S_{D_1} = \bigcap H \quad (\subset \mathbb{P}^r)$$

where the intersection runs over all hyperplanes H in $\mathbb{P}^r$ such that $D_1 \leq D_H$. Put

$$s(D_1) = \dim S_{D_1}$$

Lemma 3.4.16. For any $D_1 \in |D|$, (1) $s(D_1) + \dim |D'| = r - 1$ and (2) $s(D_1) = s(D)$.

Proof. (1) can be proved as follows:

$$\dim \{H \mid D_H \geq D_1\} = \dim \{E' \in |E| \,\big|\, E' \geq D_1\}$$

$$= \dim \{D_2 > 0 \mid D_1 + D_2 \sim E\}$$

$$= \dim \{D_2 > 0 \mid D_2 \sim D'\} = \dim |D'|$$

(2) follows from (1). Q.E.D.

Now, suppose that the equality in (*) holds, that is,

$$\dim |D| + \dim |D'| = r$$

Then, by the above lemma,

$$s(D) + s(D') = r - 2 \tag{**}$$

We freely use the previous notations. Let H_0 be a general hyperplane in $\mathbb{P}^r$. The equality in (*) implies that the map

$$\phi\colon |D| \times |D'| \to \Lambda = |E|$$

is surjective. Hence,

$$D_{H_0} = D_1 + D_2$$

for some $D_1 \in |D|$ and $D_2 \in |D'|$. By the first theorem by Bertini,

$$\dim W = 1$$

where $W = \Phi_\Lambda(M)$. Moreover,

$$D_{H_0} = D(q_1) + \cdots + D(q_e)$$

$$= D_1(q_1) + \cdots + D_m(q_1) + \cdots + D_1(q_e) + \cdots + D_m(q_e)$$

where $e = \deg W$ and

$$H_0 \cap W = \{q_1, \ldots, q_e\}$$

(see the proof of (1) of Theorem 3.4.14). Put

$$C_{\nu j} = D_j(q_\nu) \quad \text{for } 1 \leq \nu \leq e \quad \text{and} \quad 1 \leq j \leq m$$

Then $C_{\nu j}$ are mutually distinct prime divisors on M. Since $D_{H_0} = D_1 + D_2$, D_1 is a sum of some $C_{\nu j}$'s. We may put

$$\{\nu \mid \text{there is } j \text{ such that } C_{\nu j} \leq D_1\} = \{1, \ldots, c\}$$

for some integer c with $1 \leq c \leq e$.

Lemma 3.4.17. $S_{D_1} = \bigcap H$, where the intersection runs over all hyperplanes H such that $q_\nu \in H$ for $1 \leq \nu \leq c$.

Proof. Let H_1 be a hyperplane in $\mathbb{P}^r$ such that $H_1 \cap W$ consists of distinct e points

$$q_1, \ldots, q_c, q'_{c+1}, \ldots, q'_e$$

of W - Sing W - Z (see the proof of (1) of Theorem 3.4.14). Then

$$\sum_{\nu=1}^{c} \sum_{j=1}^{m} \tilde{D}_j(q_\nu) \leq \beta^* H$$

Hence,

$$D_1 \leq \sum_{\nu=1}^{c} \sum_{j=1}^{m} C_{\nu j} \leq D_{H_1}$$

so

$$S_{D_1} \subset H_1$$

It is clear that such H_1's span in $\mathbb{P}^{r*}$ the linear subspace of all hyperplanes H such that $q_\nu \in H$ for $1 \leq \nu \leq c$. Hence,

$$S_{D_1} \subset \bigcap H$$

Conversely, if $D_1 \leq D_H$, then, for any ν with $1 \leq \nu \leq c$, there is $j = j(\nu)$ such that $C_{\nu j} \leq D_H$. Since $\alpha^* D_H = \beta^* H + \tilde{F}_0$ and $\tilde{D}_j(q_\nu) \not\leq \tilde{F}_0$, we have $\tilde{D}_j(q_\nu) \leq \beta^* H$. Hence, $q_\nu \in H$ for $1 \leq \nu \leq c$. Q.E.D.

Lemma 3.4.18. $D_1 = \sum_{\nu=1}^{c} \sum_{j=1}^{m} C_{\nu j}$.

Proof. Suppose that, say, $C_{11} \not\leq D_1$. Take any $D_3 \in |D'|$. By Lemma 3.4.17, the hyperplane H in $\mathbb{P}^r$ such that $D_H = D_1 + D_3$ contains q_1. Hence, $C_{11} \leq D_H$, so $C_{11} \leq D$. This means that C_{11} is a fixed component of $|D'|$. Since every $E' \in |E|$ can be written as $E' = D_4 + D_5$ for $D_4 \in |D|$ and $D_5 \in |D'|$, C_{11} is also a fixed component of $\Lambda = |E|$, a contradiction.
Q.E.D.

Lemma 3.4.19. $s(D) = c - 1$ and $s(D') = e - c - 1$.

Proof. By the General Position Theorem (Theorem 1.6.16), we may assume that $q_1, \ldots, q_e$ are in general position. In particular, $q_1, \ldots, q_c$ are in general position. By Lemma 3.4.17, S_{D_1} is the linear subspace in $\mathbb{P}^r$ spanned by $q_1, \ldots, q_c$. Note that $c \leq r$. In fact, if $c > r$, then $S_{D_1} = \mathbb{P}^r$. Hence, by Lemma 3.4.16, D' is empty, a contradiction. Thus $c \leq r$, so

$$s(D) = s(D_1) = \dim S_{D_1} = c - 1$$

The second equality can be shown in a similar way. Q.E.D.

Lemma 3.4.20. (1) $e = r$, so W is a rational normal curve. (2) $\dim |D| = c$ and $\dim |D'| = r - c$.

Proof. This follows from Lemma 3.4.19, (**), and Lemma 3.4.16.
Q.E.D.

Now we complete the proof of Theorem 3.4.15. Fix a general point $q_0 \in W$ and put

$$D_0 = D(q_0) = D_1(q_0) + \cdots + D_m(q_0)$$

Since W is a rational normal curve,

$$D(q) \sim D_0 \text{ (linearly equivalent) for any general } q \in W$$

Now

$$D_{H_0} = D(q_1) + \cdots + D(q_r) \sim rD_0$$

Hence, $|E| = |rD_0|$. In particular, $\dim |rD_0| = r$.

Let x be a rational function on W which gives a holomorphic isomorphism $x: W \to \mathbb{P}^1$ such that $q_0 = D_\infty(x)$, the polar divisor of x. Consider the rational function

$$u = x \cdot \Phi_\Lambda$$

on M. Then $D_0 = D_\infty(u)$, so $|D_0|$ has no fixed component. Note that $L(rD_0)$ contains 1, u, ..., u^r, which are clearly linearly independent. Hence, $\{1, u, \ldots, u^r\}$ is a basis of $L(rD_0)$.

We show that, for any j with $1 \leq j \leq r$, $\{1, u, \ldots, u^j\}$ is a basis of $L(jD_0)$. In fact, since $L(jD_0) \subset L(rD_0)$, every $f \in L(jD_0)$ can be written as

$$f = c_0 + c_1 u + \cdots + c_r u^r \quad (c_\nu \in \mathbb{C})$$

If there were k such that $k > j$ and $c_k \neq 0$, then $D_\infty(f) \geq kD_0$, a contradiction. Hence, $\{1, u, \ldots, u^j\}$ is a basis of $L(jD_0)$. In particular,

$$\dim |jD_0| = j \quad \text{for } 1 \leq j \leq r$$

Finally, by Lemma 3.4.18,

$$D \sim D_1 = D(q_1) + \cdots + D(q_c) \sim cD_0$$

$$D' \sim D_2 = D(q_{c+1}) + \cdots + D(q_r) \sim (r - c)D_0$$

Thus we have proved the "only if" part of Theorem 3.4.15. The "if" part is trivial. This completes the proof of Theorem 3.4.15.

Another application of Bertini's Theorem is the important

Theorem 3.4.21. For any line bundle L on a projective manifold M, there is a divisor D such that $[D] = L$. In other words, the following sequence is exact:

$$1 \to \mathbb{C}^* \to \mathbb{C}(M)^* \to \mathrm{Div}(M) \to \mathrm{Pic}(M) \to 1$$

(compare Lemma 3.4.2).

But the proof of the theorem uses also cohomology of sheaves, so cannot be given here. See Griffiths-Harris [33, p. 161].

Corollary 3.4.22. For any line bundle L on a projective manifold M, there is a meromorphic section ξ ($\not\equiv 0$) of L on M.

Proof. Take a Cartier divisor $D = \{f_\alpha\}$ such that $[D] = L$. Since $L = \{f_\alpha/f_\beta\}$, the collection $\{f_\alpha\}$ defines a meromorphic section ξ ($\not\equiv 0$) of L on M. Q.E.D.

Corollary 3.4.23. There is a meromorphic n-form ω ($\not\equiv 0$) on any n-dimensional projective manifold.

Exercises

1. Prove Lemmas 3.4.5, 3.4.6, and Proposition 3.4.8.

2. Let M be a projective manifold in $\mathbb{P}^N$. Then (1) the set of all homogeneous polynomials F of degree d with $F \not\equiv 0$ on M defines a linear system $\Lambda_M(d)$ on M, called the linear system cut out on M by the hypersurfaces of degree d. (2) If d is sufficiently large, then $\Lambda_M(d)$ is complete (compare Mumford [70, p. 102].

3. Let H be a hyperplane in $\mathbb{P}^n$. Then (1) $\mathrm{Pic}(\mathbb{P}^n)$ is isomorphic to $\mathbb{Z}$ and is generated by [H], (2) $\Lambda_{\mathbb{P}^n}(d) = |dH|$ (complete) for any $d \geq 1$, (3) $\Phi_{|dH|}$ is a holomorphic imbedding for any $d \geq 1$, called the d-th Veronese map, and (4) a biholomorphic map of $\mathbb{P}^n$ onto itself is a projective transformation. (Hint: A biholomorphic map $f: \mathbb{P}^n \to \mathbb{P}^n$ induces a group automorphism $\mathrm{Pic}(\mathbb{P}^n) \to \mathrm{Pic}(\mathbb{P}^n)$.)

4. A birational transformation $\phi: \mathbb{P}^n \to \mathbb{P}^n$ is called an n-dimensional Cremona transformation. ϕ can be written as

$$\phi(p) = (F_0(p): \cdots : F_n(p))$$

where $F_0, \ldots, F_n$ are homogeneous polynomials of the same degree k such that they are coprime. k is called the degree of ϕ. A typical example of two-dimensional Cremona transformations is

$$\phi_0: (X_0 : X_1 : X_2) \in \mathbb{P}^2 \to (X_1X_2 : X_2X_0 : X_0X_1) \in \mathbb{P}^2$$

(1) What is the base locus of ϕ_0? (2) Prove that, for any Cremona transformation $\phi: \mathbb{P}^2 \to \mathbb{P}^2$ of degree 2, there are $\alpha, \beta \in \mathrm{Aut}(\mathbb{P}^2)$ such that $\phi = \alpha\phi_0\beta$. (In general, it is a classical theorem that the group of all two-dimensional Cremona transformations can be generated by ϕ_0 and $\mathrm{Aut}(\mathbb{P}^2)$. See Coolidge [22].)

5. Give an example of compact complex manifolds M such that the homomorphism $\mathrm{Div}(M) \to \mathrm{Pic}(M)$ is not surjective.

6. Let M be a projective manifold of dimension ≥ 2 in $\mathbb{P}^N$ and H be a general hyperplane in $\mathbb{P}^N$. Then (the pull back by the inclusion map $i: M \hookrightarrow \mathbb{P}^N$ of) the hyperplane cut D_H on M by H is a nonsingular prime divisor.

7. (i) $K_{\mathbb{P}^n} = [-(n+1)H]$, where H is a hyperplane in $\mathbb{P}^n$.

   (ii) (<u>Adjunction formula</u>). Let N be an (n + 1)-dimensional compact complex manifold and M be an n-dimensional complex submanifold of N. Then

$$K_M = (K_N \otimes [M])|_M$$

   where the right-hand side is the restriction to M of the line bundle $K_N \otimes [M]$ on N.

   (iii) If M is a nonsingular hypersurface of degree d in $\mathbb{P}^{n+1}$, then

$$K_M = [(d - n - 2)D_H]$$

   where D_H is a hyperplane cut on M.

   (iv) Let ℓ_0 be a line in $\mathbb{P}^3$. Then

$$S = \{\ell \in G(1, 3) \mid \ell \cap \ell_0 \neq \phi\}$$

   is a nonsingular irreducible hypersurface in G(1, 3) such that

$$K_{G(1,3)} = [-4S]$$

   (The homology class $\sigma \in H_6(G(1, 3), \mathbb{Z})$ determined by S is a <u>Schubert cycle</u>. See Griffiths-Harris [33, p. 197].)

4

COMPACT RIEMANN SURFACES

4.1 COMPACT RIEMANN SURFACES

It is known that a two-dimensional compact orientable topological manifold M is a sphere with g-handles. M is topologically determined by the number g of handles, which is called the genus of M. (See Figs. 4.1-4.4.)

$g = 0$

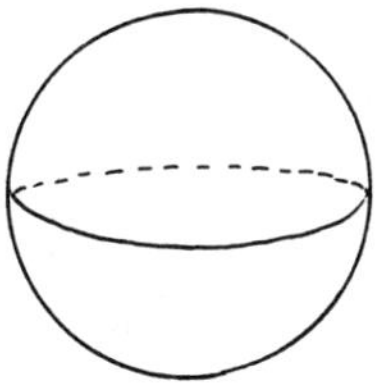

sphere

FIGURE 4.1

$g = 1$

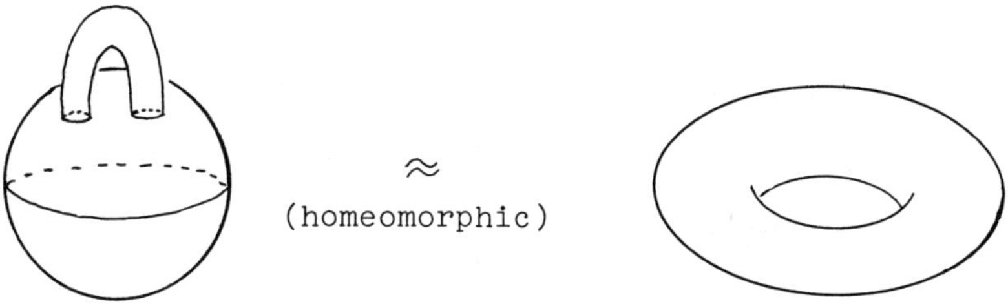

torus

FIGURE 4.2

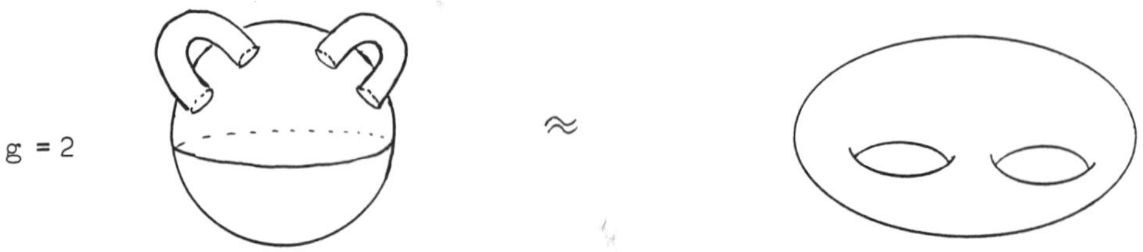

FIGURE 4.3

FIGURE 4.4

M has the Betti numbers

$$b_0 = 1, \quad b_1 = 2g, \quad b_2 = 1$$

(b_j = the j-th Betti number). Hence, the Euler-Poincaré characteristic is

$$\chi = b_0 - b_1 + b_2 = 2 - 2g$$

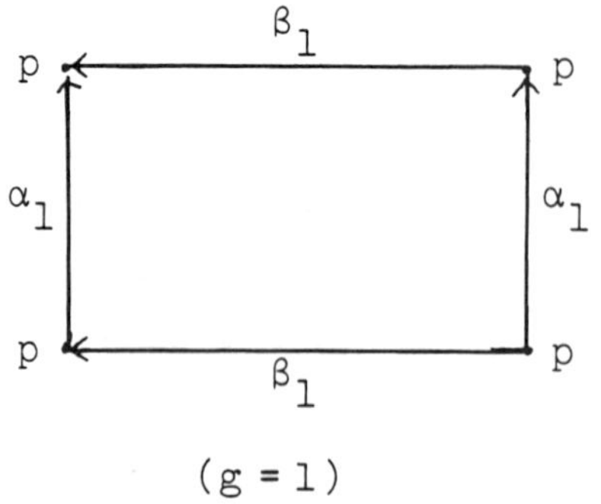

FIGURE 4.5

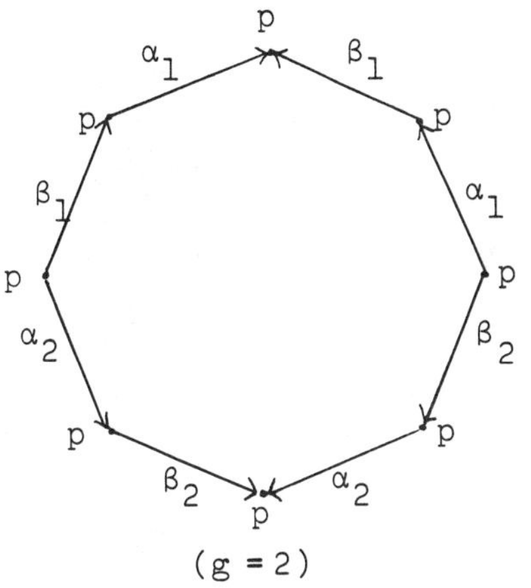

FIGURE 4.6

M can be regarded as a 4g-sided polygon whose oriented sides are identified in Figs. 4.5–4.7.

In fact, from such a polygon, M can be realized (and vice versa) as in Figs. 4.8–4.11.

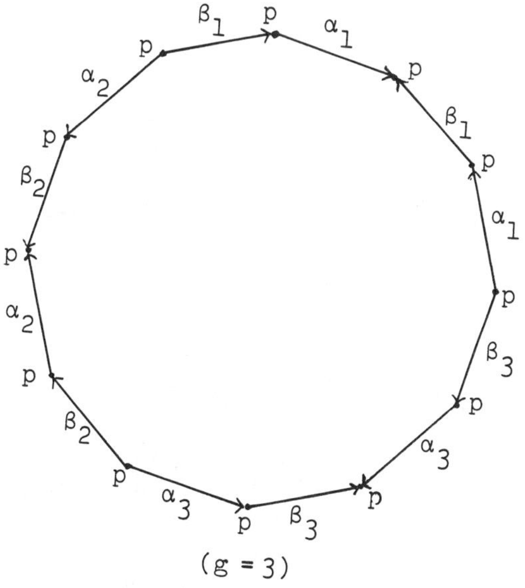

FIGURE 4.7

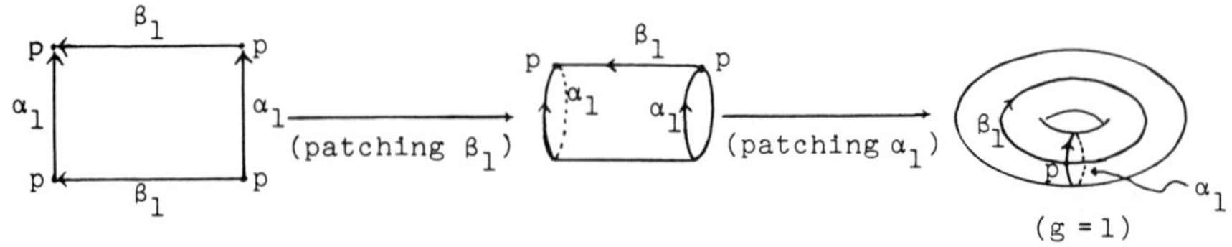

FIGURE 4.8

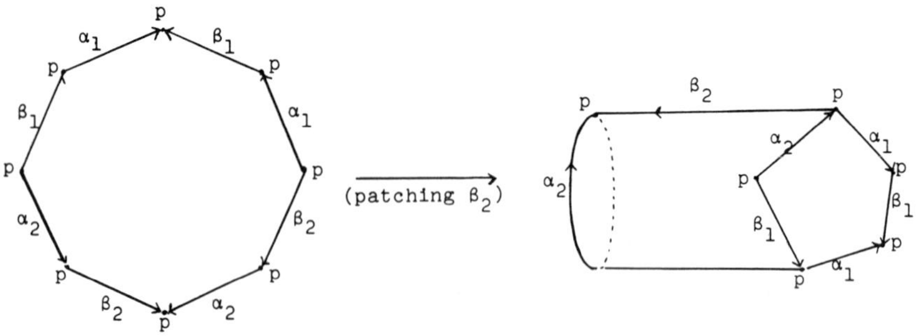

FIGURE 4.9

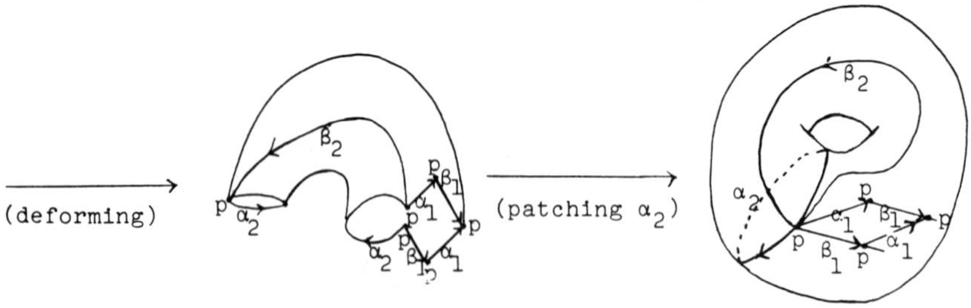

FIGURE 4.10

See, for example, Massey [63] for details.

By abuse of notation, the homotopy (respectively homology) class of a closed path γ on M is denoted by the same letter γ. Then the <u>fundamental group</u> $\pi_1(M) = \pi_1(M, p)$ <u>of</u> M is generated by $\alpha_1, \ldots, \alpha_g, \beta_1, \ldots, \beta_g$ which have the unique relation

$$\alpha_1\beta_1\alpha_1^{-1}\beta_1^{-1} \cdots \alpha_g\beta_g\alpha_g^{-1}\beta_g^{-1} = 1$$

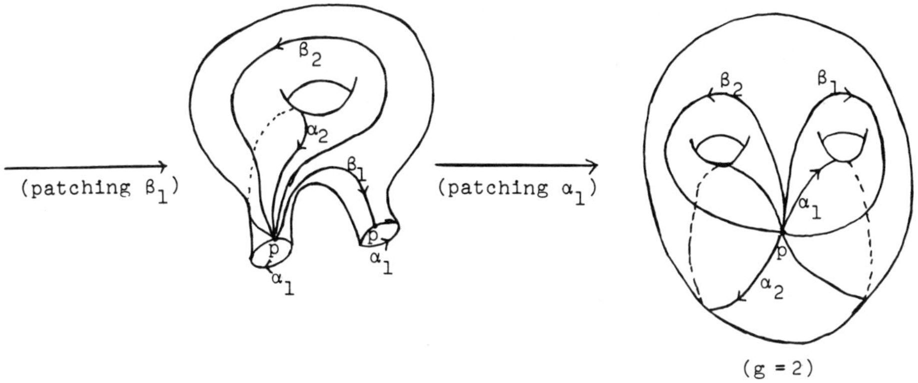

FIGURE 4.11

The first homology group $H_1(M, \mathbb{Z})$ is hence the direct sum

$$H_1(M, \mathbb{Z}) = \mathbb{Z}\alpha_1 + \cdots + \mathbb{Z}\alpha_g + \mathbb{Z}\beta_1 + \cdots + \mathbb{Z}\beta_g$$

(torsion free). The intersection numbers are

$$\alpha_j \alpha_k = \beta_j \beta_k = 0$$

$$\alpha_j \beta_k = -\beta_k \alpha_j = \delta_{jk} \qquad \text{(Kronecker's } \delta\text{)}$$

for all j and k. In general, a basis $\{\alpha_1, \ldots, \alpha_g, \beta_1, \ldots, \beta_g\}$ of $H_1(M, \mathbb{Z})$ which satisfies these relations is called a symplectic basis.

Now, a one-dimensional complex manifold is traditionally called a Riemann surface. Henceforth, we mainly treat compact Riemann surfaces.

A compact Riemann surface M is a two-dimensional compact orientable topological manifold. Hence, its genus is defined.

The projective line $\mathbb{P}^1$ (respectively a complex 1-torus) is an example of compact Riemann surfaces of genus 0 (respectively 1). Later on, we will prove that, conversely, a compact Riemann surface of genus 0 (respectively 1) is biholomorphic to $\mathbb{P}^1$ (respectively a complex 1-torus) (see Chapter 5).

The following proposition is a special case of the proper mapping theorem (Theorem 3.2.14).

Proposition 4.1.1. Let M and N be compact Riemann surfaces and $f: M \to N$ be a holomorphic map. Then f is either surjective or a constant map.

Proof. If f is not a constant map, then, by Lemma 4.1.2 below, it is an open map. Hence, f(M) is open in N and is compact. Since N is connected, f(M) = N. Q.E.D.

Lemma 4.1.2. Let M and N be Riemann surfaces and $f: M \to N$ be a nonconstant holomorphic map. Then, for any point $p \in M$, f can be locally written as

$$f: z \to w = z^k \quad (k \geq 1)$$

for a suitable local coordinate z (respectively w) around p (repectively f(p)) such that z(p) = 0 (respectively w(f(p)) = 0). In particular, f is an open map. (See Fig. 4.12.)

The proof of the lemma is left to the reader (Exercise 1).

A coordinate z around $p \in M$ such that z(p) = 0 is traditionally called a local uniformizing parameter at p. The positive integer k in Lemma 4.1.2 is called the index of ramification at p. If $k \geq 2$, then p (respectively f(p)) is called a point of ramification (respectively a branch point) of f. The set of all points of ramification (respectively branch points) is clearly a discrete set in M (respectively in N).

If M and N are compact Riemann surfaces and $f: M \to N$ is a surjective holomorphic map, then f is a ramified covering map. That is to say

$$f: M - f^{-1}(B) \to N - B$$

is a usual (unramified) covering map, where B is the set of all branch points of f, called the branch locus of f. Moreover, around every point of ramification, f can be given as in Lemma 4.1.2 for $k \geq 2$. (See Fig. 4.13.)

The degree of the covering

$$f: M - f^{-1}(B) \to N - B$$

is called the degree, deg f, of f. The following lemma is easy to prove.

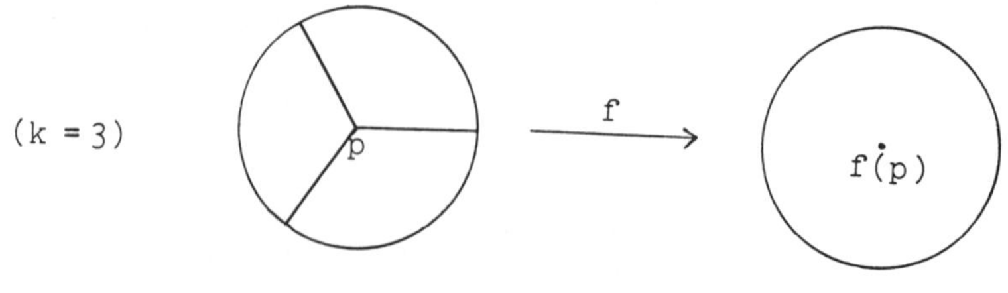

FIGURE 4.12

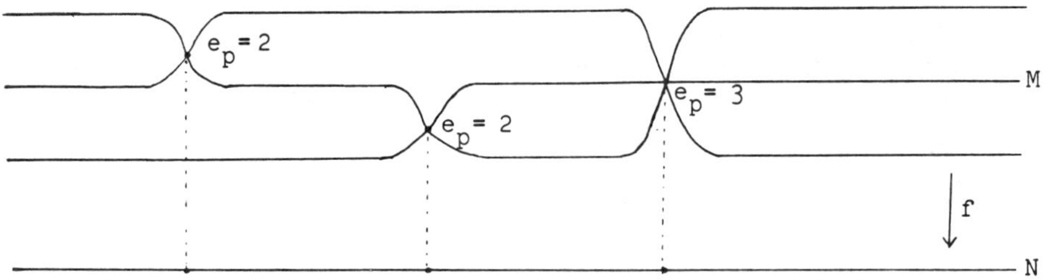

FIGURE 4.13

Lemma 4.1.3. Let M and N be compact Riemann surfaces and f: M → N be a surjective holomorphic map. Then (1) the set of all points of ramification (respectively the branch locus B) of f is a finite set in M (respectively in N), (2) for every point $q \in N$,

$$\sum_p e_p = \deg f$$

where e_p is the index of ramification at p and Σ is extended over all points $p \in f^{-1}(q)$, and (3) if deg f = 1, then f is a biholomorphic map.

Now, we prove

Theorem 4.1.4 (Riemann-Hurwitz Formula). Let M and N be compact Riemann surfaces of genus g and g', respectively. Let f: M → N be a surjective holomorphic map of degree d. Then

$$2g - 2 = d(2g' - 2) + \Sigma (e_p - 1)$$

where Σ is extended over all points $p \in M$ of ramification and e_p is the index of ramification at p. In particular, $g \geq g'$.

Proof. Let Γ be a triangulation of N such that every branch point of f is a vertex of Γ. Let s_0, s_1, and s_2 be the numbers of vertices, sides, and faces of Γ, respectively. Let $f^*\Gamma$ be the pull-back of Γ by f. It is a triangulation of M, whose numbers of vertices, sides, and faces are clearly

$$ds_0 - \Sigma (e_p - 1), \quad ds_1, \quad \text{and} \quad ds_2$$

respectively. Hence, the Euler-Poincaré characteristics of M and N are related as

$$\chi(M) = ds_0 - \Sigma(e_p - 1) - ds_1 + ds_2 = d\chi(N) - \Sigma(e_p - 1) \qquad \text{Q.E.D.}$$

For a compact Riemann surface M, a meromorphic function f is nothing but a meromorphic function in the usual sense. That is, f is holomorphic outside a finite set A in M and is expanded into a Laurent series

$$f(z) = \frac{c_{-k}}{z^k} + \cdots + \frac{c_{-1}}{z} + c_0 + c_1 z + \cdots$$

for a local uniformizing parameter z around every point $p \in A$. It is then easy to see that

Proposition 4.1.5. A meromorphic function on a compact Riemann surface M can be regarded as a holomorphic map $f\colon M \to \mathbb{P}^1$ such that $f(M) \neq \{\infty\}$, and vice versa. In particular, a nonconstant meromorphic function f on M is a (ramified) covering map of M onto $\mathbb{P}^1$. (Its degree is called the degree, deg f, of the meromorphic function f.) (See Fig. 4.14.)

Since $\mathbb{P}^1$ is simply connected, the branch locus of $f\colon M \to \mathbb{P}^1$ is nonempty, unless $M = \mathbb{P}^1$ and $f \in \mathrm{Aut}(\mathbb{P}^1)$. This is intuitively clear and follows from the Riemann-Hurwitz Formula.

It is remarkable that the following proposition holds (compare Mumford [70, p. 127]).

Proposition 4.1.6. Let M be a Riemann surface and N be a complex manifold. Let $f\colon M \to N$ be a meromorphic map. Then f is necessarily a holomorphic map.

Proof. The set S_f of all points of indeterminacy of f is an analytic set in M such that $\dim S_f \leq \dim M - 2 = -1$ (see Proposition 3.2.16). Hence, S_f is empty. Q.E.D.

Corollary 4.1.7. Let M and N be Riemann surfaces and $f\colon M \to N$ be a bimeromorphic map. Then f is a biholomorphic map.

Now, for a positive number δ, put

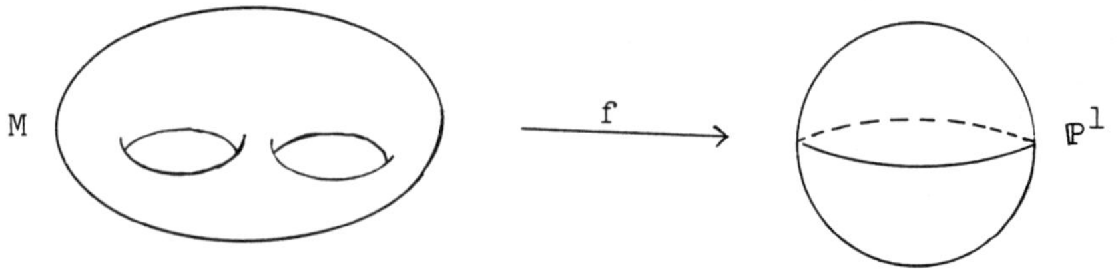

FIGURE 4.14

$$\Delta^*(0,\delta) = \{t \in \mathbb{C} \mid 0 < |t| < \delta\}$$

Then the holomorphic map

$$\pi_0 \colon t \in \Delta^*(0, \delta^{1/k}) \to t^k \in \Delta^*(0,\delta)$$

is an (unramified) covering map of degree k. The proof of the following lemma is left to the reader (Exercise 1).

Lemma 4.1.8. A k-fold (unramified) covering of $\Delta^*(0,\delta)$ is essentially unique. More precisely, let W be a Riemann surface and $\pi\colon W \to \Delta^*(0,\delta)$ be a holomorphic (unramified) covering of degree k. Then there is a biholomorphic map $\phi\colon \Delta^*(0, \delta^{1/k}) \to W$ such that the diagram

$$\begin{array}{ccc} \Delta^*(0,\delta^{1/k}) & \xrightarrow{\ \phi\ } & W \\ & {\scriptstyle \pi_0}\searrow \quad \swarrow{\scriptstyle \pi} & \\ & \Delta(0,\delta) & \end{array}$$

is commutative.

Now let

$$\Omega = \{(x,y) \in \mathbb{C}^2 \,\big|\, |x| < \epsilon, \quad |y| < \epsilon\}$$

be a small polydisc in $\mathbb{C}^2$. Consider an analytic curve

$$C = \{(x,y) \in \Omega \mid f(x,y) = 0\}$$

in Ω passing through $0 = (0,0)$. Assume that C is irreducible at 0. We may assume that

$$f(x,y) = y^k + a_1(x)y^{k-1} + \cdots + a_k(x)$$

is a Weierstrass polynomial. Then there is δ with $0 < \delta < \epsilon$ such that, for every $x \in \mathbb{C}$ with $0 < |x| < \delta$, there are just k distinct roots

$$y_1(x), \ldots, y_k(x)$$

of the equation $f(x,y) = 0$ for y such that $|y_\nu(x)| < \epsilon$ for $1 \leq \nu \leq k$. Put

$$W = \{(x,y) \in C \mid 0 < |x| < \delta, \quad |y| < \epsilon\}$$

Then, by the Implicit Mapping Theorem (see Sec. 3.1),

$$\pi: (x,y) \in W \to x \in \Delta^*(0,\delta)$$

is clearly a holomorphic (unramified) covering map of degree k. Hence, by Lemma 4.1.8, there is a biholomorphic map $\phi: \Delta^*(0, \delta^{1/k}) \to W$ such that $\pi_0 = \pi\phi$. Hence, ϕ can be written as

$$\phi(t) = (t^k, h(t))$$

where h(t) is a holomorphic function on $\Delta^*(0, \delta^{1/k})$. Since

$$f(t^k, h(t)) = 0 \quad \text{for } 0 < |t| < \delta^{1/k}$$

h(t) must satisfy

$$h(t) \to 0 \quad \text{as } t \to 0$$

Hence, h(t) can be extended holomorphically on

$$\Delta(0, \delta^{1/k}) = \{t \in \mathbb{C} \mid |t| < \delta^{1/k}\}$$

by putting h(0) = 0. Put

$$\Omega' = \{(x,y) \in \mathbb{C}^2 \mid |x| < \delta, |y| < \epsilon\}$$

$$C' = \{(x,y) \in C \mid |x| < \delta, |y| < \epsilon\}$$

Then

$$\phi: t \in \Delta(0, \delta^{1/k}) \to (t^k, h(t)) \in \Omega'$$

is a holomorphic map whose image is C'. Moreover,

$$\phi: \Delta(0, \delta^{1/k}) \to C'$$

is a homeomorphism. Note that C' is a neighborhood of 0 in C.

The above argument works for an analytic curve C in a small polydisc in $\mathbb{C}^r$ ($r \geq 3$), after a modification. (The reader may check the detail.)

Definition 4.1.9. Let $\Omega = \Omega(0,\epsilon)$ be a small polydisc in $\mathbb{C}^r$ ($r \geq 2$) and C be an analytic curve in Ω passing through 0 and irreducible at 0. A holomorphic map $\phi: \Delta(0, \rho) \to \Omega$ ($\rho > 0$) is called a local uniformizing parameter of C at 0 if (1) $\phi(0) = 0$, (2) $\phi(\Delta(0, \rho)) = C'$ is a neighborhood of 0 in C, (3) $\phi: \Delta(0, \rho) \to C'$ is a homeomorphism, and (4) $\phi: \Delta^*(0, \rho) \to C' - \{0\}$ is a biholomorphic map.

Proposition 4.1.10. A local uniformizing parameter of C at 0 exists and is unique up to holomorphic isomorphisms.

Proof. The existence has been shown already. The uniqueness follows from the Riemann Extension Theorem (see Sec. 3.1). Q.E.D.

It should be noted that local uniformizing parameters do not necessarily exist for higher dimensional analytic sets in Ω.

The following theorem is a special case of (the analytic version of) Hironaka's Theorem (Theorem 3.3.11).

Theorem 4.1.11. Let N be a complex manifold and C be a compact irreducible analytic curve in N. Then there is a nonsingular model $\phi: M \to C$ of C. That is, there are a compact Riemann surface M and a holomorphic map $\phi: M \to N$ such that (1) $\phi(M) = C$ and (2) $\phi: M - \phi^{-1}(\text{Sing } C) \to C - \text{Sing } C$ is biholomorphic. Moreover, if $\phi_1: M_1 \to C$ is another nonsingular model of C, then there is a biholomorphic map $\psi: M \to M_1$ such that $\phi_1\psi = \phi$.

Proof. Put

$$\text{Sing } C = \{p_1, \dots, p_s\}$$

Let U_j be a small neighborhood of p_j in N such that

$$C \cap U_j = C_{j1} \cup \cdots \cup C_{jk_j}$$

is the irreducible decomposition of C at p_j. We may assume that (1) $U_j \cap U_k = \phi$ for $j \neq k$ and (2) $C_{j\nu} \cap C_{j\mu} = \{p_j\}$ for $\nu \neq \mu$. Let

$$\phi_{j\nu}: \Delta_{j\nu} \to U_j$$

be a local uniformizing parameter of $C_{j\nu}$ at p_j such that $\phi_{j\nu}(\Delta_{j\nu}) = C_{j\nu}$ for $1 \leq \nu \leq k_j$. Consider the disjoint union

$$\tilde{M} = (C - \text{Sing } C) \cup (\bigcup_{j,\nu} \Delta_{j\nu})$$

Then $\tilde{M}$ is a not necessarily connected complex manifold of dimension 1. Consider the equivalence relation $\sim$ on M defined by

$$p \in C - \text{Sing } C \sim q \in \Delta_{j\nu}$$

if $p \in C_{j\nu}$ and $\phi_{j\nu}(q) = p$. Put

$$M = \tilde{M}/\sim,$$

$$\pi : \tilde{M} \to M, \text{ the natural projection}$$

Then M is a compact Hausdorff space under the induced topology. Moreover, M is a compact Riemann surface; the local coordinates on $\pi(C - \text{Sing } C)$ are those on $C - \text{Sing } C$ and the local coordinates on $\pi(\Delta_{j\mu})$ are those on $\Delta_{j\nu}$.

Now the map $\tilde{\phi} : M \to N$ defined by

$$p \in C - \text{Sing } C \to p \in C,$$

$$p \in \Delta_{j\nu} \to \phi_{j\nu}(p) \in C_{j\nu}$$

is a holomorphic map and induces a holomorphic map

$$\phi : M \to N$$

which satisfies the condition. (The reader may check the details.) Thus, the existence is proved.

The last statement of the theorem follows from Corollary 4.1.7.

Q.E.D.

Definition 4.1.12. The genus of a compact irreducible analytic curve in a complex manifold is, by definition, the genus of its nonsingular model. If it is 0 (respectively 1), then the curve is called a rational (respectively elliptic) curve.

Next, let M be a compact Riemann surface. A divisor D on M is, in this case, a formal finite sum

$$D = a_1 p_1 + \cdots + a_m p_m$$

of points p_j, $1 \leq j \leq m$, on M with the coefficients $a_j \in \mathbb{Z}$. Recall that D is said to be positive (or effective), $D > 0$, if $a_j \geq 0$ for $1 \leq j \leq m$ and $a_k > 0$ for some k. $D \geq 0$ if $D > 0$ or $D = 0$. $D_1 \geq D_2$ if $D_1 - D_2 \geq 0$.

The degree, deg D, of D is, by definition

$$\deg D = a_1 + \cdots + a_m$$

This defines a homomorphism

$$\deg\colon \mathrm{Div}(M) \to \mathbb{Z}$$

A point $p \in M$ can be regarded as a positive divisor of degree 1 on M, called the point divisor. We write the point divisor by (p) instead of p, when it is necessary to distinguish them.

Recall that divisors D_1 and D_2 are linearly equivalent, $D_1 \sim D_2$, if and only if

$$D_1 - D_2 = (f) = D_0(f) - D_\infty(f)$$

for a meromorphic function $f\colon M \to \mathbb{P}^1$, where (f), $D_0(f)$, and $D_\infty(f)$ are the principal divisor, the zero divisor, and the polar divisor of f, respectively. Note that

$$\deg D_0(f) = \deg D_\infty(f) = \deg f \quad \text{(compare Lemma 4.1.3)}$$

so

$$\deg (f) = 0$$

$\deg D_0(f)$ (respectively $\deg D_\infty(f)$) is called the order of zeros (respectively poles) of f. The equality $\deg D_0(f) = \deg D_\infty(f)$ can be also proved by the Argument Principle (see Ahlfors [3, p. 151]) and the Residue Theorem (Theorem 5.1.3). Hence,

$$\deg D_1 = \deg D_2, \quad \text{if } D_1 \sim D_2$$

so the homomorphism deg induces a homomorphism

$$\deg\colon \mathrm{Div}(M)/\!\sim\; \to \mathbb{Z}$$

where $\mathrm{Div}(M)/\!\sim$ is the divisor class group of M.

The following theorem is a special case of Theorem 3.4.4.

Theorem 4.1.13. Let D be a divisor on a compact Riemann surface M. Then $L(D)$ ($\simeq \Gamma([D])$) is finite dimensional. If $D \geq 0$, then $\dim L(D) \leq \deg D + 1$.

Proof. We may write $D = D^+ - D^-$, where D^+ and D^- are positive (or zero) divisors whose supports have no point in common. Then, by the definition of $L(D)$,

$$L(D) \subset L(D^+)$$

If $D = 0$, then $\dim L(0) = 1$.

Hence, we may assume that D is a positive divisor. Write

$$D = \nu_1 p_1 + \cdots + \nu_s p_s$$

where ν_j are positive integers and $p_j \neq p_k$ for $j \neq k$.

Take $f \in L(D)$. We expand f into the Laurent series

$$f(t) = a^{(j)}_{-\nu_j} t^{-\nu_j} + \cdots + a^{(j)}_{-1} t^{-1} + a^{(j)}_0 + \cdots$$

around every p_j, where t is a local uniformizing parameter at p_j. Then

$$A_j: f \in L(D) \rightarrow (a^{(j)}_{-\nu_j}, \ldots, a^{(j)}_{-1}) \in \mathbb{C}^{\nu_j} \quad (1 \leq j \leq s),$$

$$A: f \in L(D) \rightarrow (A_1(f), \ldots, A_k(f)) \in \mathbb{C}^d \quad (d = \deg D)$$

are linear maps. The kernel of A is clearly $\mathbb{C} = \{\text{constants}\}$. Hence,

$$\dim L(D) \leq \deg D + 1 \qquad \text{Q.E.D.}$$

A linear system on a compact Riemann surface M was classically called a <u>linear series</u>. In this book, however, we do not use this word.

Given a linear system Λ on M, the <u>degree</u>, $\deg \Lambda$, of Λ is defined by

$$\deg \Lambda = \deg D \quad \text{for } D \in \Lambda$$

This is well defined as noted above.

A linear system Λ of dimension r and degree d on M is traditionally denoted by g^r_d. When a linear system is fixed, this notation is convenient and will be sometimes used.

Let $\Lambda = g^r_d$ be a linear system on M. A <u>fixed point</u> of Λ is, by definition, a point $p \in M$ such that

$$(p) \leq D \quad \text{for all } D \in \Lambda$$

A meromorphic map $f: M \rightarrow \mathbb{P}^r$ is a holomorphic map by Proposition 4.1.6. If f is nondegenerate, then it defines a linear system $\Lambda = g^r_d$ on M <u>without fixed point</u> such that

$$f \sim \Phi_\Lambda \quad (\text{mod } \operatorname{Aut}(\mathbb{P}^r))$$

and vice versa (see Theorem 3.4.7). The degree $d = \deg \Lambda$ of Λ is also called the degree, $\deg f$, of $f: M \to \mathbb{P}^r$. The image $C = f(M)$ is a nondegenerate irreducible curve in $\mathbb{P}^r$. Let

$$\phi: M_1 \to C$$

be a nonsingular model of C. Then $f: M \to C \subset \mathbb{P}^r$ clearly induces a meromorphic map $f_1: M \to M_1$. By Proposition 4.1.6, f_1 is a surjective holomorphic map such that the diagram

$$\begin{array}{ccc} M & \xrightarrow{\;f_1\;} & M_1 \\ & {\scriptstyle f}\searrow \qquad \swarrow{\scriptstyle \phi} & \\ & C \subset \mathbb{P}^r & \end{array}$$

is commutative. The following lemma is then easy to see.

Lemma 4.1.14. $\deg f = (\deg f_1)(\deg C)$.

Let $\Lambda = g_d^r$ be a linear system on M. For a positive divisor $E = p_1 + \cdots p_k$ on M, put

$$\Lambda - E = \{D \in \Lambda \mid D \geq E\}$$

Then the following lemma is easy to see.

Lemma 4.1.15. (1) $\Lambda - E$ is a linear subsystem of Λ and (2) if Λ has no fixed point, then $\Phi_{\Lambda - E} = \pi_S \cdot \Phi_\Lambda$, where S is a linear subspace in $\mathbb{P}^r$ ($r = \dim \Lambda$) spanned by $\Phi_\Lambda(p_1), \ldots, \Phi_\Lambda(p_k)$, and π_S is the projection with the center S (compare Proposition 3.4.8).

The following theorem will be useful.

Theorem 4.1.16. Let $\Lambda = g_d^r$ be a linear system on a compact Riemann surface M.

1. A point $p \in M$ is a fixed point of Λ if and only if $\dim(\Lambda - p) = r$.
2. Λ has no fixed point and $\Phi_\Lambda: M \to \mathbb{P}^r$ is a (holomorphic) bimeromorphic map onto its image if and only if there is a point $p \in M$ such that

$$\dim(\Lambda - p - q) = r - 2 \quad \text{for all } q \in M$$

3. Λ has no fixed point and $\Phi_\Lambda: M \to \mathbb{P}^r$ is a holomorphic imbedding if and only if

$$\dim(\Lambda - p - q) = r - 2 \quad \text{for all } p, q \in M$$

1 of the theorem is trivial. The proofs of 2 and 3 are left to the reader (Exercise 1).

Note that a similar theorem holds for a linear system on a higher dimensional projective manifold which the reader may formulate.

If 3 of the theorem is the case, Λ is said to be <u>very ample</u>. A divisor on M is said to be very ample if $|D|$ is very ample. D is said to be <u>ample</u> if kD is very ample for some positive integer k. Later (in Sec. 5.1), we show that D is ample if and only if deg D > 0. Hence, this word "ample" is unnecessary. But this notion is very useful for the study of higher dimensional projective manifolds.

Finally, we state one of the deepest and most surprising theorems.

<u>Theorem 4.1.17 (Riemann's Existence Theorem)</u>. Let M be a compact Riemann surface and p and q are distinct points on M. Then there is a meromorphic function f on M such that $f(p) \neq f(q)$.

<u>Note 4.1.18</u>

1. The proof of Theorem 4.1.17 uses a lot of analysis, so cannot be given here. See, for example, Weyl [101], Ahlfors-Sario [4], Bers [11], and Springer [95]. Historically, Riemann's original proof, using the <u>Dirichlet Principle</u>, was criticized by Weierstrass. Rigorous proofs were obtained by Neumann and Schwarz (the <u>alternating method</u>) and by Hilbert (the Dirichlet Principle). Weyl [101] simplified Hilbert's method. Today, Weyl's method has been extended by Hodge, Kodaira, and so on, to <u>Kähler manifolds</u>, know as the <u>Hodge Theory</u> (see, for example, Griffiths-Harris [33]).
2. Theorem 4.1.17 is an easy consequence of the Riemann-Roch Theorem (see Sec. 5.1). Hence, if the Riemann-Roch Theorem is proved analytically for a compact Riemann surface, then Theorem 4.1.17 is also proved. Gunning [35] took this way. He proved the Riemann-Roch Theorem by a cohomological method plus the <u>Serre duality</u>. Another interesting analytic proof of the Riemann-Roch Theorem was given by Kotake [58] using the <u>heat equation</u>.

<u>Exercises</u>

1. Prove Lemmas 4.1.2, 4.1.8, and Theorem 4.1.16.

2. (1) If D is a divisor on M such that deg $D < 0$, then $L(D) = \{0\}$. (2) If deg $D = 0$, then dim $L(D) \leq 1$. Moreover, dim $L(D) = 1$ if and only if D is a principal divisor.

3. A meromorphic function f on $\mathbb{P}^1$ is a rational function in the usual sense (and vice versa): $f(z) = v(z)/u(z)$, where $u(z)$ and $v(z)$ are polynomials. (1) The degree of $f: \mathbb{P}^1 \to \mathbb{P}^1$ is $\max\{\deg u, \deg v\}$. (2) f is itself a polynomial of degree n if and only if $D_\infty(f) = n(\infty)$. Hence, the support of $D_0(f)$ is a nonempty set in $\mathbb{C}$ (the Fundamental Theorem of Algebra). (3) For any (usual) rational function f(z) of degree 2, there are $\sigma_1, \sigma_2 \in \mathrm{Aut}(\mathbb{P}^1)$ such that $f(z) = \sigma_2(\sigma_1(z)^2)$. (4) Does a similar assertion hold for (usual) rational functions of degree 3?

4. Let M and N be compact Riemann surfaces of the same genus g, and $f: M \to N$ be a surjective holomorphic map. (1) If $g = 1$, then f is an (unramified) covering map. (2) If $g \geq 2$, then f is necessarily a biholomorphic map.

4.2 BLOWING UP

Let N be an n-dimensional manifold ($n \geq 2$) and p be a point of N. We define the blowing up $Q_p(N)$ of N at p.

Let $(x_1, \ldots, x_n)$ be a connected neighborhood U of p in N such that $p = (0, \ldots, 0)$. Put

$$\tilde{U} = \{((x_1, \ldots, x_n), (\xi_1 : \cdots : \xi_n)) \in U \times \mathbb{P}^{n-1} \mid \xi_j x_k = \xi_k x_j,\ 1 \leq j, k \leq n\}$$

where $(\xi_1 : \cdots : \xi_n)$ is a homogeneous coordinate system in $\mathbb{P}^{n-1}$. Then $\tilde{U}$ is an n-dimensional complex manifold and the projection

$$\pi: (x, \xi) \in \tilde{U} \to x \in U$$

is a bimeromorphic holomorphic map such that

1. $\pi: \tilde{U} - \pi^{-1}(p) \simeq U - p$ and

2. $\pi^{-1}(p) \simeq \mathbb{P}^{n-1}$.

(See Fig. 4.15.)

Note that the tangent directions at p and points of $\pi^{-1}(p)$ are in 1-1 correspondence.

Next, consider the disjoint union

$$\tilde{N} = (N - p) \cup \tilde{U}$$

and the following equivalence relation on $\tilde{N}$:

$$(x_1, \ldots, x_n) \in N - p \sim ((x_1, \ldots, x_n), (\xi_1 : \cdots : \xi_n)) \in \tilde{U}$$

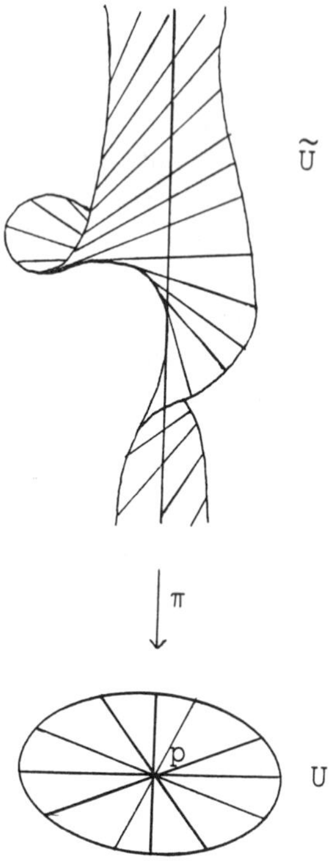

FIGURE 4.15

Then the quotient space

$$Q_p(N) = \tilde{N}/\sim$$

is naturally a n-dimensional complex manifold and the projection

$$\pi: Q_p(N) \to N$$

induced by $\pi: \tilde{U} \to U$, is a bimeromorphic holomorphic map such that (1) $\pi: Q_p(N) - \pi^{-1}(p) \simeq N - p$ and (2) $\pi^{-1}(p) \simeq \mathbb{P}^{n-1}$.

It can be easily shown that $Q_p(N)$ does not depend on the choice of local coordinate system $(x_1, \ldots, x_n)$, up to holomorphic isomorphisms.

The complex manifold $Q_p(N)$ (or the projection $\pi: Q_p(N) \to N$) is called the blowing up of N at p. $E = \pi^{-1}(p)$ is called the exceptional submanifold of $Q_p(N)$. (Conversely, N is called the blowing down of $Q_p(N)$ along E, or simply the blowing down E to one point.)

The proof of the following lemma is left to the reader (Exercise 1).

Lemma 4.2.1. Let N, p, and $\pi: Q_p(N) \to N$ be as above. Let W be a complex submanifold of N such that $p \in W$. Then the closure of $\pi^{-1}(W - p)$ in $Q_p(N)$ is (biholomorphic to) $Q_p(W)$.

Put, in particular, $N = \mathbb{P}^n$ and $p \in \mathbb{P}^n$. Take a homogeneous coordinate system $(X_0: \cdots : X_n)$ such that $p = (1: 0: \cdots : 0)$. Put

$$\tilde{\mathbb{P}}^n = \{((X_0: \cdots : X_n), (\xi_1: \cdots : \xi_n)) \in \mathbb{P}^n \times \mathbb{P}^{n-1} \mid \xi_j X_k = \xi_k X_j, \ 1 \le j, k \le n\}$$

Then the following lemma is easy to show:

Lemma 4.2.2. $\tilde{\mathbb{P}}^n = Q_p(\mathbb{P}^n)$.

Proposition 4.2.3. Let N be a projective manifold and $p \in N$. Then $Q_p(N)$ is also a projective manifold.

Proof. Suppose that N is a complex submanifold in $\mathbb{P}^r$. Then, by Lemma 4.2.1, $Q_p(N)$ is a complex submanifold of $Q_p(\mathbb{P}^r)$, which is a complex submanifold of the projective manifold $\mathbb{P}^r \times \mathbb{P}^{r-1}$ by Lemma 4.2.2. Q.E.D.

Now, let S be a two-dimensional complex manifold and C be an analytic curve on S. For a point $p \in C$, consider the blowing up

$$\pi: Q_p(S) \to S$$

of S at p. Then $\pi^{-1}(C)$ has the exceptional curve

$$E = \pi^{-1}(p)$$

as an irreducible component. Write

$$\pi^{-1}(C) = \bar{C} \cup E$$

where $\bar{C}$ is the union of other irreducible components than E. Then it is clear that (1) $\pi: \bar{C} - \bar{C} \cap E \to C - \{p\}$ is homeomorphic and (2) $\bar{C}$ is the closure in $Q_p(S)$ of $\pi^{-1}(C - p)$.

The analytic curve $\bar{C}$ in $Q_p(S)$ is called the strict (or proper) transform of C.

The proof of the following lemma is left to the reader (Exercise 1).

Lemma 4.2.4. If C is nonsingular at p, then (1) $\bar{C}$ and E meet at a unique point $\bar{p}$ transversally (that is, have distinct tangent lines at $\bar{p}$) and (2) $\bar{C}$ is nonsingular at $\bar{p}$. (See Fig. 4.16.)

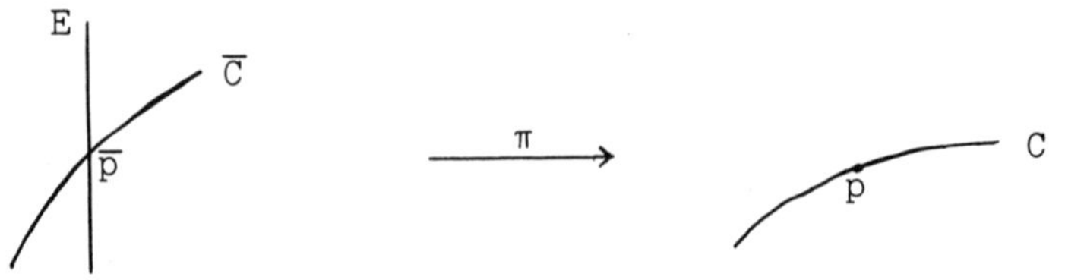

FIGURE 4.16

Now we prove

Theorem 4.2.5. Let S be a two-dimensional compact complex manifold and C be an analytic curve in S. Take a point $p \in \text{Sing } C$. Consider the blowing up

$$\pi = \pi_0 : S_1 = Q_p(S) \to S$$

at p and the strict transform $C_1 = \bar{C}$ of C. Next, take a point $p_1 \in \text{Sing } C_1$. Consider the blowing up

$$\pi_1 : S_2 = Q_{p_1}(S_1) \to S_1$$

at p_1 and the strict transform $C_2 = \bar{C}_1$ of C_1, and so on. Then, after a finite number of this process, C_k becomes a nonsingular analytic curve in S_k. Moreover, the composition of the projections

$$\begin{array}{ccccccccc} S_k & \to & S_{k-1} & \to & \cdots & \to & S_1 & \to & S \\ \cup & & \cup & & & & \cup & & \cup \\ \phi : C_k & \to & C_{k-1} & \to & \cdots & \to & C_1 & \to & C \end{array}$$

gives a nonsingular model $\phi : C_k \to C$ of C, provided C is irreducible.

Proof. Let U be a small neighborhood of p in S and $C \cap U = D_1 \cup \cdots \cup D_s$ be the irreducible decomposition of C at p. Then, clearly,

$$C_1 \cap \pi_0^{-1}(U) = \bar{D}_1 \cup \cdots \cup \bar{D}_s$$

where $\bar{D}_\nu$ is the strict transform of D_ν under the blowing up

$$\pi_0 : \pi_0^{-1}(U) = Q_p(U) \to U$$

We call the correspondence

$$D_\nu \to \bar{D}_\nu$$

the <u>strict transformation</u>.

Now, it is enough to prove that

1. every D_ν becomes nonsingular after a finite number of strict transformations and
2. nonsingular irreducible branches D_ν can be mutually disjoint after a finite number of strict transformations.

To prove 1, fix a D_ν and let

$$t \to (x, y) = (t^m, h(t))$$

be a local uniformizing parameter of D_ν at p, where h(t) is a holomorphic function such that

$$\mathrm{ord}_{t=0}\, h(t) \geq m + 1 \text{ (compare Lemma 2.1.1)}$$

(By changing y by $y - ax^\ell$ for some integer ℓ) we may assume that $k = \mathrm{ord}_{t=0}\, h(t)$ is not divisible by m. Put

$$k = qm + r, \qquad 0 < r < m$$

Let

$$h(t) = a_k t^k = a_{k+1} t^{k+1} + \cdots \qquad (a_k \neq 0)$$

be the power series expansion of h(t). Note that $\bar{D}_\nu$ passes through $\bar{p} = ((0,0), (1:0)) \in Q_p(S)$ and $\bar{D}_\nu \cap E = \{\bar{p}\}$. Around $\bar{p}$, we may take (x, ξ) as a local coordinate system in $Q_p(S)$, where $\xi = \xi_1/\xi_0$. Then, it is clear that $\bar{D}_\nu$ is given by the image of the map

$$t \to (x, \xi) = (t^m, a_k t^{k-m} + a_{k+1} t^{k+1-m} + \cdots)$$

Next, we blow up $Q_p(S)$ at $\bar{p}$, and so on. By q-times of blowings up, we can decrease the multiplicity min (m, k) of the branch. By the induction on min (m, k), we finally get the case min (m, k) = 1, which implies that the branch is nonsingular. (See Fig. 4.17.)

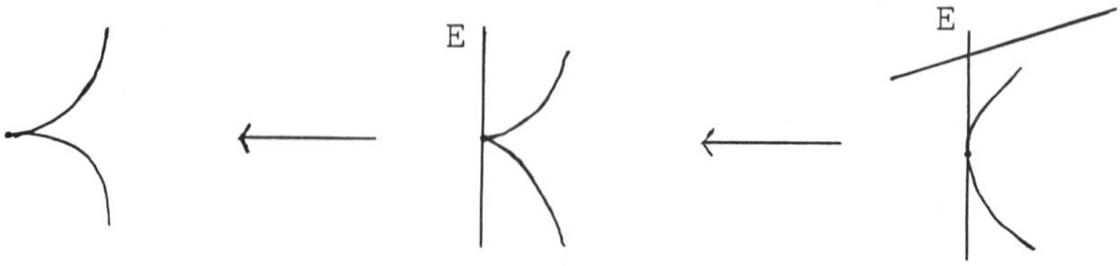

FIGURE 4.17

To prove 2, fix D_1 and D_2 such that they are nonsingular at p.

Case 1. Suppose that D_1 and D_2 have distinct tangent lines at p. Then their strict transforms $\bar{D}_1$ and $\bar{D}_2$ are clearly disjoint. (See Fig. 4.18.)

Case 2. Suppose that D_1 and D_2 have the same tangent line at p. We may take a local coordinate system (x, y) around p in S such that D_1 and D_2 are given by the equations

$$D_1: y = h_1(x) = a_2x^2 + \cdots + a_kx^k + a_{k+1}x^{k+1} + \cdots,$$

$$D_2: y = h_2(x) = a_2x^2 + \cdots + a_kx^k + b_{k+1}x^{k+1} + \cdots$$

where $a_{k+1} \neq b_{k+1}$. (x-axis is the common tangent lines to D_1 and D_2.) Then $\bar{D}_1$ and $\bar{D}_2$ pass through $\bar{p} = ((0,0), (1: 0)) \in Q_p(S)$, and

$$\bar{D}_1 \cap E = \bar{D}_2 \cap E = \{\bar{p}\}$$

Around $\bar{p}$, we may take (x, ξ) as a local coordinate system in $Q_p(S)$, where $\xi = \xi_1/\xi_0$. Then $\bar{D}_1$ and $\bar{D}_2$ are clearly given by the equations

$$\bar{D}_1: \xi = a_2x + \cdots + a_kx^{k-1} + a_{k+1}x^k + \cdots,$$

$$\bar{D}_2: \xi = a_2x + \cdots + a_kx^{k-1} + b_{k+1}x^k + \cdots$$

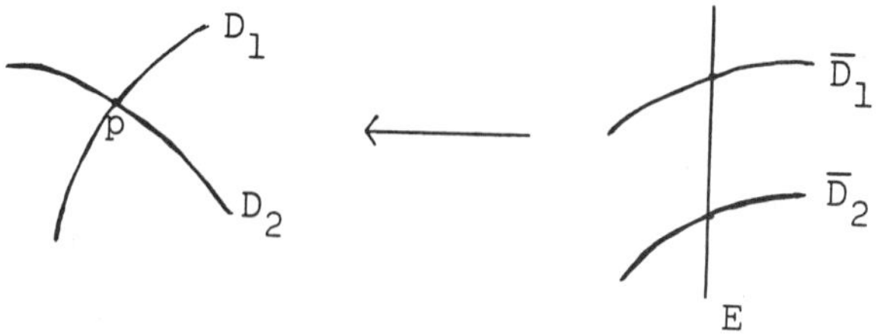

FIGURE 4.18

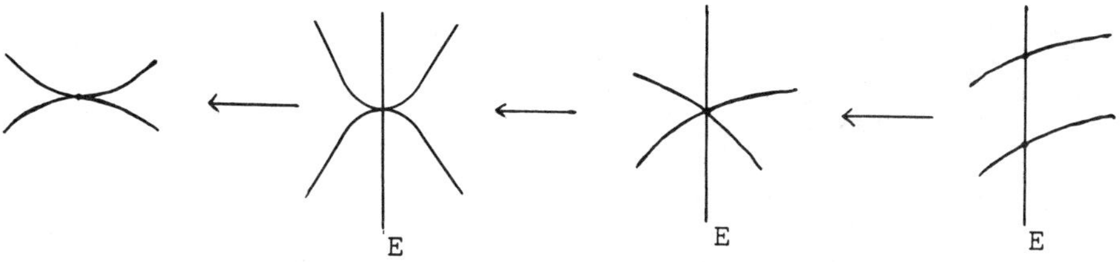

FIGURE 4.19

Next, we change (x, ξ) by $(x, \xi - ax)$, and repeat this process.

By the induction on k, this case reduces to Case 1. (See Fig. 4.19.)

Q.E.D.

Example 4.2.6

1. Let p be a cusp of C. p is a simple cusp if and only if the strict transform $\bar{C}$ is nonsingular at $\bar{p} = \bar{C} \cap E$.
2. Let p be a cusp of C of multiplicity 2. Then p is a double cusp (respectively a ramphoid cusp) (see Sec. 2.1), if and only if C becomes nonsingular at the point after two times (respectively three times) of strict transformations.
3. If p is an ordinary m-ple point of C, then $\bar{D}_1, \ldots, \bar{D}_m$ are disjoint.
4. Let p be a double point of C with two nonsingular irreducible branches D_1 and D_2. Then p is a tacnode (respectively osnode) (see Sec. 3.1), if and only if D_1 and D_2 become disjoint after two times (respectively three times) of strict transformations.

Remark 4.2.7. For the proof of the theorem without using local uniformizing parameters, see Mumford [70, p. 161].

Theorem 4.2.8. Let C be an irreducible projective curve. Then its nonsingular model M is also a projective curve.

Proof. Let C be an irreducible projective curve in $\mathbb{P}^r$ $(r \geq 2)$, and $f: M \to C$ be a nonsingular model of C. By taking generic projections, we get an irreducible plane curve $C_0 \subset \mathbb{P}^2$ and a birational regular map

$$h: C \to C_0$$

(see Theorem 1.6.22'). Let

$$\phi: C_k \to C_{k-1} \to \cdots \to C_1 \to C_0$$

be the nonsingular model of C_0 given in Theorem 4.2.5. By Proposition

4.2.3, C_k is a projective curve. Since C_k and M are bimeromorphic, they are biholomorphic by Corollary 4.1.7. Q.E.D.

Now we can prove

Theorem 4.2.9 (The trinity). The following three categories are equivalent:

i. compact Riemann surfaces (··· analysis),
ii. algebraic function fields of one variable (··· algebra),
iii. nonsingular irreducible projective curves (··· geometry).

Proof (Sketch). We construct the functors

$$\alpha\colon \text{i} \to \text{ii},\ \beta\colon \text{ii} \to \text{iii},\ \gamma\colon \text{iii} \to \text{i}$$

α: i → ii. Let M be a compact Riemann surface. By Riemann's Existence Theorem (Theorem 4.1.17), there is a nonconstant meromorphic function f on M. Put

$$n = \deg f \quad \text{and}$$

$$B = \text{the branch locus of } f\colon M \to \mathbb{P}^1$$

For any point $x \in \mathbb{P}^1 - B$, $f^{-1}(x)$ consists of distinct n points

$$p_1(x), \ldots, p_n(x)$$

For any $h \in \mathbb{C}(M)$,

$$a_1(x) = h(p_1(x)) + \cdots + h(p_n(x)),$$

$$a_2(x) = h(p_1(x)h(p_2(x)) + \cdots + h(p_{n-1}(x))h(p_n(x)),$$

$$\cdots$$

$$a_n(x) = h(p_1(x)) \cdots h(p_n(x))$$

are holomorphic maps of $\mathbb{P}^1 - B$ into $\mathbb{P}^1$. It is then easy to see that they can be extended to holomorphic maps of $\mathbb{P}^1$ into $\mathbb{P}^1$, that is, rational functions on $\mathbb{P}^1$. Consider now the equation

$$F(x,y) = y^n - a_1(x)y^{n-1} + \cdots + (-1)^n a_n(x) = 0$$

on the field $\mathbb{C}(x)$. Then

$$F(f, h) = 0 \quad \text{in} \quad \mathbb{C}(M)$$

Hence, $\mathbb{C}(M)$ is an algebraic function field of one variable such that the degree $[\mathbb{C}(M): \mathbb{C}(f)]$ of the extension satisfies

$$[\mathbb{C}(M): \mathbb{C}(f)] \leq n$$

We show in fact that

$$[\mathbb{C}(M): \mathbb{C}(f)] = n$$

Let $h \in \mathbb{C}(M)$ be such that $[\mathbb{C}(f, h): \mathbb{C}(f)]$ takes the maximal value. Then, for any $h_1 \in \mathbb{C}(M)$, there is $h_2 \in \mathbb{C}(M)$ such that

$$\mathbb{C}(f, h, h_1) = \mathbb{C}(f, h_2)$$

Hence,

$$[\mathbb{C}(f, h, h_1): \mathbb{C}(f, h)][\mathbb{C}(f, h): \mathbb{C}(f)] = [\mathbb{C}(f, h_2): \mathbb{C}(f)] \leq [\mathbb{C}(f, h): \mathbb{C}(f)]$$

Hence, $\mathbb{C}(f, h, h_1) = \mathbb{C}(f, h)$, so $h_1 \in \mathbb{C}(f, h)$. This means that

$$\mathbb{C}(M) = \mathbb{C}(f, h)$$

Let

$$G(x, y) = y^m + b_1(x)y^{m-1} + \cdots + b_m(x) \qquad (m \leq n)$$

be the irreducible polynomial over $\mathbb{C}(x)$ such that $G(f, h) = 0$ in $\mathbb{C}(M)$. Using the above notation, we claim that

$$h(p_j(x)) \neq h(p_k(x)) \quad \text{for } j \neq k, \quad \text{and} \quad x \in \mathbb{P}^1 - B \tag{*}$$

In fact, if $h(p_1(x)) = h(p_2(x))$, say, then

$$u(p_1(x)) = u(p_2(x)) \quad \text{for all } u \in \mathbb{C}(M) = \mathbb{C}(f, h)$$

This contradicts Theorem 4.1.17. Hence, (*) holds. Since every $h(p_j(x))$ satisfies $G(x, h(p_j(x)) = 0$, we have $m \geq n$. Thus $m = n$, so $[\mathbb{C}(M): \mathbb{C}(f)] = n$.

Now, consider the (contravariant) functor

$$\alpha: M \to \mathbb{C}(M)$$

$$\alpha: h \to h^*$$

where $h: M_1 \to M_2$ is a surjective holomorphic map and $h^*: \mathbb{C}(M_2) \to \mathbb{C}(M_1)$ is the injective $\mathbb{C}$-homomorphism induced by h.

β: ii $\to$ iii. Let $k = \mathbb{C}(x,y)$ be an algebraic function field of one variable such that y satisfies the irreducible equation

$$y^n + a_1(x)y^{n-1} + \cdots + a_n(x) = 0$$

over $\mathbb{C}(x)$. Write $a_j(x) = b_j(x)/b_0(x)$, where $b_j(x)$, $0 \leq j \leq n$, are polynomials such that $(b_0, b_1, \ldots, b_n) = 1$. Consider the affine curve

$$C': b_0(x)y^n + b_1(x)y^{n-1} + \cdots + b_n(x) = 0$$

Let C be the closure in $\mathbb{P}^2$ of C'. Then C is clearly irreducible. Let M be a nonsingular model of C. It is uniquely determined up to holomorphic isomorphisms. By Theorem 4.2.8, M is projective.

Let

$$i: k_1 = \mathbb{C}(x,y) \to k_2 = \mathbb{C}(z,w)$$

be an injective $\mathbb{C}$-homomorphism. Suppose that k_1 (respectively k_2) defines as above an irreducible affine curve

$$C_1': b_0(x)y^n + b_1(x)y^{n-1} + \cdots + b_n(x) = 0$$

$$(\text{respectively } C_2': c_0(z)w^m + c_1(z)w^{m-1} + \cdots + c_m(z) = 0)$$

Let C_1 (respectively C_2) be its closure in $\mathbb{P}^2$ and M_1 (respectively M_2) be a nonsingular model of C_1 (respectively C_2). Then

$$x = x(z,w) \quad \text{and} \quad y = y(z,w)$$

are rational functions of (z,w). Thus we get a dominating rational map

$$(z,w) \in C_2 \to (x,y) = (x(z,w),\ y(z,w)) \in C_1$$

Hence we have a surjective holomorphic map

$$\hat{i}: M_2 \to M_1$$

Now, consider the (contravariant) functor

$\beta: k \to M,$

$\beta: i \to \hat{i}$

$\underline{\gamma: iii \to i}$. The (covariant) functor γ is the trivial one

$\gamma: M \to M,$

$\gamma: h \to h$

where $h: M_1 \to M_2$ is a surjective regular map.

Now, we have

$$\gamma \cdot \beta \cdot \alpha = 1, \quad \alpha \cdot \gamma \cdot \beta = 1, \quad \beta \cdot \alpha \cdot \gamma = 1$$

In fact, using the above notations, consider the composition of functors

$$k = \mathbb{C}(x, y) \to M = \gamma\beta(k) \to \mathbb{C}(M) = \alpha(M)$$

Note that

$$x: (x, y) \in C \to x \in \mathbb{P}^1, \quad \text{and}$$

$$y: (x, y) \in C \to y \in \mathbb{P}^1$$

define meromorphic functions on M. Hence,

$$k \subset \mathbb{C}(M)$$

By the above argument, we get

$$[\mathbb{C}(M): \mathbb{C}(x)] = \deg x = [k: \mathbb{C}(x)]$$

Hence,

$$k = \mathbb{C}(M), \quad \text{so} \quad \alpha \cdot \gamma \cdot \beta = 1$$

The other equalities are easy to see. Q.E.D.

Corollary 4.2.10. Any compact Riemann surface is projective.

Remark 4.2.11. (1) The trinity is, of course, a peculiar phenomenon for the one-dimensional case; $\mathbb{P}^2$ and $\mathbb{P}^1 \times \mathbb{P}^1$ have the isomorphic rational function fields, while they are not biholomorphic (see Sec. 3.3). There is a complex 2-torus which is not projective (see Sec. 3.2).

(2) Chow-Kodaira [21] says that if M is a compact complex manifold of dimension 2 and alg · dim M = 2 (compare Sec. 3.2), then M is projective. But a similar assertion does not hold for dim M $\geq$ 3. See Moishezon [66] for detail.

Let

$$f(x,y) = b_0(x)y^n + b_1(x)y^{n-1} + \cdots + b_n(x)$$

be an irreducible polynomial in x and y over $\mathbb{C}$. Consider the irreducible affine curve

$$C = \{(x,y) \in \mathbb{C}^2 \mid f(x,y) = 0\}$$

For a point $(x_0, y_0) \in C$ such that $b_0(x_0) \neq 0$ and $(\partial f/\partial y)(x_0, y_0) \neq 0$, there is a unique holomorphic function $y = y(x)$ on a neighborhood of x_0 such that $y(x_0) = y_0$ and $f(x, y(x)) = 0$ (by the Implicit Mapping Theorem). By the Principle of Analytic Continuation, we get a Global Analytic Function

$$y = y(x)$$

on $\mathbb{P}^1 - B$ (B: a finite set) such that $f(x, y(x)) = 0$, in the sense that B is the natural boundary of y(x). The function y(x) does not depend on the choice of the initial point (x_0, y_0) and is a n-valued holomorphic function, called an algebraic function.

The finite set B is contained in

$$\{\infty\} \cup \{b_0(x) = 0\} \cup \{D(x) = 0\}$$

where D(x) is the discriminant of the polynomial f(x, y) of y over the field $\mathbb{C}(x)$. Every point of B is either a branch point or a pole of y(x). No essential singular point appears. Conversely, this characterizes algebraic functions (see Ahlfors [3]).

Consider the closure $\bar{C}$ in $\mathbb{P}^1 \times \mathbb{P}^1$ of C. This means that, putting $x = X_1/X_0$ and $y = Y_1/Y_0$,

$$\bar{C} = \{((X_0 : X_1), (Y_0 : Y_1)) \in \mathbb{P}^1 \times \mathbb{P}^1 \mid F(X_0, X_1; Y_0, Y_1) = 0\}$$

where $F(X_0, X_1; Y_0, Y_1) = X_0^m Y_0^n f(X_1/X_0, Y_1/Y_0)$ ($m = \deg_x f$, the degree of f with respect to x). (See Fig. 4.20.) (The lines $\overline{(\infty, 0)(\infty, \infty)}$ and $\overline{(0, \infty)(\infty, \infty)}$ in $\mathbb{P}^1 \times \mathbb{P}^1$ are called the lines of infinity.)

A nonsingular model M of $\bar{C}$ is called the Riemann surface of the algebraic function y(x). Note that the projections

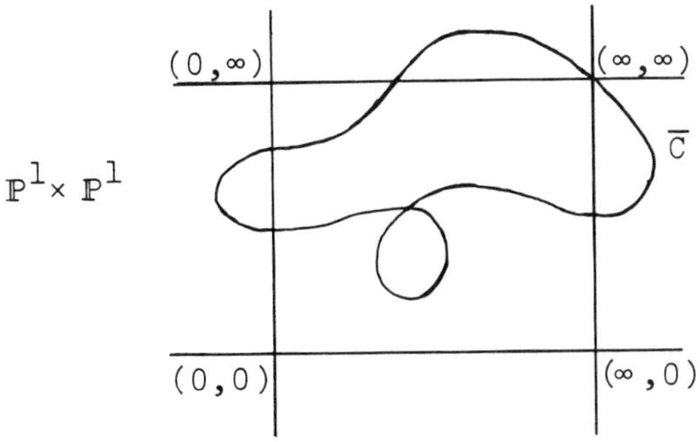

FIGURE 4.20

$$x: (x,y) \in \bar{C} \to x \in \mathbb{P}^1$$

$$y: (x,y) \in \bar{C} \to y \in \mathbb{P}^1$$

define meromorphic functions x and y, respectively, on M such that $\mathbb{C}(M) = \mathbb{C}(x,y)$. The relation of the functions x and y is given by $f(x,y) = 0$. so

$$[\mathbb{C}(M): \mathbb{C}(x)] = n = \deg_y f = \deg x$$

If we take a nonsingular model M' of the closure in $\mathbb{P}^2$ of C, then we have $\mathbb{C}(M') = \mathbb{C}(x,y)$, so M and M' are biholomorphic. Thus there is no essential distinction between the closures of C in $\mathbb{P}^1 \times \mathbb{P}^1$ and in $\mathbb{P}^2$.

From the field theoretic point of view, the simplest algebraic function field of one variable is the field $\mathbb{C}(x)$ of the (usual) rational functions. The compact Riemann surface corresponding to $\mathbb{C}(x)$ is $\mathbb{P}^1$.

The next simplest one is a quadratic extension $\mathbb{C}(x,y)$ of $\mathbb{C}(x)$, where x and y are related by

$$y^2 + a_1(x)y + a_2(x) = 0, \qquad (a_j(x) \in \mathbb{C}(x))$$

By replacing y by $y + \frac{1}{2}a_1(x)$, and so on, we may assume that the relation is given by

$$y^2 - (x - \alpha_1) \cdots (x - \alpha_n) = 0 \qquad (*)$$

where α_j, $1 \leq j \leq n$, are mutually distinct complex numbers. Eq. (*) defines an algebraic function $y(x) = \pm\sqrt{(x - \alpha_1) \cdots (x - \alpha_n)}$. Its Riemann surface M corresponds to $\mathbb{C}(x, y)$. By applying the Riemann-Hurwitz Formula to the function x, we get (compare Exercise 1)

Lemma 4.2.12. The genus of M is $\frac{1}{2}(n - 2)$ for even n and is $\frac{1}{2}(n - 1)$ for odd n.

(In the computation of g, it is more convenient to consider the closure of (*) in $\mathbb{P}^1 \times \mathbb{P}^1$ than that in $\mathbb{P}^2$.)

Definition 4.2.13. A compact Riemann surface M of genus g (≥ 2) is said to be hyperelliptic if there is a meromorphic function f of degree 2 on M.

In this case, $\mathbb{C}(M)$ is a quadratic extension of $\mathbb{C}(f)$, so M is given by the equation (*), (f = x), for some $\alpha_1, \ldots, \alpha_n$, where either $n = 2g + 2$ or $n = 2g + 1$. We may assume that (∞, ∞) is a branch point of the function f = x. Then $n = 2g + 1$. Thus,

Proposition 4.2.14. A hyperelliptic M of genus g (≥ 2) can be given by the equation

$$y^2 - (x - \alpha_1) \cdots (x - \alpha_{2g+1}) = 0$$

where $\alpha_1, \ldots, \alpha_{2g+1}$ are mutually distinct complex numbers, and vice versa.

Exercises

1. Prove Lemmas 4.2.1, 4.2.4, and 4.2.12.
2. The Cremona transformation

$$(X_0 : X_1 : X_2) \in \mathbb{P}^2 \to (X_1 X_2 : X_0 X_2 : X_0 X_1) \in \mathbb{P}^2$$

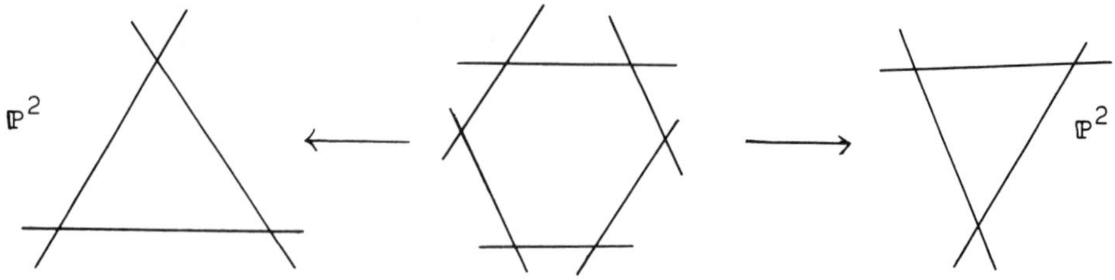

FIGURE 4.21

(see Exercise 4 of Sec. 3.4) can be obtained by (1) blowing up three points in general position and then (2) blowing down three (exceptional) curves to points. (See Fig. 4.21.)

3. Put, for n = 0, 1, 2, ...,

$$F_n = \{((X_0 : X_1 : X_2), (\xi_1 : \xi_2)) \in \mathbb{P}^2 \times \mathbb{P}^1 \mid \xi_1^n X_2 = \xi_2^n X_1\}$$

Show that F_n are $\mathbb{P}^1$-bundles over $\mathbb{P}^1$. They are called the rational ruled surfaces (or Hirzebruch surfaces). Note that

$$F_0 = \mathbb{P}^1 \times \mathbb{P}^1 \quad \text{and} \quad F_1 = Q_p(\mathbb{P}^2)$$

4. Let S be a two-dimensional compact complex manifold and C be an analytic curve on S. Let $p \in C$ be a point of multiplicity m (≥ 1) and $\pi : Q_p(S) \to S$ be the blowing up of S at p. Put $E = \pi^{-1}(p)$ and $\bar{C}$ = the strict transform of C. Regard C (respectively $\bar{C}$ and E) a divisor on S (respectively divisors on $Q_p(S)$). Then

$$\pi^*C = \bar{C} + mE$$

5. Under the same notations as in Exercise 4, suppose that p is a singular point of C. If $p_1 \in \bar{C} \cap E$ is a singular point of $C_1 = \bar{C}$, then p_1 is called an infinitely near singular point of C over p. (Note that p_1 is not a point of C but a point of C_1.) In a similar way, the infinitely near singular points of C_1 over p_1 are also called infinitely near singular points of C over p, and so on. For convenience, p is itself regarded as an infinitely near singular point of C over p. (See Fig. 4.22.) If $S = \mathbb{P}^2$ and C is irreducible, then (1)

$$g = \frac{1}{2}(n-1)(n-2) - \sum_p \sum_q \frac{1}{2} m_q(m_q - 1)$$

where g (respectively n) is the genus (respectively the degree) of C,

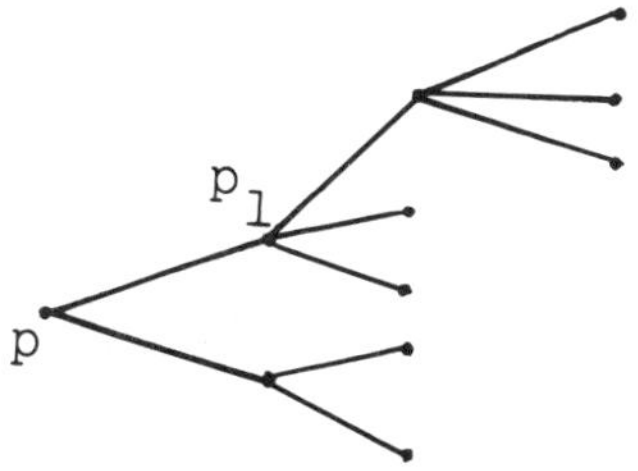

FIGURE 4.22

Σ_p runs over all singular points p of C, Σ_q runs over all infinitely near singular points q of C over p, and m_q is the multiplicity at q of C_k ($q \in C_k$) and (2)

$$\sum_q \frac{1}{2} m_q (m_q - 1) = \delta_p \qquad \text{(see Sec. 2.1)}$$

so δ_p is an integer which is analytic invariant (compare Iitaka [49]).

6. Compute the genus of the curve

$$y^n - x^m + 1 = 0$$

4.3 ELLIPTIC FUNCTIONS

Algebraic geometry has, no doubt, its origin in the theory of elliptic functions, developed by Legendre, Gauss, Abel, Jacobi, Riemann, Weierstrass, and so on. In this section, we do not intend to talk about this huge classical theory, but want to give a very short introduction to it.

Let ω_1 and ω_2 be nonzero complex numbers such that the imaginary part of ω_2/ω_1 is positive, that is, $\omega_2/\omega_1 \in \mathbb{H}$, the upper half plane. As in 3.1, put

$$\Gamma = \mathbb{Z}\omega_1 + \mathbb{Z}\omega_2$$

$$M = \mathbb{C}/\Gamma$$

$$\pi: \mathbb{C} \to M, \text{ the projection}$$

Then M is a complex 1-torus. By abuse of notation, we sometimes identify a point $z \in \mathbb{C}$ with $\pi(z) \in M$.

A meromorphic function on M can be <u>identified with</u> a meromorphic function f on $\mathbb{C}$ with periods in Γ

$$f(z + \omega) = f(z) \quad \text{for } z \in \mathbb{C} \quad \text{and} \quad \omega \in \Gamma$$

Such a function is called an <u>elliptic function with periods in</u> Γ. It is a doubly periodic function on $\mathbb{C}$, while e^z, sinz, and so on, are singly periodic functions on $\mathbb{C}$.

A (nonconstant) elliptic function can be regarded as a holomorphic ramified covering map $f: M \to \mathbb{P}^1$. Its degree is called the <u>degree</u> (or <u>order</u>) <u>of the elliptic function</u> f.

We have easily

Proposition 4.3.1

1. An entire elliptic function is a constant.
2. There is no elliptic function of degree 1.
3. All elliptic functions with periods in Γ form a field, which can be identified with $\mathbb{C}(M)$.
4. If f is an elliptic function with periods in Γ, then $f' = df/dz$ is also an elliptic function with periods in Γ.

Hence, the degree of a (nonconstant) elliptic function is at least 2. In fact, Weierstrass constructed an elliptic function of degree 2 with periods in a given Γ. We explain this as follows. Put

$$\Gamma^* = \Gamma - \{0\}$$

Lemma 4.3.2. For $k \geq 3$, the series

$$\sum_{\omega \in \Gamma^*} \frac{1}{|\omega|^k}$$

converges.

Proof. Consider the increasing sequence of parallelograms P_m, $m = 1, 2, \ldots$, as in Fig. 4.23. Put

$$\partial P_m = \text{the boundary of } P_m,$$

$$r = \inf\{|z| \mid z \in \partial P_1\}, \quad \text{so}$$

$$mr = \inf\{|z| \mid z \in \partial P_m\}$$

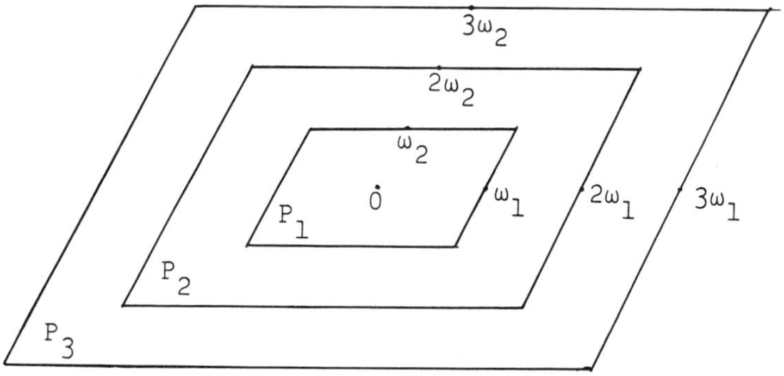

FIGURE 4.23

Put

$$S_m = \sum_{\omega \in \Gamma^* \cap \partial P_m} \frac{1}{|\omega|^k}$$

Then

$$S_m \le \frac{8m}{(mr)^k} = \frac{8}{r^k} \cdot \frac{1}{m^{k-1}}$$

Hence,

$$\sum_{\omega \in \Gamma^*} \frac{1}{|\omega|^k} \le \frac{8}{r^k} \sum_{m=1}^{\infty} \frac{1}{m^{k-1}} < +\infty$$

for $k - 1 \ge 2$. Q.E.D.

Now, consider the series

$$\sum_{\omega \in \Gamma} \frac{1}{(z - \omega)^3}$$

Take a real number R (> 0) and let $z \in \mathbb{C}$ be such that $|z| \le R$. Then

$$|z - \omega| \ge |\omega| - |z| \ge |\omega| - \frac{1}{2}|\omega| = \frac{1}{2}|\omega|$$

for any $\omega \in \Gamma$ such that $|\omega| \ge 2R$. Hence,

$$\sum_{\omega \in \Gamma, |\omega| \ge 2R} \frac{1}{|z - \omega|^3} \le 8 \sum_{\omega \in \Gamma, |\omega| \ge 2R} \frac{1}{|\omega|^3} < +\infty$$

by Lemma 4.3.2. Hence, the series

$$\phi(z) = \sum_{\omega \in \Gamma, |\omega| \ge 2R} \frac{1}{(z - \omega)^3}$$

represents a holomorphic function on $|z| < R$. Thus

$$f(z) = \sum_{\omega \in \Gamma} \frac{1}{(z - \omega)^3} = \phi(z) + \sum_{\omega \in \Gamma, |\omega| < 2R} \frac{1}{(z - \omega)^3}$$

is a meromorphic function on $|z| < R$. Since R is arbitrary, f(z) is a meromorphic function on $\mathbb{C}$ with the poles of order 3 at every $\omega \in \Gamma$.

f(z) is an elliptic function of degree 3 with periods in Γ. In fact,

$$f(z + \omega) = \sum_{\omega' \in \Gamma} \frac{1}{(z + \omega - \omega')^3} = f(z) \quad \text{for any } \omega \in \Gamma$$

In order to get an elliptic function of degree 2, we consider the integration of -2f(z) as follows

$$\wp(z) = \int_{\infty}^{z} \frac{-2}{z^3}\, dz + \sum_{\omega \in \Gamma^*} \int_{0}^{z} \frac{-2}{(z - \omega)^3}\, dz$$

$$= \frac{1}{z^2} + \sum_{\omega \in \Gamma^*} \left(\frac{1}{(z - \omega)^2} - \frac{1}{\omega^2}\right)$$

(Note that the integrations do not depend on the choice of paths, for $-2/(z - \omega)^3$ has no residue at $z = \omega$.) Then this series converges to a meromorphic function $\wp(z)$ with the poles of order 2 at every $\omega \in \Gamma$. It is easy to see that $\wp(z)$ is an elliptic function of degree 2 with periods in Γ. $\wp(z)$ is called the Weierstrass $\wp$-function. Its derivative $\wp'(z) = d\wp/dz = -2f(z)$ is also an elliptic function with periods in Γ.

We list some properties of $\wp$ and $\wp'$.

1. $\wp(z)$ (respectively $\wp'(z)$) is an elliptic function with periods in Γ and the poles of order 2 (respectively 3) at every $\omega \in \Gamma$.
2. $\wp(z)$ (respectively $\wp'(z)$) is an even (respectively odd) function

$$\wp(-z) = \wp(z), \qquad \wp'(-z) = -\wp'(z)$$

3. $\wp'(z) = 0$ if and only if $2z \in \Gamma$ and $z \notin \Gamma$.

Proposition 4.3.3. $\wp'(z)^2 = 4\wp(z)^3 - g_2\wp(z) - g_3$, where $g_2 = 60 \sum_{\omega \in \Gamma^*} \frac{1}{\omega^4}$ and $g_3 = 140 \sum_{\omega \in \Gamma^*} \frac{1}{\omega^6}$.

(In general, $\sum_{\omega \in \Gamma^*} \frac{1}{\omega^{2k}}$ $(k \geq 2)$ is called an Eisenstein series, which converges absolutely by Lemma 4.3.2.)

Proof. We expand $\wp(z)$ and $\wp'(z)$ into the Laurent series around 0:

$$\wp(z) = \frac{1}{z^2} + c_2 z^2 + c_3 z^4 + \cdots + c_k z^{2k-2} + \cdots,$$

$$\wp'(z) = -\frac{2}{z^3} + 2c_2 z + 4c_3 z^3 + \cdots + (2k-2)c_k z^{2k-3} + \cdots$$

where $c_k = (2k-1)\Sigma_{\omega \in \Gamma^*} \frac{1}{\omega^{2k}}$ $(k \geq 2)$. Hence,

$$\wp'(z)^2 - 4\,\wp(z)^3 + 20c_2\,\wp(z) = -28c_3 + \cdots$$

The right-hand side represents a holomorphic function around 0, while the left-hand side represents an elliptic function with the poles at, at most, the points in Γ. Hence, by Proposition 4.3.1, this function must be a constant $-28c_3$. Q.E.D.

From this proposition, we can easily get (compare Exercise 1)

Proposition 4.3.4. The map

$$\Phi: z \in M = \mathbb{C}/\Gamma \to (1: \wp(z): \wp'(z)) \in \mathbb{P}^2$$

is a holomorphic imbedding of the complex one-torus M into $\mathbb{P}^2$. Its image is the nonsingular plane cubic curve

$$C: y^2 = 4x^3 - g_2 x - g_3$$

(See Fig. 4.24.)

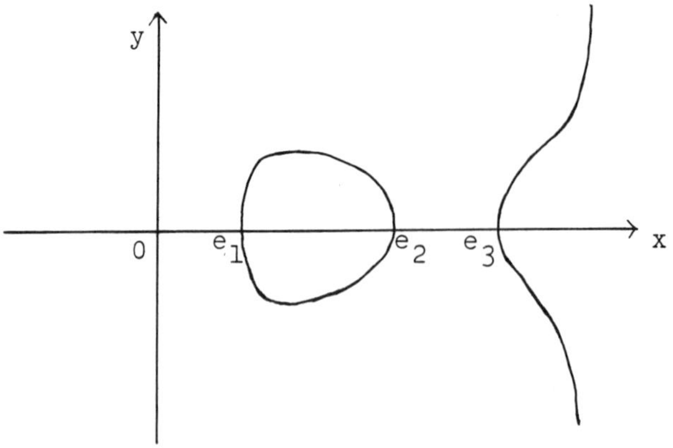

FIGURE 4.24

(Here we identify the affine curve with its closure in $\mathbb{P}^2$. $\Phi(0) = (0:0:1)$.)

Note that, if M and C are identified through Φ, then the functions $\wp$ and $\wp'$ are nothing but the projections

$$\wp: (x, y) \in C \to x \in \mathbb{P}^1$$

$$\wp': (x, y) \in C \to y \in \mathbb{P}^1$$

Write

$$4x^3 - g_2 x - g_3 = 4(x - e_1)(x - e_2)(x - e_3)$$

Then e_1, e_2, and e_3 are mutually distinct, for C is not a rational curve. We may put

$$e_1 = \wp(\omega_1/2), \quad e_2 = \wp(\omega_2/2), \quad \text{and} \quad e_3 = \wp((\omega_1 + \omega_2)/2)$$

By the Riemann-Hurwitz Formula, $\wp(z)$ has just four points of ramification (with the index 2) on $M = \mathbb{C}/\Gamma$. They are 0, $\omega_1/2$, $\omega_2/2$, and $(\omega_1 + \omega_2)/2$. On C, they are

$$p_\infty = (0:0:1), \quad p_1 = (e_1, 0), \quad p_2 = (e_2, 0), \quad \text{and} \quad p_3 = (e_3, 0)$$

respectively. This is geometrically clear. (p_∞ is a flex of C.)

Using the projection $(x, y) \to x$, the topological structure of the nonsingular plane cubic curve C can be seen as follows: (1) take two copies of $\mathbb{P}^1$, (2) cut each $\mathbb{P}^1$ from ∞ to e_1 and from e_2 to e_3, and (3) glue two copies along the cuts. (See Fig. 4.25.)

Proposition 4.3.5. $\wp$ and $\wp'$ generate $\mathbb{C}(M)$, that is, $\mathbb{C}(M) = \mathbb{C}(\wp, \wp')$.

Proof. $[\mathbb{C}(M): \mathbb{C}(\wp, \wp')]$ divides both $[\mathbb{C}(M): \mathbb{C}(\wp)] = 2$ and $[\mathbb{C}(M): \mathbb{C}(\wp')] = 3$ (see Sec. 4.2). Hence, $[\mathbb{C}(M): \mathbb{C}(\wp, \wp')] = 1$, so $\mathbb{C}(M) = \mathbb{C}(\wp, \wp')$. Q.E.D.

In particular, $\wp''$, $\wp'''$, ... can be expressed as rational functions of $\wp$ and $\wp'$. It can be, in fact, shown that

$$\wp'' = 6\wp^2 - \frac{1}{2} g_2, \quad \wp''' = 12\wp\wp', \ \ldots$$

The complex 1-torus M is a complex Lie group. Its addition reflects to $\wp(z)$ as follows

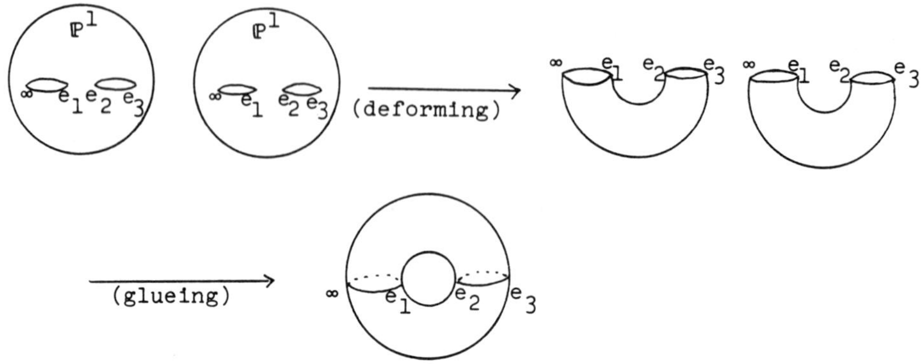

FIGURE 4.25

Theorem 4.3.6 (Addition Formula for $\wp$-function)

$$\wp(z+w) = \frac{1}{4}\left(\frac{\wp'(z) - \wp'(w)}{\wp(z) - \wp(w)}\right)^2 - \wp(z) - \wp(w)$$

Proof. Fix $w \in \mathbb{C}$ such that $w \notin \Gamma$. Put

$$\phi(z) = \frac{1}{2}\left(\frac{\wp'(z) - \wp'(w)}{\wp(z) - \wp(w)}\right)$$

Around $z = 0$, $\phi(z)$ is expanded into the Laurent series

$$\phi(z) = -\frac{1}{z} - \wp(w)z + \cdots$$

Hence, $z = 0$ is a pole of $\phi(z)$ of order 1. Other possible poles are the solutions of

$$\wp(z) - \wp(w) = 0$$

It has only two solutions $z = w$ and $z = -w \pmod{\Gamma}$. Suppose that

$$2w \neq 0 \pmod{\Gamma}$$

Then $z = w$ is also a solution of

$$'(z) - \wp'(w) = 0$$

Hence, $z = w$ is not a pole of $\phi(z)$. On the other hand, around $z = -w$, $\phi(z)$ can be expanded into the Laurent series

$$\phi(z) = \frac{1}{z + w} + b_1 (z + w) + b_2 (z + w)^2 + \cdots$$

where $b_1, b_2, \ldots$ are constants. Hence, $\phi(z)$ has the poles of order 1 at $z = 0$ and $z = -w \pmod{\Gamma}$. The case when

$$2w = 0 \pmod{\Gamma}$$

can be treated in a similar way and the same assertion holds.

Now, consider the elliptic function

$$h(z) = \wp(z + w) + \wp(z) - \phi(z)^2$$

with periods in Γ. Then $h(z)$ has the poles at, at most, 0 and $-w \pmod{\Gamma}$. But, around $z = 0$,

$$h(z) = -\wp(w) + \cdots$$

is holomorphic. In a similar way, $h(z)$ is holomorphic around $z = -w$. Hence, $h(z)$ is the constant $-\wp(w)$.

We have assumed that $w \notin \Gamma$. But the formula is symmetric for z and w. Hence, the formula holds also for $w \in \Gamma$. Q.E.D.

Besides $\wp$-function, Weierstrass defined other functions $\zeta(z)$ and $\sigma(z)$. They are not elliptic functions. $\zeta(z)$ is defined by

$$\zeta(z) = \frac{1}{z} - \int_0^z \sum_{\omega \in \Gamma^*} \left(\frac{1}{(z - \omega)^2} - \frac{1}{\omega^2} \right) dz$$

$$= \frac{1}{z} + \sum_{\omega \in \Gamma^*} \left(\frac{1}{z - \omega} + \frac{1}{\omega} + \frac{z}{\omega^2} \right)$$

The series converges, so $\zeta(z)$ is a meromorphic function on $\mathbb{C}$ with the poles of order 1 at every $\omega \in \Gamma$. $\zeta(z)$ is not an elliptic function (by Proposition 4.3.1). Note that

$$\frac{d\zeta}{dz} = -\wp(z)$$

From this, the differences

$$\eta_1 = \zeta(z+\omega_1) - \zeta(z), \quad \text{and}$$

$$\eta_2 = \zeta(z+\omega_2) - \zeta(z)$$

are (nonzero) constants. By integrating along a parallelogram,(compare Exercise 1),

Proposition 4.3.7 (Legendre's Formula).

$$\eta_1\omega_2 - \eta_2\omega_1 = 2\pi\sqrt{-1}$$

Next, $\sigma(z)$ is defined by

$$\begin{aligned}\sigma(z) &= \exp\left\{\int_\infty^z \frac{dz}{z} + \int_0^z \sum_{\omega\in\Gamma^*}\left(\frac{1}{z-\omega}+\frac{1}{\omega}+\frac{1}{\omega^2}\right)dz\right\}\\ &= \exp\left\{\log z + \sum_{\omega\in\Gamma^*}\left(\log\left(1-\frac{z}{\omega}\right)+\frac{z}{\omega}+\frac{z^2}{2\omega^2}\right)\right\}\\ &= z\prod_{\omega\in\Gamma^*}\left\{\left(1-\frac{z}{\omega}\right)\exp\left(\frac{z}{\omega}+\frac{z^2}{2\omega^2}\right)\right\}\end{aligned}$$

The infinite product Π converges, so $\sigma(z)$ is an entire function. Note that

$$\frac{d\log\sigma(z)}{dz} = \zeta(z), \quad \text{and}$$

$$\sigma(-z) = -\sigma(z)$$

From these relations, we get easily

$$\sigma(z+\omega_j) = -\exp\left(\eta_j z + \frac{1}{2}\eta_j\omega_j\right)\sigma(z) \quad (j = 1,\ 2) \qquad (*)$$

Using (*), the following lemma can be shown (compare Exercise 1).

Lemma 4.3.8. If the complex numbers $a_1, \dots, a_n, b_1, \dots, b_n$ $(n \geq 2)$ satisfy $a_1 + \cdots + a_n = b_1 + \dots + b_n \pmod{\Gamma}$, then

$$h(z) = \frac{\sigma(z-a_1)\cdots\sigma(z-a_n)}{\sigma(z-b_1)\cdots\sigma(z-b_n)}$$

is an elliptic function with periods in Γ such that $\pi(a_1), \dots, \pi(a_n)$ (respectively $\pi(b_1)$, ..., $\pi(b_n)$) are the set of all zeros (respectively poles) of $h\colon M = \mathbb{C}/\Gamma \to \mathbb{P}^1$, counting multiplicity.

For a point $p \in M = \mathbb{C}/\Gamma$, we denote by (p) the point divisor determined by p. When we write $p_1 + p_2$, it is the sum in the complex Lie group M. Hence, $p_1 + p_2 \in M$. When we write $(p_1) + (p_2)$, it is the sum in Div(M). Hence, $(p_1) + (p_2)$ is a divisor, and so on.

Theorem 4.3.9 (Abel). The divisors $D = (p_1) + \cdots + (p_n)$ and $D' = (q_1) + \cdots + (q_n)$ are linearly equivalent if and only if

$$p_1 + \cdots + p_n = q_1 + \cdots + q_n$$

Proof. Let f be an elliptic function with periods in Γ, that is, a meromorphic function on $M = \mathbb{C}/\Gamma$. Put $n = \deg f$ and

$$D_0(f) = (p_1) + \cdots + (p_n) \quad \text{and} \quad D_\infty(f) = (q_1) + \cdots + (q_n)$$

Let P be the parallelogram as in Fig. 4.26. By operating a translation if necessary, we may assume that there are neither poles nor zeros on the boundary ∂P of P. Let $a_1, \ldots, a_n$ (respectively $b_1, \ldots, b_n$) be the zeros (respectively poles) of f in P counting multiplicity. We may assume that

$$p_j = \pi(a_j) \text{ (respectively } q_j = \pi(b_j)) \quad \text{for } 1 \le j \le n$$

Then, by integrating along ∂P,

$$\sum a_j - \sum b_j = \frac{1}{2\pi\sqrt{-1}} \int_{\partial P} \frac{zf'(z)}{f(z)}\, dz$$

$$= \frac{1}{2\pi\sqrt{-1}} \int_0^{\omega_1} \frac{zf'(z)}{f(z)}\, dz - \frac{}{2\pi\sqrt{-1}} \int_0^{\omega_1} (z + \omega_2) \frac{f'(z)}{f(z)}\, dz$$

$$+ \frac{1}{2\pi\sqrt{-1}} \int_0^{\omega_2} (z + \omega_1) \frac{f'(z)}{f(z)}\, dz - \frac{1}{2\pi\sqrt{-1}} \int_0^{\omega_2} \frac{zf'(z)}{f(z)}\, dz$$

$$= \frac{1}{2\pi\sqrt{-1}} \int_0^{\omega_2} d \log f(z) - \frac{1}{2\pi\sqrt{-1}} \int_0^{\omega_1} d \log f(z)$$

$$= \frac{\omega_1}{2\pi\sqrt{-1}} (\log f(\omega_2) - \log f(0)) - \frac{\omega_2}{2\pi\sqrt{-1}} (\log f(\omega_1) - \log f(0))$$

$$= k_1\omega_1 - k_2\omega_2 \quad \text{for some } k_1, k_2 \in \mathbb{Z}$$

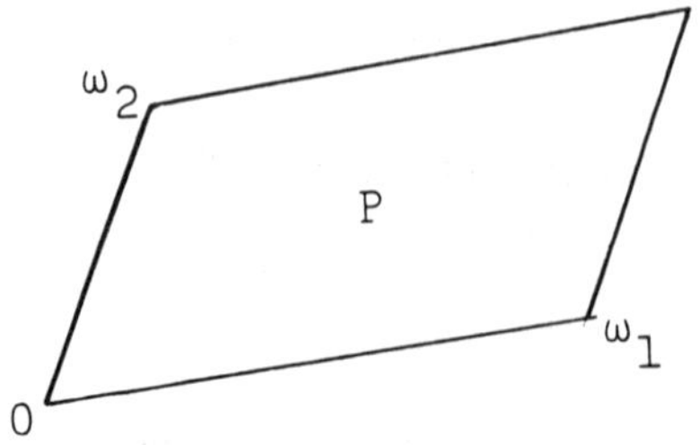

FIGURE 4.26

Hence,

$$p_1 + \cdots + p_n = q_1 + \cdots + q_n$$

The converse follows from Lemma 4.3.8. Q.E.D.

The parallelogram P which appeared in the proof of the theorem or a translation of P is called a fundamental parallelogram. A similar method to the proof of the theorem applied to f(z) itself shows 1 of the following lemma.

Lemma 4.3.10

1. For an elliptic function f with periods in Γ,

$$\Sigma \operatorname{Res}_z f = 0$$

where Σ runs over all poles z of f in a fundamental parallelogram P and $\operatorname{Res}_z f$ is the residue of f at z.

2. For a positive divisor $D = (p_1) + \cdots + (p_n)$ on $M = \mathbb{C}/\Gamma$

$$\dim L(D) \leq n$$

3. For $D = (0) + \cdots + (0) = n(0)$

$$\dim L(n(0)) = n$$

In fact, L(n(0)) has the following basis

$$\{1, \wp, \wp', \wp^2, \wp'\wp, \ldots, \wp'\wp^{(n-4)/2}, \wp^{n/2}\} \quad \text{for even n}$$

$$\{1, \wp, \wp', \wp^2, \wp'\wp, \ldots, \wp^{(n-1)/2}, \wp'\wp^{(n-3)/2}\} \quad \text{for odd n}$$

Proof. 2 follows from 1 and the proof of Theorem 4.1.13. 3 follows from 2.
Q.E.D.

This lemma shows in particular that

$$\Phi_{|2(0)|}: z \in M \to (1: \wp(z)) \in \mathbb{P}^1,$$

$$\Phi_{|3(0)|}: z \in M \to (1: \wp(z): \wp'(z)) \in \mathbb{P}^2,$$

$$\Phi_{|4(0)|}: z \in M \to (1: \wp(z): \wp'(z): \wp(z)^2) \in \mathbb{P}^3,$$

$$\Phi_{|5(0)|}: z \in M \to (1: \wp(z): \wp'(z): \wp(z)^2: \wp'(z)\,\wp(z)) \in \mathbb{P}^4$$

and so on. Note that $\Phi_{|n(0)|}$ $(n \geq 3)$ is a holomorphic imbedding of M into $\mathbb{P}^{n-1}$. In particular, the divisors in $|3(0)|$ are the line cuts of the nonsingular plane curve

$$C = \Phi_{|3(0)|}(M): y^2 = 4x^3 - g_2 x - g_3$$

Hence, by Theorem 4.3.9,

Proposition 4.3.11

1. $z_1 + z_2 + z_3 \in \Gamma$ if and only if the points

$$p_j = (1: \wp(z_j): \wp'(z_j)), \quad j = 1, 2, 3$$

on C are collinear. (See Fig. 4.27.)

2. $z_1 + z_2 + z_3 \in \Gamma$ if and only if

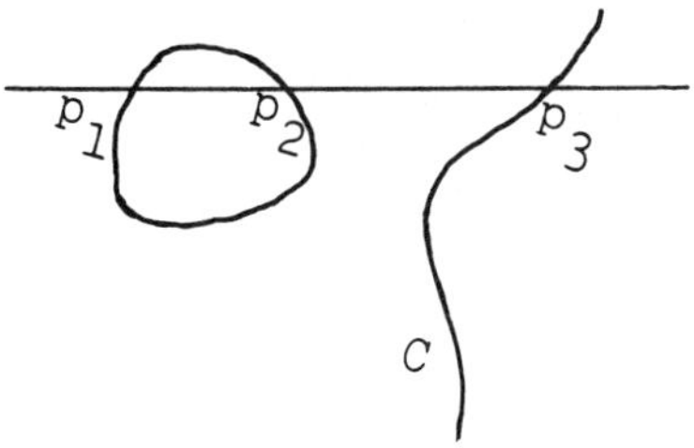

FIGURE 4.27

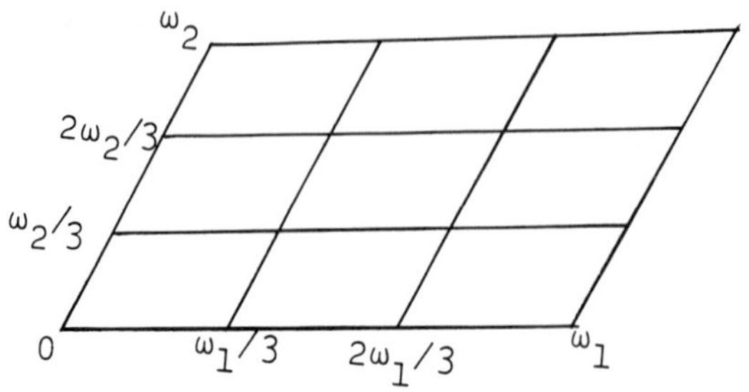

FIGURE 4.28

$$\begin{vmatrix} 1 & \wp(z_1) & \wp'(z_1) \\ 1 & \wp(z_2) & \wp'(z_2) \\ 1 & \wp(z_3) & \wp'(z_3) \end{vmatrix} = 0$$

(This is also called an Addition Formula.)

3. $3p = 0 \in M$ if and only if $p \in C$ is a flex of C.
4. There are just nine flexes on C (compare Exercise 7 of Sec. 1.3). (See Fig. 4.28.)

Now the coordinate z of $\mathbb{C}$ can be taken as a local coordinate of $M = \mathbb{C}/\Gamma$. dz is then considered as a holomorphic differential on M. dz has no zero. The proof of the following lemma is left to the reader.

Lemma 4.3.12. The vector space $\Gamma(K_M)$ of all holomorphic differentials is one-dimensional: $\Gamma(K_M) = \mathbb{C}\,dz$.

Put $x = \wp(z)$ and $y = \wp'(z)$. Then

$$y^2 = \left(\frac{dx}{dz}\right)^2 = 4x^3 - g_2x - g_3$$

$$dz = \frac{dx}{y} = \frac{dx}{\sqrt{4x^3 - g_2x - g_3}}$$

Note that $z = 0$ corresponds to $x = \infty$. Hence,

$$z = \int_\infty^x \frac{dx}{\sqrt{4x^3 - g_2x - g_3}} \tag{**}$$

This means that $\wp(z)$ is the inverse function of the (multivalued) function defined by the integral (**).

The indefinite integral

$$I_1 = \int \frac{dx}{\sqrt{4x^3 - g_2 x - g_3}}$$

is called the Weierstrass canonical form of the elliptic integral of the first kind. The Weierstrass canonical form of the elliptic integral of the second (respectively third) kind is defined by

$$I_2 = \int \frac{x\,dx}{\sqrt{4x^3 - g_2 x - g_3}}$$

(respectively

$$I_3 = \int \frac{dx}{(x - a)\sqrt{4x^3 - g_2 x - g_3}} \qquad (a \in \mathbb{C})\Big)$$

In general, if $y = y(x) = \sqrt{P(x)}$ is the algebraic function defined by the equation

$$y^2 = P(x)$$

where $P(x)$ is a cubic (or quartic) polynomial with distinct roots, and if $R(x, y)$ is a rational function of (x, y), then

$$I = \int R(x, \sqrt{P(x)})\, dx$$

is called an elliptic integral. Operating a suitable birational transformation

$$(x, y) \to \left(\frac{ax + b}{cx + d}, \psi(x, y)\right)$$

where $ad - bc \neq 0$ and ψ is a rational function of (x, y), we may assume that

$$P(x) = 4x^3 - g_2 x - g_3$$

for some g_2 and g_3 with the discriminant $g_2^3 - 27g_3^2 \neq 0$. Then, it is classically known that

Theorem 4.3.13. Any elliptic integral $I = \int R(x, \sqrt{4x^3 - g_2 x - g_3})\, dx$ can be written as a linear combination of I_1, I_2, I_3 and $Q(x, \sqrt{4x^3 - g_2 x - g_3})$, where $Q(x, y)$ is a rational function of (x, y).

Finally, consider the function

$$\delta = \delta(\omega_1, \omega_2) = \frac{g_2^3}{g_2^3 - 27g_3^2}$$

of (ω_1, ω_2), where g_2 and g_3 are defined in Proposition 4.3.3. The proof of the following lemma is left to the reader (Exercise 1).

Lemma 4.3.14.

1. $\delta(\lambda\omega_1, \lambda\omega_2) = \delta(\omega_1, \omega_2)$ for $\lambda \neq 0$.
2. $\mathbb{C}/(\mathbb{Z}\omega_1 + \mathbb{Z}\omega_2) \simeq \mathbb{C}/(\mathbb{Z}\omega_1' + \mathbb{Z}\omega_2')$ if and only if $\delta(\omega_1, \omega_2) = \delta(\omega_1', \omega_2')$.

We define a function J on the upper half plane $\mathbb{H}$ by

$$J(\tau) = \delta(1, \tau) \qquad (\tau \in \mathbb{H})$$

Then, by the lemma,

Theorem 4.3.15.

1. J is a holomorphic function on $\mathbb{H}$.
2. $J\left(\frac{a\tau + b}{c\tau + d}\right) = J(\tau)$ for all $\tau \in \mathbb{H}$ and $\begin{pmatrix} a & b \\ c & d \end{pmatrix} \in SL(2, \mathbb{Z})$.
3. J induces a bijective map of $\mathbb{H}/SL(2, \mathbb{Z})$ onto $\mathbb{C}$.

Note that $\mathbb{H}/SL(2, \mathbb{Z})$ is the moduli space of complex 1-tori (see Sec. 3.1). The function $J(\tau)$ is called the elliptic modular function.

Note 4.3.16. (1) There are many books on the classical theory of elliptic functions and elliptic integrals. See, for example, Cayley [18], Tannery-Molk [96], Hurwitz-Courant [46], Siegel [94, Vol. 1]. (2) Historically, Gauss first found an elliptic function as the inverse function of the arclength

$$\int_0^x \frac{dx}{\sqrt{1 - x^4}}$$

of the lemniscate (see Sec. 2.2). But he did not publish his results. Then, Abel and Jacobi independently found elliptic functions as the inverse functions of the elliptic integrals

$$u = \int_0^x \frac{dx}{\sqrt{(1 - c^2x^2)(1 + e^2x^2)}} \qquad \text{(c, e: nonzero real numbers)},$$

$$u = \int_0^x \frac{dx}{\sqrt{(1 - x^2)(1 - k^2x^2)}} \qquad (0 < k < 1)$$

respectively. The latter function is denoted by $x = sn(u) = sn(u, k)$, which has the periods 4K and $2\sqrt{-1}\,K'$, where

$$K = \int_0^1 \frac{dx}{\sqrt{(1 - x^2)(1 - k^2x^2)}} \quad \text{and} \quad K' = \int_0^1 \frac{dx}{\sqrt{(1 - x^2)(1 - k'^2x^2)}}$$

where $k' = \sqrt{1 - k^2}$.

Exercises

1. Prove Propositions 4.3.4 and 4.3.7, and Lemmas 4.3.8 and 4.3.14.
2. Prove the following formula

$$\wp(2z) = -2\,\wp(z) + \frac{(3\,\wp(z)^2 - g_2/4)^2}{4\,\wp(z)^3 - g_2\,\wp(z) - g_3}$$

3. Give the addition formula for $\wp'$.
4. (1) Any elliptic function f with periods in Γ can be uniquely written as

$$f = R_1(\wp) + R_2(\wp)\,\wp'$$

where R_1 and R_2 are rational functions. (2) If f is an even (respectively odd) function, then f can be written as

$$f = R_1(\wp) \qquad (\text{respectively } f = R_2(\wp)\,\wp')$$

5. Prove the following formula:

$$\wp\left(z + \frac{\omega_1}{2}\right) = e_1 + \frac{(e_1 - e_2)(e_1 - e_3)}{\wp(z) - e_1},$$

$$\wp\left(z + \frac{\omega_2}{2}\right) = e_2 + \frac{(e_2 - e_1)(e_2 - e_3)}{\wp(z) - e_2},$$

$$\wp\left(z + \frac{\omega_1 + \omega_2}{2}\right) = e_3 + \frac{(e_3 - e_1)(e_3 - e_2)}{\wp(z) - e_3}$$

where $e_1 = \wp(\omega_1/2)$, $e_2 = \wp(\omega_2/2)$, and $e_3 = \wp((\omega_1 + \omega_2)/2)$.

6. For any elliptic function $f: M = \mathbb{C}/(\mathbb{Z}\omega_1 + \mathbb{Z}\omega_2) \to \mathbb{P}^1$ of degree 2, there are $\sigma_1 \in \mathrm{Aut}(M)$ and $\sigma_2 \in \mathrm{Aut}(\mathbb{P}^1)$ such that $f = \sigma_2 \cdot \wp \cdot \sigma_1$. (This means that an holomorphic map $M \to \mathbb{P}^1$ of degree 2 is essentially equal to $\wp$.)
7. For a complex 1-torus $M = \mathbb{C}/(\mathbb{Z} + \mathbb{Z}\tau)$ ($\tau \in \mathbb{H}$), the following two conditions are equivalent: (1) there is a holomorphic map $f: M \to \mathbb{P}^1$ of degree 3 with three distinct points p, q, r of ramification such that $e_p = e_q = e_r = 3$, and (2) $J(\tau) = 0$.

5

RIEMANN-ROCH THEOREM

5.1 THE RIEMANN-ROCH THEOREM

Let M be a compact Riemann surface of genus g. Let $L = \{f_{jk}\}$ be a line bundle on M. By Corollary 4.2.10, M can be imbedded into a projective space. Hence, by Theorem 3.4.21, there is a divisor D on M such that $[D] = L$. If we regard $D = \{h_j\}$ as a Cartier divisor, then

$$h_j = f_{jk} h_k \tag{*}$$

This means that $\xi = \{h_j\}$ is a meromorphic section of L. Conversely, a meromorphic section $\xi = \{h_j\}$ of L satisfies (*), so ξ defines a Cartier divisor $(\xi) = \{h_j\}$ on M such that $[(\xi)] = L$. (ξ) is called the divisor of the meromorphic section ξ.

Lemma 5.1.1. For any meromorphic sections ξ_1 and ξ_2 of L, (1) there is a unique meromorphic function f on M such that $\xi_2 = f\xi_1$ (f is denoted by ξ_2/ξ_1), (2) $(\xi_1) \sim (\xi_2)$, linearly equivalent, and (3) $\deg(\xi_1) = \deg(\xi_2)$.

Proof. Let $\{U_j\}$ be an open covering of M. Put $\xi_1 = \{h_j^{(1)}\}$ and $\xi_2 = \{h_j^{(2)}\}$, where $h_j^{(1)}$ and $h_j^{(2)}$ are meromorphic functions on U_j. Then

$$\frac{h_j^{(2)}}{h_j^{(1)}} = \frac{f_{jk} h_k^{(2)}}{f_{jk} h_k^{(1)}} = \frac{h_k^{(2)}}{h_k^{(1)}}$$

on $U_j \cap U_k$. Hence, $f = h_j^{(2)}/h_j^{(1)}$ is a meromorphic function on M. The uniqueness of f is clear. (2) and (3) are then trivial. Q.E.D.

By (3) of the lemma, deg (ξ) is independent of the choice of meromorphic sections ξ of L. Hence, we may put

$$\deg L = \deg(\xi)$$

and call it the degree of the line bundle L.

A meromorphic section ω of the canonical bundle $K = K_M$ of M is a meromorphic 1-form on M. It is traditionally called an Abelian differential. In a local coordinate z_j on U_j, ω can be written as

$$\omega = f_j(z_j)dz_j$$

Here, the meromorphic functions $f_j(z_j)$ on U_j satisfy

$$f_j(z_j) = \frac{dz_k}{dz_j} f_k(z_k) \quad \text{for } z_j = z_k \in U_j \cap U_k$$

The divisor (ω) of ω is the (Cartier) divisor $\{f_j(z_j)\}$, which is called a canonical divisor (see Sec. 3.4).

Let f be a meromorphic function on M. Then

$$\left\{ \frac{df}{dz_j} dz_j \right\}$$

is a meromorphic differential on M, as is easily checked. This is called the total differential of f and is denoted by df. Using this, we can prove

Proposition 5.1.2. $\deg K = 2g - 2$, where g is the genus of M.

Proof. It is enough to show that

$$\deg(df) = 2g - 2$$

for a meromorphic function f. Around a pole p of f, f can be expanded into the Laurent series

$$f(z) = a_{-\nu}z^{-\nu} + \cdots + a_{-1}z^{-1} + a_0 + a_1 z + a_2 z^2 + \cdots \qquad (a_{-\nu} \neq 0)$$

where z is a local uniformizing parameter at p. Then

$$df = (-\nu a_{-\nu}z^{-\nu-1} - \cdots - a_{-1}z^{-2} + a_1 + 2a_2 z + \cdots)dz$$

Around a point q which is not a pole, f can be expanded into the power series

$$f(z) = b_0 + b_k z^k + b_{k+1} z^{k+1} + \cdots \qquad (b_k \neq 0)$$

for some $k \geq 1$. Then

$$df = (kb_k z^{k-1} + (k+1)b_{k+1} z^k + \cdots)dz$$

Hence,

$$\deg(df) = \Sigma'_q(e_q - 1) - \Sigma_p(\nu_p + 1)$$

where Σ'_q runs over all points q of ramification (with the ramification index e_q) of $f: M \to \mathbb{P}^1$ such that q is not a pole, and Σ_p runs over all poles p (with the order ν_p) of f. Then, by the Riemann-Hurwitz Formula (Theorem 4.1.4),

$$\begin{aligned} \deg(df) &= \Sigma'_q(e_q - 1) + \Sigma_p(\nu_p - 1) - \Sigma_p(\nu_p - 1) - \Sigma_p(\nu_p + 1) \\ &= 2g - 2 + 2n - 2\Sigma_p \nu_p \\ &= 2g - 2 \end{aligned}$$

Q.E.D.

A holomorphic 1-form ω on M is called a homomorphic differential or traditionally an Abelian differential of the first kind. The set $\Gamma(K)$ of all holomorphic differentials form a vector space and

$$|K| = \{(\omega) \mid \omega \in \Gamma(K)\}$$

is the canonical linear system.

Now, let $\omega = \{f_j dz_j\}$ be an Abelian differential on M. Take a path

$$\gamma: [0, 1] \to M$$

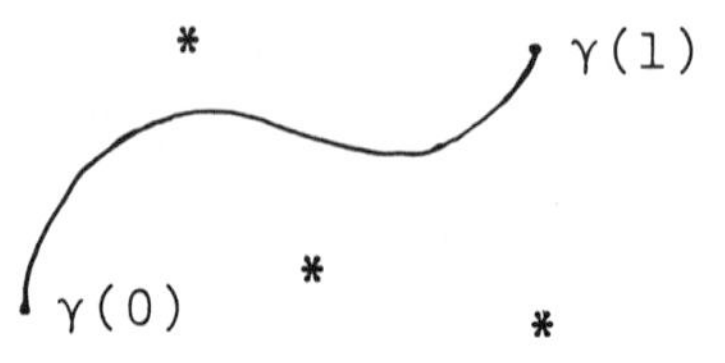

FIGURE 5.1

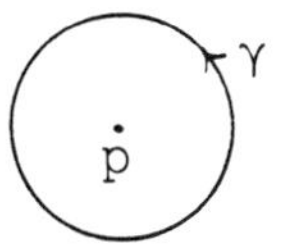

FIGURE 5.2

(a piecewise differentiable continuous map) on M, on which there is no pole of ω. (See Fig. 5.1.) Then the integral

$$\int_\gamma \omega = \int_0^1 \gamma^*\omega$$

is called the Abelian integral of ω along γ.

Around a pole p of ω, ω can be expanded into the Laurent series

$$\omega = (a_{-\nu} z^{-\nu} + \cdots + a_{-1} z^{-1} + a_0 + a_1 z + \cdots)\, dz \qquad (a_{-\nu} \neq 0)$$

where z is a local uniformizing parameter at p. ν is called the order of ω at the pole p. Put

$$\operatorname{Res}_p \omega = a_{-1} = \frac{1}{2\pi\sqrt{-1}} \int_\gamma \omega$$

(see Fig. 5.2), where γ is a small oriented circle around p. $\operatorname{Res}_p \omega$ is independent of the choice of z and is called the residue of ω at p.

Theorem 5.1.3 (Residue Theorem). For an Abelian differential ω,

$$\sum_p \operatorname{Res}_p \omega = 0$$

where Σ runs over all poles p of ω.

Proof. Let $\{p_1, \ldots, p_m\}$ be the set of all poles of ω. Let Δ_j be a small disc around p_j such that $\bar{\Delta}_j \cap \bar{\Delta}_k = \phi$ for $j \neq k$. Put

$$\gamma_j = \partial\bar{\Delta}_j = \text{the oriented boundary of } \bar{\Delta}_j$$

$$W = \bigcup \Delta_j$$

Then the oriented boundary of M − W is

FIGURE 5.3

$$\partial(M - W) = (-\gamma_1) + \cdots + (-\gamma_m)$$

(see Fig. 5.3). By Stokes' Theorem,

$$\sum_p \mathrm{Res}_p\, \omega = \sum_j \frac{1}{2\pi\sqrt{-1}} \int_{\gamma_j} \omega = \frac{-1}{2\pi\sqrt{-1}} \int_{M-W} d\omega = 0 \qquad \text{Q.E.D.}$$

For a divisor D on M, put

$$A(D) = \{\omega \mid \omega \text{ is an Abelian differential on M such that } (\omega) \geq D \text{ or } \omega = 0\}$$

Lemma 5.1.4.

1. $A(D)$ is a vector space over $\mathbb{C}$ and is isomorphic to $L((\omega_0) - D)$, where ω_0 is a (fixed) Abelian differential.
2. $A(0) = \Gamma(K)$.
3. If $D \geq 0$, then $A(D)$ is the set of all holomorphic differentials vanishing on D.
4. If $D_1 \geq D_2$, then $A(D_1) \subset A(D_2)$.
5. If $\deg D \geq 2g - 1$, then $A(D) = \{0\}$.

The proof is easy and is left to the reader (Exercise 1).
By abuse of rotation, we write

$$(\omega_0) = K \quad \text{and} \quad (\omega_0) - D = K - D$$

For a divisor D on M, put

$$\ell(D) = \dim L(D) = \dim \Gamma([D]),$$

$$r(D) = \ell(D) - 1 = \dim |D|,$$

$$i(D) = \dim A(D) = \ell(K - D)$$

These are classical notations. $i(D)$ is called the index of speciality of D.

Theorem 5.1.5 (Riemann-Roch Theorem). Let M be a compact Riemann surface of genus g and D be a divisor on M. Then

1. $\dim \Gamma(K_M) = g$, and
2. $\ell(D) - i(D) = \deg D + 1 - g$.

Remark 5.1.6.

1. The left-hand side of 1 of the theorem is an analytic quantity, while the right-hand side is a topological quantity. This is one of the most primitive forms of the index theorem.
2. 1 of the theorem follows from 2 (and Proposition 5.1.2) by putting $D = (\omega)$ for an Abelian differential ω. But we prove 1 first.
3. 2 of the theorem can also be written as (2)' $\ell(D) - \ell(K - D) = \deg D + 1 - g$, or (2)'' $r(D) - i(D) = \deg D - g$, or (2)''' $\ell r(D) - r(K - D) = \deg D + 1 - g$.
4. For the proof of the theorem, we follow Mumford [70], whose proof looks very beautiful geometrically. There are many other proofs. See, for example, Walker [100], Chevalley [20], Serre [87], Fulton [29].

For the proof of the theorem, we imbed M into a projective space $\mathbb{P}^N$ $(N \geq 2)$. If $N \geq 3$, then the composition of generic projections

$$\pi: M \to \mathbb{P}^2$$

is a birational map of M onto a plane nodal curve C (see Theorem 1.6.22). If $N = 2$, then we put $M = C$ and π = the identity map. Put

$$n = \deg C \quad \text{and}$$

$$s = \text{the number of nodes of } C$$

Then, by the genus formula (see Sec. 2.1),

$$g = \tfrac{1}{2}(n - 1)(n - 2) - s$$

(Note that we proved the genus formula in Sec. 2.1 by using Proposition 5.1.2.)

If $n = 1$ or 2, then C is a line or an irreducible conic, so M is biholomorphic to $\mathbb{P}^1$. In this case, the Riemann-Roch Theorem can be proved directly (compare Exercise 2). Hence, we may assume that

$$n \geq 3$$

Suppose that C is defined by the equation

$$C = \{F = 0\}$$

where F is an irreducible homogeneous polynomial of degree n in (X_0, X_1, X_2). Consider the affine curve

$$\{f(x,y) = F(1,x,y) = 0\}$$

where $(x,y) = (X_1/X_0, X_2/X_0)$. Then C is the closure of it in $\mathbb{P}^2$. By abuse of notation, we write

$$C = \{f = 0\}$$

By a suitable choice of $(X_0 : X_1 : X_2)$, we may assume that

1. C does not pass through (0: 0: 1) nor (0: 1: 0),
2. there is no node of C on the line $\{X_0 = 0\}$ of infinity, and
3. the line $\{X_0 = 0\}$ is not tangent to C.

Then the projections

$$x: (x,y) \in C \to x \in \mathbb{P}^1,$$

$$y: (x,y) \in C \to y \in \mathbb{P}^1$$

can be regarded as meromorphic functions x and y, respectively, on M of degree n.

Consider the Abelian differential

$$\omega = \frac{p(x,y)\,dx}{(\partial f/\partial y)} = -\frac{p(x,y)\,dy}{(\partial f/\partial x)}$$

on M, where p(x,y) is a polynomial of (x,y). (Note that $(\partial f/\partial x)\,dx + (\partial f/\partial y)\,dy = df = 0$ on M.) Using 1–3,

Lemma 5.1.7. If $\deg p(x,y) \leq n - 3$, and the curve $C_1 = \{p(x,y) = 0\}$ passes through every node of C, then $\omega = (p(x,y)\,dx)/(\partial f/\partial y)$ is a holomorphic differential on M.

The proof of the lemma is left to the reader (Exercise 1). By the lemma,

Lemma 5.1.8. $\dim \Gamma(K_M) \geq g$.

Proof. The vector space of all polynomials p(x,y) of degree at most n − 3 has the dimension $\frac{1}{2}(n-1)(n-2)$. The condition that the curve $\{p(x,y)=0\}$ passes through every node of C imposes s linear conditions on the coefficients of $p(x,y)$. Hence,

$$\dim \Gamma(K) \geq \tfrac{1}{2}(n-1)(n-2) - s = g \qquad \text{Q.E.D.}$$

Lemma 5.1.9. Let D be a divisor on M such that $D \geq 0$. Then

$$\ell(D) \leq \deg D - \dim \Gamma(K) + 1 + i(D)$$

Proof. Put $h = \dim \Gamma(K)$. If $h = 0$, then $i(D) = 0$. In this case, the lemma follows from Theorem 4.1.13. Suppose that $h \geq 1$. Let

$$\{\omega_1, \ldots, \omega_h\}$$

be a basis of $\Gamma(K)$. The assertion is trivial for $D = 0$. Put

$$D = \nu_1 p_1 + \cdots + \nu_m p_m$$

where $p_j \neq p_k$ for $j \neq k$, and $\nu_j > 0$ for $1 \leq j \leq m$. Take $f \in L(D)$. We expand ω_j and f around p_k into the power (respectively Laurent) series as follows:

$$\omega_j = (b^{(k)}_{j,0} + b^{(k)}_{j,1} t + \cdots)\, dt,$$

$$f(t) = a^{(k)}_{-\nu_k} t^{-\nu_k} + \cdots + a^{(k)}_{-1} t^{-1} + a^{(k)}_0 + \cdots$$

where t is a local uniformizing parameter at p_k. Then

$$\operatorname{Res}_{p_k}(f\omega_j) = b^{(k)}_{j,0} a^{(k)}_{-1} + b^{(k)}_{j,1} a^{(k)}_{-2} + \cdots + b^{(k)}_{j,\nu_k-1} a^{(k)}_{-\nu_k}$$

Hence, by the Residue Theorem (Theorem 5.1.3),

$$0 = \sum_k \operatorname{Res}_{p_k}(f\omega_j) = \sum_{k=1}^{m} \sum_{i=0}^{\nu_k-1} b^{(k)}_{j,i} a^{(k)}_{-(i+1)} \quad \text{for } 1 \leq j \leq h \qquad (*)$$

As in the proof of Theorem 4.1.13, consider the linear maps

$$A_k: f \in L(D) \to (a^{(k)}_{-\nu_k}, \ldots, a^{(k)}_{-1}) \in \mathbb{C}^{\nu_k}$$

$$A: f \in L(D) \to (A_1(f), \ldots, A_m(f)) \in \mathbb{C}^d$$

where $d = \deg D$. Note that

$$\operatorname{rank} A = \ell(D) - 1 = r(D)$$

Next, consider the linear map

$$B: (\ldots, (a^{(k)}_{-\nu_k}, \ldots, a^{(k)}_{-1}), \ldots) \in \mathbb{C}^d$$

$$\to \left(\sum_{k=1}^{m} \sum_{i=0}^{\nu_k - 1} b^{(k)}_{1,i} a^{(k)}_{-(i+1)}, \ldots, \sum_{k=1}^{m} \sum_{i=0}^{\nu_k - 1} b^{(k)}_{h,i} a^{(k)}_{-(i+1)} \right) \in \mathbb{C}^h$$

Then (*) means that $BA = 0$. Hence,

$$\ell(D) - 1 = \operatorname{rank} A \leq \dim\ker B = d - \operatorname{rank} B$$

On the other hand,

$$\omega = c_1\omega_1 + \cdots + c_h\omega_h \in \Gamma(K)$$

belongs to $A(D)$ if and only if

$$c_1 b^{(k)}_{1,i} + c_2 b^{(k)}_{2,i} + \cdots + c_h b^{(k)}_{h,i} = 0 \quad \text{for } 0 \leq i \leq \nu_k - 1 \quad \text{and} \quad 1 \leq k \leq m$$

This means that

$$(c_1, \ldots, c_h) \in \ker {}^tB$$

where ${}^tB: \mathbb{C}^h \to \mathbb{C}^d$ is the transpose of B. Hence,

$$i(D) = \dim\ker {}^tB = h - \operatorname{rank} {}^tB = h - \operatorname{rank} B$$

Hence,

$$\ell(D) \leq d + 1 - \operatorname{rank} B = d + 1 - h + i(D) \qquad \text{Q.E.D.}$$

Corollary 5.1.10. dim $\Gamma(K) = g$. In particular,

$$\Gamma(K) = \left\{ \frac{p(x,y)\, dx}{\partial f/\partial y} \,\middle|\, \deg p(x,y) \leq n - 3 \text{ and } \{p(x,y) = 0\} \text{ passes through every node of } C \right\}$$

Proof. By Lemma 5.1.8, $h = \dim \Gamma(K) \geq g$. Hence, if $h = 0$, then $g = 0$. Suppose now that $h \geq 1$. Take a nonzero $\omega \in \Gamma(K)$ and put $D = (\omega)$. Then, $A(D) = \mathbb{C}\omega$ by the definition of $A(D)$. Note also that $L(D)$ is isomorphic to $\Gamma(K)$. Hence, by applying Lemma 5.1.9 to $D = (\omega)$,

$$h = \ell(D) \leq (2g - 2) - h + 1 + 1$$

Hence, $h \leq g$, so $h = g$. Q.E.D.

Next, we prove

Lemma 5.1.11. Let D be a divisor on M such that $D \geq 0$. Then

$$\ell(D) \geq \deg D - g + 1 + i(D)$$

Proof.

(1°) Special case. We use the same notations as above. Let $\{\hat{p}_1, \ldots, \hat{p}_s\}$ be the set of all nodes of C. Put $\pi^{-1}(\hat{p}_j) = \{p_j, p_j'\}$ for $1 \leq j \leq s$. Let m ($\geq s$) be a sufficiently large positive integer and $L_1, \ldots, L_m$ be lines in $\mathbb{P}^2$ such that

$$\hat{p}_j \in L_j \text{ and } \hat{p}_j \notin L_k \text{ for } 1 \leq j \leq s \text{ and } 1 \leq k \leq m \text{ such that } j \neq k \qquad (*)_1$$

and

$$\text{no } L_k \text{ is tangent to C for } 1 \leq k \leq m \qquad (*)_2$$

Put

$$G = L_1 \cdots L_m,$$

$$D_1 = C \cdot G,$$

$$D = D_1 - (p_1 + p_1' + \cdots + p_s + p_s')$$

where $C \cdot G$ is the intersection zero-cycle (see Sec. 1.4) and is regarded as a positive divisor on M. By Bezout's Theorem (Theorem 1.3.5),

$\deg D_1 = mn, \quad \text{so} \quad \deg D = mn - 2s$

Consider the vector space

$$W = \{H \mid H \text{ is a homogeneous polynomial of degree } m \text{ such that } H(\hat{p}_j) = 0 \text{ for } 1 \leq j \leq s\}$$

Then, it is clear that

$$\dim W \geq \tfrac{1}{2}(m+2)(m+1) - s$$

For any $H \in W$, H/G can be regarded as a meromorphic function on M. The principal divisor (H/G) satisfies

$$\left(\frac{H}{G}\right) + D \geq 0$$

Hence, $H/G \in L(D)$. Consider the linear map

$$H \in W \to \frac{H}{G} \in L(D)$$

Its kernel consists of such H that

$$H = FH'$$

where F (respectively H') is a homogeneous polynomial of degree n (respectively $m - n$) such that $C = \{F = 0\}$. Hence,

$$\begin{aligned}\dim L(D) &\geq \dim W - \tfrac{1}{2}(m-n+2)(m-n+1) \\ &\geq \tfrac{1}{2}(m+2)(m+1) - \tfrac{1}{2}(m-n+2)(m-n+1) - s \\ &= (mn - 2s) - (\tfrac{1}{2}(n-1)(n-2) - s) + 1 \\ &= \deg D - g + 1\end{aligned}$$

Note that $i(D) = 0$ for $m >> 0$ by (5) of Lemma 5.1.4.

(2°) General case. For a given D such that $D \geq 0$, we may assume that no node of C is contained in the support of D. In fact, we imbed M into $\mathbb{P}^3$ and take a generic projection

$$\pi_p : M \to C \subset \mathbb{P}^2$$

with the center $p \in \mathbb{P}^3 - M$ such that $p \notin \overline{qr}$ for any points q and r in the support of D.

Take a sufficiently large positive integer m and lines $L_1, \ldots, L_m$ which satisfy $(*)_1$ and $(*)_2$ above and

$$G = L_1 \cdots L_m = 0 \text{ on } D \qquad (*)_3$$

This means that, if we put

$$C \cdot G - (p_1 + P_1' + \cdots + P_s + p_s') = D + E$$

then $E \geq 0$. We may assume further that

$$\text{the supports of D and E have no point in common} \qquad (*)_4$$

By (1°) and Lemma 5.1.9,

$$\ell(D + E) = \deg D + \deg E - g + 1$$

If $E = 0$, then the lemma is proved. Suppose that $E > 0$. Write

$$E = \mu_1 q_1 + \cdots + \mu_r q_r$$

where $\mu_j > 0$ and $q_j \neq q_k$ for $j \neq k$.

Take $f \in L(D + E)$. In a neighborhood of q_j, f can be written as

$$f(t) = a^{(j)}_{-\mu_j} t^{-\mu_j} + \cdots + a^{(j)}_{-1} t^{-1} + a^{(j)}_0 + \cdots$$

where t is a local uniformizing parameter at q_j.

Consider the linear maps

$$\psi_j: f \in L(D + E) \to (a^{(j)}_{-\mu_j}, \ldots, a^{(j)}_{-1}) \in \mathbb{C}^{\mu_j}$$

$$\psi: f \in L(D + E) \to (\psi_1(f), \ldots, \psi_r(f)) \in \mathbb{C}^e \qquad (e = \deg E)$$

Then it is clear that the sequence

$$0 \to L(D) \to L(D + E) \xrightarrow{\psi} \mathbb{C}^e$$

is exact.

On the other hand, consider the bilinear map

$$\text{Res}: ((\cdots, (a^{(j)}_{-\mu_j}, \ldots, a^{(j)}_{-1}), \ldots), \omega) \in \mathbb{C}^e \times A(D)$$

$$\to \sum_j \text{Res}_{q_j}\left(\sum_{\nu=-\mu_j}^{-1} a^{(j)}_\nu t^\nu\right)\omega \in \mathbb{C}$$

Then, by the residue theorem (Theorem 5.1.3),

$$\text{Res}\,(\psi(f), \omega) = \sum_j \text{Res}_{q_j}(f\omega) = 0 \quad \text{for } f \in L(D+E) \quad \text{and} \quad \omega \in A(D)$$

(Note that $(f\omega) + E = (f) + (\omega) + E \geq (f) + D + E \geq 0$.) Put

$$W = \mathbb{C}^e \text{ and } W^* = \text{the dual vector space to } W$$

Then the map

$$R: \omega \in A(D) \to \text{Res}\,(\cdot, \omega) \in W^*$$

is linear. We show that R is injective for $m >> 0$. Suppose that

$$\text{Res}\,(z, \omega) = 0 \quad \text{for all } z \in W$$

Put

$$\omega = (b^{(j)}_0 + b^{(j)}_1 t + \cdots)\, dt$$

Then, for $z = (0, \ldots, 0, (0, \ldots, 0, \overset{(k)}{1}, 0, \ldots, 0), 0, \ldots, 0) \in \mathbb{C}^e$,

$$\text{Res}\,(z, \omega) = b^{(j)}_{\mu_j - k} = 0$$

This means that $\omega \in A(D+E)$. But, if $m >> 0$, then $\deg(D+E) = mn - 2s \geq 2g - 1$, so $A(D+E) = 0$ by (5) of Lemma 5.1.4. Hence, $\omega = 0$, so R is injective for $m >> 0$.

Now, since $\text{Res}\,(\psi(L(D+E)), A(D)) = 0$, we have

$$\dim \psi(L(D+E)) + \dim A(D) \leq \dim W = \deg E, \quad \text{for } m >> 0$$

Hence,

$$\begin{aligned}\ell(D) &= \ell(D + E) - \dim \psi(L(D + E)) \\ &\geq \deg D + \deg E - g + 1 - (\deg E - i(D)) \\ &= \deg D - g + 1 + i(D)\end{aligned}$$ Q.E.D.

By Lemma 5.1.9, Lemma 5.1.11, and Corollary 5.1.10, the Riemann-Roch Theorem is proved for D such that $D \geq 0$.

Finally, we prove the theorem for a general D. The idea is similar to the proof of Lemma 5.1.11. Write

$$D = D^{+} - D^{-}$$

where $D^{+} \geq 0$, $D^{-1} \geq 0$ and the supports of D^{+} and D^{-} have no point in common. Let E_1 be a positive divisor such that

1. $\deg E_1 > 2g - 2$ and
2. the supports of E_1 and D^{+} (respectively E_1 and D^{-}) have no point in common.

Put

$$E = E_1 + D^{-}$$

Then

$$D + E = D^{+} + E_1 > 0 \quad \text{and} \quad \deg (D + E) > 2g - 2$$

Hence, by Lemmas 5.1.9 and 5.1.11,

$$\ell(D + E) = \deg (D + E) - g + 1$$

Write

$$D^{-} = \sum_j \mu_j p_j \quad \text{and} \quad E_1 = \sum_k \nu_k q_k$$

For $f \in L(D + E) = L(D^{+} + E_1)$, put

$$f(t) = a_0^{(j)} + a_1^{(j)} t + \cdots,$$

$$f(u) = b_{-\nu_k}^{(k)} u^{-\nu_k} + \cdots + b_{-1}^{(k)} u^{-1} + b_0^{(k)} + \cdots$$

where t (respectively u) is a local uniformizing parameter at p_j (respectively q_k). Consider the linear maps

$$\psi_j^{(1)}: f \in L(D+E) \to (a_0^{(j)}, \ldots, a_{\mu_j-1}^{(j)}) \in \mathbb{C}^{\mu_j},$$

$$\psi_k^{(2)}: f \in L(D+E) \to (b_{-\nu_k}^{(k)}, \ldots, b_{-1}^{(k)}) \in \mathbb{C}^{\nu_k}, \quad \text{and}$$

$$\psi: f \in L(D+E) \to (\ldots, \psi_j^{(1)}(f), \ldots; \ldots, \psi_k^{(2)}(f), \ldots) \in \mathbb{C}^e$$

where $e = \deg E$. Then it is clear that the sequence

$$0 \to L(D) \to L(D+E) \xrightarrow{\psi} \mathbb{C}^e$$

is exact. Put $W = \mathbb{C}^e$. Consider the bilinear map

$$\text{Res}: ((\ldots, (a_0^{(j)}, \ldots, a_{\mu_j-1}^{(j)}), \ldots; \ldots, (b_{-\nu_k}^{(k)}, \ldots, b_{-1}^{(k)}), \ldots), \omega) \in W \times A(D)$$

$$\to \sum_j \text{Res}_{p_j} \Big(\sum_{\nu=0}^{\mu_j-1} a_\nu^{(j)} t^\nu \Big) \omega + \sum_k \text{Res}_{q_k} \Big(\sum_{\nu=-\nu_k}^{-1} b_\nu^{(k)} u^\nu \Big) \omega \in \mathbb{C}$$

Then the linear map

$$R: \omega \in A(D) \to \text{Res}(\cdot, \omega) \in W^*$$

is injective. In fact, if $R(\omega) = 0$, then, as in the proof of Lemma 5.1.11, we have $\omega \in A(D+E)$. But $A(D+E) = 0$, for $\deg(D+E) \geq 2g-1$.

On the other hand, for $f \in L(D+E)$ and $\omega \in A(D)$, we have

$$\text{Res}(\psi(f), \omega) = \sum_j \text{Res}_{p_j}(f\omega) + \sum_k \text{Res}_{q_k}(f\omega) = 0$$

Hence,

$$\begin{aligned} \ell(D) &= \ell(D+E) - \dim \psi(L(D+E)) \\ &\geq \deg D + \deg E - g + 1 - (\deg E - i(D)) \\ &= \deg D - g + 1 + i(D) \end{aligned}$$

Thus,

$$\ell(D) - \ell(K - D) \geq \deg D - g + 1$$

Using K - D instead of D, we get

$$\ell(K - D) - \ell(D) \geq \deg (K - D) - g + 1 = g - 1 - \deg D$$

Hence,

$$\ell(D) - \ell(K - D) = \deg D - g + 1$$

This completes the proof of the Riemann-Roch Theorem.

The Riemann-Roch Theorem contains a lot of information on curves. In the rest of this section, we give some direct applications of the theorem.

Henceforth, let M be a compact Riemann surface of genus g.

Proposition 5.1.12. Let D be a divisor on M.

1. If $\deg D \geq 2g + 1$, then D is very ample (see Sec. 4.1).
2. D is ample if and only if $\deg D > 0$.
3. If $g > 0$ and $\deg D \geq 2g$, then $|D|$ has no fixed point.

Proof. If $\deg D \geq 2g + 1$, then, by the Riemann-Roch Theorem,

$$r(D - p - q) = \deg (D - p - q) - g = r(D) - 2$$

for any points p and q of M. Hence, D is very ample by Theorem 4.1.16.

2 follows from 1. The proof of 3 is similar to that of 1. Q.E.D.

In this proposition, putting $g = 0$ (respectively $g = 1$) and $\deg D = 1$ (respectively 3), we get

Corollary 5.1.13. (1) M is biholomorphic to $\mathbb{P}^1$ if and only if $g = 0$. (2) M is biholomorphic to a nonsingular plane cubic curve if and only if $g = 1$.

The following two propositions are easy to prove (compare Exercise 1).

Proposition 5.1.14. Suppose that $g \geq 1$. Let D be a divisor on M such that $\deg D = 2g - 2$. Then r(D) is either $g - 2$ or $g - 1$. Moreover, $r(D) = g - 1$ if and only if D is a canonical divisor.

Proposition 5.1.15. If $g \geq 2$ and $m \geq 2$, then $\ell(mK) = (2m - 1)(g - 1)$.

Recall that a compact Riemann surface M of genus g (≥ 2) is said to be hyperelliptic if there is a meromorphic function on M of degree 2. Otherwise, M is said to be <u>nonhyperelliptic</u>.

The proof of the following lemma is left to the reader (Exercise 1).

<u>Lemma 5.1.16.</u> Let M be hyperelliptic and be given by the equation

$$y^2 = (x - \alpha_1) \cdots (x - \alpha_{2g+1}) \qquad (\alpha_j \neq \alpha_k \text{ for } j \neq k)$$

(see Proposition 4.2.14). Then

1. $\Gamma(K) = \left\{ \dfrac{h(x)}{y}\, dx \mid h(x) \text{ is a polynomial of degree} \leq g - 1 \right\}$, and
2. Φ_K is a (2 - 1)-map of M onto a rational normal curve in $\mathbb{P}^{g-1}$.

<u>Proposition 5.1.17.</u> If g = 2, then M is necessarily hyperelliptic.

<u>Proof.</u> The canonical linear system |K| is a g_2^1 without fixed point. Q.E.D.

However, a "general" M with $g \geq 3$ is nonhyperelliptic (see Sec. 5.4).

<u>Theorem 5.1.18.</u> Suppose that $g \geq 2$. Then (1) $|K_M|$ has no fixed point and (2) $|K_M|$ is very ample if and only if M is nonhyperelliptic.

<u>Proof.</u> We prove (2) ((1) can be shown in a simpler way). If M is hyperelliptic, then $|K_M|$ is not very ample by Lemma 5.1.16. Suppose that M is nonhyperelliptic. By the Riemann-Roch Theorem,

$$\ell(K) = g,$$

$$\ell(K - p) = \ell(p) + g - 2 \qquad (p \in M), \tag{*}$$

$$\ell(K - p - q) = \ell(p + q) + g - 3 \qquad (p,\ q \in M)$$

If $\ell(p + q) \geq 2$, then there is a nonconstant meromorphic function f such that $D_\infty(f) \leq p + q$. If $D_\infty(f) = p$, say, then the degree of f is 1, so f: $M \simeq \mathbb{P}^1$, a contradiction. If $D_\infty(f) = p + q$, then the degree of f is 2, so M is hyperelliptic, a contradiction.

Hence, $\ell(p + q) = 1$, so $\ell(K - p - q) = g - 2$. Thus |K| is very ample by Theorem 4.1.16. Q.E.D.

The image

$$C_K = \Phi_K(M)$$

of the canonical map $\Phi_K: M \to \mathbb{P}^{g-1}$ of a nonhyperelliptic M into $\mathbb{P}^{g-1}$ is called the canonical curve (of M). C_K has the degree 2g - 2. Conversely (compare Exercise 1)

Proposition 5.1.19. If C is a nonsingular, nondegenerate irreducible curve in $\mathbb{P}^{g-1}$ of degree 2g - 2 and genus g, then C is nonhyperelliptic and is (projectively equivalent to) its canonical curve C_K.

Canonical curves are very important. Any projective geometric property of C_K reflects some intrinsic property of M.

Example 5.1.20. Let g = 3 and M be nonhyperelliptic. Then C_K is a nonsingular plan quartic curve (and vice versa). Let ℓ be a bitangent line of C_K. Put

$$C_K \cdot \ell = 2(p_1 + p_2)$$

(See Fig. 5.4.) Then $2(p_1 + p_2) \sim K$, so $p_1 + p_2$ is a half-canonical divisor such that $p_1 \neq p_2$. (In general, a positive divisor D on M is called a half-canonical divisor if $2D \sim K_M$. Half-canonical divisors are important in the theory of theta-functions on the Jacobian variety J(M) of M (see Sec. 5.2). In a similar way, a flex of order 2 (see Sec. 1.3) gives a half-canonical divisor of the form 2p. Hence, by Exercise 3 of Sec. 1.5, the number of half-canonical divisors on M = 28.

Example 5.1.21. Let g = 4 and M be nonhyperelliptic. Then C_K is a nonsingular space sextic curve in $\mathbb{P}^3$. C_K is the complete intersection of a quadric surface Q and a cubic surface R in $\mathbb{P}^3$

$$C_K = Q \cap R$$

In fact, put

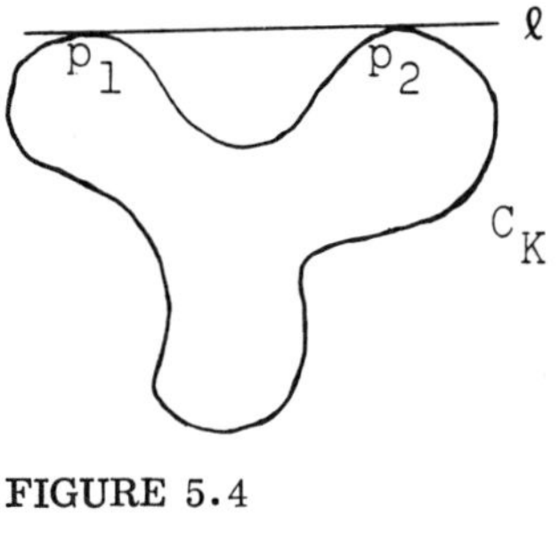

FIGURE 5.4

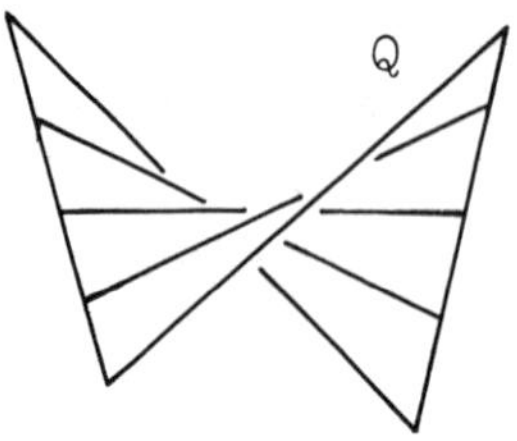

FIGURE 5.5

W_m = the vector space of all homogeneous polynomials of degree m in the variables X_0, X_1, X_2, X_3

Then, dim $W_2 = 10$ and dim $W_3 = 20$. Consider the linear map

$$\alpha_m : F \in W_m \to F \mid C_K \in \Gamma(mK) \quad (m \geq 2)$$

where $F \mid C_K$ is the restriction of F to C_K. Since dim $\Gamma(mK) = (2m - 1)(g - 1)$,

$$\dim \ker \alpha_2 \geq 1 \quad \text{and} \quad \dim \ker \alpha_3 \geq 5$$

Take a nonzero $F_0 \in \ker \alpha_2$ and put $Q = \{F_0 = 0\}$. Then Q contains C_K. Put

$$W_3(F_0) = \{G \in W_3 \mid G = F_0 L \text{ for a linear form } L\}$$

Then dim $W_3(F_0) = 4$. Take $G_0 \in \ker \alpha_3 - W_3(F_0)$ and put $R = \{G_0 = 0\}$. Then $Q \cap R$ is a curve in $\mathbb{P}^3$ containing C_K. But

$$\deg(Q \cap R) = 6 = \deg(C_K), \quad \text{so } C_K = Q \cap R$$

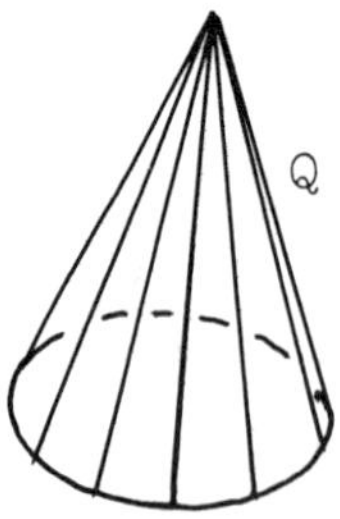

FIGURE 5.6

There are two cases.

1. Suppose that Q is a nonsingular quadric surface. In this case, there are two one-parameter families $\{\ell_\lambda\}$ and $\{\ell'_\lambda\}$ of lines on Q. C_K meets every ℓ_λ (respectively ℓ'_λ) at three points. The projections π and π' with the centers ℓ_λ and ℓ'_λ, respectively, give distinct linear pencils of degree 3. (Note that the projections with the center lines in the same family give the same linear pencil.) No other linear pencil of degree 3 exists on M. (See Fig. 5.5.)
2. Suppose that Q is a quadric cone. In this case, there is a unique linear pencil of degree 3 on M. (See Fig. 5.6.)

Finally, we talk Weierstrass points on M.

Lemma 5.1.22. Let p be a point of M.

1. If $m \geq g + 1$, then there is a nonconstant meromorphic function f on M such that $D_\infty(f) \leq mp$.
2. If $g \geq 1$ (respectively $g = 0$) and $m \geq 2g$ (respectively $m \geq 1$), then there is a meromorphic function f on M such that $D_\infty(f) = mp$.

Proof. By the Riemann-Roch Theorem,

$$\ell(mp) = m + 1 - g + i(mp)$$

If $m \geq g + 1$, then $\ell(mp) \geq 2$. This shows 1. If $g \geq 1$ and $m \geq 2g$, then $i(mp) = i((m - 1)p) = 0$. Hence,

$$\ell(mp) - \ell((m - 1)p) = 1$$

Hence, 2 follows. The assertion in the case $g = 0$ is trivial. Q.E.D.

In the rest of this section, we suppose that $g \geq 1$. For a point $p \in M$, note that

$$1 = \ell(p) \leq \ell(2p) \leq \cdots \leq \ell(2gp) = g + 1$$

Hence, there are just g-values of m with $2 \leq m \leq 2g$ such that $\ell(mp) - \ell((m - 1)p) = 1$. Other g-values of m with $1 \leq m \leq 2g$ satisfy $\ell(mp) = \ell((m - 1)p)$.

Definition 5.1.23. An integer m with $1 \leq m \leq 2g - 1$ is called a gap value at p if there is no meromorphic function f such that $D_\infty(f) = mp$. There are just g gap values

$$n_1 = 1 < n_2 < \cdots < n_g \leq 2g - 1$$

at p, called the gap sequence at p. If the gap sequence at p is

$$1, 2, \ldots, g$$

then p is called a non-Weierstrass point. Otherwise, p is called a Weierstrass point.

Example 5.1.24.

m	1	2	3	4	5	6	7	8	9	...	2g - 1	2g
$\ell(mp)$	1	1	2	2	2	3	3	3	4	...	g	g + 1
g.v.	n_1	n_2		n_3	n_4		n_5	n_6		...		

(g.v. = gap values).

The proof of the following lemma is left to the reader (Exercise 1).

Lemma 5.1.25. The following three conditions are equivalent:

1. p is a non-Weierstrass point.
2. $\ell(gp) = 1$.
3. $i(gp) = 0$.

Now we prove

Theorem 5.1.26.

1. If $g = 1$, then there is no Weierstrass point.
2. If $g \geq 2$, then the number w of Weierstrass points is finite and satisfies

$$1 \leq w \leq (g - 1)g(g + 1)$$

Proof. 1 is trivial. To prove 2, let $g \geq 2$ and $\{\omega_1, \ldots, \omega_g\}$ be a basis of $\Gamma(K)$. For a point $p \in M$, ω_j can be written around p as

$$\omega_j = h_j(t)\, dt$$

where t is a local coordinate around p. For $\omega = a_1\omega_1 + \cdots + a_g\omega_g \in \Gamma(K)$, $\omega \in A(gp)$ if and only if

$$a_1h_1(p) + \cdots + a_gh_g(p) = 0,$$

$$a_1h'_1(p) + \cdots + a_gh'_g(p) = 0,$$

$$\cdots$$

$$a_1h_1^{(g-1)}(p) + \cdots + a_gh_g^{(g-1)}(p) = 0$$

where $h' = dh/dt$, $h'' = d^2h/dt^2$, and so on. Put

$$W(t) = \begin{vmatrix} h_1(t) & \cdots & h_g(t) \\ h'_1(t) & \cdots & h'_g(t) \\ \cdots & & \\ h_1^{(g-1)}(t) & \cdots & h_g^{(g-1)}(t) \end{vmatrix}$$

Then $i(gp) \geq 1$ if and only if $W(p) = 0$.

Note that, if s is another local coordinate around p, then

$$W(t) = \left(\frac{ds}{dt}\right)^{g(g+1)/2} W(s)$$

This means that

$$W(\omega_1, \ldots, \omega_g) = \{W(t)\}$$

is a holomorphic section of $\frac{1}{2}g(g + 1)K$. It is not difficult to check that $W(\omega_1, \ldots, \omega_g)$ is not identically zero (compare Exercise 9). $W(\omega_1, \ldots, \omega_g)$ is called the <u>Wronskian form</u> of $\omega_1, \ldots, \omega_g$.

Now, a point $p \in M$ is a Weierstrass point if and only if p is a zero of $W(\omega_1, \ldots, \omega_g)$. But

$$\deg\left(\tfrac{1}{2}g(g + 1)K\right) = (g - 1)g(g + 1)$$

Hence,

$$1 \leq w \leq (g - 1)g(g + 1)$$ Q.E.D.

As an application of Theorem 5.1.26,

<u>Theorem 5.1.27 (Schwarz)</u>. If $g \geq 2$, then Aut (M) is a finite group.

Proof (Segre). Let $\{p_1, \cdots, p_w\}$ be the set of all Weierstrass points on M. Let G_1 be the group of all permutations of $p_1, \cdots, p_w$. Every $\sigma \in \mathrm{Aut}(M)$ clearly induces a permutation $\phi_1(\sigma)$ of $p_1, \cdots, p_w$. The map

$$\phi_1 : \sigma \in \mathrm{Aut}(M) \to \phi_1(\sigma) \in G_1$$

is then a homomorphism. Hence, it suffices to show that ker ϕ_1 is a finite group. Put

$$m = \min\{k \mid \ell(kp_1) = 2\}$$

Then $2 \leq m \leq g$, for p_1 is a Weierstrass point. Let $\{1, f\}$ be a basis of $L(mp_1)$. Then $D_\infty(f) = mp_1$. Let $\{p_1, q_2, \cdots, q_s\}$ be the set of all points of ramification of $f: M \to \mathbb{P}^1$. By the Riemann-Hurwitz Formula,

$$2g - 2 = -m - 1 + (e_2 - 1) + \cdots + (e_s - 1)$$

where e_j $(2 \leq j \leq s)$ is the ramification index of f at q_j. Note that $e_j \leq m$ for $2 \leq j \leq s$. Hence, we get

1. $s \geq 3$ and
2. the number t $(\leq s)$ of branch points

$$\{f(p_1) = \infty, f(q_2), \cdots, f(q_s)\} = \{\infty, \lambda_2, \cdots, \lambda_t\}$$

is greater than or equal to 3.

Now, take $\sigma \in \ker \phi_1$. The composition $f \cdot \sigma$ clearly belongs to $L(mp_1)$. Hence,

3. $f \cdot \sigma = af + b$

for some constants a $(\neq 0)$ and b. Hence, at $q = q_j$,

$$\left(\frac{df}{dw}\right)_{\sigma(q)} \left(\frac{d\sigma}{dz}\right)_q = a\left(\frac{df}{dz}\right)_q = 0$$

where z (respectively w) is a local coordinate around q (respectively $\sigma(q)$). Hence, $(df/dw)_{\sigma(q)} = 0$, so $\sigma(q)$ is a point of ramification of f. In other words, σ induces a permutation $\phi_2(\sigma)$ of the set $\{q_2, \cdots, q_s\}$. Then

$$\phi_2 : \sigma \in \ker \phi_1 \to \phi_2(\sigma) \in G_2$$

is clearly a homomorphism, where G_2 is the group of all permutations of $q_2, \ldots, q_s$. Hence, it suffices to show that $\ker \phi_2$ is a finite group.

3 shows that the diagram

$$\begin{array}{ccc} M & \xrightarrow{\sigma} & M \\ {\scriptstyle f}\downarrow & & \downarrow{\scriptstyle f} \\ \mathbb{P}^1 & \xrightarrow{\tau} & \mathbb{P}^1 \end{array}$$

is commutative, where τ is the automorphism of $\mathbb{P}^1$ defined by

$$\tau: x \in \mathbb{P}^1 \to ax + b \in \mathbb{P}^1$$

If $\sigma \in \ker \phi_2$, then $\lambda_2, \ldots, \lambda_t$ are fixed by τ. By 2, $t \geq 3$. Hence, τ is the identity map, so $f \cdot \sigma = f$.

Take $\lambda \in \mathbb{C} - \{\lambda_2, \ldots, \lambda_t\}$ and put

$$f^{-1}(\lambda) = \{r_1, \ldots, r_m\}$$

Then every $\sigma \in \ker \phi_2$ induces a permutation $\phi_3(\sigma)$ of $r_1, \ldots, r_m$. The map

$$\phi_3: \sigma \in \ker \phi_2 \to \phi_3(\sigma) \in G_3$$

is clearly a homomorphism, where G_3 is the group of all permutations of $r_1, \ldots, r_m$. It is now clear that ϕ_3 is injective. Q.E.D.

Note 5.1.28. (1) For the Riemann-Roch Theorem for higher dimensional manifolds, see Hirzebruch [42], Palais [79] and Borel-Serre [14]. (2) For the automorphism groups of compact Riemann surfaces, see Accola [1] and the references in it.

Exercises

1. Prove Lemmas 5.1.4, 5.1.7, 5.1.16, 5.1.19, and 5.1.25, and Propositions 5.1.14 and 5.1.15.
2. Prove the Riemann-Roch Theorem for $M = \mathbb{P}^1$. Prove then any $r + 1$ points on a rational normal curve in $\mathbb{P}^r$ are in general position. (This can be shown also directly.)
3. Prove the following Brill-Noether's Formula: If D and E are divisors on M such that $D + E \sim K$, then $i(D) - i(E) = \frac{1}{2}(\deg D - \deg E)$.
4. If M is hyperelliptic and is given by the equation

$$y^2 = (x - \alpha\ ) \cdots (x - \alpha_{2g+1}) \quad (\alpha_j \neq \alpha_k \quad \text{for } j \neq k)$$

(see Proposition 4.2.14), then (1) $\{p_1 = (\alpha_1, 0), \cdots, p_{2g+1} = (\alpha_{2g+1}, 0), p_\infty = (\infty, \infty)\}$ is the set of all Weierstrass points of M, and (2) the gap sequence at every p_j is $\{1, 3, 5, \ldots, 2g - 1\}$.

5. Compute a basis of $\Gamma(K)$ for (a nonsingular model of the closure in $\mathbb{P}^2$ of) the curves (1) $C = \{y^3 = x^7 - 1\}$ and (2) $\hat{C} = \{y^3 = x^8 - 1\}$.

6. (1) Prove that the curve $C = \{y^3(x^3 + 1) = x^3 - 1\}$ (respectively $\hat{C} = \{y^3 = x^6 - 1\}$) is nonhyperelliptic and genus 3. (2) Compute a basis of $\Gamma(K)$ of C (respectively $\hat{C}$). (3) Prove that the canonical curve C_K of C (respectively $\hat{C}$) can be written as $C_K = Q \cap R$, where Q is a nonsingular quadric surface (respectively a quadric cone) and R is a cubic surface.

7. Let C_K be the canonical curve ($\subset \mathbb{P}^4$) of a nonhyperelliptic M of genus 5. Suppose that M has no meromorphic function of degree 3. Then, C_K is the complete intersection of 3 quadric hypersurfaces: $C_K = Q_1 \cap Q_2 \cap Q_3$.

8. Let $n_1 = 1 < n_2 < \cdots < n_g$ be the gap sequence at a point $p \in M$. Then $j + \ell(n_j p) = n_j + 1$ for $1 \leq j \leq g$.

9. Let $\{\omega_1, \ldots, \omega_g\}$ and $\{\phi_1, \ldots, \phi_g\}$ be two bases of $\Gamma(K)$. Then (1) the Wronskian form $W(\omega_1, \ldots, \omega_g)$ is not the zero section of $\frac{1}{2}g(g + 1)K$ and (2) $W(\omega_1, \ldots, \omega_g) = \det(a_{jk})W(\phi_1, \ldots, \phi_g)$, where $(\omega_1, \ldots, \omega_g) = (\phi_1, \ldots, \phi_g)(a_{jk})$.

10. Prove the following <u>theorem of Hurwitz</u>: Let $g \geq 2$. Let w (respectively #Aut (M)) be the number of Weierstrass points on M (respectively the order of Aut (M)). Then (1) $2g + 2 \leq w \leq (g - 1)g(g + 1)$, (2) $w = 2g + 2$ if and only if M is hyperelliptic, and (3) $\#\mathrm{Aut}(M) \leq 84(g - 1)$.

5.2 JACOBIAN VARIETY AND ABEL'S THEOREM

We first recall de Rham's Theorem. Let M be a paracompact differentiable manifold. For an integer $p \geq 0$, put

$$Z^p(M) = \{\text{complex valued differentiable closed p-forms on M}\},$$

$$B^p(M) = \{\text{complex valued differentiable exact p-forms on M}\},$$

$$H^p_{DM}(M) = \frac{Z^p(M)}{B^p(M)}$$

$H^p_{DM}(M)$ is called the <u>p-th de Rham group of</u> M.

Theorem 5.2.1 (de Rham). $H^p_{DM}(M)$ is canonically isomorphic to $H^p(M, \mathbb{C})$. The exterior product

$$(\phi, \psi) \in H^p_{DM}(M) \times H^q_{DM}(M) \to \phi \wedge \psi \in H^{p+q}_{DM}(M)$$

corresponds to the cup product $H^p(M, \mathbb{C}) \times H^q(M, \mathbb{C}) \to H^{p+q}(M, \mathbb{C})$.

For the proof of this theorem, see de Rham [26] or Kodaira-Morrow [57]. The isomorphism in the theorem is given as follows: Let ϕ be a closed p-form. Then, by Stokes' Theorem, the map

$$\gamma \in \{\text{p-cycles on M}\} \to \int_\gamma \phi \in \mathbb{C}$$

induces a homomorphism

$$\hat{\phi}\colon H_p(M, \mathbb{Z}) \to \mathbb{C}$$

Then

$$\phi(\text{mod } B^p(M)) \in H^p_{DM}(M) \to \hat{\phi} \in \text{Hom}\,(H_p(M, \mathbb{Z}), \mathbb{C}) = H^p(M, \mathbb{C})$$

gives the above isomorphism, provided M is compact.

Lemma 5.2.2. Let M be a compact Riemann surface of genus g. Then

1. a holomorphic 1-form on M is a closed form,
2. if $\omega \in \Gamma(K)$ satisfies that $\int_\gamma \omega = 0$ for any 1-cycle γ, then $\omega = 0$,
3. if $\omega \in \Gamma(K)$ satisfies that $\text{Re}\left(\int_\gamma \omega\right) = 0$ for any 1-cycle, then $\omega = 0$, ($\text{Re}(\xi)$ is the real part of $\xi \in \mathbb{C}$),
4. the map $\omega \in \Gamma(K) \to \left(\gamma \to \text{Re}\left(\int_\gamma \omega\right)\right) \in \text{Hom}\,(H_1(M, \mathbb{Z}), \mathbb{R}) = H^1(M, \mathbb{R})$ is a $\mathbb{R}$-linear isomorphism, and
5. if $\{\omega_1, \ldots, \omega_g\}$ is a basis of $\Gamma(K)$, then $\{\omega_1, \ldots, \omega_g, \bar{\omega}_1, \ldots, \bar{\omega}_g\}$ is a basis of $H^1_{DM}(M)$, ($\bar{\omega}$ is the complex conjugate of ω).

Proof. To prove 1, $d\omega = \partial\omega + \bar{\partial}\omega = \partial\omega$ is of type (2, 0), so $d\omega = 0$, for $\dim M = 1$. To prove 2, take a point $p_0 \in M$. If $\int_\gamma \omega = 0$ for any 1-cycle γ, then

$$h(p) = \int_{p_0}^{p} \omega$$

is independent of the choice of the path from p_0 to p and is a holomorphic function on M, so is the constant 0 ($=h(p_0)$). Hence, $\omega = dh = 0$. To prove 5, it suffices to show that, for ω and η in $\Gamma(K)$, if $\int_\gamma(\omega + \bar{\eta}) = 0$ for any 1-cycle γ, then $\omega = \eta = 0$. Take a point $p_0 \in M$. Then

$$h(p) = \int_{p_0}^{p} (\omega + \bar{\eta})$$

is a well-defined differentiable function on M. Since $\int_{p_0}^{p} \omega$ $\left(\text{respectively } \int_{p_0}^{p} \bar{\eta}\right)$ is locally well defined and holomorphic (respectively anti-holomorphic), we have $\partial\bar{\partial}h = 0$, so Re (h) and Im (h) are <u>harmonic functions</u>. By the Maximum Principle (see Ahlfors [3]), h is the constant 0 ($=h(p_0)$). Hence, $dh = \omega + \bar{\eta} = 0$. Since ω (respectively $\bar{\eta}$) is of type (1, 0) (respectively (0, 1)), we get $\omega = \eta = 0$. 3 can be proved in a similar way to 5. 4 is an easy consequence of 3. Q.E.D.

Henceforth, let M be a compact Riemann surface of genus g. Let ω be an Abelian differential on M. For a 1-cycle γ on M not passing through the poles of ω,

$$\int_\gamma \omega$$

is called the <u>period of</u> ω <u>along</u> γ. This depends only on the homology class $[\gamma] \in H_1(M, \mathbb{Z})$ to which γ belongs.

Let $\{\omega_1, \ldots, \omega_g\}$ be a basis of $\Gamma(K)$ and $\{\gamma_1, \ldots, \gamma_{2g}\}$ be a (free) basis of $H_1(M, \mathbb{Z})$. The $(2g \times g)$-matrix

$$\Omega = \begin{pmatrix} \int_{\gamma_1} \omega_1 & \cdots & \int_{\gamma_1} \omega_g \\ \cdots & & \\ \int_{\gamma_{2g}} \omega_1 & \cdots & \int_{\gamma_{2g}} \omega_g \end{pmatrix}$$

is called a <u>period matrix</u>.

<u>Lemma 5.2.3</u>. The 2g row vectors

$$\left(\int_{\gamma_k} \omega_1, \ldots, \int_{\gamma_k} \omega_g\right) \in \mathbb{C}^g, \qquad 1 \leq k \leq 2g$$

are linearly independent over $\mathbb{R}$.

This lemma follows from de Rham's Theorem, 5 of Lemma 5.2.2, and the following lemma, whose proof is straightforward.

Lemma 5.2.4. The 2g row vectors of a $(2g \times g)$-matrix A are linearly independent over $\mathbb{R}$ if and only if $\det(A\bar{A}) \neq 0$.

By Lemma 5.2.3,

$$J(M) = \mathbb{C}^g/\Omega$$

is a complex g-torus, called the Jacobian variety of M.

If $\{\omega_1, \ldots, \omega_g\}$ is changed to another basis $\{\omega_1', \ldots, \omega_g'\}$ of $\Gamma(K)$, where

$$(\omega_1', \ldots, \omega_g') = (\omega_1, \ldots, \omega_g)B \qquad (B \in GL(g, \mathbb{C}))$$

and $\{\gamma_1, \ldots, \gamma_{2g}\}$ is changed to another basis $\{\gamma_1, \ldots, \gamma_{2g}\}$ of $H_1(M, \mathbb{Z})$, where

$$(\gamma_1', \ldots, \gamma_{2g}') = (\gamma_1, \ldots, \gamma_{2g})\,{}^tA \qquad (A \in GL(2g, \mathbb{Z}))$$

then the period matrix Ω' corresponding to these new bases satisfies

$$\Omega' = A\Omega B$$

Hence, $\mathbb{C}^g/\Omega'$ is biholomorphic to $\mathbb{C}^g/\Omega$ (see Exercise 13 of Sec. 3.1). Thus the Jacobian variety is uniquely determined up to holomorphic isomorphisms.

Let $\{\alpha_1, \ldots, \alpha_g, \beta_1, \ldots, \beta_g\}$ be a symplectic basis of $H_1(M, \mathbb{Z})$ (see Sec. 4.1). Let Ω be the period matrix with respect to this basis and a basis $\{\omega_1, \ldots, \omega_g\}$ of $\Gamma(K)$. Put

$$J = \begin{pmatrix} 0 & I_g \\ -I_g & 0 \end{pmatrix}$$

where I_g is the $(g \times g)$-identity matrix. Then

Theorem 5.2.5 (Riemann's Bilinear Relations).

$$ {}^t\Omega J \Omega = 0 \qquad \text{(I)}$$

$$\sqrt{-1}\,{}^t\Omega J \bar{\Omega} > 0 \text{ (positive definite hermitian matrix)} \qquad \text{(II)}$$

Proof. Let $\{\alpha_1^*, \ldots, \alpha_g^*, \beta_1^*, \ldots, \beta_g^*\}$ be the basis of $H^1(M, \mathbb{C})$ dual to $\{\alpha_1, \ldots, \alpha_g, \beta_1, \ldots, \beta_g\}$. Then

$$\begin{aligned} &\alpha_j^* \cdot \alpha_k^* = \beta_k^* \cdot \beta_k^* = 0, \\ &\alpha_j^* \cdot \beta_k^* = -\beta_k^* \cdot \alpha_j^* = \delta_{jk} \end{aligned} \tag{1}$$

where the product is the cup product. Note that, for $\omega \in \Gamma(K)$,

$$\omega \sim c_1\alpha_1^* + \cdots + c_g\alpha_g^* + d_1\beta_1^* + \cdots + d_g\beta_g^* \text{ (cohomologous)} \tag{2}$$

where

$$c_j = \int_{\alpha_j} \omega \quad \text{and} \quad d_j = \int_{\beta_j} \omega, \qquad 1 \leq j \leq g$$

Now we have the relations

$$\int_M \omega_j \wedge \omega_k = 0 \tag{3}$$

$$\sqrt{-1} \int_M \omega \wedge \bar{\omega} > 0 \quad \text{for any nonzero } \omega \in \Gamma(K) \tag{4}$$

Then, (3) (respectively (4)), de Rham's Theorem, (1), and (2) imply I (respectively II). Q.E.D.

Corollary 5.2.6. Let Ω be the period matrix in Theorem 5.2.5. Write $\Omega = \begin{pmatrix} \Omega_1 \\ \Omega_2 \end{pmatrix}$, where Ω_1 and Ω_2 are $(g \times g)$-matrices. Then $\det \Omega_1 \neq 0$.

Proof. For $x = (x_1, \ldots, x_g) \in \mathbb{C}^g$, suppose that $\Omega_1 {}^t x = 0$. Then

$$\sqrt{-1}\, x\, {}^t\Omega J \bar{\Omega}\, {}^t\bar{x} = \sqrt{-1}\, x\, {}^t\Omega_1 \bar{\Omega}_2\, {}^t\bar{x} - \sqrt{-1}\, x\, {}^t\Omega_2 \bar{\Omega}_1\, {}^t\bar{x} = 0$$

Hence, $x = 0$ by the bilinear relation II. Q.E.D.

Using this corollary, put

$$Z = \Omega_2 \Omega_1^{-1}$$

Then $\binom{I_g}{Z}$ is the period matrix with respect to the basis $\{\alpha_1, \ldots, \alpha_g, \beta_1, \ldots, \beta_g\}$ and $\{\omega'_1, \ldots, \omega'_g\}$, where

$$(\omega'_1, \ldots, \omega'_g) = (\omega_1, \ldots, \omega_g)\Omega_1^{-1}$$

In this case, the bilinear relations can be written as

$$^tZ = Z \tag{I}$$

$$\mathrm{Im}\, Z > 0 \quad \text{(positive definite)} \tag{II}$$

where Im Z is the imaginary part of the matrix Z.

Definition 5.2.7. $\mathbb{H}_g = \{Z \mid Z$ is a $(g \times g)$-matrix such that $^tZ = Z$ and $\mathrm{Im}\, Z > 0\}$ is called the g-dimensional Siegel's upper half space.

$\mathbb{H}_g$ is a $\frac{1}{2}g(g+1)$-dimensional complex manifold. Note that the period matrix $\binom{I}{Z}$ is uniquely determined by $\{\alpha_1, \ldots, \alpha_g, \beta_1, \ldots, \beta_g\}$ and is independent of the choice of a basis $\{\omega'_1, \ldots, \omega'_g\}$ of $\Gamma(K)$ as above. But, if the symplectic basis $\{\alpha_1, \ldots, \alpha_g, \beta_1, \ldots, \beta_g\}$ is changed to another one $\{\alpha'_1, \ldots, \alpha'_g, \beta'_1, \ldots, \beta'_g\}$, then $\binom{I}{Z}$ is changed. Put

$$(\alpha'_1, \ldots, \alpha'_g, \beta'_1, \ldots, \beta'_g) = (\alpha_1, \ldots, \alpha_g, \beta_1, \ldots, \beta_g)X$$

where $X \in GL(2g, \mathbb{Z})$. Then X clearly satisfies

$$^tXJX = J$$

That is, X is an element of

$$Sp(g, \mathbb{Z}) = \{X \in GL(2g, \mathbb{Z}) \mid {}^tXJX = J\}$$

(Note that, if $X \in Sp(g, \mathbb{Z})$, then $\det X = 1$.) Write

$$X = \begin{pmatrix} A & B \\ C & D \end{pmatrix} \quad \text{(A, B, C, D: } (g \times g)\text{-matrices)}$$

The period matrix $\binom{I}{Z}$ is then changed to $\binom{I}{Z'}$, where

$$Z' = (C + DZ)(A + BZ)^{-1} \tag{1}$$

Conversely, for any $X \in Sp(g, \mathbb{Z})$, if Z' is defined by (1), then $\begin{pmatrix} I \\ Z' \end{pmatrix}$ is a period matrix of J(M).

Thus, for a compact Riemann surface M, a point

$$i(M) = Z (\text{mod } Sp(g, \mathbb{Z})) \in \mathbb{H}_g / Sp(g, \mathbb{Z})$$

is uniquely determined, where the action of $Sp(g, \mathbb{Z})$ on $\mathbb{H}_g$ is given by (1). But not every point of $\mathbb{H}_g / Sp(g, \mathbb{Z})$ corresponds to some compact Riemann surface of genus g, if $g \geq 4$.

A beautiful theorem by Torelli says

Theorem 5.2.8 (Torelli). Compact Riemann surfaces M and M' of genus g are biholomorphic if and only if $i(M) = i(M') \in \mathbb{H}_g / Sp(g, \mathbb{Z})$.

For the proof, see, for example, Griffiths-Harris [33, p. 359].

It is known that, for every point $Z \in \mathbb{H}_g$, the complex g-torus

$$\frac{\mathbb{C}^g}{\begin{pmatrix} I \\ Z \end{pmatrix}}$$

is an Abelian variety (that is, is projective), so J(M) is an Abelian variety. In general,

Theorem 5.2.9. A complex g-torus $\mathbb{C}^g/\Omega$ is an Abelian variety if and only if there is a nonsingular skew symmetric integral $(2g \times 2g)$-matrix Q such that (1) ${}^t\Omega Q^{-1} \Omega = 0$ and (2) $\sqrt{-1}\, {}^t\Omega Q^{-1} \bar{\Omega} > 0$ (positive definite hermitian).

This beautiful theorem is fundamental in the theory of Abelian varieties. For the proof, one needs the theory of theta functions (see Siegel [93]). For another proof using Kodaira's imbedding theorem, see, for example, Griffiths-Harris [33].

The nonsingular skew symmetric integral $(2g \times 2g)$-matrix Q in the theorem is called a polarization of the complex torus $\mathbb{C}^g/\Omega$. The pair $(\mathbb{C}^g/\Omega, Q)$ is called a polarized Abelian variety. If

$$Q = J = \begin{pmatrix} 0 & I \\ -I & 0 \end{pmatrix}$$

then $(\mathbb{C}^g/\Omega, J)$ is called a principally polarized Abelian variety. J(M) is such an example.

The Riemann theta function $\vartheta = \vartheta(Z, \zeta)$ on $\mathbb{H}_g \times \mathbb{C}^g$ is defined by

$$\vartheta(Z, \zeta) = \sum_{\eta \in \mathbb{Z}^g} \exp \left\{ 2\pi\sqrt{-1} \left(\tfrac{1}{2} \eta Z {}^t\eta + \eta {}^t\zeta \right) \right\}$$

for $(Z, \zeta) \in \mathbb{H}_g \times \mathbb{C}^g$. Then $\vartheta(Z, \zeta)$ is a holomorphic function on $\mathbb{H}_g \times \mathbb{C}^g$ and satisfies the following identity

$$\vartheta(Z, \zeta + mZ + n) = \exp\{2\pi\sqrt{-1}(-\tfrac{1}{2}mZ{}^tm - m{}^t\zeta)\}\vartheta(Z, \zeta) \qquad (*)$$

for $m, n \in \mathbb{Z}^g$. For a fixed $Z \in \mathbb{H}_g$, the holomorphic function

$$\vartheta(\zeta) = \vartheta(Z, \zeta), \qquad \zeta \in \mathbb{C}^g$$

on $\mathbb{C}^g$ is also called the Riemann theta function. Note that by (*),

$$\Theta = \{\zeta \in \mathbb{C}^g/\Omega \mid \vartheta(\zeta) = 0\} \quad \left(\Omega = \binom{I}{Z}\right)$$

is a well-defined irreducible hypersurface of $\mathbb{C}^g/\Omega$ and is called the theta divisor of $(\mathbb{C}^g/\Omega, J)$.

Next, we talk about Abel's theorem. For points $(p_1, \ldots, p_n)$ and $(q_1, \ldots, q_n)$ of M^n, we define the following equivalence relation:

$$(p_1, \ldots, p_n) \sim (q_1, \ldots, q_n)$$

if there is a permutation σ of $1, \ldots, n$ such that $q_j = p_{\sigma(j)}$ for $1 \leq j \leq n$. $M^n/\sim\, = M^n/\mathfrak{S}_n$, where $\mathfrak{S}_n$ is the symmetric group of n letters $1, \ldots, n$.

$$S^nM = \frac{M^n}{\mathfrak{S}_n}$$

is called the n-th symmetric product of M. S^nM can be naturally identified with the set of all positive divisors on M of degree n:

$$(p_1, \ldots, p_n)(\text{mod } \mathfrak{S}_n) \longleftrightarrow D = p_1 + \cdots + p_n$$

Proposition 5.2.10. S^nM is a n-dimensional compact complex manifold.

Proof (Sketch). For a divisor $D_0 \in S^nM$, write

$$D_0 = a_1p_1 + \cdots + a_sp_s \qquad (p_j \neq p_k \text{ for } j \neq k,\ \Sigma a_j = n)$$

Let U_j be a small neighborhood of p_j in M and t_j be a local uniformizing parameter at p_j. We may suppose that

$$U_j = \{t_j \in \mathbb{C} \mid |t_j| < \epsilon\}$$

Consider a_j-copies $(U_{j\alpha}, t_{j\alpha})$, $1 \leq \alpha \leq a_j$, of (U_j, t_j). Put

$$z_{j1} = t_{j1} + \cdots + t_{ja_j},$$

$$z_{j2} = (t_{j1})^2 + \cdots + (t_{ja_j})^2,$$

$$\cdots$$

$$z_{ja_j} = (t_{j1})^{a_j} + \cdots + (t_{ja_j})^{a_j}$$

Then the map

$$\psi_j: (t_{j1}, \ldots, t_{ja_j}) \in U_{j1} \times \cdots \times U_{ja_j} \to (z_{j1}, \ldots, z_{ja_j}) \in \mathbb{C}^{a_j}$$

induces a homeomorphism $\hat{\psi}_j$ of $S^{a_j}U_j$ (the a_j-th symmetric product of U_j) onto an open set of $\mathbb{C}^{a_j}$. We take $\hat{\psi}_1 \times \cdots \times \hat{\psi}_s$ as a local chart of S^nM. Then S^nM becomes an n-dimensional compact complex manifold. (The details are left to the reader.) Q.E.D.

Lemma 5.2.11. If D is a divisor on M such that $\deg D = n$, then the natural injection

$$i: E \in |D| \to E \in S^nM$$

is a holomorphic map.

Proof. Let $\{\xi_0, \ldots, \xi_r\}$ be a basis of $\Gamma([D])$. Every $\xi \in \Gamma([D])$ can be written as

$$\xi = X_0\xi_0 + \cdots + X_r\xi_r$$

Then $(X_0 : \cdots : X_r)$ is the homogeneous coordinate of $E = (\xi) \in |D|$. Take

$$\xi^0 = X_0^0\xi_0 + \cdots + X_r^0\xi_r \quad \text{and} \quad D_0 = (\xi^0)$$

We use the notations in the proof of Proposition 5.2.10. Let $\xi_\nu(t_j)$ be the holomorphic function on U_j representing ξ_ν, $0 \leq \nu \leq r$. Let Γ be a small oriented circle around p_j. Then, there is a neighborhood U of $(X_0^0 : \cdots : X_r^0)$ in $|D|$ such that

$$X_0\xi_0(t_j) + \cdots + X_r\xi_r(t_j) \neq 0 \quad \text{for } t_j \in \Gamma \quad \text{and} \quad (X_0 : \cdots : X_r) \in U$$

Then, for k = 1, 2, ...,

$$z_{jk} = (t_{j1})^k + \cdots + (t_{ja_j})^k = \frac{1}{2\pi\sqrt{-1}} \int_\Gamma \frac{t_j^k(X_0\xi_0'(t_j) + \cdots + X_r\xi_r'(t_j))}{X_0\xi_0(t_j) + \cdots + X_r\xi_r(t_j)} dt_j$$

is a holomorphic function of $(X_1/X_0, \ldots, X_r/X_0) \in U$. Q.E.D.

We show later that i: $|D| \to S^nM$ is in fact a holomorphic imbedding (see Corollary 5.2.23).

Now, we fix a point $p_0 \in M$ once for all. Let $\{\omega_1, \ldots, \omega_g\}$ be a basis of $\Gamma(K)$. Then

$$\Gamma = \left\{ \left(\int_\gamma \omega_1, \ldots, \int_\gamma \omega_g \right) \middle| \ \gamma \text{ is a 1-cycle on M} \right\}$$

is a free additive group such that $J(M) = \mathbb{C}^g/\Gamma$. Consider the map

$$\phi = \phi_n : D = p_1 + \cdots + p_n \in S^nM$$

$$\to \left(\int_{p_0}^{p_1} \omega_1 + \cdots + \int_{p_0}^{p_n} \omega_1, \ldots, \int_{p_0}^{p_1} \omega_g + \cdots + \int_{p_0}^{p_r} \omega_g \right) \bmod \Gamma$$

$$\in J(M)$$

ϕ is called the (n-th) Jacobi map. It is easy to see (compare the proof of Proposition 5.2.22 below) that

Proposition 5.2.12. $\phi : S^nM \to J(M)$ is a holomorphic map.

Now we can state Abel's Theorem.

Theorem 5.2.13 (Abel). For D and E in S^nM, $D \sim E$ (linearly equivalent) if and only if $\phi(D) = \phi(E)$.

For the proof of this important theorem, we need some preparations. An Abelian differential η on M is said to be of the second kind (respectively the third kind) if (1) η has a pole and (2) $\mathrm{Res}_p(\eta) = 0$ for any pole p of η

(respectively (2) η has the order 1 at every pole). For a nonconstant meromorphic function f on M, its total differential df is of the second kind.

Lemma 5.2.14. For any point $p \in M$ and any integer m with $m \geq 2$, there is an Abelian differential $\eta = \eta_p^{(m)}$ of the second kind such that $D_\infty(\eta) = mp$, where $D_\infty(\eta)$ is the polar divisor of η. Moreover, such $\eta = \eta_p^{(m)}$ can be chosen so that

$$\eta = \left(\frac{1}{t^m} + a_0 + a_1 t + \cdots\right) dt$$

where t is a given local uniformizing parameter at p.

Proof. By the definition,

$$A(-mp) = \{\eta \mid \eta \text{ is an Abelian differential on M such that } D_\infty(\eta) \leq mp\}$$

Note that $\Gamma(K) \subset A(-mp)$. By the Riemann-Roch Theorem,

$$i(-mp) = \dim A(-mp) = m + g - 1 + \ell(-mp) = m + g - 1$$

for $m \geq 1$. If $m = 1$, then $i(-mp) = g$, so $A(-p) = \Gamma(K)$. This means that there is no Abelian differential η such that $D_\infty(\eta) = p$. (This is also clear from the Residue Theorem, Theorem 5.1.3.)

If $m \geq 2$, then

$$i(-mp) - i(-(m-1)p) = 1$$

Hence, there is η such that $D_\infty(\eta) = mp$. η is of the second kind by the residue theorem.

The last assertion is trivial. Q.E.D.

The proof of the lemma implies

Corollary 5.2.15. For any point $p \in M$ and any integer m (≥ 2), take $\eta_p^{(m)} \in A(-mp)$ such that $D_\infty(\eta_p^{(m)}) = mp$. Then the vector space $A(-mp)$ is spanned by $\eta_p^{(m)}, \eta_p^{(m-1)}, \ldots, \eta_p^{(2)}$ and $\Gamma(K)$.

In a similar way,

Lemma 5.2.16. For any distinct points p and q on M, there is an Abelian differential $\eta = \eta_{p,q}$ of the third kind such that $D_\infty(\eta) = p + q$.

Proof. Note that

$$A(-p - q) = \{\eta \mid D_\infty(\eta) \leq p + q\} \supset \Gamma(K)$$

By the Riemann-Roch Theorem

$$i(-p - q) = \dim A(-p - q) = g + 1 > g = \dim \Gamma(K) \qquad \text{Q.E.D.}$$

Corollary 5.2.17. For any distinct points p and q on M, take $\eta_{p,q} \in A(-p - q)$ such that $D_\infty(\eta_{p,q}) = p + q$. Then the vector space $A(-p - q)$ is spanned by $\eta_{p,q}$ and $\Gamma(K)$. Moreover, such $\eta_{p,q}$ can be uniquely chosen so that

1. $\mathrm{Res}_p\, \eta_{p,q} = 1$,
2. $\mathrm{Res}_q\, \eta_{p,q} = -1$, and
3. $\mathrm{Re}\left(\int_\gamma \eta_{p,q}\right) = 0$ for any 1-cycle γ passing through neither p nor q.

(Re (z) is the real part of $z \in \mathbb{C}$.)

Proof. The first assertion is clear from the proof of the lemma. Let us prove the second assertion. Take any $\hat{\eta} \in A(-p - q)$ such that $D_\infty(\hat{\eta}) = p + q$. Put $a = \mathrm{Res}_p\, \hat{\eta}$. Then, by the residue theorem, $\mathrm{Res}_q\, \hat{\eta} = -a$. Since $\hat{\eta} \notin \Gamma(K)$, $a \neq 0$. By replacing $\hat{\eta}$ by $\hat{\eta}/a$, we may assume that

$$\mathrm{Res}_p\, \hat{\eta} = 1 \quad \text{and} \quad \mathrm{Res}_q\, \hat{\eta} = -1$$

By 4 of Lemma 5.2.2, the map

$$\alpha\colon \omega \in \Gamma(K) \to \left(\gamma \to \mathrm{Re}\left(\int_\gamma \omega\right)\right) \in H^1(M, \mathbb{R})$$

is a $\mathbb{R}$-linear isomorphism. Hence, there is $\omega \in \Gamma(K)$ such that

$$\mathrm{Re}\left(\int_\gamma \omega\right) = \mathrm{Re}\left(\int_\gamma \hat{\eta}\right)$$

for any 1-cycle γ passing through neither p nor q. Put

$$\eta_{p,q} = \hat{\eta} - \omega$$

Then $\eta_{p,q}$ satisfies conditions (1)-(3). The uniqueness of such $\eta_{p,q}$ follows from the fact that the above map α is a $\mathbb{R}$-linear isomorphism. Q.E.D.

Now, we are ready to prove Abel's Theorem. By Lemma 5.2.11, the natural injection $i: |D| \to S^n M$ is holomorphic. Hence,

$$\phi \cdot i : |D| = \mathbb{P}^r \to J(M)$$

is holomorphic, so is a constant map by Exercise 14 of Sec. 3.1.

Conversely, suppose that $\phi(D) = \phi(E)$. We show that there is a meromorphic function f on M such that

$$D - E = (f)$$

We may suppose that the supports of D and E are disjoint. Put

$$D = p_1 + \cdots + p_n \quad \text{and} \quad E = q_1 + \cdots + q_n$$

We may assume that $p_0 \neq p_1, \ldots, p_n, q_1, \ldots, q_n$. Let $\eta_{(j)} = \eta_{p_j, q_j}$ be the Abelian differential of the third kind satisfying (1)-(3) of Corollary 5.2.17 with respect to $p = p_j$ and $q = q_j$. Put

$$\eta = \eta_{(1)} + \cdots + \eta_{(n)} \quad \text{and}$$

$$f(p) = \exp \int_{p_0}^{p} \eta$$

where the path of the integration passes through none of $p_1, \ldots, p_n, q_1, \ldots, q_n$. Then f is a meromorphic function on M which satisfies the condition.

To show this, we first prove that f is single valued. It suffices to show that

$$\frac{1}{2\pi\sqrt{-1}} \int_{\gamma} \eta \in \mathbb{Z}$$

for any 1-cycle γ passing through none of $p_1, \ldots, p_n, q_1, \ldots, q_n$.

Let $\{\alpha_1, \ldots, \alpha_g, \beta_1, \ldots, \beta_g\}$ be a symplectic basis of $H_1(M, \mathbb{Z})$ as in Sec. 4.1. (See Fig. 5.7.) Let Δ be the simply connected domain as in the picture. We may suppose that none of $p_0, p_1, \ldots, p_n, q_1, \ldots, q_n$ is on the boundary

$$\partial\Delta = \alpha_1 \cup \cdots \cup \alpha_n \cup \beta_1 \cup \cdots \cup \beta_n$$

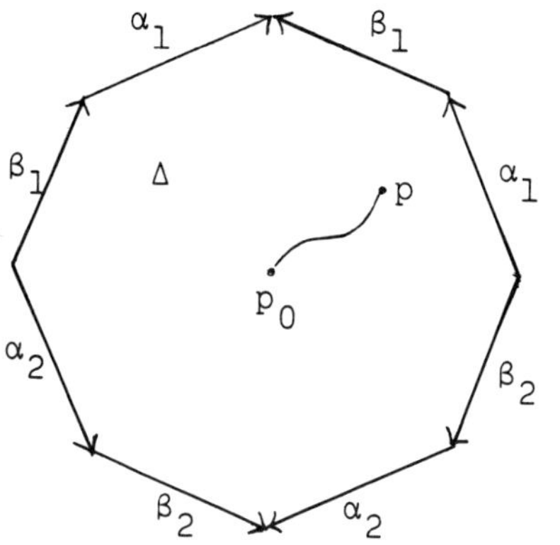

FIGURE 5.7

of Δ. Take $\omega \in \Gamma(K)$ and consider the function

$$h(p) = \int_{p_0}^{p} \omega$$

Since Δ is simply connected, h is a single-valued holomorphic function on Δ.

We apply the Residue Theorem for $h\eta$ and get

$$\frac{1}{2\pi\sqrt{-1}} \int_{\partial\Delta} h\eta = \sum_j h(p_j) - \sum_j h(q_j) = \sum_j \int_{p_0}^{p_j} \omega - \sum_j \int_{p_0}^{q_j} \omega$$

On the other hand, if $p \in \alpha_k$ and $p' \in \alpha_k$ as in Fig. 5.8 correspond to the same point in M,

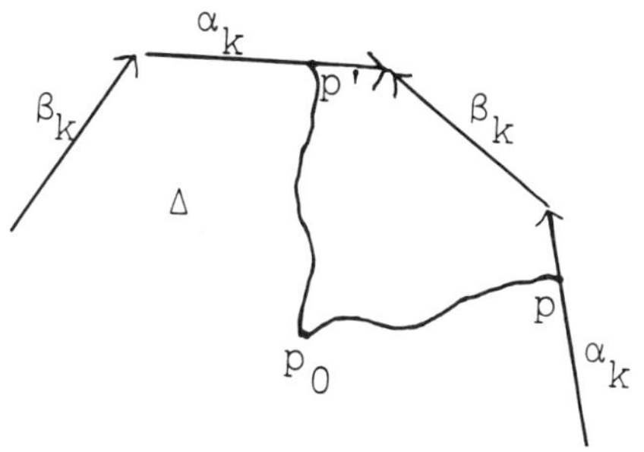

FIGURE 5.8

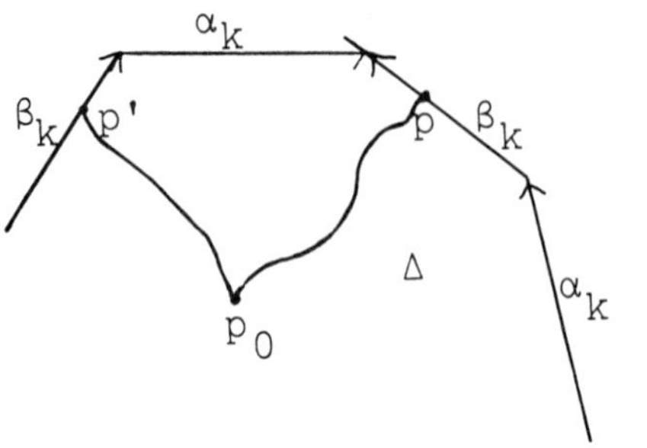

FIGURE 5.9

then

$$h(p') - h(p) = \int_{p_0}^{p'} \omega - \int_{p_0}^{p} \omega = \int_{\beta_k} \omega, \text{ a constant}$$

In a similar way, if $p \in \beta_k$ and $p' \in \beta_k$ as in Fig. 5.9 correspond to the same point in M, then

$$h(p') - h(p) = -\int_{\alpha_k} \omega, \text{ a constant}$$

Hence,

$$\int_{\partial\Delta} h\eta = \sum_k \int_{\alpha_k + \alpha_k^{-1}} h\eta + \sum_k \int_{\beta_k + \beta_k^{-1}} h\eta$$

$$= \sum_k \int_{\alpha_k} (-h(p') + h(p))\eta + \sum_k \int_{\beta_k} (-h(p') + h(p))\eta$$

$$= \sum_k \left(\int_{\alpha_k} \omega \int_{\beta_k} \eta - \int_{\beta_k} \omega \int_{\alpha_k} \eta \right)$$

Thus, we get

$$\sum_j \int_{p_0}^{p_j} \omega - \sum_j \int_{p_0}^{q_j} \omega = \frac{1}{2\pi\sqrt{-1}} \sum_k \left(\int_{\alpha_k} \omega \int_{\beta_k} \eta - \int_{\beta_k} \omega \int_{\alpha_k} \eta \right)$$

By the assumption that $\phi(D) = \phi(E)$, there is a 1-cycle

$$\gamma = c_1\alpha_1 + \cdots + c_g\alpha_g + d_1\beta_1 + \cdots + d_g\beta_g \quad (c_k, d_k \in \mathbb{Z})$$

such that

$$\left(\sum_j \int_{p_0}^{p_j} \omega_1 - \sum_j \int_{p_0}^{q_j} \omega_1, \ldots, \sum_j \int_{p_0}^{p_j} \omega_g - \sum_j \int_{p_0}^{q_j} \omega_g\right)$$

$$= c_1\left(\int_{\alpha_1} \omega_1, \ldots, \int_{\alpha_1} \omega_g\right) + \cdots + c_g\left(\int_{\alpha_g} \omega_1, \ldots, \int_{\alpha_g} \omega_g\right)$$

$$+ d_1\left(\int_{\beta_1} \omega_1, \ldots, \int_{\beta_1} \omega_g\right) + \cdots + d_g\left(\int_{\beta_g} \omega_1, \ldots, \int_{\beta_g} \omega_g\right)$$

$$= \left(\int_\gamma \omega_1, \ldots, \int_\gamma \omega_g\right)$$

This means that

$$\int_\gamma \omega = \sum_j \int_{p_0}^{p_j} \omega - \sum_j \int_{p_0}^{q_j} \omega = \frac{1}{2\pi\sqrt{-1}} \sum_k \left(\int_{\alpha_k} \omega \int_{\beta_k} \eta - \int_{\beta_k} \omega \int_{\alpha_k} \eta\right) \qquad (*)$$

for any $\omega \in \Gamma(K)$.

By 4 of Lemma 5.2.2, the map

$$\omega \in \Gamma(K) \to \left(\gamma \to \operatorname{Re}\left(\int_\gamma \omega\right)\right) \in H^1(M, \mathbb{R})$$

is a $\mathbb{R}$-linear isomorphism. Hence, we can take ω such that, say,

$$\operatorname{Re}\left(\int_{\alpha_1} \omega\right) = 1, \quad \operatorname{Re}\left(\int_{\alpha_k} \omega\right) = 0, \quad 2 \le k \le g, \quad \text{and}$$

$$\operatorname{Re}\left(\int_{\beta_j} \omega\right) = 0, \quad 1 \le j \le g$$

Then, (*) implies that

$$c_1 + \sqrt{-1}\,x = \frac{1}{2\pi\sqrt{-1}}\left(y + \int_{\beta_1} \eta\right)$$

for some real numbers x and y. (Note that $\int_{\alpha_k} \eta$ and $\int_{\beta_k} \eta$ are pure imaginary.) Comparing the real parts

$$\frac{1}{2\pi\sqrt{-1}} \int_{\beta_1} \eta = c_1 \in \mathbb{Z}$$

In a similar way, we can show that

$$\frac{1}{2\pi\sqrt{-1}} \int_{\alpha_k} \eta \in \mathbb{Z} \quad \text{and} \quad \frac{1}{2\pi\sqrt{-1}} \int_{\beta_k} \eta \in \mathbb{Z}$$

for $1 \leq k \leq g$. This proves that f is single valued.

Finally, we show that f is a meromorphic function on M such that $(f) = D - E$. It is clear that f is holomorphic on $M - \{p_1, \ldots, p_n, q_1, \ldots, q_n\}$. Write D as

$$D = ap_1 + D'$$

where a is a positive integer and D' is a divisor with $D' \geq 0$ whose support does not contain p_1. Let t be a local uniformizing parameter at p_1. Then η can be locally written as

$$\eta = \left(\frac{a}{t} + b_0 + b_1 t + \cdots\right) dt$$

around p_1. Hence, around p_1,

$$f(t) = \left(\exp \int_{p_0}^{t_0} \eta\right)\left(\exp \int_{t_0}^{t} \left(\frac{a}{t} + b_0 + b_1 t + \cdots\right) dt\right)$$

where $t_0 \neq 0$ is fixed. Note that $\exp \int_{p_0}^{t_0} \eta$ is a nonzero constant with respect to t. On the other hand,

$$\exp \int_{t_0}^{t} \left(\frac{a}{t} + b_0 + b_1 t + \cdots\right) dt = \left(\frac{t}{t_0}\right)^a \exp \int_{t_0}^{t} (b_0 + b_1 t + \cdots)\, dt$$

Hence, in a neighborhood of p_1, f(t) has a unique zero, t = 0, of order a.

A similar argument works for other points of the supports of D and E, so we get (f) = D - E.

This completes the proof of Abel's Theorem.

Example 5.2.18. First, if g = 0, then $J(\mathbb{P}^1)$ = one point. Abel's Theorem in this case says that any D and E in $S^n\mathbb{P}^1$ are linearly equivalent, so

$$S^n\mathbb{P}^1 = |n(\infty)|$$

In fact, if $D = (z_1) + \cdots + (z_n)$ and $E = (w_1) + \cdots + (w_n)$, $(z_j, w_j \in \mathbb{C})$, say, then D - E = (f), where

$$f(t) = \frac{(t - z_1) \cdots (t - z_n)}{(t - w_1) \cdots (t - w_n)}$$

Note that $D = (z_1) + \cdots + (z_n) \in S^n\mathbb{P}^1 = |n(\infty)|$ has the homogeneous coordinate $(a_0 : \cdots : a_n)$, where the equation

$$a_0 X^n + a_1 X^{n-1} + \cdots + a_n = 0$$

has the roots $z_1, \ldots, z_n$. Aut $(\mathbb{P}^1)$ acts on $|n(\infty)|$ as follows:

$$(\sigma, (z_1) + \cdots + (z_n)) \in \text{Aut}(\mathbb{P}^1) \times |n(\infty)| \to (\sigma z_1) + \cdots + (\sigma z_n) \in |n(\infty)|$$

Hence, Aut $(\mathbb{P}^1)$ also acts on $\mathbb{P}^n = |n(\infty)|^*$. Note that, for $\sigma \in \text{Aut}(\mathbb{P}^1)$, the diagram

$$\begin{array}{ccc} \mathbb{P}^1 & \xrightarrow{\sigma} & \mathbb{P}^1 \\ \Phi \downarrow & & \downarrow \Phi \\ \mathbb{P}^n & \xrightarrow{\sigma} & \mathbb{P}^n \end{array}$$

is commutative, where $\Phi = \Phi_{|n(\infty)|}$. For example, the homomorphism Aut $(\mathbb{P}^1) \to$ Aut $(\mathbb{P}^3)$ can be given by

$$\begin{pmatrix} a & b \\ c & d \end{pmatrix} \to \begin{pmatrix} a^3 & 3a^2b & 3ab^2 & b^3 \\ a^2c & 2abc + a^2d & 2abd + b^2c & b^2d \\ ac^2 & 2acd + bc^2 & 2bcd + ad^2 & bd^2 \\ c^3 & 3c^2d & 3cd^2 & d^3 \end{pmatrix}$$

Next, let $M = \mathbb{C}/(\mathbb{Z}\omega_1 + \mathbb{Z}\omega_2)$ be a complex 1-torus. If z is the standard coordinate in $\mathbb{C}$, then dz is in $\Gamma(K)$ and span $\Gamma(K)$. (See Fig. 5.10.) Taking one-cycles α and β as in the picture, the period matrix is

$$\Omega = \begin{pmatrix} \int_\alpha dz \\ \int_\beta dz \end{pmatrix} = \begin{pmatrix} \omega_1 \\ \omega_2 \end{pmatrix}$$

Hence, $J(M) = \mathbb{C}/\Omega = M$. Put $p_0 = 0$, the zero of the additive group M. The Jacobi map

$$\phi \colon S^n M \to J(M) = M$$

is then given by

$$\phi((p_1) + \cdots + (p_n)) = \sum_j \int_0^{p_j} dz \quad (\mathrm{mod}(\mathbb{Z}\omega_1 + \mathbb{Z}\omega_2))$$

$$= p_1 + \cdots + p_n$$

(the sum in the additive group M). Hence, Abel's Theorem in this case is nothing but Theorem 4.3.9. It can be shown that ϕ is a $\mathbb{P}^{n-1}$-bundle over M.

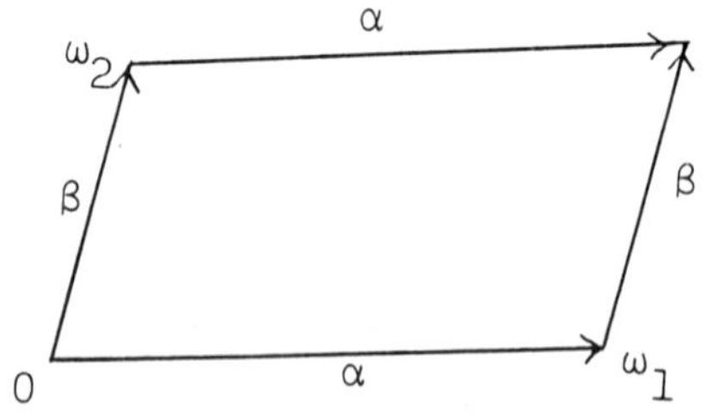

FIGURE 5.10

Theorem 5.2.19. If $g \geq 1$, then $\phi = \phi_1: S^1M = M \to J(M)$ is a holomorphic imbedding.

Proof. For distinct points p and q on M, if $\phi(p) = \phi(q)$, then $p \sim q$ by Abel's Theorem. Hence, $g = 0$, a contradiction. Hence, ϕ is injective. Note that

$$(d\phi)_p = ((\omega_1)_p, \dots, (\omega_g)_p) \quad \text{for } p \in M$$

Since $|K|$ has no fixed point, $(d\phi)_p \neq 0$ for all $p \in M$. Q.E.D.

Corollary 5.2.20. If a compact Riemann surface M has genus 1, then M is (biholomorphic to) a complex 1-torus.

Proof. By the theorem, $\phi: M \to J(M)$ is a holomorphic imbedding. Since $\dim J(M) = 1$, ϕ is in fact a holomorphic isomorphism. Q.E.D.

The equality

$$(d\phi)_p = ((\omega_1)_p, \dots, (\omega_g)_p) \quad \text{for } p \in M$$

in the proof of Theorem 5.2.19 can be expressed as

Corollary 5.2.21. The canonical map $\Phi_K: M \to \mathbb{P}^{g-1}$ is the Gauss map of $\phi: M \to J(M)$.

Here, we identify $\mathbb{P}^{g-1}$ with the projective space $\mathbb{P}(T_0 J(M))$ of all one-dimensional vector subspaces of the tangent space $T_0 J(M)$ at the zero $0 \in J(M)$. $T_p J(M)$ and $T_0 J(M)$ are also identified by the translation $t_p: q \in M \to p + q \in M$.

For a point $p \in M$, let t be a local uniformizing parameter at p. Every $\omega \in \Gamma(K)$ can be locally written as

$$\omega = h(t)\, dt$$

where h(t) is a holomorphic function of t. Put

$$\omega^{(k)}(p) = \frac{d^k h}{dt^k}(0) \quad \text{for } k = 0, 1, \dots$$

Proposition 5.2.22. For $D \in S^n M$, write

$$D = a_1 p_1 + \cdots + a_s p_s$$

where a_j are positive integers such that $\Sigma a_j = n$, and $p_j \neq p_k$ for $j \neq k$. Then

$$(d\phi)_D = \begin{pmatrix} \omega_1^{(0)}(p_1) \cdots \omega_1^{(a_1-1)}(p_1) & \cdots & \omega_1^{(0)}(p_s) \cdots \omega_1^{(a_s-1)}(p_s) \\ \cdots & & \\ \omega_g^{(0)}(p_1) \cdots \omega_g^{(a_1-1)}(p_1) & \cdots & \omega_g^{(0)}(p_s) \cdots \omega_g^{(a_s-1)}(p_s) \end{pmatrix} \tag{1}$$

up to nonzero constant. (This matrix is called the Brill-Noether matrix of D.)

$$\text{rand } (d\phi)_D = n - \dim |D| \tag{2}$$

Proof. To prove (1), let t be a local uniformizing parameter at P_1 in a neighborhood U in M. For $\omega \in \Gamma(K)$, the Abelian integral

$$w(t) = \int_{p_0}^{t} \omega$$

can be expanded as

$$w(t) = w(0) + w'(0)t + \frac{1}{2!} w''(0)t^2 + \cdots + \frac{1}{a_1!} w^{(a_1)}(0)t^{a_1} + \cdots$$

$$= w(0) + \omega^{(0)}(p_1)t + \frac{1}{2!}\omega^{(1)}(p_1)t^2 + \cdots + \frac{1}{a_1!}\omega^{(a_1-1)}(p_1)\, t^{a_1} + \cdots$$

Let $(U_1, t_1), \ldots, (U_{a_1}, t_{a_1})$ be a_1-copies of (U, t). Then

$$w(t_1) + \cdots + w(t_{a_1})$$

$$= a_1 w(0) + \omega^{(0)}(p_1)z_1 + \frac{1}{2!}\omega^{(1)}(p_1)z_2 + \cdots + \frac{1}{a_1!}\omega^{(a_1-1)}(p_1)z_{a_1} + \cdots$$

where $z_k = t_1^k + \cdots + t_{a_1}^k$. (Here, $\cdots$ indicates the higher order terms with respect to $(z_1, \ldots, z_{a_1})$.)

Note that $(z_1, \ldots, z_{a_1})$ is a part of the local coordinate at D in $S^n M$ (see the proof of Proposition 5.2.10). Hence, $(d\phi)_D$ can be given by the above matrix up to nonzero constant.

To prove (2), take $(c_1, \ldots, c_g) \in \mathbb{C}^g$. Then, by (1),

$$^t(d\phi)_D\,^t(c_1, \ldots, c_g) = 0$$

if and only if

$$c_1\omega_1 + \cdots + c_g\omega_g \in A(D)$$

Hence, by the Riemann-Roch Theorem,

$$\text{rank}\,(d\phi)_D = \text{rank}\,^t(d\phi)_D = g - i(D) = n - \dim |D| \qquad \text{Q.E.D.}$$

Corollary 5.2.23. The natural injection i: $|D| \to S^nM$ is a holomorphic imbedding. Its image is a fiber of ϕ: $S^nM \to J(M)$, that is, $|D| = \phi^{-1}\phi(D)$.

Proof. The injection i is holomorphic by Lemma 5.2.11, so $i(|D|)$ is an irreducible analytic set of dimension $r = \dim |D|$ in S^nM. By Abel's Theorem, $i(|D|)$ is a fiber $\phi^{-1}(x)$ of ϕ: $S^nM \to J(M)$ $(x = \phi(D))$. By Proposition 5.2.22 and Exercise 12 of Sec. 3.2, $\phi^{-1}(x)$ is nonsingular. Hence, by Exercise 13 of Sec. 3.2, i: $|D| \to \phi^{-1}(x)$ is biholomorphic. Q.E.D.

Lemma 5.2.24. (1) If $n \leq g$, then $\dim |D| = 0$ for a general $D \in S^nM$. (2) If $n \geq g$, then $\dim |D| = n - g$ for a general $D \in S^nM$.

Proof. To prove (1), by the Riemann-Roch Theorem, $\dim |D| = 0$ if and only if $i(D) = g - n$. If $D = p_1 + \cdots + p_n$ is such that $p_j \neq p_k$ for $j \neq k$, then $i(D) = g - n$ if and only if the points $\Phi_K(p_j)$, $1 \leq j \leq n$, are in general position in $\mathbb{P}^{g-1}$. Since the canonical map Φ_K: $M \to \mathbb{P}^{g-1}$ is nondegenerate, this occurs for a general D.

The proof of (2) is similar to that of (1). Q.E.D.

Put

$$W_n(M) = \phi(S^nM)$$

Then $W_n(M)$ is an irreducible analytic set in $J(M)$.

Theorem 5.2.25.

1. If $n \leq g$, then $\dim W_n(M) = n$ and ϕ: $S^nM \to W_n(M)$ is generically bijective, that is, $\phi^{-1}(x)$ is one point for a general $x \in W_n(M)$.
2. (Jacobi inversion). If $n \geq g$, then ϕ: $S^nM \to J(M)$ is surjective.

Proof. 1 follows from (1) of Lemma 5.2.24 and the Proper Mapping Theorem (Theorem 3.2.14). 2 follows from (2) of Lemma 5.2.24 and Exercise 9 of Sec. 3.2. Q.E.D.

Theorem 5.2.26. $W_{g-1}(M)$ is a translation of the theta divisor Θ on J(M).

For the proof of this beautiful theorem, see, for example, Griffiths-Harris [33].

More generally, we put

$$G_n^r(M) = \{D \in S^nM \mid \dim |D| \geq r\}$$

$$W_n^r(M) = \phi(G_n^r(M))$$

for r = 0, 1, ... Then, by Proposition 5.2.22 and Exercise 8 of Sec. 3.2,

$$S^nM = G_n^0(M) \supset G_n^1(M) \supset \cdots$$

$$(\text{respectively } W_n(M) = W_n^0(M) \supset W_n^1(M) \supset \cdots)$$

is a decreasing sequence of analytic sets in S^nM (respectively J(M)).

Proposition 5.2.27. Let $n \leq g$ and $r \geq g$.

1. If $G_n^{r+1}(M)$ is nonempty, then $G_n^r(M) - G_n^{r+1}(M)$ is nonempty.
2. $G_n^r(M) - G_n^{r+1}(M)$ is open dense in $G_n^r(M)$.
3. $W_n^r(M) - W_n^{r+1}(M)$ is open dense in $W_n^r(M)$.

Proof. 3 is an easy consequence of 2, so we prove 1 and 2. It suffices to show that, for any $D_0 = p_1 + \cdots + p_n$ such that $\dim |D_0| = r + 1$ ($r \geq 0$), there is $D \in S^nM$ in any small neighborhood of D_0 in S^nM such that $\dim |D| = r$. We assume that $|D_0|$ has a fixed point. (The case when $|D_0|$ has no fixed point can be treated in a similar and simpler way.) Let

$$E_0 = p_1 + \cdots + p_s \qquad 1 \leq s \leq n - 1$$

be the fixed part of $|D_0|$. Then the linear system

$$|D_0 - E_0| = |D_0| - E_0 = \{D' - E_0 \mid D' \in |D_0|\}$$

has the dimension r + 1. By Theorem 3.4.15,

$$\dim |E_0| + r + 1 = \dim |E_0| + \dim |D_0 - E_0| \leq \dim |D_0| = r + 1$$

Hence, $\dim |E_0| \leq 0$, so $\dim |E_0| = 0$. By Proposition 5.2.22,

$$\operatorname{rank} (d\phi_s)_{E_0} = s$$

This means that the first s column vectors in the Brill-Noether matrix $(d\phi_n)_{D_0}$ are linearly independent.

We may suppose that

$$p_j \neq p_k \quad \text{for } 1 \leq j \leq n \quad \text{and} \quad s + 1 \leq k \leq n \quad \text{with } j \neq k$$

In fact, if D' is a general member of $|D_0 - E_0|$, then $E_0 + D'$ satisfies the condition by Bertini's Theorem (Theorem 3.4.15).

Note that $\dim |D_0| = r + 1 \geq 1$. Hence, by Proposition 5.2.22, we may assume that the last column vector

$${}^t(\omega_1(p_n), \ldots, \omega_g(p_n))$$

of $(d\phi_n)_{D_0}$ is linearly dependent on the other column vectors.

Let U be any small neighborhood of p_n in M. For $q \in U$, put

$$D = D(q) = p_1 + \cdots + p_{n-1} + q$$

Then $\operatorname{rank} (d\phi_n)_D$ is either $n - (r + 1)$ or $n - r$.

Suppose that

$$\operatorname{rank} (d\phi_n)_D = n - (r + 1) \quad \text{for all } q \in U$$

Then ${}^t(\omega_1(q), \ldots, \omega_g(q))$ belongs to the vector subspace of $\mathbb{C}^g$ spanned by the other n - 1 column vectors of $(d\phi_n)_D$, which do not depend on $q \in U$. Since $n - 1 \leq g - 1$, ${}^t(\omega_1(q), \ldots, \omega_g(q))$ belongs to a fixed hyperplane in $\mathbb{C}^g$, so there is $(a_1, \ldots, a_g) \in \mathbb{C}^g - \{0\}$ such that

$$a_1\omega_1(q) + \cdots + a_g\omega_g(q) = 0 \quad \text{for all } q \in U$$

Hence,

$$a_1\omega_1 + \cdots + a_g\omega_g = 0 \quad \text{in } \Gamma(K)$$

a contradiction. Thus, there is $q \in U$ such that rank $(d\phi_n)_D = n - r$, that is, $\dim |D| = r$. Q.E.D.

Note 5.2.28. (1) Abel's Theorem can be generalized to higher dimensional projective manifolds. See the very interesting paper by Griffiths [32]. (2) For deeper analysis on $G_n^r(M)$ and $W_n^r(M)$, see Martens [61] and Gunning [35]. (3) as for theta functions, see, for example, Rauch-Farkas [82].

Exercises

1. Let $A_2(M)$ be the vector space of all Abelian differentials of the first and second kinds. Let $B(M)$ be the vector subspace of $A_2(M)$ of all total differentials df ($f \in \mathbb{C}(M)$). (1) Let p be a non-Weierstrass point of M. Then every $\eta \in A_2(M)$ can be uniquely written as a linear combination of $\eta_p^{(2)}, \ldots, \eta_p^{(g+1)}, \omega_1, \ldots, \omega_g$, modulo $B(M)$, where $\{\omega_1, \ldots, \omega_g\}$ is a basis of $\Gamma(K)$ and $\eta_p^{(k)}$ are the Abelian differentials of the second kind in Lemma 5.2.14. (2) $\dim (A_2(M)/B(M)) = 2g$. (3) (Algebraic de Rham's Theorem). The pairing

$$(\eta, \gamma) \in \left(\frac{A_2(M)}{B(M)}\right) \times H_1(M, \mathbb{C}) \to \int_\gamma \eta \in \mathbb{C}$$

is a nondegenerate bilinear form and gives a linear isomorphism

$$\frac{A_2(M)}{B(M)} \simeq H^1(M, \mathbb{C})^*$$

(For a generalization of this result to higher dimensional manifolds, see, for example, Griffiths-Harris [33, p. 454].)

2. Prove the following theorem:

Theorem (Galois-Riemann). Every Abelian differential η on M can be written as a linear combination of Abelian differentials of the first, second, and third kinds. More precisely, if $\{q_1, \ldots, q_s\}$ is the set of all poles of η and p is a fixed non-Weierstrass point of M such that $p \neq q_j$, $1 \leq j \leq s$, then η can be written as a linear combination of

$$\omega_1, \ldots, \omega_g, \eta_p^{(2)}, \ldots, \eta_p^{(g+1)}, \eta_{q_1,p}, \ldots, \eta_{q_s,p}$$

module B(M) (see Exercise 1), where $\eta_{q_j,p}$ are the Abelian differentials of the third kind in Corollary 5.2.17.

3. Let η (respectively ω) be an Abelian differential of the third (respectively first) kind with poles $p_1, \ldots, p_n$. Prove the reciprocity law:

$$\frac{1}{2\pi\sqrt{-1}}\left\{\sum_{k=1}^{g}\left(\int_{\alpha_k}\omega\int_{\beta_k}\eta - \int_{\beta_k}\omega\int_{\alpha_k}\eta\right)\right\} = \sum_{j=1}^{n}(\mathrm{Res}_{p_j}\,\eta)\int_{p_0}^{p_j}\omega$$

4. Nonsingular plane cubic curves C_1 and C_2 are biholomorphic if and only if they are projectively equivalent.

5. For a complex number λ such that $\lambda \neq 0, 1$, let C_λ be the (closure in $\mathbb{P}^2$ of the affine) curve

$$C_\lambda : y^2 = x(x-1)(x-\lambda)$$

(1) $C_\lambda \simeq C_\mu$ (biholomorphic) if and only if μ is one of

$$\lambda,\ \frac{1}{\lambda},\ 1-\lambda,\ \frac{1}{1-\lambda},\ \frac{\lambda-1}{\lambda},\ \frac{\lambda}{\lambda-1}$$

(2) Put

$$J(\lambda) = \frac{4}{27}\,\frac{(\lambda^2-\lambda+1)^3}{\lambda^2(\lambda-1)^2}$$

Then $C_\lambda \simeq C_\mu$ if and only if $J(\lambda) = J(\mu)$.

(3) Let γ be a closed arc in x-plane as in Fig. 5.11.

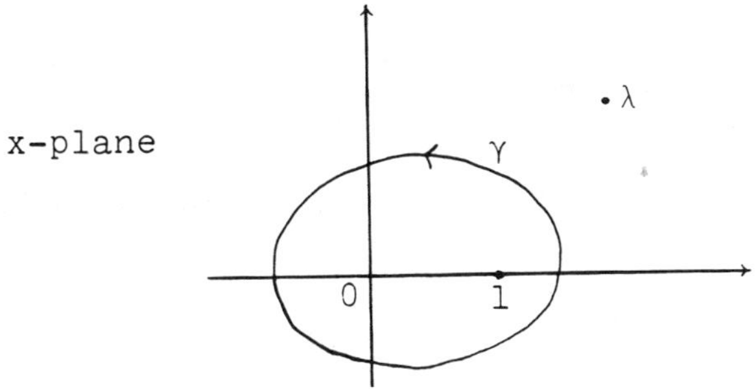

FIGURE 5.11

Put $\omega(\lambda) = \int_\gamma \frac{dx}{y}$. Then $\omega(\lambda)$ is a multivalued holomorphic function on $\mathbb{C} - \{0, 1\}$ and satisfies the Gauss hypergeometric differential equation

$$\lambda(\lambda - 1)\frac{d^2\omega}{d\lambda^2} + (2\lambda - 1)\frac{d\omega}{d\lambda} + \frac{1}{4}\omega = 0 \tag{*}$$

(4) Let D be the universal covering space of $\mathbb{C} - \{0, 1\}$. Let $\omega_1(\lambda)$ and $\omega_2(\lambda)$ ($\lambda \in D$) be (suitably chosen) linearly independent solutions of (*). Put $\tau(\lambda) = \omega_2(\gamma)/\omega_1(\lambda)$. Then

$$C_\lambda \simeq \frac{\mathbb{C}}{\mathbb{Z}\omega_1(\lambda) + \mathbb{Z}\omega_2(\lambda)} \simeq \frac{\mathbb{C}}{\mathbb{Z} + \mathbb{Z}\tau(\lambda)}$$

and

$$\tau: \lambda \in D \to \tau(\lambda) \in \mathbb{H} \text{ (the upper half plane)}$$

is biholomorphic. (Its inverse map

$$\lambda: \tau \in \mathbb{H} \to \lambda \in D$$

is also called an elliptic modular function.)
(5) $J(\lambda) = J(\tau(\lambda))$ for all $\lambda \in D$ (see Theorem 4.3.15).

5.3 HOLOMORPHIC MAPS INTO PROJECTIVE SPACES

Let M be a compact Riemann surface of genus g. By Theorem 3.4.7, there is a natural one-to-one correspondence

$$\Lambda \longleftrightarrow \Phi_\Lambda \ (\text{mod Aut}\,(\mathbb{P}^r))$$

between linear systems $\Lambda = g_n^r$ on M without fixed point and nondegenerate holomorphic maps of degree n of M into $\mathbb{P}^r$ module Aut $(\mathbb{P}^r)$.

For g = 0 or g = 1, more precisely,

Example 5.3.1. (1) If g = 0, then $M = \mathbb{P}^1$ and $S^n\mathbb{P}^1 = |n(\infty)|$ (see Example 5.2.18). Hence, the set

$$\mathbb{G}_n^r(\mathbb{P}^1) = \{\Lambda \mid \Lambda \text{ is a } g_n^r \text{ on } \mathbb{P}^1\}$$

can be identified with the Grassman variety $G(r, |n(\infty)|)$. This implies that every nondegenerate $f: \mathbb{P}^1 \to \mathbb{P}^r$ of degree n can be uniquely written as

$$f = \pi_S \cdot \Phi_{|n(\infty)|}$$

where π_S is the projection with the center an $(n - r - 1)$-plane S in $\mathbb{P}^n = |n(\infty)|^*$ such that $S \cap C_n = \phi$. Here, $C_n = \Phi_{|n(\infty)|}(\mathbb{P}^1)$ is the rational normal curve in $\mathbb{P}^n$.

(2) If g = 1, then M is a complex 1-torus by Corollary 5.2.20:

$$M = \frac{\mathbb{C}}{\mathbb{Z}\omega_1 + \mathbb{Z}\omega_2} \qquad \left(\tau = \frac{\omega_2}{\omega_1} \in \mathbb{H}\right)$$

For $n \geq 1$, the Jacobi map $\phi: S^nM \to J(M) = M$ is given by

$$\phi: (p_1) + \cdots + (p_n) \to p_1 + \cdots + p_n$$

(see Example 5.2.18). ϕ is a $\mathbb{P}^{n-1}$-bundle over M. Aut (M) acts on S^nM as follows:

$$(\sigma, (p_1) + \cdots + (p_n)) \in \mathrm{Aut}(M) \times S^nM \to (\sigma p_1) + \cdots + (\sigma p_n) \in S^nM$$

Put

$$\mathbb{G}^r_n(M) = \{\Lambda \mid \Lambda \text{ is a } g^r_n \text{ on } M\}$$

Then $\mathbb{G}^r_n(M)$ is the set of all r-planes in some fiber $\phi^{-1}(p) = |D|$ of ϕ. It can be easily shown that $\mathbb{G}^r_n(M)$ is a fiber bundle over M with the standard fiber the Grassmann variety $G(r, n - 1)$. Aut (M) naturally acts on $\mathbb{G}^r_n(M)$. If $\Lambda \subset |(p_1) + \cdots + (p_n)|$, then $T_q(\Lambda) \subset |n(0)|$, where $q \in M$ satisfies

$$p_1 + \cdots + p_n + nq = 0$$

and T_q is the translation

$$T_q: p \in M \to p + q \in M$$

Hence every $\Lambda \in \mathbb{G}^r_n(M)$ is equivalent to a linear subsystem in $|n(0)|$ under the action of Aut (M). This implies that, for every nondegenerate $f: M \to \mathbb{P}^r$, there are $q \in M$ and an $(n - r - 2)$-plane S with $S \cap C_n = \phi$ ($C_n = \Phi_{|n(0)|}(M) \subset \mathbb{P}^{n-1}$) such that

$$f = \pi_S \cdot \Phi_{|n(0)|} \cdot T_q$$

where π_S is the projection with the center S. The quotient space $\mathbb{G}_n^r(M)/\mathrm{Aut}(M)$ can be then identified with $G(n - r - 2, n - 1)/B$:

$$\frac{\mathbb{G}_n^r(M)}{\mathrm{Aut}(M)} = \frac{G(n - r - 2, n - 1)}{B}$$

Here B is the finite subgroup of Aut (M) generated by

$$\{T_q \mid q \text{ is a point of M such that } nq = 0\}$$

and

$$S_\zeta : p \in M \rightarrow \zeta p \in M$$

where

$$\zeta = -1, \text{ if } \tau = \frac{\omega_2}{\omega_1} \text{ is equivalent under } SL(2, \mathbb{Z}) \text{ to neither } \sqrt{-1} \text{ nor } \frac{1 + \sqrt{-3}}{2},$$

$$\zeta = \sqrt{-1}, \text{ if } \tau = \sqrt{-1},$$

$$\zeta = \frac{1 + \sqrt{-3}}{2}, \text{ if } \tau = \frac{1 + \sqrt{-3}}{2}$$

In fact, every element $\sigma \in B$ maps $|n(0)|$ to $|n(0)|$, so acts on $\mathbb{P}^{n-1} = |n(0)|^*$ such that the diagram

$$\begin{array}{ccc} M & \xrightarrow{\sigma} & M \\ \Phi_{|n(0)|} \Big\downarrow & & \Big\downarrow \Phi_{|n(0)|} \\ \mathbb{P}^{n-1} & \xrightarrow{\sigma} & \mathbb{P}^{n-1} \end{array}$$

is commutative.

Next, we prove

Proposition 5.3.2. Let M be a compact Riemann surface of genus g. Let n and r be positive integers such that $n \geq g + r$. Then there is a g_n^r on M without fixed point.

Proof. Let p be a non-Weierstrass point of M. Then

$$\dim |np| = n - g \quad \text{and} \quad \dim |(n-1)p| = n - 1 - g$$

for $n \geq g + 1$. This implies that $|np|$ has no fixed point for $n \geq g + 1$.

If $r = n - g$, then $|np|$ satisfies the condition. Suppose that $r < n - g$. Take any $(n - g - r - 1)$-plane S in $\mathbb{P}^{n-g}$ such that $S \cap \Phi_{|np|}(M) = \phi$. Then the linear system

$$\Lambda = \{(H) \mid H \text{ is a hyperplane in } \mathbb{P}^{n-g} \text{ such that } S \subset H\}$$

((H) = the pullback by $\Phi_{|np|}$ of the hyperplane cut by H) satisfies the condition. Q.E.D.

Lemma 5.3.3. Let $f: M \to \mathbb{P}^r$ be a nondegenerate holomorphic map of degree $n \leq g + r - 1$. Then there are linearly independent $\omega_0, \cdots, \omega_r$ in $\Gamma(K)$ such that $f = (\omega_0 : \cdots : \omega_r)$. In other words, there is a $(g - r - 2)$-plane S in $\mathbb{P}^{g-1}$ such that $f = \pi_S \cdot \Phi_{|K|}$, where π_S is the projection with the center S.

Proof. Let Λ be the fixed point free linear system on M corresponding to f. For $D \in \Lambda$,

$$i(D) = g - n + \dim |D| \geq g - n + r \geq 1$$

by the Riemann-Roch theorem. Hence, there is a nonzero $\eta \in \Gamma(K - [D])$. Take linearly independent $\xi_0, \cdots, \xi_r \in \Gamma([D])$ such that $f = (\xi_0 : \cdots : \xi_r)$. Put $\omega_j = \xi_j \eta \in \Gamma(K)$ for $1 \leq j \leq r$. (Here $\xi_j \eta$ is, by definition, the section $\{\xi_{j\alpha}\eta_\alpha\}$ of $[D] + (K - [D]) = K$, where $\xi_j = \{\xi_{j\alpha}\}$ and $\eta = \{\eta_\alpha\}$ on an open covering $\{U_\alpha\}$ of M.) Then

$$(\omega_0 : \cdots : \omega_r) = (\xi_0 : \cdots : \xi_r) = f$$

Let H_j be the hyperplane in $\mathbb{P}^{g-1}$ corresponding to ω_j. Then $S = H_0 \cap \cdots \cap H_r$ satisfies

$$f = \pi_S \cdot \Phi_{|K|}$$

Q.E.D.

This lemma shows that projective geometric properties of the canonical curve $C_K = \Phi_K(M)$ give information in order to look for nondegenerate $f: M \to \mathbb{P}^r$ of lower degree.

Proposition 5.3.4. Let M be hyperelliptic and $\phi: M \to \mathbb{P}^1$ be a holomorphic map of degree 2.

1. For any nondegenerate holomorphic map $f: M \to \mathbb{P}^r$ of degree $n \leq g + r - 1$, there is a nondegenerate $h: \mathbb{P}^1 \to \mathbb{P}^r$ such that $f = h \cdot \phi$. In particular, n must be even.
2. A holomorphic map $M \to \mathbb{P}^1$ of degree 2 is unique up to Aut $(\mathbb{P}^1)$.
3. For $1 \leq r \leq g - 1$, put $\Lambda_r = \{\phi^{-1}(E) \mid E \in |r(\infty)|\}$. Then Λ_r is a unique g^r_{2r} on M. Moreover, Λ_r is complete: $\Lambda_r = |rD_0|$ $(D_0 \in \Lambda_1)$.
4. If $1 \leq r \leq g - 1$ and $s > r$, then there is no g^s_{2r} on M.

Proof. To prove 1, note that since $\mathbb{C}(M)$ is a quadratic extension of $\mathbb{C}(\phi)$, M is birational to the curve

$$y^2 = (x - \alpha_1) \cdots (x - \alpha_{2g+1}) \qquad (\alpha_j \neq \alpha_k \text{ for } j \neq k)$$

and ϕ is identified with the projection

$$(x, y) \to x$$

(compare Proposition 4.2.14). Moreover,

$$\Gamma(K) = \left\{ \frac{u(x)}{y} dx \mid u(x) \text{ is a polynomial of degree} \leq g - 1 \right\}$$

(see Exercise 4 of Sec. 5.1). Hence, by Lemma 5.3.3,

$$f = (u_0(x): \cdots : u_r(x))$$

where every $u_j(x)$ is a polynomial of degree $\leq g - 1$. This means that $f = h \cdot \phi$, where $h = (u_0: \cdots : u_r): \mathbb{P}^1 \to \mathbb{P}^r$.

2 follows from 1.

To prove 3, let Λ be a g^r_{2r} on M. Let E_0 be the fixed part of Λ. Put $k = \deg E_0$ (≥ 0). Then $\Lambda - E_0$ is a g^r_{2r-k} without fixed point. Note that $2r - k \leq g + r - 1$. By 1,

$$\Phi_{\Lambda - E_0} = h \cdot \phi$$

for some $h: \mathbb{P}^1 \to \mathbb{P}^r$. In particular, k must be even: $k = 2k'$. Since $\deg(\Lambda - D_0) = 2(r - k')$, we have $\deg h = r - k'$. Note that h is nondegenerate. Hence, $k' = 0$ and $h = \Phi_{|r(\infty)|}$. Thus

$$\Lambda = \Lambda_r = \{\phi^{-1}(E) \mid E \in |r(\infty)|\}$$

Λ_r is complete. In fact, if $|D| \supset \Lambda_r$ and $|D| \neq \Lambda_r$, then $|D|$ has a linear subsystem g^r_{2r} such that $g^r_{2r} \neq \Lambda_r$, a contradiction. Hence, $\Lambda_r = |rD_0|$ $(D_0 \in \Lambda_1)$.

This last argument also shows 4. Q.E.D.

A positive divisor D on M is said to be special if $i(D) \geq 1$, that is, if there is $\omega \in \Gamma(K)$ such that $(\omega) \geq D$. Otherwise, D is said to be nonspecial. A linear system Λ on M is said to be special (respectively nonspecial) if one (hence, all) divisor D in Λ is special (respectively nonspecial). Special linear systems appeared in Lemma 5.3.3 and Proposition 5.3.4. By the Riemann-Roch Theorem,

Lemma 5.3.5. Let D be a positive divisor on M.

1. If $\deg D \geq 2g - 1$, then D is nonspecial.
2. If $\deg D \leq g + \dim |D| - 1$, then D is special.
3. If $\deg D \leq g - 1$, then D is special.

For a positive divisor D on M, put

$$S_D = \bigcap_{(H) \geq D} H$$

where the intersection runs over all hyperplanes H in $\mathbb{P}^{g-1}$ such that the hyperplane sections (H) (canonical divisors) contain D. If there is no such H, put $S_D = \mathbb{P}^{g-1}$. Clearly,

Lemma 5.3.6.

1. D is special if and only if $S_D \neq \mathbb{P}^{g-1}$.
2. $\dim S_D = g - 1 - i(D) = \deg D - \ell(D)$.

One of the most famous theorems on special divisors is

Theorem 5.3.7 (Clifford). Let D be a special divisor on M. Then

$$\dim |D| \leq \tfrac{1}{2} \deg D$$

The equality holds if and only if (1) $D \in |K|$, or (2) M is hyperelliptic and $|D| = |kD_0|$ for some k such that $1 \leq k \leq g - 1$, where D_0 is a positive divisor such that $|D_0| = g^1_2$.

Proof. If $D \in |K|$, then

$$\dim |D| = g - 1 = \tfrac{1}{2}\deg D$$

Suppose that $D \notin |K|$. Then there is a positive divisor E such that $D + E \in |K|$. By Theorem 3.4.15 and the Riemann-Roch Theorem,

$$\dim |D| + \dim |E| \leq g - 1$$

$$\dim |D| - \dim |E| = \deg D + 1 - g$$

Adding them, we have

$$\dim |D| \leq \tfrac{1}{2}\deg D$$

By Theorem 3.4.15 again, the equality holds if and only if there is a positive divisor D_0 such that (i) $|D_0|$ has no fixed point, (ii) $\dim |jD_0| = j$ for $1 \leq j \leq g - 1$ and (iii) $D \sim kD_0$ and $E \sim \ell D_0$ for some k, $\ell \geq 1$. By (iii),

$$(k + \ell)D_0 \in |K|$$

By (ii) and the first part of Theorem 3.4.15,

$$k + \ell = (k + \ell) \dim |D_0| \leq \dim |(k + \ell)D_0| = \dim |K| = g - 1$$

Hence, by (ii) again,

$$\dim |(k + \ell)D_0| = k + \ell$$

Hence, $k + \ell = g - 1$, so

$$(g - 1)D_0 \in |K|$$

Hence, $\deg D_0 = 2$, so $|D_0| = g^1_2$. This means that M is hyperelliptic.
The converse follows from 3 of Proposition 5.3.4. Q.E.D.

The following two theorems can be obtained from the Clifford theorem and a deep analysis on $\dim W^r_n(M)$.

Theorem 5.3.8. Let $g \geq 2$. There is no $f: M \to \mathbb{P}^1$ of degree g if and only if g is odd and M is hyperelliptic.

Theorem 5.3.9 (G. Martens [62], Horiuchi [44]. Let $g \geq 4$. There is no $f: M \to \mathbb{P}^1$ of degree $g - 1$ if and only if g is even and M is hyperelliptic.

Definition 5.3.10. Let n be a positive integer. M is said to be n-gonal if (1) there is $f: M \to \mathbb{P}^1$ of degree n and (2) there is no $h: M \to \mathbb{P}^1$ of degree $\leq n - 1$. A three-gonal M is said to be trigonal.

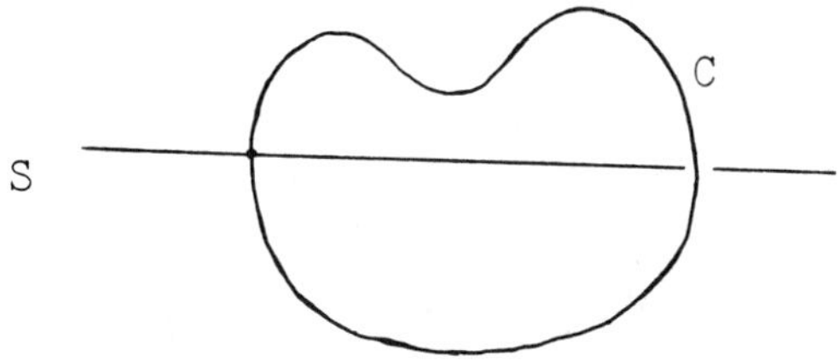

FIGURE 5.12

A two-gonal M with $g \geq 2$ is nothing but a hyperelliptic compact Riemann surface. If $g = 3$ or 4, then M is trigonal if and only if M is nonhyperelliptic. But, if $g \geq 5$, then a general M is neither hyperelliptic nor trigonal.

Lemma 5.3.11. Let M be n-gonal and $\Lambda = g_n^1$ be a linear pencil of degree n on M. Then Λ is complete.

Proof. Suppose that Λ is not complete. Then Λ is a linear subsystem of a of a complete linear system $|D|$ of dimension $r \geq 2$. Put $C = \Phi_{|D|}(M)$ $(\subset \mathbb{P}^r)$. Take an $(r-2)$-plane S such that $S \cap C \neq \phi$. Then $\deg \pi_S \cdot \Phi_{|D|} < n$, a contradiction. (See Fig. 5.12.)

Theorem 5.3.12 (Accola [2], Namba [71]).

1. Let $\phi: M \to \mathbb{P}^1$ and $f: M \to \mathbb{P}^1$ be of degree m and n, respectively. If $(m, n) = 1$ (coprime), then $(m-1)(n-1) \geq g$.
2. Let p be a prime number and $\phi: M \to \mathbb{P}^1$ be of degree p. (i) If $f: M \to \mathbb{P}^1$ is of degree n such that $(p-1)(n-1) \leq g-1$, then $n \equiv 0 \pmod p$ and there is $h: \mathbb{P}^1 \to \mathbb{P}^1$ such that $f = h \cdot \phi$. (ii) If $(p-1)^2 \leq g-1$, then M is p-gonal and ϕ is unique up to $\mathrm{Aut}(\mathbb{P}^1)$.

Proof. Let $\phi: M \to \mathbb{P}^1$ and $f: M \to \mathbb{P}^1$ be of degree m and n, respectively. By changing ϕ by $\sigma \cdot \phi$ $(\sigma \in \mathrm{Aut}(\mathbb{P}^1))$ if necessary, we may assume that the supports of $D_\infty(\phi)$ and $D_\infty(f)$ have no point in common. Consider the holomorphic map

$$\Phi: p \in M \to (1: \phi(p): f(p)) \in \mathbb{P}^2$$

Let C be its image and be defined by the irreducible equation

$$F(x,y) = 0$$

Let $m' = \deg_y F$ (respectively $n' = \deg_x F$) be the degree of F with respect to y (respectively x). Then, since ϕ (repectively f) is the pull-back by Φ of

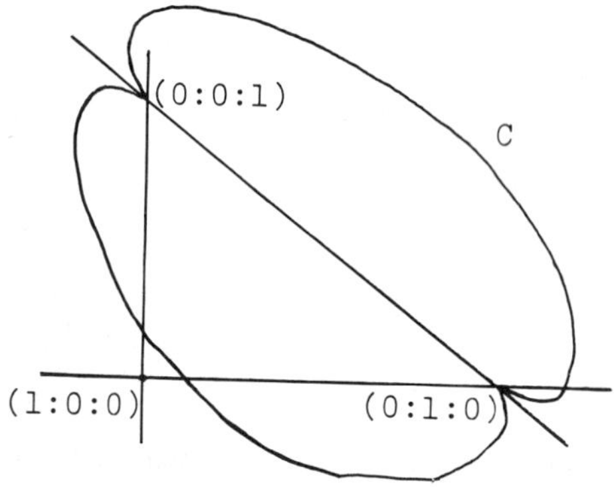

FIGURE 5.13

$$(x, y) \in C \to x \in \mathbb{P}^1 \quad \text{(respectively } (x, y) \in C \to y \in \mathbb{P}^1\text{)}$$

we have

$$m = em' \quad \text{and} \quad n = en' \tag*{$(*)_1$}$$

where e is the mapping degree of $\Phi: M \to C$. Note that $\deg C = m' + n'$ and the points (0: 0: 1) and (0: 1: 0) are (singular) points of C with the multiplicity n' and m', respectively. (See Fig. 5.13.)

Let g_0 be the genus of C. Then

$$g_0 \le g \quad \text{and} \tag*{$(*)_2$}$$

$$g_0 \le \tfrac{1}{2}(m' + n' - 1)(m' + n' - 2) - \tfrac{1}{2}n'(n' - 1) - \tfrac{1}{2}m'(m' - 1)$$

$$= (m' - 1)(n' - 1) \tag*{$(*)_3$}$$

(see Corollary 2.1.10).

Now, suppose that $(m, n) = 1$. Then, by $(*)_1$, $e = 1$. Hence, $m' = m$ and $n' = n$ and $\Phi: M \to C$ is birational, so $g_0 = g$. Hence, by $(*)_3$, $g \le (m - 1)(n - 1)$. This proves 1 of the theorem.

Next, let $m = p$ be a prime number and $(p - 1)(n - 1) \le g - 1$. If $e = 1$, then $g \le (p - 1)(n - 1)$ by the same reason as above, a contradiction. Hence, $e = p$, so $m' = \deg_y F = 1$. Hence,

$$F(x, y) = B_0(x) + B_1(x)y$$

where B_0 and B_1 are polynomials of x. Hence, on M,

$$0 = F(\phi, f) = B(\phi) + B_1(\phi)f$$

so

$$f = -\frac{B_0(\phi)}{B_1(\phi)}$$

This proves (i). (ii) is a special case of (i). Q.E.D.

For example, a trigonal M with $g \geq 5$ has a unique g_3^1. (On a nonhyperelliptic M with $g = 4$, there may be two g_3^1's (see Example 5.1.21).)

As applications to Theorem 5.3.12, we state the following theorems on the equivalence problem of compact Riemann surfaces.

Theorem 5.3.13. Let M and N be trigonal and be defined by the equations

$$M: \; y^3 + a_2(x)y + a_3(x) = 0$$

$$N: \; y^3 + b_2(x)y + b_3(x) = 0$$

where $a_j(x)$ and $b_j(x)$ $(j = 2, 3)$ are rational functions of x. Suppose that the genera of M of N are greater than 4. Then $M \simeq N$ (biholomorphic) if and only if there are rational functions u(x) and v(x) of x and $\sigma \in \mathrm{Aut}\,(\mathbb{P}^1)$ such that

$$b_2 \cdot \sigma = D(3a_2u^2 + 9a_3uv - a_2^2v^2)$$

$$b_3 \cdot \sigma = D^2(u^3 + a_2uv^2 + a_3v^3)$$

where $D = -4a_2^3 - 27a_3^2$.

Theorem 5.3.14. Let p be a prime number and n be an integer such that $n \equiv -1 \pmod{p}$ and $n \geq 2p + 1$. Let M and N be defined by the equations

$$M: \; y^p = (x - \alpha_1) \cdots (x - \alpha_n)$$

$$N: \; y^p = (x - \beta_1) \cdots (x - \beta_n)$$

where α_j and β_j $(1 \leq j \leq n)$ are complex numbers such that $\alpha_j \neq \alpha_k$ and $\beta_j \neq \beta_k$ for $j \neq k$. Then $M \simeq N$ if and only if there is $\sigma \in \mathrm{Aut}\,(\mathbb{P}^1)$ such that

$$\{\sigma(\alpha_1), \ldots, \sigma(\alpha_n), \sigma(\infty)\} = \{\beta_1, \ldots, \beta_n, \infty\}$$

Theorem 5.3.15. Let p be a prime number and n be an integer such that $n \not\equiv 0 \pmod{p}$ and $n \geq 2p + 1$. Let M and N be defined by the equations

$$M: y^n = (x - \alpha_1) \cdots (x - \alpha_p),$$

$$N: y^n = (x - \beta_1) \cdots (x - \beta_p)$$

where α_j and β_j $(1 \leq j \leq p)$ are complex numbers such that $\alpha_j \neq \alpha_k$ and $\beta_j \neq \beta_k$ for $j \neq k$. Then $M \simeq N$ if and only if there is $\sigma \in \mathrm{Aut}(\mathbb{C})$ such that

$$\{\sigma(\alpha_1), \ldots, \sigma(\alpha_p)\} = \{\beta_1, \ldots, \beta_p\}$$

Corollary 5.3.16. Let n be an integer such that $n \not\equiv 0 \pmod{3}$ and $n \geq 2$. For $\lambda \in \mathbb{C} - \{0, 1\}$, let M_λ be defined by the equation

$$M_\lambda: y^n = x(x - 1)(x - \lambda)$$

Then $M_\lambda \simeq M_\mu$ if and only if μ is equal to one of the following values:

$$\lambda, \frac{1}{\lambda}, 1 - \lambda, \frac{1}{1 - \lambda}, \frac{\lambda - 1}{\lambda}, \frac{\lambda}{\lambda - 1}$$

Theorem 5.3.13 can be shown directly. For the proofs of Theorem 5.3.14-Corollary 5.3.16, see Namba [72].

Theorem 5.3.17 (compare Namba [71]). Let C be a nonsingular plane curve of degree n.

1. If $n \geq 2$, then C is (n - 1)-gonal and every g^1_{n-1} on C can be obtained by the projection π_p with the center a point $p \in C$. Moreover, if $n \geq 3$, then such a point p is uniquely determined.

2. If $n \geq 5$, $n \leq d \leq 2n - 5$ and $f: C \to \mathbb{P}^1$ is of degree d, then $d = n$ and $f = \pi_p$ for a unique point $p \in \mathbb{P}^2 - C$.

3. If $n \geq 4$, $1 \leq d \leq 2n - 4$ and a nondegenerate holomorphic map $f: C \to \mathbb{P}^2$ is of degree d, then $d = n$ and f = the identity map. In particular, {line cuts of C} is a unique g^2_n on C for $n \geq 4$.

4. The linear system {line cuts of C} is complete for $n \geq 1$.

5. If $n \geq 4$, then every g^2_{2n-3} can be obtained in the following way: Take any (not necessarily distinct) three points p, q, and r on C. Then

$$g^2_{2n-3} = |2D - p - q - r|$$

where D is a line cut of C. This g^2_{2n-3} has no fixed point if and only if p, q, and r are not collinear. (If p = q = r, then this condition means that p is not a flex of C.) In this case

$$\Phi = \Phi_{|2D-p-q-r|} : C \to \mathbb{P}^2$$

is birational onto the image curve C', which has at least one and at most three singular points. Every singular point has the multiplicity n - 2.

6. If $n \geq 2$, $r \geq 3$, and $1 \leq d \leq 2n - 3$, then there is no g^r_d on C.

Proof. To prove 1, note that the genus of C is $\frac{1}{2}(n-1)(n-2)$. The assertion is trivial for n = 2. If n = 3, then C is an elliptic curve. Every $\Lambda = g^1_2$ on C is complete and can be written as $|(q_1) + (q_2)|$. Take $q_0 \in C$ such that q_0, q_1, and q_2 are collinear. Then Λ can be obtained by π_{q_0}.

Let $n \geq 4$. In this case, $n - 1 \leq g$. Hence, by Lemma 5.3.3, every $f: C \to \mathbb{P}^1$ of degree $\leq n - 1$ can be written as

$$f = \frac{\omega_1}{\omega_0} \quad \text{for some } \omega_0, \omega_1 \in \Gamma(K)$$

Hence, by Corollary 5.1.10,

$$f = \frac{F}{G}$$

for some homogeneous polynomials F and G of degree n - 3. Put

$$H = (F, G), \quad (\text{G.C.D.}), \quad F = F_1H \quad \text{and} \quad G = G_1H$$

Then H, F_1, and G_1 are homogeneous and

$$f = \frac{F_1}{G_1}$$

Put k = deg H. Let

$$C.F_1 = a_1p_1 + \cdots + a_sp_s$$

$$(\text{respectively } C.G_1 = b_1p_1 + \cdots + b_sp_s)$$

be the intersection zero-cycle of the plane curves C and F_1 (respectively C and G_1), where a_j and b_j $(1 \leq j \leq s)$ are nonnegative integers such that

$$\Sigma a_j = \Sigma b_j = n(n - 3 - k)$$

We regard $C \cdot F_1$ and $C \cdot G_1$ as positive divisors on C. Put

$$c_j = \min\{a_j, b_j\} \quad \text{for } 1 \leq j \leq s, \quad \text{and}$$

$$D = c_1 p_1 + \cdots + c_s p_s$$

Then the linear pencil

$$\{C \cdot (\lambda F_1 + \mu G_1)\}_{(\lambda:\mu) \in \mathbb{P}^1}$$

on C has the fixed part D. Hence,

$$\deg f = n(n - 3 - k) - \Sigma c_j$$

By Lemma 1.3.8 and Bezout's Theorem (Theorem 1.3.5),

$$\Sigma c_j \leq \Sigma I_{p_j}(F_1, G_1) \leq (n - 3 - k)^2$$

Hence,

$$\deg f \geq n(n - 3 - k) - (n - 3 - k)^2 = (n - 3 - k)(3 + k)$$

Note that $0 \leq k \leq n - 3$. The function $\psi(k) = (n - 3 - k)(3 + k)$ with $0 \leq k \leq n - 3$ takes its minimal positive value $n - 1$ at $k = n - 4$. (See Fig. 5.14.) Hence,

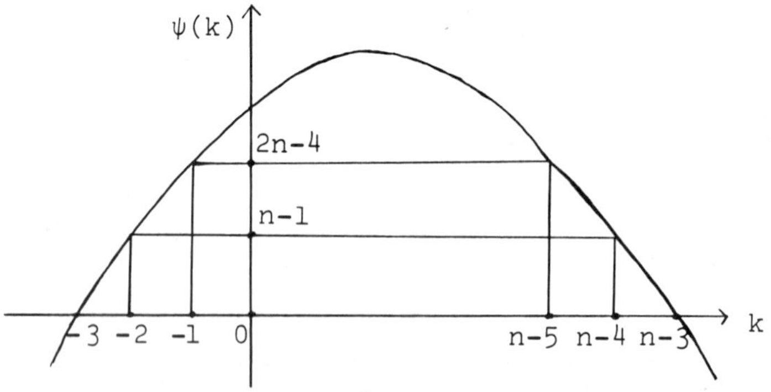

FIGURE 5.14

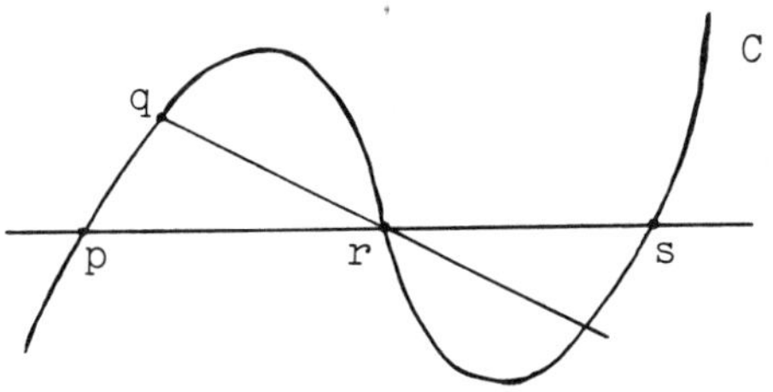

FIGURE 5.15

deg $f \geq n - 1$. The equality holds if and only if F_1 and G_1 are lines and $p = F_1 \cap G_1$ is a point of C. Thus $f = \pi_p$ with $p \in C$.

Suppose that, for distinct points p and q on C, π_p and π_q give the same linear pencil g^1_{n-1} on C. This means that, by choosing a fixed coordinate on $\mathbb{P}^1$, there is $\sigma \in \mathrm{Aut}(\mathbb{P}^1)$ such that $\pi_q = \sigma \cdot \pi_p$. Take a point $r \in C$ such that (1) p, q, and r are mutually distinct, (2) p, q, and r are not collinear, and (3) $r \notin T_pC$ and $p \notin T_rC$. (See Fig. 5.15.) If $n \geq 3$, then there is a point $s \in C$ such that p, r, and s are mutually distinct and collinear. Then $\pi_p(r) = \pi_p(s)$. Hence,

$$\pi_q(r) = \sigma\pi_p(r) = \sigma\pi_p(s) = \pi_q(s)$$

so q, r, and s are collinear, which is impossible.

To prove 2, first we note that

$$2n - 4 \leq g \quad \text{for } n \geq 5$$

Hence, as in the proof of 1, every $f: C \to \mathbb{P}^1$ of degree $\leq 2n - 4$ can be written as $f = F/G = F_1/G_1$. The function $\psi(k)$ of k in the proof of 1 takes its second minimal positive value $2n - 4$ at $k = n - 5$, if $n \geq 5$. Hence, if $n \geq 5$, $n \leq d \leq 2n - 5$ and $f: C \to \mathbb{P}^1$ is of degree d, then $k = n - 4$, F_1 and G_1 are lines and $p = F_1 \cap G_1 \in \mathbb{P}^2 - C$. Thus $d = n$ and $f = \pi_p$ with $p \in \mathbb{P}^2 - C$.

The uniqueness of the point p can be shown in a similar way to proof of 1.

To prove 3, if $n = 4$, then C is the canonical curve of itself. Hence, $\{\text{line cuts of } C\} = |K|$ is the unique g^2_4 (see Proposition 5.1.14), and there is no g^2_d on C such that $1 \leq d \leq 3$.

Suppose that $n \geq 5$. Let C' be the image of $f: C \to \mathbb{P}^2$ and put

$$d' = \deg C' \quad \text{and}$$

$$e = \text{the mapping degree of } f: C \to C'$$

Then $d = d'e$, the degree of f. Take a nonsingular point p' on C'. Then

$$\deg(\pi_{p'} \cdot f) = (d' - 1)e = d - e \leq 2n - 5 \quad \text{for } d \leq 2n - 4$$

Hence, by 1 and 2, either $d - e = n$ or $d - e = n - 1$.

Suppose that $d - e = n$. Then there is a point $p \in \mathbb{P}^2 - C$ such that $\pi_{p'} \cdot f = \pi_p$. This implies that

$$E - f^{-1}(p') \sim D$$

where D is a line cut of C and E is the pull-back by f of a line cut of C'. Take another nonsingular point q' on C'. Then, by the same reason as above,

$$E - f^{-1}(q') \sim D$$

Hence,

$$f^{-1}(p') \sim f^{-1}(q')$$

This contradicts 1, for

$$\deg f^{-1}(p') = e = d - n \leq n - 4 < n - 1$$

Hence, we have

$$d - e = n - 1$$

By 1, there is a point $p \in C$ such that $\pi_{p'} \cdot f = \pi_p$. This implies that

$$E - f^{-1}(p') \sim D - p$$

where D and E are as above. Take another nonsingular point $q' \in C'$. Then, by the same reason as above, there is $q \in C$ such that

$$E - f^{-1}(q') \sim D - q$$

Hence,

$$f^{-1}(p') + q \sim f^{-1}(q') + p$$

Note that

$$\deg(f^{-1}(p') + q) = e + 1 = d - n + 2 \leq n - 2$$

Hence, by 1 again,

$$f^{-1}(p') + q = f^{-1}(q') + p$$

Note that the supports of $f^{-1}(p')$ and $f^{-1}(q')$ are disjoint. Hence,

$$e = 1, \quad d = n, \quad f^{-1}(p') = p \quad \text{and} \quad f^{-1}(q') = q$$

Note that C' is nonsingular. In fact,

$$\deg (\pi_{p'} \cdot f) \leq n - 2 \quad \text{for a singular point } p' \in C'$$

which contradicts 1. Hence, $f: C \to C'$ is biholomorphic and

$$\pi_{f(p)} \cdot f = \pi_p \quad \text{for all } p \in C$$

This implies that

$$\{\text{the pull-backs by f of line cuts of } C'\} \subset \{\text{line cuts of } C\}$$

These linear systems are both two-dimensional, so they are equal. Thus

$$f = \text{the identity map of } C, \text{ up to Aut } (\mathbb{P}^2)$$

To prove 4, note that the assertion is trivial for $n \leq 3$. The assertion for $n \geq 4$ follows from 3.

To prove 6, note that the assertion is trivial for n = 2 and 3. Suppose that $n \geq 4$. Let $\Lambda = g_d^r$ be a linear system on C such that $r \geq 3$ and $1 \leq d \leq 2n - 3$. We may assume that Λ has no fixed point. Let $\Lambda' = g_d^3$ be a linear subsystem of Λ without fixed point. Take a point $p' \in \Phi_{\Lambda'}(C) \subset \mathbb{P}^3$. Then $\pi_{p'} \cdot \Phi_{\Lambda'}$ gives a $g_{d'}^2$ on C such that $1 \leq d' \leq d - 1 \leq 2n - 4$. By 3, $d' = n$ and $\pi_{p'} \cdot \Phi_{\Lambda'}$ = the identity map of C (up to Aut $(\mathbb{P}^2)$). In particular, $\Phi_{\Lambda'}$ is birational, so $d' = n = d - 1$ and $\Phi_{\Lambda'}(C)$ is nonsingular. Hence $\Phi_{\Lambda'}$: $C \to \Phi_{\Lambda'}(C)$ is biholomorphic. Put $p = \Phi_{\Lambda'}^{-1}(p')$. Take another point $q' \in \Phi_{\Lambda'}(C)$ and put $q = \Phi_{\Lambda'}^{-1}(q')$. Then

$$p + D = D_H \sim q + D$$

where D (respectively D_H) is a line cut of C (respectively a plane cut of $\Phi_{\Lambda'}(C)$). Hence,

$$p \sim q$$

which contradicts $g = \frac{1}{2}(n - 1)(n - 2) \neq 0$ for $n \geq 4$.

To prove 5, note that the case n = 4 was treated in Sec. 2.3. Suppose that $n \geq 5$. The idea is similar to the case n = 4.

Take a $\Lambda = g_{2n-3}^2$ on C without fixed point. By 6, Λ is complete. Put

$$C' = \Phi_\Lambda(C) \quad (\subset \mathbb{P}^2)$$
$$d' = \deg C'$$
$$e = \text{the mapping degree of } \Phi_\Lambda\colon C \to C'$$

Then

$$d'e = 2n - 3$$

We first show that $e = 1$, so $\Phi_\Lambda\colon C \to C'$ is birational. In fact, suppose that $e \geq 2$. Take a nonsingular point $p' \in C'$. Then

$$\deg \pi_{p'} \cdot \Phi_\Lambda = e(d' - 1) = 2n - 3 - e \leq 2n - 5$$

Hence, by 1 and 2,

$$\deg \pi_{p'} \cdot \Phi_\Lambda = 2n - 3 - e = n \text{ or } n - 1$$

First, suppose that $2n - 3 - e = n$, that is, $e = n - 3$. But e must divide $2n - 3$. Hence, e must divide $n = (2n - 3) - e$. Since $n \geq 5$, this occurs if and only if $n = 6$, $e = 3$ and $2n - 3 = 9$. By 2, there is $p \in \mathbb{P}^2 - C$ such that $\pi_{p'} \cdot \Phi_\Lambda = \pi_p$. Hence,

$$D + \Phi_\Lambda^{-1}(p') \in \Lambda$$

where D is a line cut of C. Take another nonsingular $q' \in C'$. Then, by the same reason,

$$D + \Phi_\Lambda^{-1}(q') \in \Lambda$$

Hence,

$$\Phi_\Lambda^{-1}(p') \sim \Phi_\Lambda^{-1}(q')$$

But $\deg \Phi_\Lambda^{-1}(p') = e = 3$. This contradicts 1 for $n = 6$.

Second, suppose that $2n - 3 - e = n - 1$, that is, $e = n - 2$. But e must divide $2n - 3$, so e must divide $n - 1 = (2n - 3) - e$, which contradicts $n \geq 5$.

Thus $e = 1$, $d' = 2n - 3$, and $\Phi_\Lambda\colon C \to C'$ is birational. By the Genus Formula, C' must have a singular point p'. Put

$$\Phi_\Lambda^{-1}(p') = p_1 + \cdots + p_m$$

where $m = m_{p'}$ (≥ 2) is the multiplicity of C' at p'. Then

$$\deg \pi_{p'} \cdot \Phi_\Lambda = 2n - 3 - m \leq 2n - 5$$

Hence, by 1 and 2,

$$2n - 3 - m = n \text{ or } n - 1$$

Suppose that $2n - 3 - m = n$, that is, $m = n - 3$. By 2, there is $p \in \mathbb{P}^2 - C$ such that $\pi_{p'} \cdot \Phi_\Lambda = \pi_p$. Hence

$$D + p_1 + \cdots + p_m \in \Lambda$$

where D is a line cut of C. But Λ is complete and of dimension 2. Hence,

$$\Lambda = \{\text{line cuts of } C\} + (p_1 + \cdots + p_m)$$

This means that $p_1 + \cdots + p_m$ is the fixed part of Λ, a contradiction.

Thus $2n - 3 - m = n - 1$, that is, $m = n - 2$. By 1, there is $p \in C$ such that $\pi_{p'} \cdot \Phi_\Lambda = \pi_p$. Hence,

$$\Lambda - \Phi_\Lambda^{-1}(p') \sim D - p$$

where D is a line cut of C. Hence,

$$D - p + p_1 + \cdots + p_{n-2} \in \Lambda$$

so

$$\Lambda = |D - p + p_1 + \cdots p_{n-2}|$$

Note that $p \neq p_1, \ldots, p_{n-2}$. In fact, if $p = p_{n-2}$, say, then $\Lambda = |D + p_1 + \cdots + p_{n-3}|$, which has clearly the fixed part $p_1 + \cdots + p_{n-3}$, a contradiction.

If there is another singular point $q' \in C'$, then, by the same reason, there are $q, q_1, \ldots, q_{n-2}$ on C such that

$$\Lambda = |D - q + q_1 + \cdots + q_{n-2}|$$

Hence

$$p_1 + \cdots + p_{n-2} + q \sim q_1 + \cdots + q_{n-2} + p$$

Note that $p \neq q$. By 1 again, there is $r \in C$ such that

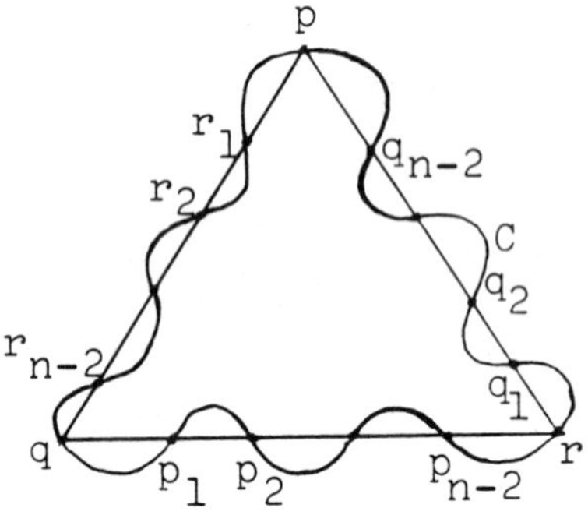

FIGURE 5.16

$$p_1 + \cdots + p_{n-2} + q + r \quad \text{and}$$

$$q_1 + \cdots + q_{n-2} + p + r$$

are line cuts of C. Put

$$C.\overline{pq} = p + q + r_1 + \cdots + r_{n-2}$$

(See Fig. 5.16.) Then

$$\Lambda = |D - r + r_1 + \cdots + r_{n-2}| = |2D - p - q - r|$$

and

$$r' = \Phi_\Lambda(r_1) = \cdots = \Phi_\Lambda(r_{n-2})$$

is another singular point of C'.

It is clear that there is no other singular point on C' than p', q', and r'. The equality $\Lambda = |2D - p - q - r|$ holds even if

$$p \neq q = r \ (p' \neq q' = r') \quad \text{or} \quad p = q = r \ (p' = q' = r')$$

(See Fig. 5.17.)

Conversely, take (not necessarily distinct) three points p, q, and r and consider the linear system

$$\Lambda = |2D - p - q - r|$$

on C of degree 2n - 3, where D is a line cut of C.

Note that, for conics G and H, if C.G = C.H, then G = H by the Max Noether Theorem (Theorem 1.4.12) for $\ell = m = 2 < n$. Hence, by Proposition 1.4.11,

$$\dim \Lambda \geq \dim \Lambda(2; C; p + q + r) = 2$$

By 6,

$$\dim \Lambda = 2 \quad \text{so} \quad \Lambda = g^2_{2n-3}$$

We show that Λ has no fixed point if and only if p, q, and r are not collinear. In fact, if s is a fixed point of Λ, then

$$\dim |2D - p - q - r - s| = 2$$

so $|2D - p - q - r - s|$ is a g^2_{2n-4}. By 3, this linear system has the fixed part $s_1 + \cdots + s_{n-4}$ and

$$|2D - p - q - r - s| = |D| + s_1 + \cdots + s_{n-4}$$

Hence,

$$D \sim p + q + r + s + s_1 + \cdots + s_{n-4} \tag{*}$$

This means that p, q, and r are collinear. Conversely, if p, q, and r are collinear, then there are points s, $s_1, \ldots, s_{n-4}$ on C such that (*) holds. Then

$$2D - p - q - r - s \sim D + s_1 + \cdots + s_{n-4}$$

Hence, $\dim |2D - p - q - r - s| = 2$, so s is a fixed point of $|2D - p - q - r|$.

Q.E.D.

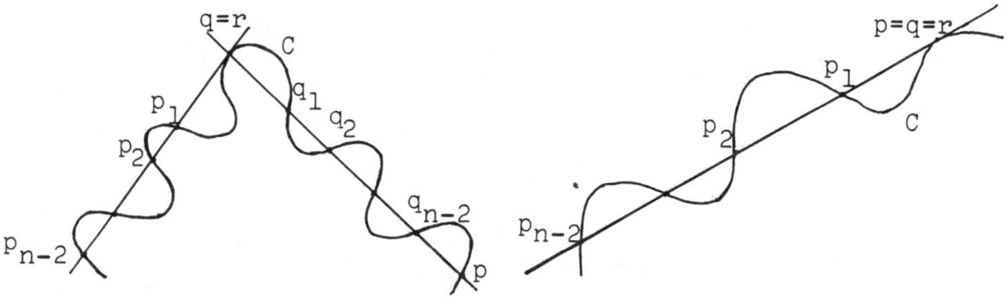

FIGURE 5.17

Remark 5.3.18. $\Phi_{|2D-p-q-r|}: C \to C'$ in 5 of the theorem is nothing but the Cremona transformation ϕ_0 in Exercise 4 of Sec. 3.4, if p, q, and r are mutually distinct.

Corollary 5.3.19. Let C_1 and C_2 be nonsingular plane curves of degree n. Then C_1 and C_2 are biholomorphic if and only if they are projectively equivalent, that is, there is $\sigma \in \mathrm{Aut}(\mathbb{P}^2)$ such that $\sigma(C_1) = C_2$.

Proof. The assertion is trivial for n = 1 and 2. See Exercise 4 of Sec. 5.2 for n = 3. If $n \geq 4$, then the assertion follows from 3 of the theorem. Q.E.D.

As an application, we state

Theorem 5.3.20. Let V and W be nonsingular hypersurfaces in $\mathbb{P}^{r+1}$ of degree n defined by the equations

$$V: X_{r+1}^n = F(X_0, \dots, X_r)$$

$$W: X_{r+1}^n = G(X_0, \dots, X_r)$$

where F and G are homogeneous polynomials in $X_0, \dots, X_r$ of degree n. Suppose that $(n, r) \neq (4, 2)$. Then V and W are biholomorphic if and only if there is $\sigma \in \mathrm{Aut}(\mathbb{P}^r)$ such that

$$\sigma\{F = 0\} = \{G = 0\}$$

For the proof of the theorem, we use Corollary 5.3.19 for r = 1. For $r \geq 2$, we use a theorem in Matsumura-Monsky [64]. See Namba [72] for details.

Next, let $\Lambda = g_n^r$ be a fixed point free linear system on a compact Riemann surface M of genus g. Let W be the (r + 1)-dimensional vector subspace of $\Gamma([D])$ corresponding to Λ, where $D \in \Lambda$. Put

$$S(W) = \bigoplus_{m=0}^{\infty} S^m(W), \quad \text{the symmetric tensor algebra of W}$$

Consider the ring homomorphism

$$\alpha: \xi_1 \circ \cdots \circ \xi_m \in S(W) \to \xi_1 \cdots \xi_m \in \bigoplus_{m=0}^{\infty} \Gamma([mD])$$

Then every element of the kernel of the linear map

$$\alpha_m = \alpha | S^m(W) : S^m(W) \to \Gamma([mD])$$

defines a hypersurface in $\mathbb{P}^r$ of degree m containing the curve $C = \Phi_\Lambda(M)$, and vice versa.

Theorem 5.3.21 (M. Noether-Enriques-Petri). Let $g \geq 4$ and M be nonhyperelliptic.

1. $\alpha: S(\Gamma(K)) \to \bigoplus \Gamma(mK)$ is surjective.
2. The ideal ker (α) is generated by ker (α_2) and ker (α_3).
3. The ideal ker (α) is generated by ker (α_2), except the following two cases: (a) M is trigonal, (b) M is biholomorphic to a nonsingular plane quintic curve ($g = 6$).

For the proof of the theorem, see Saint-Donat [84] or Shokurov [91].

The theorem implies that the canonical curve C_K is an intersection of quadric and cubic hypersurfaces in $\mathbb{P}^{g-1}$. Moreover, C_K is an intersection of quadric hypersurfaces, except cases a and b. See Example 5.1.21 and Exercise 7 of Sec. 5.1.

If M is trigonal, then the intersection of all quadric hypersurfaces in ker (α_2) is a rational ruled surface (see Exercise 3 of Sec. 4.2), containing C_K (see [84] or [91]). If M is biholomorphic to a nonsingular plane quintic curve, then the intersection is the Veronese surface $\Phi_{|2H|}(\mathbb{P}^2)$ in $\mathbb{P}^5$, where H is a line in $\mathbb{P}^2$.

A quadric hypersurface Q in $\mathbb{P}^r$ is said to be of rank k ($1 \leq k \leq r + 1$) if Q is defined by the equation

$$X_0^2 + \cdots + X_{k-1}^2 = 0$$

for a suitable homogeneous coordinate system $(X_0 : \cdots : X_r)$ in $\mathbb{P}^r$. We write $k = \operatorname{rank} Q$. The linear subspace

$$V_Q = \{X_0 = \cdots = X_{k-1} = 0\}$$

of $\mathbb{P}^r$ is contained in Q and is called the vertex of Q.

Lemma 5.3.22. Let $\Lambda = g_d^r$ ($r \geq 3$) be a fixed point free linear system on M. Let $\Lambda_1 = g_a^1$ and $\Lambda_2 = g_b^1$ ($a + b = d$) be linear pencils on M such that

$$D_1 + D_2 \in \Lambda \quad \text{for all } D_1 \in \Lambda_1 \quad \text{and } D_2 \in \Lambda_2 \tag{*}$$

Then the pair (Λ_1, Λ_2) induces a quadric hypersurface $Q = Q(\Lambda_1, \Lambda_2)$ in $\mathbb{P}^r$ of rank 3 or 4 such that $C = \Phi_\Lambda(M) \subset Q$. Q is of rank 3 if and only if

$$\Lambda_1 - F_1 = \Lambda_2 - F_2$$

where F_1 (respectively F_2) is the fixed part of Λ_1 (respectively Λ_2). Conversely, any quadric hypersurface Q in $\mathbb{P}^r$ of rank 3 or 4 such that $C \subset Q$ can be constructed by a pair (Λ_1, Λ_2) of linear pencils Λ_1 and Λ_2 satisfying (*).

Proof. Let $\{\xi_0, \xi_1\}$ (respectively $\{\eta_0, \eta_1\}$ be a basis of the vector subspace of $\Gamma([D_1])$ (respectively $\Gamma([D_2])$) corresponding to Λ_1 (respectively Λ_2). ($D_1 \in \Lambda_1$ and $D_2 \in \Lambda_2$). Then, by (*),

$$(\xi_j \eta_k) = (\xi_j) + (\eta_k) \in \Lambda \quad \text{for } j, k = 0, 1$$

$Q = Q(\Lambda_1, \Lambda_2)$ is defined by the equation

$$(\xi_0 \eta_0) \circ (\xi_1 \eta_1) - (\xi_0 \eta_1) \circ (\xi_1 \eta_0) = 0$$

It can be easily shown that Q does not depend on the choice of $\{\xi_0, \xi_1\}$ and $\{\eta_0, \eta_1\}$.

Q is clearly of rank ≤ 4. Q cannot be of rank ≤ 2, for $C = \Phi_\Lambda(M)$ is irreducible and nondegenerate. Q is of rank 3 if and only if either

$$\xi_0 \eta_0 = \xi_1 \eta_1 \quad \text{or} \quad \xi_0 \eta_1 = \xi_1 \eta_0$$

that is, if and only if

$$\Lambda_1 - F_1 = \Lambda_2 - F_2$$

The converse is easy to prove. Q.E.D.

The following lemma is easy to show.

Lemma 5.3.23. Let $Q = Q(\Lambda_1, \Lambda_2)$ be the quadric hypersurface in $\mathbb{P}^r$ defined in Lemma 5.3.22. If rank Q = 4, then there are just two one-parameter families $\{S_\lambda\}_{\lambda \in \mathbb{P}^1}$ and $\{T_\mu\}_{\mu \in \mathbb{P}^1}$ of (r - 2)-planes which are contained in Q and contain the vertex V_Q of Q. Moreover, for every $\lambda \in \mathbb{P}^1$, the projection $\pi_{S_\lambda} \cdot \Phi_\Lambda : M \to \mathbb{P}^1$ (respectively $\pi_{T_\lambda} \cdot \Phi_\Lambda : M \to \mathbb{P}^1$) corresponds to $\Lambda_1 - F_1$ (respectively $\Lambda_2 - F_2$). $F_1 + F_2$ is contained in V_Q. If rank Q = 3, then there is a unique one-parameter family $\{S_\lambda\}_{\lambda \in \mathbb{P}^1}$ of (r - 2)-planes which are contained in Q and contain V_Q. Moreover, for every $\lambda \in \mathbb{P}^1$, the

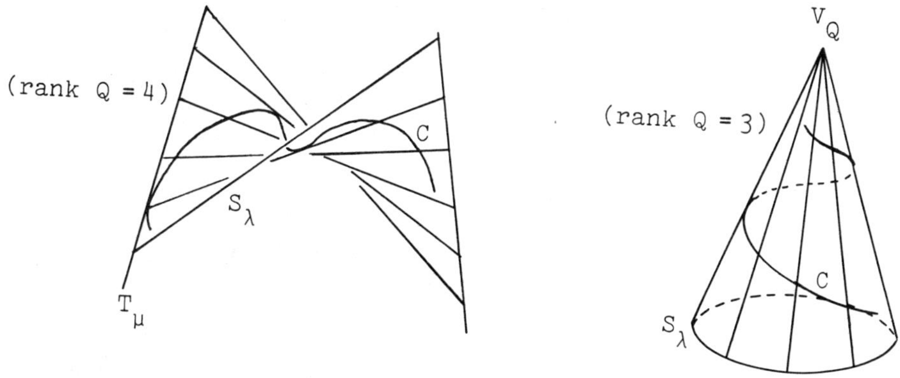

FIGURE 5.18

projection $\pi_{S_\lambda} \cdot \Phi_\Lambda : M \to \mathbb{P}^1$ (respectively $\pi_{T_\lambda} \cdot \Phi_\Lambda : M \to \mathbb{P}^1$) corresponds to $\Lambda_1 - F_1$. F_1 is contained in V_Q. (See Fig. 5.18.)

Conversely,

Lemma 5.3.24. Let $\Lambda = g_n^r$ ($r \geq 3$) and $C = \Phi_\Lambda(M)$ be as in Lemma 5.3.22. Let S and S' be distinct $(r - 2)$-planes in $\mathbb{P}^r$ such that $\pi_S \cdot \Phi_\Lambda: M \to \mathbb{P}^1$ and $\pi_{S'} \cdot \Phi_\Lambda: M \to \mathbb{P}^1$ correspond to the same linear pencil Λ_1. Then there is a linear pencil Λ_2 on M such that (1) (Λ_1, Λ_2) satisfies the condition (*) in Lemma 5.3.22 and (2) S and S' are members of the same one-parameter family $\{S_\lambda\}_{\lambda \in \mathbb{P}^1}$ of $(r - 2)$-planes in $\mathbb{P}^r$ which are contained in $Q = Q(\Lambda_1, \Lambda_2)$ and contain the vertex V_Q of Q.

Proof. Let S and S' be defined by the equations

$$S = \{\omega_1 = \omega_2 = 0\}$$

$$S' = \{\omega_3 = \omega_4 = 0\}$$

where $\omega_j \in \Gamma([D])$ for $1 \leq j \leq 4$ ($D \in \Lambda_1$). Then, by choosing ω_1 and ω_2 suitably, we may assume that

$$\frac{\omega_2}{\omega_1} = \frac{\omega_4}{\omega_3} : M \to \mathbb{P}^1$$

Let Q be the quadric hypersurface in $\mathbb{P}^r$ defined by

$$Q: \omega_1 \circ \omega_4 - \omega_2 \circ \omega_3 = 0$$

The vertex V_Q of Q is given by

$$V_Q = \{\omega_1 = \omega_2 = \omega_3 = \omega_4 = 0\}$$

Put $v = \dim V_Q$ and let F_0 be the pull-back by Φ_Λ of the v-plane cut of C by V_Q (see Sec. 1.6). Now let Λ' be the linear pencil corresponding to

$$\frac{\omega_3}{\omega_1} = \frac{\omega_4}{\omega_2} : M \to \mathbb{P}^1$$

Put $\Lambda_2 = \Lambda' + F_0$. (The fixed part of Λ_2 is F_0.) Then, clearly,

$$Q = Q(\Lambda_1, \Lambda_2)$$

and Q satisfies the condition. Q.E.D.

Example 5.3.25. (1) Consider the linear system $\Lambda = |3(\infty)|$ on $\mathbb{P}^1$. Then $C = \Phi_\Lambda(\mathbb{P}^1)$ is the twisted cubic in $\mathbb{P}^3$. Let Λ_1 and Λ_2 be linear pencils on $\mathbb{P}^1$ such that $D_1 + D_2 \in \Lambda$ for all $D_1 \in \Lambda_1$ and $D_2 \in \Lambda_2$. We may assume that $\Lambda_1 = |(\infty)|$ and Λ_2 is a linear subsystem of $|2(\infty)|$. If Λ_2 has no fixed point, then $Q(|(\infty)|, \Lambda_2)$ is a nonsingular quadric surface in $\mathbb{P}^3$ which contains C. If Λ_2 has a fixed point p, then $\Lambda_2 = |(\infty)| + p$ and $Q(|(\infty)|, |(\infty)| + p)$ is a quadric cone in $\mathbb{P}^3$ which contains C and has the vertex point p. The set of all quadric surfaces in $\mathbb{P}^3$ which contain C forms a two-parameter family

$$\{Q(|(\infty)|, \Lambda_2) \mid \Lambda_2 \subset |2(\infty)|\}$$

(2) Let $M = \mathbb{C}/(\mathbb{Z}\omega_1 + \mathbb{Z}\omega_2)$ be a complex 1-torus and $\Lambda = |4(0)|$. Let Λ_1 and Λ_2 be the linear pencils on M such that $D_1 + D_2 \in \Lambda$ for all $D_1 \in \Lambda_1$ and $D_2 \in \Lambda_2$. Then $\deg \Lambda_1 = \deg \Lambda_2 = 2$, so

$$\Lambda_1 = |(p_1) + (p_2)| \quad \text{and} \quad \Lambda_2 = |(p_3) + (p_4)|$$

where

$$p_1 + p_2 + p_3 + p_4 = 0$$

Put

$$\lambda = \wp(p_1 + p_2) = \wp(p_3 + p_4) \qquad (\wp\text{: Weierstrass } \wp\text{-function})$$

Then, clearly, $Q(\Lambda_1, \Lambda_2)$ depends only on $\lambda \in \mathbb{P}^1$, so we may put

$$Q(\Lambda_1, \Lambda_2) = Q_\lambda$$

The set of all quadric surfaces in $\mathbb{P}^3$ which contain $C = \Phi_\Lambda(M)$ forms a one-parameter family $\{Q_\lambda\}_{\lambda \in \mathbb{P}^1}$. Note that $\Lambda_1 = \Lambda_2$ if and only if λ is a branch point of $\mathfrak{p}$, so there are just four quadric cones in $\{Q_\lambda\}_{\lambda \in \mathbb{P}^1}$.

A similar assertion holds for any linear system g_4^3 on M.

(3) Let M be a genus g = 2 and $\Lambda = g_5^3$ be a linear system on M. Then Λ is complete and very ample. Let Λ_1 and Λ_2 be linear pencils such that $D_1 + D_2 \in \Lambda$ for all $D_1 \in \Lambda_1$ and $D_2 \in \Lambda_2$. Then, we may assume that

$$\Lambda_1 = |K| \quad \text{and} \quad \Lambda_2 = g_3^1$$

Λ_2 is uniquely determined by Λ. ($\Lambda_2 = |D - K|$ for $D \in \Lambda$.) Hence there is a unique quadric surface $Q = Q(|K|, \Lambda_2)$ in $\mathbb{P}^3$ which contains $C = \Phi_\Lambda(M)$. Q is a cone if and only if $|D - 2K|$ is nonempty ($D \in \Lambda$).

Proposition 5.3.26. Let $f: M \to \mathbb{P}^1$ be of degree $\leq g - 1$ (respectively $\leq \frac{1}{2}(g+1)$). Then there is a quadric hypersurface Q of rank ≤ 4 (respectively rank 3) in $\mathbb{P}^{g-1}$ which contains the canonical curve $C_K = \Phi_K(M)$ such that $f = \pi_S$ for a $(g-3)$-plane S contained in Q.

Proof. Let Λ_1 be the linear pencil on M corresponding to f. If deg f $\leq g - 1$, then, by the Riemann-Roch Theorem, dim $|K - D_\infty(f)| \geq 1$. Take a linear pencil $\Lambda_2 \subset |K - D_\infty(f)|$. Then $Q = Q(\Lambda_1, \Lambda_2)$ satisfies the condition.

If deg f $\leq \frac{1}{2}(g+1)$, then, by the Riemann-Roch Theorem, $|K - 2D_\infty(f)|$ is nonempty. Take $D_0 \in |K - 2D_\infty(f)|$ and put $\Lambda_2 = \Lambda_1 + D_0$. Λ_2 is then a linear pencil with the fixed part D_0. Then $Q = Q(\Lambda_1, \Lambda_2)$ satisfies the condition. Q.E.D.

Problem. Let $g \geq 4$ and M be nonhyperelliptic. Let

$$\alpha_2: S^2(\Gamma(K)) \to \Gamma(2K)$$

be the surjective linear map in Theorem 5.3.21. Is ker (α_2) spanned by quadric hypersurfaces in P^{g-1} of rank < 4?

If M is trigonal, then the problem is affirmative (Andreotti-Mayer [5]). If $g \leq 6$, then the problem is affirmative (Arbarello-Harris [7]). See also Mumford [69] and Namba [71].

Note 5.3.27.

1. Many variations and modifications of the Clifford theorem are known. See, for example, Martens [61] and Gunning [35].
2. n-gonal compact Riemann surfaces have many interesting properties. See, for example, Coolidge [22] and Coppens [23].
3. A good refinement of Theorem 5.3.14 was obtained by Kato [50].
4. For deep analysis on defining equations of projective varieties, see Mumford [68] and Fujita [28].

Exercises

1. Prove the following theorem by Accola.

Theorem (Accola [2]). Let M and N be compact Riemann surfaces of genus g and g_0, respectively. Let ϕ: M $\to$ N be a (possibly ramified) double covering. Then, for any f : M $\to$ $\mathbb{P}^1$ of degree $n \leq g - 2g_0$, there is h: N $\to$ $\mathbb{P}^1$ such that $f = h \cdot \phi$. In particular, n is even.

2. Let α and β be distinct complex numbers in $\mathbb{C} - \{0, 1\}$. Let $M_{\alpha,\beta}$ be the compact Riemann surface defined by the equation

$$M_{\alpha,\beta}: y^3 = x(x-1)(x-\alpha)(x-\beta)$$

(1) $M_{\alpha,\beta} \simeq M_{\gamma,\delta}$ (biholomorphic) if and only if there is $\sigma \in \mathrm{Aut}(\mathbb{C})$ such that $\{\sigma(0), \sigma(1), \sigma(\alpha), \sigma(\beta)\} = \{0, 1, \gamma, \delta\}$. (Hint Use Corollary 5.3.19.) (2) Let γ be a closed arc in x-plane as in Fig. 5.19. Put

$$\omega(\alpha, \beta) = \int_\gamma \frac{dx}{y}$$

Then $\omega = \omega(\alpha, \beta)$ is a multivalued holomorphic function on

$$D = \{(\alpha, \beta) \in \mathbb{C}^2 \mid \alpha\beta(\alpha - 1)(\beta - 1)(\alpha - \beta) \neq 0\}$$

and satisfies the following Appell's hypergeometric differential equation:

$$9\alpha(1-\alpha)(\alpha-\beta)r = 3(5\alpha^2 - 4\alpha\beta - 3\alpha + 2\beta)p + 3\beta(1-\beta)q + (\alpha-\beta)\omega,$$

$$3(\alpha-\beta)s = p - q, \qquad (*)$$

$$9\beta(1-\beta)(\beta-\alpha)t = 3\alpha(1-\alpha)p + 3(5\beta^2 - 4\alpha\beta - 3\beta + 2\alpha)q + (\beta-\alpha)\omega$$

x-plane

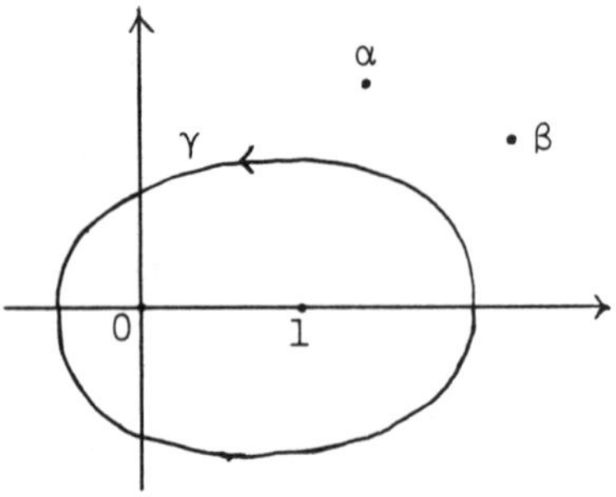

FIGURE 5.19

where $p = \partial\omega/\partial\alpha$, $q = \partial\omega/\partial\beta$, $r = \partial^2\omega/\partial\alpha^2$, $s = \partial^2\omega/\partial\alpha\,\partial\beta$, and $t = \partial^2\omega/\partial\beta^2$. (3) Let $\tilde{D}$ be the universal covering space of D. Let $\omega_j(z)$ ($z \in \tilde{D}$, $j = 0, 1, 2$) be (suitably chosen) linearly independent solutions of (*). Then the image of the holomorphic map

$$\Psi: z \in \tilde{D} \to (\omega_0(z): \omega_1(z): \omega_2(z)) \in \mathbb{P}^2$$

is contained in the hyperball $B = \{(1: \lambda_1: \lambda_2) \in \mathbb{P}^2 \mid |\lambda_1|^2 + |\lambda_2|^2 < 1\}$. Moreover, the inverse map

$$\Psi^{-1}: B \to \tilde{D}$$

can be defined and gives automorphic functions, called <u>Picard's modular functions</u> (see Picard [81] and Shiga [90]).

3. Let C be a nonsingular plane curve of degree $n \geq 5$. Then every linear pencil $\Lambda = g^1_{2n-4}$ on C without fixed point can be obtained as

$$\Lambda = |2D - p - q - r - s| = \{C\cdot(\lambda F + \mu G)\}_{(\lambda:\mu) \in \mathbb{P}^1} - (p + q + r + s)$$

where (1) D is a line cut of C, (2) p, q, r, and s are (not necessarily distinct) points on C in general position, that is, no three points of them are collinear, and (3) F and G are distinct conics passing through p, q, r, and s.

5.4 RECENT TOPICS ON LINEAR SYSTEMS ON A CURVE

In this last section, we talk about some recent topics on linear systems on a compact Riemann surface without giving proofs, which are often based on various new methods of modern algebraic geometry. To do this, however, we adhere to our point of view.

Let M be a compact Riemann surface of genus g. For integers $r \geq 0$ and $n \geq 1$, put

$$\mathbb{G}^r_n(M) = \{g^r_n \text{ on } M\} \quad \text{and}$$

$$\phi: \Lambda \in \mathbb{G}^r_n(M) \to \phi(D) \in J(M) \qquad (D \in \Lambda)$$

where $\phi(D)$ is the image of D of the Jacobi map ϕ defined in Sec. 5.2. ϕ is called the <u>Jacobi map</u> again. Note that $\phi(\mathbb{G}^r_n(M)) = W^r_n(M)$. (Do not confuse $\mathbb{G}^r_n(M)$ and $G^r_n(M)$ in Sec. 5.2!)

Theorem 5.4.1. $\mathbb{G}_n^r(M)$ is a compact complex analytic space such that $\phi\colon \mathbb{G}_n^r(M) \to J(M)$ is holomorphic.

Here, a complex analytic space, or simply a complex space is, roughly speaking, a Hausdorff space which is locally identified with analytic sets in polydiscs in $\mathbb{C}^N$'s. See Gunning-Rossi [36] or Narasimhan [74] for the rigorous definition.

$\mathbb{G}_n^0(M)$ can be identified with S^nM.

A linear system $\Lambda \in \mathbb{G}_n^r(M)$ is said to be semi-regular if the linear map

$$\xi \otimes \eta \in W \otimes \Gamma([K - D]) \to \xi\eta \in \Gamma(K) \qquad (D \in \Lambda)$$

is injective, where W is the $(r + 1)$-dimensional vector subspace of $\Gamma([D])$ corresponding to Λ. Clearly, every element of $\mathbb{G}_n^0(M) = S^nM$ is semi-regular.

Theorem 5.4.2 (Semi-regularity theorem). If $\Lambda \in \mathbb{G}_n^r(M)$ is semi-regular, then Λ is a nonsingular point of $\mathbb{G}_n^r(M)$ and $\dim_\Lambda \mathbb{G}_n^r(M) = (r + 1)(n - r) - rg$.

Corollary 5.4.3. If $\Lambda = |D| \in \mathbb{G}_n^r(M)$ is complete and semi-regular, then $\phi(\Lambda) = \phi(D)$ is a nonsingular point of $W_n^r(M)$ and $\dim_{\phi(D)} W_n^r(M) =$ $(r + 1)(n - r) - rg$.

In contrast to the semi-regularity,

Theorem 5.4.4. For $D \in S^nM$ with $\dim |D| = r$, suppose that the linear map

$$\xi \otimes \eta \in \Gamma([D]) \otimes \Gamma([K - D]) \to \xi\eta \in \Gamma(K)$$

is surjective. Then $\phi(D)$ is an isolated point of $W_n^r(M)$.

Put

$$\mathbb{F}_n^r(M) = \{\Lambda \in \mathbb{G}_n^r(M) \mid \Lambda \text{ has a fixed point}\}$$

Then

Lemma 5.4.5. $\mathbb{F}_n^r(M)$ is a (closed) analytic subset of $\mathbb{G}_n^r(M)$.

Put

$$\mathrm{Hol}\,(M, \mathbb{P}^r) = \{\text{holomorphic maps of } M \text{ into } \mathbb{P}^r\}$$

$$\mathrm{Hol}_n\,(M, \mathbb{P}^r) = \{f \in \mathrm{Hol}\,(M, \mathbb{P}^r) \mid f \text{ is nondegenerate and } \deg f = n\}$$

$\mathrm{Hol}\,(M, \mathbb{P}^r)$ is then a complex space which is called a Douady space (Douady

[27]). $\mathrm{Hol}_n(M, \mathbb{P}^r)$ is an open (and closed) subspace of $\mathrm{Hol}(M, \mathbb{P}^r)$. The correspondence between linear systems g_n^r and nondegenerate holomorphic maps into $\mathbb{P}^r$ can be summarized as follows:

Theorem 5.4.6. $\mathrm{Hol}_n(M, \mathbb{P}^r)$ is a principal $\mathrm{Aut}(\mathbb{P}^r)$-bundle over $\mathbb{G}_n^r(M) - \mathbb{F}_n^r(M)$.

For the proofs of the above theorems, see Namba [71], in which a deformation theory of linear systems on projective manifolds is developed.

Next, we ask when $\mathbb{G}_n^r(M)$ is nonempty. One of the most important theorems in this direction is

Theorem 5.4.7 (Brill-Noether [15], Kleiman-Laksov [54, 55], Kempf [52]. If $\rho = (r+1)(n-r) - rg \geq 0$, then there is a g_n^r on M. Moreover, $\dim W_n^r(M) \geq \rho$.

The point of the proof of this theorem by Kleiman-Laksov [55] is to use Porteous' Formula in the Schubert calculus.

Corollary 5.4.8. (1) There is $f: M \to \mathbb{P}^1$ such that $\deg f \leq (g+3)/2$.
(2) There is a nondegenerate $f: M \to \mathbb{P}^2$ such that $\deg f \leq (2g+8)/3$.

Theorem 5.4.9 (Griffiths-Harris [34]). (1) If $\rho = (r+1)(n-r) - rg$ is negative and M is "general" (in the sense of moduli), then there is no g_n^r on M.
(2) If $\rho \geq 0$ and M is general, then $\dim W_n^r(M) = \rho$.

In the proof of the theorem, Griffiths-Harris [34] noted that $\dim W_n^r(M)$ does not decrease by specializations and constructed a singular curve which satisfies a similar condition. Their idea was used to prove

Theorem 5.4.10 (Gieseker [31]). Let M be general. Then, for any line bundle L on M, the linear map

$$\xi \otimes \eta \in \Gamma(L) \otimes \Gamma(K - L) \to \xi\eta \in \Gamma(K)$$

is injective.

This theorem was first stated in Petri [80], so was called Petri's conjecture. See Arbarello-Cornalba-Griffiths-Harris [6] for further discussion. Note that, if a linear system Λ is semi-regular, then every linear subsystem of Λ is also semi-regular. Hence, by Theorem 5.4.9 and 5.4.10,

Corollary 5.4.11. Let M be general. If $\rho = (r+1)(n-r) - rg \geq 0$, then $\mathbb{G}_n^r(M)$ is nonsingular and $\dim \mathbb{G}_n^r(M) = \rho$.

Another interesting recent result is

Theorem 5.4.12 (Fulton-Lazarsfeld [30]). If $\rho \geq 1$, then $W_n^r(M)$ is connected.

Their proof is also very interesting.

Corollary 5.4.13. If $\rho \geq 1$ and M is general, then $W_n^r(M)$ is irreducible.

Next, on the structure of $W_n(M) = \phi(S^n M)$, a beautiful theorem by Kempf says

Theorem 5.4.14 (Kempf [53]). Let $n \leq g - 1$ and $D \in S^n M$ be a divisor with $\dim |D| = r$.

1. $\phi(D)$ is a (singular) point with multiplicity $\binom{g-n+r}{r}$ of $W_n(M)$.
2. The tangent cone $TC_{\phi(D)}(W_n(M))$ at $\phi(D)$ to $W_n(M)$ is equal to

$$\cup_{E \in |D|} (d\phi)_E(T_E S^n M)$$

3. There are a neighborhood U of $\phi(D)$ in J(M) and a $(g - n + r) \times (r + 1)$-matrix valued holomorphic function u(x) on U with $u(\phi(D)) = 0$ such that (i) $W_n(M) \cap U$ is the set of zeros of all $(r + 1) \times (r + 1)$-minors of u(x) and (ii) the tangent cone $TC_{\phi(D)}(W_n(M))$ is the set of zeros of all $(r + 1) \times (r + 1)$-minors of the linear term of u(x) (in the power series expansion).

Corollary 5.4.15 (Weil). $W_{g-1}^1(M)$ is the singular locus of $W_{g-1}(M)$.

Note that $W_{g-1}(M)$ is a translation of the theta divisor Θ on J(M) (Theorem 5.2.26).

The tangent cone $TC_x(W_n(M))$ to $W_n(M)$ at $x \in J(M)$ is contained in the tangent space $T_x J(M)$ to J(M) at x. Consider the projectivized tangent cone $\widehat{TC}_x(W_n(M))$ to $W_n(M)$ at $x \in J(M)$. This is an algebraic set in the projective space $\mathbb{P}(T_x J(M))$, which can be, by translation, identified with

$$\mathbb{P}^{g-1} = |K|^* = \mathbb{P}(T_0 J(M)) \qquad \text{(compare Corollary 5.2.21)}$$

Corollary 5.4.16. Under the above identification,

$$\widehat{TC}_x(W_n(M)) = \cup_{D \in \phi^{-1}(x)} S_D$$

where S_D is the linear subspace in $\mathbb{P}^{g-1}$ spanned by D (see Sec. 5.3). Moreover, if $\dim \phi^{-1}(x) \geq 1$, then $\widehat{TC}_x(W_n(M))$ contains the canonical curve $C_K = \Phi_K(M)$.

If $|D|$ is a g^1_{g-1}, then $|K - D|$ is also a g^1_{g-1} by the Riemann-Roch Theorem. By Lemma 5.3.22, a quadric hypersurface $Q(|D|, |K - D|)$ in $\mathbb{P}^{g-1}$ of rank ≤ 4 which contains the canonical curve C_K can be constructed.

Corollary 5.4.17. If $|D|$ is a g^1_{g-1}, then

$$\widehat{TC}_{\phi(D)}(W_{g-1}(M)) = Q(|D|, |K - D|)$$

See Andreotti-Mayer [5] for detail and further discussion.

By Clifford's Theorem (Theorem 5.3.7), we can prove

Theorem 5.4.18. Let $g \geq 4$ and M be nonhyperelliptic.

1. (Namba [71]) $\mathbb{G}^1_{g-1}(M)$ is pure $(g - 4)$-dimensional.
2. (Martens [61], Andreotti-Mayer [5]) $W^1_{g-1}(M)$ is pure $(g - 4)$-dimensional.

Proposition 5.4.19. If M is hyperelliptic, then $\mathbb{G}^1_n(M)$ and $W^1_n(M)$ are pure $(n - 2)$-dimensional for $2 \leq n \leq g$.

Next, let $g \geq 2$ and T_g, $\{M_t\}_{t \in T_g}$, Γ_g be the Teichmüller space, the Teichmüller family, and the Teichmüller modular group, respectively. They have the following properties:

1. T_g is a complex manifold which is biholomorphic to a holomorphically convex bounded domain in $\mathbb{C}^{3g-3}$,
2. for any compact Riemann surface M of genus g, there is a point $t \in T_g$ such that M is biholomorphic to M_t,
3. for any point $o \in T_g$, there is a neighborhood U of o in T_g such that M_t is not biholomorphic to M_o for any $t \in U - \{0\}$,
4. Γ_g acts on T_g and on $\{M_t\}_{t \in T_g}$ properly discontinuously,
5. for points s and t in T_g, M_s and M_t are biholomorphic if and only if the there is $\sigma \in \Gamma_g$ such that $t = \sigma(s)$, and
6. for every point $t \in T_g$, the isotropy subgroup

$$\Gamma_g(t) = \{\sigma \in \Gamma_g \mid \sigma(t) = t\}$$

at t is canonically isomorphic to Aut (M_t) (see, for example, Bers [12]).

By H. Cartan's theorem (Cartan [17]),

Theorem 5.4.20. The quotient space $\mathbb{M}_g = T_g/\Gamma_g$ is an irreducible complex space of dimension 3g - 3 for $g \geq 2$.

$\mathbb{M}_g$ is the set of all holomorphically isomorphism classes of compact Riemann surfaces of genus g and is called the moduli space of compact Riemann surfaces of genus g.

$\mathbb{M}_g$ is in fact a quasi-projective variety, that is, a Zariski open set of a projective variety (Baily [8], Mumford [67]), and has canonical compactifications (Satake [85], Deligne-Mumford [24]).

It is a difficult problem to look at the structure of $\mathbb{M}_g$. The structure of $\mathbb{M}_2$ was analyzed completely by Igusa [48]. $\mathbb{M}_2$ is a rational variety. But it is an unsolved problem if $\mathbb{M}_3$ and $\mathbb{M}_4$ are rational varieties. For bigger g ($g \geq 25$ for odd g and $g \geq 40$ for even g), $\mathbb{M}_g$ is of general type (Harris-Mumford [38]).

Torelli's Theorem (Theorem 5.2.8) says that the natural map

$$i : \mathbb{M}_g \to \frac{\mathbb{H}_g}{S_p(g, \mathbb{Z})}$$

is injective. $S_p(g, \mathbb{Z})$ acts on $\mathbb{H}_g$ properly discontinuously. The irreducible complex space $\mathbb{H}_g/S_p(g, \mathbb{Z})$ of dimension $\frac{1}{2}g(g + 1)$ is called the moduli space of principally polarized Abelian varieties of dimension g. The map i is holomorphic. See Oort-Steenbrink [77] for more.

If $g \geq 4$, then i is not surjective. The Schottky problem asks to characterize $i(\mathbb{M}_g)$ (or its closure $\overline{i(\mathbb{M}_g)}$) by some special properties. See Mumford [69] and Beauville [10].

Now,

Theorem 5.4.21 (Namba [71]). For $g \geq 2$ (respectively g = 1), the disjoint union

$$\mathbb{G}^r_{g,n} = \cup_{t \in T_g} \mathbb{G}^r_n(M_t)$$

$$\text{(respectively } \mathbb{G}^r_{1,n} = \cup_{\tau \in H} \mathbb{G}^r_n\left(\frac{\mathbb{C}}{\mathbb{Z} + \mathbb{Z}\tau}\right) \quad (\mathbb{H} = \text{the upper half plane}))$$

is a complex space such that the natural projection

$$\pi : \mathbb{G}^r_{g,n} \to T_g$$

$$\text{(respectively } \pi : \mathbb{G}^r_{1,n} \to \mathbb{H})$$

is a proper holomorphic map.

It can be shown that $\mathbb{G}^1_{g,n}$ is nonsingular and of dimension 2g + 2n - 5 for $g \geq 2$ (Namba [71]).

Consider pairs (M, Λ) of compact Riemann surfaces M of genus g and linear systems $\Lambda = g^r_n$ on M. Two such pairs (M, Λ) and (M', Λ') are said to be equivalent if there is a holomorphic isomorphism $\alpha: M \to M'$ such that $\alpha(\Lambda) = \Lambda'$, that is, $\alpha(D) \in \Lambda'$ for all $D \in \Lambda$, and vice versa. An equivalence class is denoted by [M, Λ]. The set of all equivalence classes is denoted by $\mathbb{L}^r_{g,n}$.

Theorem 5.4.22. For $g \geq 1$, $\mathbb{L}^r_{g,n}$ is a complex space such that the natural projection $\pi: \mathbb{L}^r_{g,n} \to \mathbb{M}_g$ is a proper holomorphic map. (The image $\pi(\mathbb{L}^r_{g,n})$ is denoted by $\mathbb{M}^r_{g,n}$.)

Proof (Sketch). (1) Suppose that $g \geq 2$. Then Γ_g acts properly discontinuously on $\mathbb{G}^r_{g,n}$. Hence,

$$\mathbb{L}^r_{g,n} = \frac{\mathbb{G}^r_{g,n}}{\Gamma_g}$$

is a complex space which satisfies the condition.

(2) Suppose that g = 1. Let $M_\tau = \mathbb{C}/(\mathbb{Z} + \mathbb{Z}\tau)$ ($\tau \in \mathbb{H}$) be a complex 1-torus and $\Lambda \in \mathbb{G}^r_n(M_\tau)$. Then, by Example 5.3.1, there is a linear subsystem Λ' in |n(0)| such that (M_τ, Λ) is equivalent to (M_τ, Λ'). Hence, we may assume that Λ is itself a linear subsystem of |n(0)|. Then, we can find a group $\tilde{B}$ acting properly discontinuously on the fiber bundle

$$\hat{\mathbb{G}}^r_{1,n} = \bigcup_{\tau \in \mathbb{H}} G_\tau(r, |n(0)|)$$

where $G_\tau(r, |n(0)|)$ is the Grassmann variety of all r-planes in the linear system |n(0)| on M_τ. Then

$$\mathbb{L}^r_{1,n} = \frac{\hat{\mathbb{G}}^r_{1,n}}{\tilde{B}}$$

is a complex space which satisfies the condition (see Namba [71] for details). Note that every fiber $\pi^{-1}(\tau)$ of

$$\pi: \mathbb{L}^r_{1,n} \to \mathbb{M}_1 = \frac{\mathbb{H}}{SL(2, \mathbb{Z})}$$

is biholomorphic to the complex space G(n - r - 2, n - 1)/B in Example 5.3.1.

Q.E.D.

Remark 5.4.23. $\mathbb{L}^r_{0,n} = G(n - r - 1, n)/\mathrm{Aut}(\mathbb{P}^1)$ is not even Hausdorff in general.

Finally, let C and C' be nondegenerate irreducible curves in $\mathbb{P}^r$ ($r \geq 2$) of degree n and genus g. C and C' are said to be projectively equivalent, $C \sim C'$, if there is $\sigma \in \mathrm{Aut}(\mathbb{P}^r)$ such that $\sigma(C) = C'$. Put

$$\mathscr{C}^r_{g,n} = \frac{\{C\}}{\sim}$$

Theorem 5.4.24. If $g \geq 1$, then $\mathscr{C}^r_{g,n}$ is a complex space.

Proof (Sketch). For two such C and C' in $\mathbb{P}^r$, let $f: M \to C$ and $f': M' \to C'$ be nonsingular models of C and C', respectively. Then f (respectively f') gives a linear system $\Lambda = g^r_n$ on M (respectively Λ' on M') without fixed point. It is then easy to see that $C \sim C'$ if and only if (M, Λ) and (M', Λ') are equivalent. Hence $\mathscr{C}^r_{g,n}$ can be identified with a subset of $\mathbb{L}^r_{g,n}$, which is an open set as is easily seen. Q.E.D.

Note 5.4.25. For recent topics on linear systems on curves, see Arbarello-Cornalba-Griffiths-Harris [6]. For various recent topics on curve theory, the beautiful book by Mumford [69] is strongly recommended.

Exercises

1. If M has genus 2, then $\phi: S^2M \to J(M)$ is a blowing up one point.
2. Let Λ be a g^1_n on M with the fixed part D_0 (≥ 0). Then Λ is semi-regular if and only if $i(2D - D_0) = 0$ for $D \in \Lambda$.
3. Let Λ be a g^2_n on M without fixed point such that $\Phi_\Lambda: M \to \mathbb{P}^2$ is birational onto a nodal curve C of degree n with k nodes, ($g = \frac{1}{2}(n-1)(n-2) - k$). Then (M, Λ) is a nonsingular point of the complex space $\mathbb{G}^2_{g,n}$.
4. Let Λ (and M) be as in Exercise 3. Suppose n and k satisfy one of the following conditions:

   (i) $n = 4$ and $k = 0$

   (ii) $n = 5$ and $0 \leq k \leq 1$,

   (iii) $n \geq 6$ and $0 \leq k \leq n - 3$

   Then (1) Λ is complete and (2) $\phi(\Lambda)$ is an isolated point of $W^2_n(M)$.

REFERENCES

1. R. D. M. Accola, On the number of automorphisms of a closed Riemann surface, Transac. Amer. Math. Soc., 131, 398-408 (1968).
2. R. D. M. Accola, Strongly branched coverings of closed Riemann surfaces, Proc. Amer. Math. Soc., 26, 315-322 (1970).
3. L. V. Ahlfors, Complex Analysis, Third ed., McGraw-Hill, New York, 1979.
4. L. V. Ahlfors and L. Sario, Riemann Surfaces, Princeton University Press, Princeton, New Jersey, 1960.
5. A. Andreotti and A. Mayer, On period relations for abelian integrals on algebraic curves, Ann. Scu. Norm. Sup. Pisa, 21, 189-238 (1967).
6. E. Arbarello, M. Cornalba, P. Griffiths, and J. Harris, Topics in the Theory of Algebraic Curves, To appear in Princeton University Press, Princeton, New Jersey.
7. E. Arbarello and J. Harris, Canonical curves and quadrics of rank 4, Compositio Math., 43, 145-179 (1980/81).
8. W. Baily, On the theory of θ-functions, the moduli of abelian varieties and the moduli of curves, Annals of Math., 75, 342-381 (1962).
9. H. F. Baker, Principles of Geometry, Cambridge University Press, Cambridge, 1922.
10. A. Beauville, Prym varieties and the Schottky problem, Inv. Math., 41, 149-196 (1977).
11. L. Bers, Riemann Surfaces, Lec. Notes, Courant Inst. (1957/58).
12. L. Bers, Uniformization, moduli and Kleinian groups, Bull. London Math. Soc., 4, 257-300 (1972).
13. S. Bochner and D. Montgomery, Groups on analytic manifolds, Annals of Math., 48, 659-669 (1947).
14. A. Borel and J. P. Serre, Le théorèm de Riemann-Roch (d'aprés Grothendieck), Bull. Soc. Math. France, 86, 97-136 (1958).

15. A. Brill and M. Noether, Über die algebraischen Functionen und ihre Anwendungen in der Geometrie, Math. Ann., 7, 269-310 (1874).
16. H. Cartan, Sur les groupes de transformations analytiques, Act. Sc. et Indus., Hermann, Paris, 1935.
17. H. Cartan, Quotient d'un espace analytique par un groupe d'automorphisms, Algebraic Geometry and Topology, A sympos. in honor of S. Lefschetz, Princeton University Press, Princeton, New Jersey, 1957, pp. 90-102.
18. A. Cayley, Elliptic Functions, Dover, New York, 1961.
19. S. S. Chern, Complex Manifolds without Potential Theory, Van Nostrand, New York, 1967.
20. C. Chevalley, Introduction to the Theory of Algebraic Functions of One Variable, Amer. Math. Soc., Providence, Rhode Island, 1951.
21. W. L. Chow and K. Kodaira, On analytic surfaces with two independent mermorphic functions, Proc. Nat. Acad. Sci. U.S.A., 38, 319-325 (1952).
22. J. L. Coolidge, A Treatise on Algebraic Plane Curves, Dover, New York, 1959.
23. M. R. M. Coppens, A study of 4-gonal curves of genus $g \geq 7$, Univ. Utrecht, Preprint, 221 (1981).
24. P. Deligne and D. Mumford, The irreducibility of the space of curves of given genus, Publ. Math. I.H.E.S., 36, 75-110 (1969).
25. Del Pezzo, Sulla quintica con cinque punti cuspidale, Napoli Rendiconti, Ser. 2, 3, 46 (1889).
26. G. de Rham, Varietes Differentiables, Hermann, Paris, 1960.
27. A. Douady, Le problème des modules pour les sous-espaces analytiques compacts d'un espace analytiques donné, Ann. Inst. Fourier, 16, 1-98 (1966).
28. T. Fujita, Defining equations for certain types of polarized varieties, Complex Analysis and Algebraic Geometry, edited by W. L. Baily and T. Shioda, Iwanami Shoten - Cambridge University Press, New York, 1977.
29. W. Fulton, Algebraic Curves, Benjamin, Elmsford, New York, 1969.
30. W. Fulton and R. Lazarsfeld, On the connectedness of degeneracy loci and special divisors, Acta Math., 146, 271-283 (1981).
31. D. Gieseker, Stable curves and special divisors, Inv. Math., 66, 251-275 (1982).
32. P. Griffiths, Variations of a theorem of Abel, Inv. Math., 35, 321-390 (1976).
33. P. Griffiths and J. Harris, Principles of Algebraic Geometry, Wiley, New York, 1978.
34. P. Griffiths and J. Harris, On the variety of special linear systems on a general algebraic curve, Duke Math., 47, 233-272 (1980).
35. R. C. Gunning, Lectures on Riemann Surfaces, Math. Notes, 2, 6, 12, Princeton University Press, Princeton, New Jersey.

36. R. C. Gunning and H. Rossi, Analytic Functions of Several Complex Variables, Prentice-Hall, New York, 1965.
37. J. Harris, The genus of space curves, Math. Ann., 249, 191-204 (1980).
38. J. Harris and D. Mumford, On the Kodaira dimension of the moduli space of curves, Inv. Math., 67, 23-86 (1982).
39. R. Hartshorne, Algebraic Geometry, Springer-Verlag, New York, 1977.
40. H. Hironaka, Resolution of singularities of an algebraic variety over a field of characteristic zero, I, II, Annals of Math., 79, 109-326 (1964).
41. H. Hironaka, Bimeromorphic smoothing of complex spaces, Lec. Notes, Harvard University, Cambridge, Massachusetts, 1971.
42. F. Hirzebruch, Topological Methods in Algebraic Geometry, 3rd ed., Springer-Verlag, New York, 1966.
43. W. V. D. Hodge and D. Pedoe, Methods of Algebraic Geometry, Cambridge University Press, New York, 1952.
44. Horiuchi, R., On the existence of meromorphic functions with certain lower order on non-hyperelliptic Riemann surfaces, J. Math. Kyoto Univ., 21, 397-416 (1981).
45. L. Hörmander, Complex Analysis in Several Variables, Van Nostrand, New York, 1966.
46. A. Hurwitz and R. Courant, Funktionentheorie, 4th ed., Springer-Verlag, New York, 1964.
47. S. Hwang, Study on plane quintic curves (in Chinese), Master's thesis, University of Tokyo, Tokyo, 1980.
48. J. Igusa, Arithmetic variety of moduli for genus two, Annals of Math., 72, 612-649 (1960).
49. S. Iitaka, Algebraic Geometry, Springer-Verlag, New York, 1982.
50. T. Kato, Conformal equivalence of compact Riemann surfaces, Japan J. Math., 7, 281-289 (1981).
51. S. Kawai, Algebraic Geometry (in Japanese), Baifukan, 1977.
52. G. Kempf, Schubert methods with application to algebraic curves, Publ. Math. Certrum, Amsterdam, 1971.
53. G. Kempf, On the geometry of a theorem of Riemann, Annals of Math., 98, 178-185 (1973).
54. S. L. Kleiman and D. Laksov, On the existence of special divisors, Amer. J. Math., 94, 431-436 (1972).
55. S. L. Kleiman and D. Laksov, Another proof of the existence of special divisors, Acta Math., 132, 163-176 (1974).
56. K. Kodaira, On compact analytic surfaces, I, Annals of Math., 71, 111-152 (1960).
57. K. Kodaira and J. Morrow, Complex Manifolds, Holt, Rinehart and Winston, New York, 1971.
58. T. Kotake, An analytic proof of the classical Riemann-Roch theorem, Global Analysis, Amer. Math. Soc., Providence, Rhode Island, 1970, pp. 137-146.
59. M. Kuranishi, Deformations of compact complex manifolds, Proc. Internat. Sem., Univ. Montreal, Montreal, 1969.

60. S. Lefschetz, On the existence of loci with given singularities, Transac. Amer. Math. Soc., 14, 23-41 (1913).
61. H. Martens, On the varieties of special divisors on a curve, I, II, Jour. Reine Angew. Math., 227, 111-120 (1967); ibid., 233, 89-100 (1968).
62. Von G. Martens, Funktionen von vorgegebener Ordnung auf komplexen Kurven, Jour. Reine Angew. Math., 320, 68-85 (1980).
63. W. S. Massey, Algebraic Topology, Springer-Verlag, New York, 1977.
64. H. Matsumura and P. Monsky, On the automorphisms of hypersurfaces, J. Math. Kyoto Univ., 3, 347-361 (1964).
65. Y. Matsushima, Differentiable Manifolds, Marcel Dekker, New York, 1972.
66. B. G. Moishezon, On n-dimensional compact varieties with n algebraically independent meromorphic functions, I, II, III, English translation, AMS Translation Ser. 2, 63, 51-177.
67. D. Mumford, Geometric Invariant Theory, Springer-Verlag, New York, 1965.
68. D. Mumford, Varieties defined by quadratic equations, C.I.M.E. 1969-III, 29-100.
69. D. Mumford, Curves and Their Jacobians, University of Michigan Press, Ann Arbor, Michigan, 1975.
70. D. Mumford, Algebraic Geometry I, Complex Projective Varieties, Springer-Verlag, New York, 1976.
71. M. Namba, Families of Meromorphic Functions on Compact Riemann Surfaces. Lec. Notes, 767, Springer-Verlag, 1979.
72. M. Namba, Equivalence problem and automorphism groups of certain compact Riemann surfaces, Tsukuba J. Math., 5, 319-338 (1981).
73. M. Namba, Linear Systems on Compact Riemann Surfaces and Applications, Lec. Notes, I.P.N., Mexico, to appear.
74. R. Narasimhan, Introduction to the Theory of Analytic Spaces. Lec. Notes, 25, Springer-Verlag, New York, 1966.
75. R. Narasimhan, Several Complex Variables, University of Chicago Press, Chicago, 1971.
76. D. G. Northcott, Ideal Theory, Cambridge University Press, New York, 1965.
77. F. Oort and J. Steenbrink, On the local Torelli problem for algebraic curves, Jour. de geom. algébrique d'Angers, Rockville Sijthoff and Noordhoff, 1980.
78. G. Orzech and M. Orzech, Plane Algebraic Curves, Marcel Dekker, New York, 1981.
79. R. S. Palais, Seminar on the Atiyah-Singer Index Theorem, Annals of Math Studies, 57, Princeton University Press, Princeton, New Jersey, 1965.
80. K. Petri, Über die invariante Darstellung algebraischer Funktionen einer Veranderlichen, Math. Ann., 88, 242-289 (1922).

81. E. Picard, Sur les fonctions de deux variables independentes analogues aux conctions modulaires, Acta Math., 2, 114-135 (1883).
82. H. E. Rauch and H. M. Farkas, Theta Functions with Applications to Riemann Surfaces, Williams-Wilkins, Baltimore, 1974.
83. R. Remmert, Holomorphe und meromorphe Abbildungen komplexer Räume, Math. Ann., 133, 328-360 (1957).
84. B. Saint-Donat, On Petri's analysis of the linear system of quadrics through a canonical curve, Math. Ann., 206, 157-175 (1973).
85. I. Satake, On the compactification of the Siegel space, J. Indian Math. Soc., 20, 259-281 (1956).
86. J. P. Serre, Géomètrie algébrique et géomètrie analytique, Ann. Inst. Fourier, 6, 1-42 (1956).
87. J. P. Serre, Groupes Algébriques et Corps de Classes, Hermann, Paris, 1959.
88. J. P. Serre, Cours D'arithmétique, Presses Univ. de France, 1970.
89. F. Severi, Vorlesungen über Algebraische Geometrie, Teubner, Leipzig, 1921.
90. H Shiga, One attempt to the K3 modular functions, I, II. Ann. Scu. Norm. Sup. Pisa, 6, 609-635 (1979).
91. V. V. Shokurov, The Noether-Enriques theorem on canonical curves, Math. USSR Sbornik, 15, 361-403 (1971).
92. C. L. Siegel, Meromorphe Funktionen auf kompakten analytischen Mannigfaltigkeiten, Göttinger Nacher., 1955, pp. 71-77.
93. C. L. Siegel, Analytic Functions of Several Complex Variables, University of Tokyo Press, Tokyo, 1970.
94. C. L. Siegel, Topics in Complex Function Theory, I, II, III, Wiley-Interscience, New York, 1969-1973.
95. G. Springer, Introduction to Riemann Surfaces, Addison-Wesley, Reading, Massachusetts, 1957.
96. J. Tannery and J. Molk, Éléments de la Théorie des Fonctions Elliptiques, Gauthier-Villars et Fils, Paris, 1893.
97. K. Ueno, Classification Theory of Algebraic Varieties and Compact Complex Spaces, Lec. Notes, 439, Springer-Verlag, New York, 1975.
98. Van der Waerden, Algebra, 7th ed., Springer-Verlag, New York, 1966.
99. G. Veronese, Behandlung der projectivischen Verhältnisse der Räume verschiedenen Dimensionen durch das Princip des Projicirens und Schneidens, Math. Ann., 19, 161-234 (1882).
100. R. Walker, Algebraic Curves, Dover, New York, 1962.
101. H. Weyl, Die Idee der Riemannschen Flächen, Third ed., Teubner, 1955.
102. H. Yoshihara, On plane rational curves, Proc. Japan Acad., 55, 152-155 (1979).

103. O. Zariski, On the non-existence of curves of order 8 with 16 cusps, Collected Papers, Vol. III, pp. 176-185.

104. O. Zariski and P. Samuel, Commutative Algebra, I, II. Van Nostrand, New York, 1960.

INDEX